Biotechnology for Medicinal Plants

Suman Chandra · Hemant Lata
Ajit Varma
Editors

Biotechnology for Medicinal Plants

Micropropagation and Improvement

Springer

Editors
Suman Chandra
National Center for Natural Product Research
Research Institute of Pharmaceutical Sciences
School of Pharmacy
University of Mississippi
Oxford, MS
USA

Hemant Lata
National Center for Natural Product Research
Research Institute of Pharmaceutical Sciences
School of Pharmacy
University of Mississippi
Oxford, MS
USA

Ajit Varma
Amity Institute of Microbial Sciences
Amity University Uttar Pradesh
Noida, Uttar Pradesh
India

ISBN 978-3-642-29973-5 ISBN 978-3-642-29974-2 (eBook)
DOI 10.1007/978-3-642-29974-2
Springer Heidelberg New York Dordrecht London

Library of Congress Control Number: 2012942923

Printed on acid-free paper

Springer is part of Springer Science+Business Media (www.springer.com)

Suman Chandra and Hemant Lata dedicate this book to their parents and children Rishi and Riddhi for their love and support

Foreword

Plants are by far the most abundant and cost effective renewable resource of medicines since ancient times. Even today, the World Health Organization (WHO) estimates that up to 80 % of the world's population still relies mainly on traditional remedies such as herbs in one form or another for their medicine. The scientific foundation for the emergence of pharmacology in the eighteenth and nineteenth centuries was based largely on plant-derived substances. With the introduction of new botanical therapeutic agents, plants are also poised to continue with a key role in the era of modern medicine. Worldwide, the beneficial impact of plant biotechnology has been almost exclusively on medicinal plants. Plant biotechnology has been successfully employed for the large-scale production of phytopharmaceuticals, perfumes, colors, flavors, and biopesticides. Recently, we have begun to realize the ability to better exploit plant-derived bioactives as a result of dramatic advances in metabolic engineering, biochemical genomics, chemical and molecular characterization, and pharmaceutical screening. Furthermore, plant biotechnology acts as a core research tool in the basic research for plant biology and a practical tool for plant improvement.

This volume is a unique overview of medicinal plant biotechnology, describing the whole spectrum of biotechnological tools from cell-culture techniques, to the biosynthesis and accumulation of pharmaceutical compounds in the plants, genetic transformation, metabolic engineering, metabolomics, processes for the production of authenticated plant raw materials and quality control of materials for their safe and efficacious use.

Dr. Suman Chandra and Dr. Hemant Lata, along with Dr. Ajit Varma, have envisaged this special volume that aims to review the state of the art for medicinal plant biotechnology. I have been privileged to see the exemplary work of Drs. Chandra and Lata in our own program using many of these approaches, and to see their aspirations and their maturation in applying this expertise in our own highly interdisciplinary program. They are great team players and gave themselves unselfishly to this task. I am very pleased to see them working with Dr. Varma in bringing together these chapters by eminent scientists from across the globe. Their dedicated work on this project is evident in the quality, comprehensiveness, and cohesiveness of this volume. This work will be useful for researchers in the pharmaceutical and biotechnological industries, medicinal chemists, biochemists, botanists, molecular biologists, academicians and students alike. I congratulate the editors and authors for their endeavors and wish the book all success.

Prof. Dr. Larry A. Walker
Director, National Center for Natural Products Research
Professor of Pharmacology, School of Pharmacy
University of Mississippi
Associate Director for Basic Sciences
University of Mississippi Cancer Institute
USA

Preface

Plant-based medicines are used in all civilizations and cultures, and have been one of the indispensable lines of defense in maintaining health and combating diseases worldwide. With the increasing demand of medicinal plants at the global level for use in pharmaceuticals and dietary supplements, biotechnology has emerged as a powerful tool for their conservation and improvement. By adopting techniques such as in vitro propagation and genetic transformation, biotechnology plays an important role in multiplication and genetic enhancement of medicinal plants. The rapid advances in plant genomics, transcriptomics, proteonomics, and the recent emergence of metabolomics along with molecular biology and analytical chemistry will greatly facilitate and enhance the metabolic engineering of medicinal plants.

The purpose of "Biotechnology for Medicinal Plants: Micropropagation and Improvement" is to present in a single volume the comprehensive knowledge and the experience of renowned researchers and scientists in the field of medicinal plant biotechnology. The book provides an overview of modern plant biotechnology and discusses the potential applications of plant biotechnology in the improvement of medicinal plants. Each chapter is independently written by experts in their field of endeavor ranging from micropropagation protocols, advantages of different biotechnological tools in the plant micropropagation, the biosynthesis of useful secondary metabolites, the metabolomics, the transcriptomics, the metabolomic engineering, the bioinformatics, to the quality control of phytopharmaceuticals.

The book begins with introductory chapters on the overview of downstream processes of plant cell and tissue culture (Chap. 1, Yesil-Celiktas and Vardar-Sukan) and the *Agrobacterium rhizogenes*-mediated transformation in medicinal plants (Chap. 2, Roychowdhury et al.) followed by the biotechnology of *Scutellaria* propagation (Chap. 3, Joshee et al.), microbial endophytes (Chap. 4, Rathod et al.), *Cannabis* propagation for the production of phytocannabinoids (Chap. 5, Chandra et al.), and the micropropagation of medicinal *Epilobium* species (Chap. 6, Constantin et al.). In the next few chapters the discussion is focused on the advantages and amplification of photoelicitation (Chap. 7, Matsuura et al.), microspore culture

(Chap. 8, Ferrie), and cellular heterogeneity (Chap. 9, Patil and Roberts) in plant cell and tissue culture.

Hairy root cultures are a useful means for studying the biochemical and gene expression profiles of the biosynthetic pathways, and for metabolic engineering of medicinal plants. Chapter 10 (Sheludko and Gerasymenko) outlines the biosynthetic potential of hairy roots for the production of novel natural products, while in Chap. 11 (Bunsupa et al.) focus is laid on the molecular biology and biotechnology of quinolizidine alkaloid biosynthesis in leguminosaeLeguminosae plants.

Chapter 12 (Aravindaram and Yang), summarizes the various analytical techniques most frequently used in phytomedicinal research and the potential as well as the limitations of various approaches are discussed.

Free radicals are well known for their involvement in the pathogenesis of a large number of degenerative diseases such as cancer, cardiovascular diseases, neurological disorder and diabetes, in addition to their role in immunomodulation. Over the years, research on antioxidants present in medicinal plants has gained enormous popularity and emerged as an arena for providing potential therapeutic agents to prevent/quench free radical generation that may damage the biological system of the human body. Chapter 13 (Bhatt et al.), focuses on the antioxidants present in various important medicinal plants and their implication in the treatment of certain diseases.

Chapters 14–16 provide in-depth discussion on metabolic engineering in medicinal plants. In Chap. 14, Chow and Sato describe the metabolic engineeringmetabolic engineering and synthetic biology for the production of isoquinoline alkaloids, whereas in the following chapter (Chap. 15) Shoji and Hashimoto explore the role of Isoquinoline Alkaloids ‘jasmonate-responsive transcription factors’ as new tools for metabolic engineering and gene discovery in medicinal plants. In Chap. 16, Zarate et al. review methodologies and advances with the future directions of metabolic engineering of plant cellular metabolism.

Using *Hypericum* as a model, use of metabolomics and transcriptomics to analyze the regulation and biosynthesis of medicinal compounds is described in Chap. 17 (Crispin and Wurtele). Metabolomics, the comprehensive analysis of diverse metabolites, has greatly expanded metabolite fingerprinting and profiling as well as the selection and identification of marker metabolites in medicinal plants. The methodology typically employs multivariate analysis to statistically process the massive amount of analytical chemistry data, not only including the major, but also fingerprinting the minor metabolites, resulting from high-throughput and simultaneous metabolite analysis. Using the advances in computational science, Chap. 18 (Okada et al.) of this book focuses on multivariate analysis of analytical chemistry data and utility of the KNApSAcK family database to understand metabolic diversity in medicinal plants. This chapter also describes the effectiveness and potential of computational systems biology in medicinal plant research.

Authenticity and consistency in the quality of biomass are the key parameters for the safety and efficacy of any phytopharmaceutical. Somatic mutation is a common problem in micropropagation that can alter the genetics and the chemical profile of the propagated crops. The quality assurance of starting martial is therefore a prerequisite. The concluding chapter (Chap. 19, Sucher et al.) of this book discusses the use of genomic and transcriptomic profiling of propagated plants for the quality control of plant-based drugs.

It has been a pleasure to edit this book, primarily due to the splendid cooperation of contributors in preparation of the manuscript according to the guidelines provided, strict adherence to time schedules and the richness of the material provided by them, for that we are highly thankful. We express our gratitude and heartfelt thanks to each author for their generous contribution of time and effort.

We also wish to thank Dr. Jutta Lindenborn and Dr. Christina Eckey at Springer Heidelberg, for their generous assistance and patience in finalizing the volume. Suman Chandra and Hemant Lata in particular express special gratitude to Prof. Dr. Mahmoud ElSohly for his guidance, constant encouragement, and kind support.

Oxford, MS, USA Suman Chandra
Hemant Lata
Noida, UP, India Ajit Varma

Contents

Chapter 1
Downstream Processes for Plant Cell and Tissue Culture

Ozlem Yesil-Celiktas and Fazilet Vardar-Sukan

1.1 Introduction

The first attempt to cultivate plant cells in vitro was almost a century ago. The aseptic culture of cells, tissues, organs, and their components under defined physical and chemical conditions in vitro, has become an important tool in both basic and applied sciences as well as in commercial applications (Thorpe 2007). The prospect of using such culturing techniques is for obtaining secondary metabolites, such as active compounds for pharmaceutics and cosmetics, hormones, enzymes, proteins, antigens, food additives and natural pesticides from the harvest of the cultured cells, or tissues. Industrial companies spend an increasing amount of resources on new product development and are anxious for getting good return on product development investments. Generally, product development is a risky activity as it is very resource demanding and the outcome is uncertain. The majority of all product development projects do not result in a marketable product and all innovations are far from being profitable Therefore, companies look for different ways to reduce the uncertainty and by doing so remain viable and competitive. The task of product development may be divided into two major sub-tasks. Considering the business mission of the companies and the target customers, one of the tasks is to decide what to develop in terms of product functions and properties. The other is to find good technical solutions to achieve the functions and properties. The first focuses on future customer requirements, demands, and market developments while the second deals with more technical issues (Sandell 1996) which will be the focus of this chapter.

O. Yesil-Celiktas (✉) · F. Vardar-Sukan
Department of Bioengineering, Ege University, 35100 Bornova, Izmir, Turkey
e-mail: ozlem.yesil.celiktas@ege.edu.tr; ozlemyesil@gmail.com

F. Vardar-Sukan
e-mail: fazilet.vardar.sukan@ege.edu.tr; fazilet.vardar@gmail.com

S. Chandra et al. (eds.), *Biotechnology for Medicinal Plants*,
DOI: 10.1007/978-3-642-29974-2_1, © Springer-Verlag Berlin Heidelberg 2013

The design and development of simplified and cost-effective separation processes to purify and isolate active compounds is a major challenge for commercialization of biotechnological industries. Various steps of separation processes used for the purification of active compounds are commonly known as downstream processes. Considering the sequence of a plant cell and tissue culture cultivation process, two major methodologies namely, cell culture studies and clonal propagation techniques are widely used. Cell culture studies begin with callus initiation using in vitro cultures for the purpose of determining the medium that best adapts for cultivation. When calli are obtained, they can undergo somaclonal variation, generally during several subcultures (Gurel 1989). When genetic stability is reached, callus lines need to be screened in order to evaluate the productivity of each cell line so that the best performing lines can be taken to cell suspensions. Various approaches can be used to increase the production of secondary metabolites in cell suspensions, but elicitation is generally one of the most successful. The next step is the bioreactor studies leading to a possible commercial production of secondary metabolites. This is a critical step as various problems might arise when scaling-up from shake flasks to bioreactors. Subsequently, the biomass is treated with a number of steps for recovery and purification purposes. Downstream operations used are solid–liquid separation processes such as centrifugation and filtration, followed by cell disruption techniques and concentration steps such as membrane processes, finally separation and purification processes such as various types of extraction and chromatographic techniques. Particularly, dilute bioreactor streams lead to high purification costs, where downstream processing costs could be as high as 10 times the cultivation cost for value-added compounds. Hence, the application of integrated downstream processes suitable for the nature of the specific plant cell and tissue cultures are of great importance for the economic recovery and purification of secondary metabolites.

1.2 Mass Transfer Considerations

Mass transfer resistances play an important role in maximizing the separation efficiencies in downstream processes. Mass transfer occurs when compounds are transported from regions of high concentration to regions where the concentration is low or in other words from regions of high chemical potential to low. The driving force of mass transfer is molecular and bulk diffusions.

Molecular diffusion is the movement of compounds in a mixture under the influence of a concentration difference. Diffusion of compounds occurs in the direction in order to overcome the concentration gradient. On the other hand, bulk diffusion sometimes referred as convective mass transfer or turbulent diffusion is the transfer of compounds in the presence of bulk fluid motion. Molecular diffusion occurs whenever there is a concentration gradient. Additionally, if the bulk fluid is also moving, the overall rate of mass transfer will be higher due to the contribution of convective motions. Analysis of mass transfer is most important in multi-phase systems where interfacial boundary layers provide significant mass-transfer resistances. Two mass-transfer situations which occur in downstream processing of plant cell and tissue cultures are liquid–solid mass transfer for the recovery of intracellular compounds from the biomass and liquid–liquid mass transfer between the organic solvents and the broth for the recovery of extracellular compounds.

1.2.1 Liquid–Solid Mass Transfer

Mass transfer between a liquid and a solid is important in plant cell and tissue culture processes due to the presence of clumps, pellets, flocs, and films of cells. The active compounds in the solid phase must be transported into the liquid phase as in the case of leaching and supercritical fluid extraction. Moreover, adsorption of compounds onto surfaces, such as in chromatography and adsorption requires transport from liquid phase to solid. Liquid–solid mass transfer is also important in crystallization as molecules move from the liquid to the face of the growing crystal.

Considering the transport phenomena, active compounds diffuse from biomass to the solvent phase or active compounds in liquid phase diffuse into or onto the solid as in the case of various adsorbents which are illustrated in Fig. 1.1. In each case, the fluid velocity is reduced and a boundary layer develops near the interface. The local concentration of the compounds at the surface decreases as a result of adsorption and a concentration gradient is established throughout the film. The concentration difference between the bulk liquid and the phase interface is the driving force for the mass transfer of the compounds from the liquid to the solid, allowing the adsorption to continue. For higher recovery of compounds from the liquid phase, the contact area for mass transfer is an important parameter. For instance, considering adsorption characteristics of digoxin recovery, the increase of the contact area for mass transfer significantly enhanced digoxin production (Hong et al. 1998).

In the case of active compounds diffusing from biomass to the solvent phase, first of all the target compounds in solid particles of biomass solvate, then the dissolved compounds diffuse to the outer surface of the solid particle and finally diffuse from the liquid film formed on the surface of the solid particle into the bulk solvent. Main factors affecting the extraction time are the external diffusion coefficient and the difference in chemical potential between the solid–liquid interface and the bulk liquid.

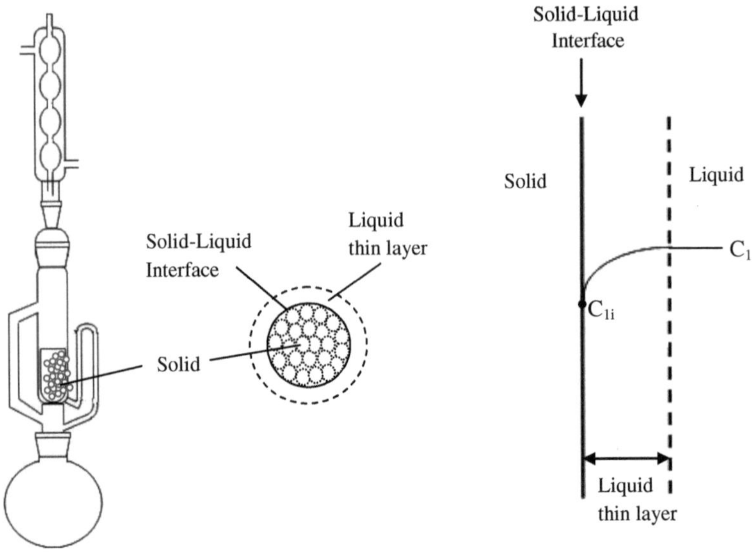

Fig. 1.1 Liquid–solid mass transfer; a Soxhlet apparatus and a schematic representation of active compound C_{1i} diffusing from biomass to the solvent forming the liquid phase

1.2.2 Liquid–Liquid Mass Transfer

Liquid–liquid mass transfer between the stream from the bioreactor and the solvent is encountered in the recovery of extracellular compounds. Organic solvents are used to isolate active compounds from the broths. Mass transfer of the active compounds from one phase to another involves transport from the bulk of one phase to the interface, and then from the interface to the bulk of the second phase. This phenomenon is expressed in the film theory as formation of a fluid film or mass-transfer boundary layer wherever there is contact between two phases.

The situation at the interface between two liquid phases is shown in Fig. 1.2. Assume that the phases are two immiscible liquids and that C_1 is initially at higher concentration in the first liquid phase than in the other liquid phase. Compounds are present at bulk concentration C_1 in the broth, subsequently, this concentration falls to C_{1i} at the interface. In the solvent phase, the concentration falls from C_{2i} at the interface to C_2 in the bulk. Although the bulk liquids may be well mixed, diffusion of the active compounds is crucial in effecting mass transfer because the local fluid velocities approach zero at the interface. Therefore, diffusion is an important mechanism of mass transfer close to the interface between liquids.

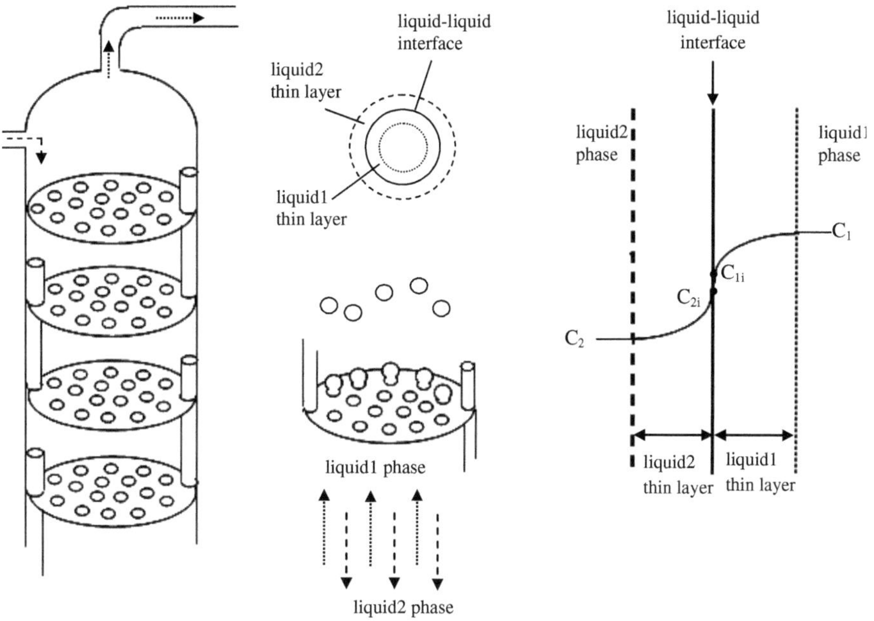

Fig. 1.2 Liquid–liquid mass transfer, schematic representations of an extraction column, and the transport phenomena between two liquid phases

1.3 Downstream Processes

Forming a mixture is a spontaneous, natural process that is accompanied by an increase in entropy or randomness. The inverse process, the separation of that mixture into its compounds is not a spontaneous process which requires an expenditure of energy (Seader and Henly 1998). The mixture to be separated may exist as a single homogeneous liquid or as two or more phases. Based on this, appropriate separation methods have to be applied with the ultimate aim of highest recovery at the lowest possible cost.

1.3.1 Solid–Liquid Separation

Removal of insolubles is generally the first step enabling the capture of the active compounds as a solute in a particulate-free phase. If the final product is a soluble compound, solid particles need to be separated from the liquid phase before it is further treated to purify the soluble compound. If the final product is biomass, then separation of cell aggregates is the major step in product recovery, leading to a significant volume reduction. Typical separation operations with respect to plant cell and tissue cultures are filtration and centrifugation which will be discussed in the scope of solid–liquid separation.

Filtration is used to separate solid particles from a fluid–solid mixture by forcing the fluid through a filter medium while the particles are retained. The solid layer deposited on the filter media is called the "filter cake." As the thickness of the filter cake increases, resistance occurs in further filtration. Therefore, filtration is performed using vacuum or positive pressure gradient, sometimes incrementally increased by time. Rate of filtration depends on the parameters such as the area of the filter cake, the viscosity of the mixture, the pressure difference, and the resistance to filtration.

As for various research studies, an attempt was to accomplish intermittent cell/medium separation of rosmarinic acid produced by *Anchusa officinalis* in the bioreactor by means of automated in situ culture filtration (Su et al. 1995). The cyanobacterium *Spirulina platensis* is an attractive source of chlorophyll, a green pigment used in food, pharmaceutical and cosmetic industries, and other high-value cell components. Filtration is applied as easy and cheap recovery approach from the cultivation medium (Soletto et al. 2005). Another study was conducted to show that auxin was required for the formation of active cyclin-dependent kinase associated complexes, promoting assembly of the complex in tobacco suspension cultures. The researchers found that kinase activity against tobacco retinoblastoma-related protein was maximal in fractions separated using get filtration (Harashima et al. 2007).

In centrifugation, separation occurs due to differences in density. In the case of the centrifugation of plant cell and tissue cultures, particles are generally small in size; the viscosities of the media can be relatively high, while the particle densities might also be similar to those of the suspending fluids. These obstacles have to be overcome in large-scale applications. In a study, a new protocol has been developed for somatic embryogenesis, plant regeneration, and transformation of a banana cultivar where liquid co-cultivation was carried out with centrifugation (Ghosh et al. 2009).

1.3.2 Cell Disruption Techniques

Cell disruption is used for the recovery of intracellular compounds after separating the cells and aggregates from the liquid broth. Cell walls can be ruptured by mechanical or nonmechanical techniques. The mechanical approaches are targeted more towards cell wall disruption while the nonmechanical techniques are mainly used for destabilizing the cell membrane.

The various mechanical disruption techniques involve using bead mills, rotor–stator mill, homogenizers, and ultrasonic vibrations. The bead mills are cylindrical tubes loaded with small beads. The beads collide with the biomaterial as the mill rotates, resulting in broken cell walls. The rotor–stator mill consists of a stationary block with a stator and a rotor. The cell suspension is fed into the gap between the rotating rotor and the fixed stator and expelled through the outlet due to centrifugal action. The generated turbulence and shear lead to disruption of cells. Homogenizer consists of a high pressure pump that injects the cell suspension through a discharge valve causing an explosion of the cells. The cells can also be subjected to ultrasonic vibrations by introducing an ultrasonic vibration emitting tip into the cell suspension. The ultrasonic vibration could be emitted continuously or in the form of short pulses. The size of the emitting tips is determined based on the volume of sample being processed, whereas the duration of ultrasound is ascertained based on the cell type, the sample size, and the cell concentration.

Nonmechanical disruption techniques involve using detergents, organic solvents enzymes, and osmotic shock. Chemical lysis is carried out by adding detergents and solvents to solubilize cell walls. The detergents and solvents need to be subsequently removed from the product and this generally involves an additional purification step in the process. An enzyme can be used to disrupt plant cells such as cellulase. The main limitation of using an enzyme is its relatively high cost but, on the other hand, the process is less energy demanding and very specific. Another technique is osmotic shock resulting from a difference in solute concentration across a semi permeable membrane. A rapid influx of water into the cell takes place when a cell is suddenly transferred from an isotonic to hypotonic medium such as distilled water. Subsequently, the cell wall is expanded and ruptured.

Among these techniques, homogenization, sonication, steam explosion (Hartonen et al. 2007; Kurosumi et al. 2006), and treatment with supercritical fluids (Castor et al. 1998) were applied to plant cell and tissue cultures. The pretreatment efficiency depends on the type of tissue extracted and the metabolite localization, so ideally, any considered procedure should be tested prior to concentration and separation. The disrupted cells need to be separated from the other cellular compounds by removing the cell debris.

1.3.3 Concentration

Membrane processes are used to concentrate streams from bioreactors by permitting some compounds to pass through while retaining others. Hence, the feed stream leaves the membrane as a permeate containing small compounds and a retentate with large compounds. Membrane technologies have speedily gained importance in processes involving plant cell and tissue cultures. Various membrane processes are available with varying pore sizes and driving forces.

Microfiltration is a pressure driven process that separates micron range particles from liquid streams of plant cell and tissue cultures. Typical operating conditions are 0.3–3.3 bar pressures, 3–6 m/s flow rates with membrane pore sizes of 0.05–3 mm. Hwang et al. (2005) developed a simple and scalable, chromatography based process to purify a type of colony stimulating factor from recombinant rice cell culture. It consisted of cell removal by microfiltration, ammonium sulfate precipitation, and anion exchange chromatography. The purification process resulted in a final purity of about 95 % and an overall recovery yield of 48 % from the culture broth.

On the other hand, there are growing efforts to facilitate in situ medium exchange particularly to eliminate possible introduction of contaminants to the process using microfiltration. A study carried out in *Cyperus aromaticus* cell suspension culture to produce a bioinsecticide known as juvenile hormone III focused on the effect of different volumes and flow rates of replenishment toward cell biomass and the hormone. Medium replenishment of 75 % through the membrane system leads to higher cell biomass and metabolite production which corresponded to values as high as 64 and 112 %, respectively. The reduction in doubling time, extended exponential growth and increment increases in specific growth rates were also demonstrated. A 50 % medium replenishment with membrane filtration system was identified to be an economical and feasible application for the production of juvenile hormone III at large scale. The results obtained showed the potential of in situ membrane filtration system for the enhancement of cell biomass and metabolite yields in plant cell culture (Chan et al. 2010).

Ultrafiltration is another technique used for concentration and fractionation purposes in plant cell and tissue cultures. This is a low pressure process separating dissolved solutes of 5–100 μm in size and is used in numerous industries for concentrating large process streams. In one of the earlier studies, ultrafiltration was used to obtain a high molecular weight fraction of tobacco cell suspension cultures (Takahashi and Warren 1994). In the other study, ultrafiltration was applied to ascertain the responsible fraction for the plant activation of m-phenylenediamine in suspension cultures of tobacco cells (Gichner et al. 1995). Ultrafiltration was also used to deproteinize cytosol in which phosphoribosylpyrophosphate synthetase from *Hevea brasiliensis* Mull. Arg. latex was present (Gallois et al. 1997). Treatment of cell-suspension cultures of *Platanus acerifolia* with a crude elicitor preparation induced the synthesis of the hydroxycoumarin phytoalexins, scopoletin and umbelliferone, and their accumulation in the growth medium. Only the protein-containing fraction of the culture filtrate was involved in cell response. By ultrafiltration of this last fraction, a major eliciting glycoprotein able to induce 80 % coumarin synthesis was isolated (Alami et al. 1998).

1.3.4 Separation and Purification

Subsequent to the concentration of the valuable fraction, various methods such as extraction and chromatography are employed to recover soluble compounds.

1.3.4.1 Solvent Extraction

Solvent extraction is used to separate active compounds from the valuable fraction where the solute is contacted with a selected solvent exhibiting favorable solubility properties.

If the valuable fraction is concentrated liquid as in the case of extracellular compounds, liquid–liquid extraction is applied. In this operation, the solute, which is dissolved in the liquid broth, is transferred into a new solvent, which is either insoluble or partially soluble in the broth. The liquid broth is named as raffinate, and the solvent as extract solvent or simply the solvent. The limited solubilities of solvents constitute the basis of this operation. Thus two liquid layers are formed, each containing different fractions of all the components depending on the respective partial solubilities. The two layers separate from each other due to the density difference. The layer which contains a higher percentage of the extract solvent is called the extract phase and the layer containing a higher percentage of the raffinate solvent is called the raffinate phase. In many cases, there is more than one solvent that can be used for a specific extraction operation. The following properties of the solvents have to be taken into account in order to select the best one among the potential solvents. Selectivity of solvent (β) is defined as a ratio at equilibrium conditions. It indicates the separation power of the selected solvent for the solute dissolved in the raffinate solvent. If $\beta = 1$, this solvent has no selectivity for the solute and hence it cannot be used for the extraction. Higher the value of β, the easier it is to separate the solute from solvent.

$$\beta = \frac{y/y_c}{x/x_c}$$

Distribution coefficient (K) is defined as $K = y/x$ which should be as high as possible to accomplish the given extraction with minimum amount of solvent usage. The selected solvent should have no or low solubility in raffinate solvent to minimize the solvent losses. Separation of the solvent from the solute is achieved by rectification. Hence, the selected solvent should not form an azeotropic solution with the solute and if possible, the relative volatility of this solution should be high. Another parameter to be considered is the density of solvent. Density difference should exist between the phases for the separation of extract and raffinate phases. Hence, the density of the selected solvent should be different from the density of raffinate solvent. Furthermore, interfacial tension has to be taken into consideration. One of the phases is first dispersed in the other phase as small droplets by intensive mixing and then coalescence occurs by allowing the settling. Interfacial tension among the liquids plays an important role in dispersion and coalescence. Low interfacial tension means easy dispersion but difficult coalescence. Since short coalescence time is preferred, the selected solvent should exhibit high interfacial tension in the presence of the raffinate solvent. As the solvent is recovered by rectification and reused many times and during which it is heated and cooled repeatedly, it should be chemically stable to prevent decomposition. Finally, whenever possible the selected solvent should have low viscosity and vapor pressure, should be nontoxic, and should have low unit cost (Alpay and Demircioglu 2006).

If the valuable fraction is a dried or lyophilized biomass, then solid–liquid extraction which is known as leaching is applied. When the two phases are in contact, the active compounds diffuse from the biomass into the liquid phase resulting in the separation of the compounds. However, the active compounds are generally intracellular, hence the rate of extraction may be slow due to the additional diffusional resistances of the cell walls. Drying the biomass leads to ruptured cell walls, thereby enhancing mass transfer.

Considering the transport phenomena, active compounds diffuse from the biomass into the solvent phase basically in three steps. Solvation of target compounds in solid particles of biomass, diffusion of dissolved compounds to the outer surface of the solid particle (internal diffusion), and finally diffusion from the liquid film around the solid particle into the bulk solvent (external diffusion). Main factors affecting the extraction time are external diffusion coefficients and differences in the chemical potentials in the liquid film.

The choice of solvent depends on the physical properties and oxidation sensitivity of the biomass produced by plant cell and tissue culture. Both nonpolar and polar solvents are used.

However, the disadvantages of solvent extraction can be listed as follows:

- Lower efficiency
- Solvent residues affecting the purity of the compounds
- High costs for high purity requirements
- Longer process durations
- Undesired side reactions affecting the quality of the product

Solvent extraction is widely used in industry due to its low investment costs, but additional separation steps have to be introduced in order to isolate the active compounds from the solvent.

In an earlier study, the recovery of capsaicin from a *Capsicum annuum* cell culture broth was reported. First, the solids were removed by filtration, and then liquid–liquid extraction was used to isolate the capsaicin. Chloroform was used to perform the extraction of the capsaicin in a reciprocating Karr column and the operating conditions determined in the pilot plant column were used to design a commercial-size reciprocating column (Tapia et al. 1993). Another study dealing with a new design was a two-liquid-phase bioreactor designed to extract indole alkaloids from *Catharanthus roseus* hairy roots with silicon oil with the aim of continuous selective extraction of secondary metabolites (Tikhomiroff et al. 2002). Such scale-up attempts and innovative designs can enhance large-scale cultivation of plant cell and tissue cultures providing cost-effective solutions. In more recent works, solvent extraction was applied to different fractions of cultures. For instance, lipids were extracted with chloroform–methanol and the remaining fraction containing vitamin D3 compounds was extracted with polar solvents from *Solanum glaucophyllum* grown in vitro in darkness (Curino et al. 2001). For the production of volatile flavor compounds from *Agastache rugosa* suspension culture, callus was obtained from the leaves of *A rugosa* and volatiles were isolated by liquid–liquid continuous extraction and identified by gas chromatography/mass spectrometry (Kim et al. 2001).

Taxol is another secondary metabolite widely used as a chemotherapeutic agent and applications of large-scale cultivation exist (Son et al. 2000). Suspension culture of *Taxus chinensis* cells was carried out in aqueous-organic two-phase systems for the production and in situ solvent extraction of taxol (paclitaxel). Three organic solvents, hexadecane, decanol, and dibutylphthalate, were tested at 5–20 % (v/v) in the culture liquid. All of these solvents stimulated taxol release and the yield per cell, though decanol and higher concentrations of the other two solvents depressed biomass growth significantly. Dibutyl phthalate at 10 % was the optimal solvent for improving taxol production and release with minimal cell growth inhibition. The time of solvent addition to the culture also affected taxol production, with the addition during the late-log growth phase being most favorable (Wang et al. 2001). In another study dealing with taxol, different organic solvents, including saturated aliphatic hydrocarbons, aliphatic alkanols, esters of dicarboxylic acids, monoterpenoids and unsaturated lipids, were investigated to test the effect of in situ paclitaxel removal from *Taxus chinensis* suspension cultures. The results indicated that dibutyl phthalate, as well as, oleic acid and terpineol were ideal solutions for in situ extraction of paclitaxel (Zhang and Xu 2001). Furthermore, a large-scale purification method which consisted of solvent extraction, synthetic adsorbent treatment, and two steps of precipitation for the isolation and purification was developed for producing paclitaxel from the biomass of *Taxus chinensis*. The biomass from plant cell cultures was mixed with methanol and stirred at room temperature for 30 min. Extraction was repeated at least four times. Each methanol extract was collected and concentrated at a temperature of 40 °C under reduced pressure to decrease the volume of the methanol extract down to 30 % of the original. As a second extraction step, the concentrated methanol extract was added to organic solvents (dichloromethane, chloroform, diethylether, and hexane) with varying polarities at a volume ratio of 25 % for liquid–liquid extraction and extraction was carried out at room temperature for 30 min. The extraction was repeated at least three times and the crude extracts thus obtained were pooled and dried at room temperature under reduced pressure (Pyo et al. 2004).

In our study, stem explants of young shoot of *Rosmarinus officinalis* were cultured in both woody plant medium and Murashige and Skoog media for callus initiation. The lyophilized calli were subjected to solvent extraction via sonicating at 35 kHz for 45 min with (50 °C) and without heat treatment at room temperature to obtain the methanolic extracts. Active constituents of eight calli extracts were analyzed by HPLC, and rosmarinic acid was determined to be the primary compound. Rosmarinic acid contents of the samples exposed to thermal treatment (50 °C) were higher than the untreated samples (Yesil-Celiktas et al. 2007) implying the importance of temperature-induced solubility. In another study, successful scale up of *Azadirachta indica* suspension culture for azadirachtin production was done in a stirred tank bioreactor with two different impellers. However, extraction of azadirachtin was not conducted on large scale, but rather at laboratory scale (Prakash and Srivastava 2007).

A variant plant of opium poppy (*Papaver somniferum* L.) containing high thebaine, which is a paramorphine, was obtained during a mutation breeding experiment. The seeds of the variant plant were subjected to in vitro studies to investigate the prospects of thebaine production through tissue culture. Alkaloid profile showed higher thebaine in stems followed by leaf callus, stem callus, and cotyledons. Such a high content of thebaine from tissue raised plant material was argued to be a new vista for the extraction of thebaine at commercial level. However, the samples were extracted in 5 ml DMSO for more than 3 months which is not an appropriate and cost-effective protocol for possible scale-up studies (Chaterjee et al. 2010). Alkaloids were of interest in another study where a treatment involving an acid-based extraction from the cell culture of *Passiflora alata* was carried out (Machado et al. 2010). A cell suspension culture was developed from calli of grape rootstock for the bioproduction of resveratrol using methyl jasmonate as an elicitor. Liquid–liquid extraction of the culture medium and a solid–liquid extraction of the cells showed that other metabolites such as trans-epsilon-viniferin, trans-delta-viniferin, and a trans-3-methylviniferin as well as trans-piceatannol were also produced in a 2 L bioreactor cell culture of grapevine. Furthermore, a one step FPLC method was developed for the purification of resveratrol and epsilon-viniferin (Donnez et al. 2011). A recent study is a good example where extraction was combined with sonication in order to enhance the recovery of secondary metabolites. Cell suspension cultures of *Cephalotaxus fortunei* were manipulated to produce secondary metabolites of pharmaceutical interest and seven abietane diterpenoids were produced by the suspension-cultured *C. fortunei* cells by complementing an elicitation strategy with an in situ product removal strategy. After subculturing was completed, methyl jasmonate solution in ethanol was added to the culture medium together with autoclaved XAD-7HP resin. After 2 weeks of culture, the cells and the resin were harvested separately based on density. The suction-filtered resin was extracted with methanol three times at room temperature. Each extraction was performed by shaking on a rotary shaker at 100 rpm for 2 h followed by sonication for 20 min. The combined methanol extracts were filtered, and the supernatant was evaporated to dryness (Xu et al. 2011).

As mentioned earlier, developing methods for quali-quantitative analysis of metabolites produced by plant cell and tissue cultures are also very important for commercial applications. The hydrophilic metabolome of heterotrophic *Arabidopsis thaliana* cells grown in suspension was targeted and water-soluble metabolites were extracted using four protocols; perchloric acid, boiling ethanol, methanol, and methanol/chloroform. They were detected and quantified using 1H nuclear magnetic resonance (NMR) spectroscopy. The quantitative studies demonstrated that the four extraction protocols commonly used do lead to quite similar molecular compositions as analyzed by 1H NMR. The methanol/chloroform method proved effective and reliable to prepare series of physiologically significant extracts from plant cells for 1H NMR analysis. Reproducibility of the detected metabolome was assessed over long periods of time by analyzing a large number of separate extracts prepared from independent cultures. Quantitatively resolved 1H NMR of cell extracts proved to be robust and reliable for routine metabolite profiling of plant cell cultures (Gromova and Roby 2010).

1.3.4.2 Supercritical Fluid Extraction

Supercritical fluid extraction (SFE) is an alternative technology for separation and purification of secondary metabolites from plant cell and tissue cultures. This technology has the potential to be an environmentally friendly "green" process. A fluid is said to be supercritical when its temperature and pressure values are slightly above the liquid–vapor critical point. Under these conditions, the fluid displays unique properties, such as high density, high diffusivity, and low viscosity. Furthermore, the extreme compressibility of supercritical fluids makes them very sensitive to minor temperature and pressure gradients causing supercritical fluids to be useful in many industrial applications. Supercritical fluid processes are applied to increase product performance to levels that cannot be achieved by traditional processes.

The critical temperatures, pressures, and densities of various fluids are provided in Table 1.1. Carbon dioxide is generally the most desirable solvent for supercritical fluid extraction due to its low critical temperature (31 °C) and pressure (73.8 bar). Additionally, being inert, nonflammable, and nontoxic and having a low surface tension and high diffusivity characters, CO_2 is attractive as a supercritical solvent for pharmaceutical and food industries. Since its diffusivity is one to two orders of magnitude greater than conventional fluids, a greater extraction rate is obtained compared to conventional solvent extraction.

CO_2 molecule is composed of two oxygen atoms covalently bonded to a single carbon atom displaying no permanent dipoles which makes CO_2 ideal for the separation of nonpolar compounds (Kaiser et al. 2001). On the other hand, the solubility of polar and high molecular weight compounds is very limited in CO_2. Hence, polar solvents with high dielectric constants are added at low concentrations in order to increase the solvating power of CO_2. These are named as modifiers, co-solvents, or entrainers (Taylor 1996; McHugh and Krukonis 1994; Dobbs et al. 1987). Therefore, adding of solvents such as methanol, ethanol, acetone, and water is required for the efficient extraction of compounds with polar properties existing in plant cell and tissue cultures.

A typical supercritical CO_2 extraction setup is comprised of a CO_2 tube, high pressure CO_2 pump, CO_2 flow-meter, temperature controllers, high pressure co-solvent pump, automated Back Pressure Regulator, cooler, and heater as illustrated in Fig. 1.3. The extractor volumes range between 10 and 1,000 ml for laboratory scale, 1.5–10 L for pilot scale and above 10 L as industrial scale. In Fig. 1.4, a 100 ml laboratory scale Thar and a 6.5 L pilot scale Natex supercritical fluid processing units can be seen. Liquefied CO_2 is introduced into the packed column through a piston pump with a cooling jacket. Both the pressure and temperature of the cartridge are automatically reached and maintained by a control unit according to the settings. After the desired pressure and temperature are reached, the dynamic extraction starts. The variables are temperature, pressure, CO_2 flow, and extraction duration and co-solvent ratio if polar compounds are of interest. Finally, extracts are collected from the separator outlet after releasing CO_2 from the system at laboratory scale applications, whereas CO_2 is recycled in

Table 1.1 The critical temperature, pressure, and density values of some solvents (Rivzi et al. 1986)

Solvent	Critical temperature (°C)	Critical pressure (atm)	Critical density (g/ml)
Methane	−83.2	45.40	0.162
Ethylene	9.3	49.64	0.218
Chlorofluoromethane	28.9	38.68	0.579
Carbon dioxide	31.06	72.83	0.468
Ethane	32.3	48.16	0.203
Propylene	91.9	45.59	0.233
Propane	96.7	41.84	0.217
Ammonia	132.5	111.52	0.235
Diethyl ether	194.6	35.92	0.265
n-pentane	196.5	33.26	0.237
Acetone	235.0	46.40	0.278
Methanol	239.5	79.84	0.272
Benzene	289.0	48.26	0.302
Toluene	318.6	40.56	0.292
Pyridine	346.9	55.56	0.312
Water	374.2	217.11	0.322
Xenon	16.6	57.63	1.113

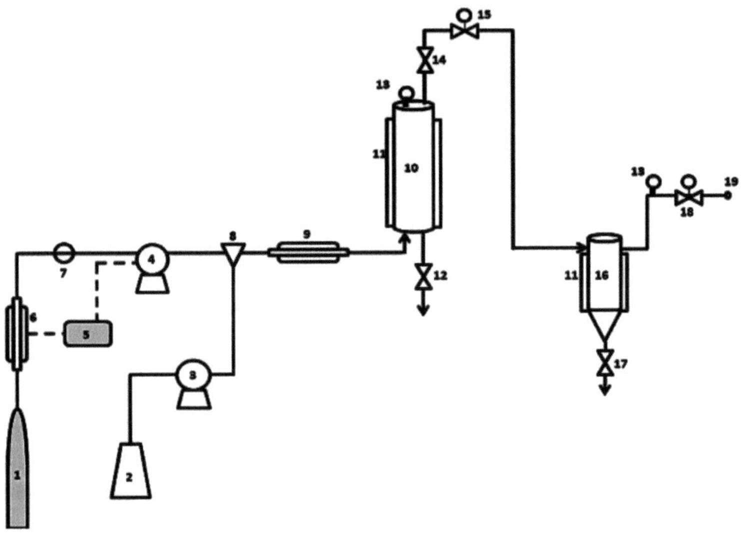

Fig. 1.3 The diagram of supercritical fluid process used (1—CO_2 supply in, 2—Co-solvent reservoir, 3—Co-solvent pump, 4—CO_2 pump, 5—Cooling bath, 6—Cooling heat exchanger, 7—Flow meter, 8—Mixer, 9—Heat exchanger, 10—Extraction vessel, 11—Heat jacket, 12—Bleed valve, 13—Gauge, 14—On–off valve, 15—Automated Back Pressure Regulator (BPR), 16—Fraction collector, 17—Drain valve, 18—Manual BPR, 19—Vent)

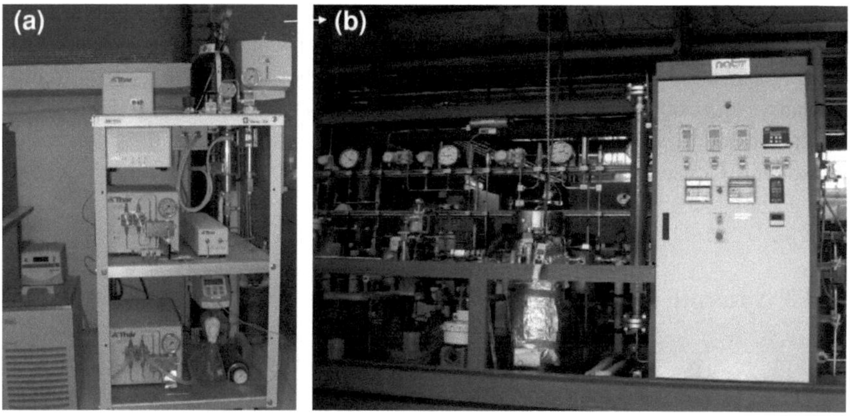

Fig. 1.4 A 100 ml laboratory scale Thar (**a**) and a 6.5 L pilot scale Natex (**b**) supercritical fluid processing units

most of the pilot scale and all industrial-scale applications. Therefore, the extracts obtained with SFE can be considered to be more sustainable and healthier, conforming with industrial requirements of healthy food and pharmaceutical products because they generally are solvent-free, even if co-solvents are applied in small amounts to enhance the extraction of high polarity active compounds.

While processing plant cell and tissue cultures, supercritical and subcritical fluids can also be used as a cell disruption technique releasing structural biomass constituents. These can further be subjected to a multiplicity of sub- and supercritical extraction steps with different solvation conditions used for each fraction (Darani and Farahani 2005), thereby conducting a two-step process in one single operation. Supercritical fluid extraction is a relatively new method and laboratory studies are still being carried out to investigate other potential applications. Only one study was identified for the supercritical fluid extraction of the biomass, where the supercritical CO_2 extraction of capsidiol from pepper fruit tissues activated with *Alternaria alternate* Keissler suspension culture as a biotic elicitor was investigated. The effects of separation parameters such as temperature, pressure, supercritical solvent flow rate, particle diameter and also initial capsidiol concentration on solubility, initial extraction rate, and extraction yield were studied. The optimal extraction conditions were obtained at a temperature of 40 °C, the pressure of 400 bar, the supercritical CO_2 flow rate of 2 cm^3/min, and the average particle diameter of 116 μm. The results showed that the ratio of the supercritical CO_2 extraction yield to the organic solvent extraction yield changed from 84 to 97 % depending on the initial capsidiol concentration (Salgin et al. 2005).

In terms of comparing the costs associated with SFE and traditional extraction plants, the capital cost of the SFE is generally higher than the traditional, while the operating costs are lower. However, it is necessary to take into account the capital costs of all the equipment required for a traditional process, such as evaporation systems for solvent recovery, costs associated with building requirements, instrumentation, and electrical connections with appropriate safety requirements. The operating costs of the traditional processes, particularly the energy costs of water evaporation are considerably high. On the other hand, SFE is an attractive alternative for obtaining value-added compounds from biomass without any solvent residues, resulting in higher sales prices implying shorter return on investment. Consequently, SFE presents a potential for commercial applications with acceptable quality.

1.3.4.3 Adsorptive Separation

Adsorption involves adsorbing the active compounds on the surface of solid particles. In adsorption operations, the substance being concentrated on the surface is called the adsorbate, whereas the material to which the adsorbate binds is the adsorbent. The ideal adsorbent material has a high surface area per unit volume which can be achieved if the solid contains a network of fine internal pores providing a large internal surface area. Silica gel, oxidic resins, activated carbons, and polymers with different surface functionalizations are commonly used for adsorption of biological molecules. As the polarity of the resins strongly influences the adsorption behavior, the preselection of applicable sorbents focuses on covering a wide range of polarities. Additionally, the resins are characterized by different specific surface areas, pore diameters, and porosities (Winkelnkemper et al. 2011). A typical adsorption operation consists of the following stages: adsorption step which loads solute onto the adsorbent, a washing step to remove residual unadsorbed material, desorption or elution of adsorbate with a suitable solvent, washing to remove residual eluant, and regeneration of the adsorption resin to its original condition.

Cell and tissue culture research focused on adsorption process in 1990s. The main drive was to test suitability and efficacy of various adsorbents for the recovery of value-added compounds from cultured plant cells and tissues. However, the findings from the studies have also highlighted the potential of adsorbents to overcome growth inhibition. The use of a polycarboxyl ester resin, for the in situ separation of berberine, isoquinoline alkaloid, produced from plant cell culture of *Thalictrum rugosum* was studied (Choi 1992). The suitability of adsorbent polymeric resins was investigated for the accumulation of sanguinarine from *Papaver somniferum*cell cultures. In addition to sanguinarine, the resins were found to adsorb growth regulators and vitamins from the culture medium. Growth inhibition was overcome by delaying resin addition for approximately 4 days after cell inoculation in fresh medium. Resin addition to actively growing unelicited cultures led to increases in sanguinarine production and release of 30 to 40 and 60 %, respectively. The addition of resins to elicited cultures led to increases in alkaloid production of up to 50–85 % with similar increases in alkaloid release as observed for unelicited cells. Overall yields of sanguinarine increased from 21 mg/g biomass dry weight for elicited cultures to more than 39 mg/g when elicitation was combined with resin addition (Williams et al. 1992). Another study focused on the integrated recovery of solavetivone from fungus elicited hairy root cultures of *Hyoscyamus muticus* using volatile organic solvents and solid-phase adsorbents in an external loop extraction configuration. Hexane and pentane were shown to be toxic when added directly to the culture; however, growth of roots was not inhibited when cultures were exposed to media saturated with these hydrocarbons. Solid-phase neutral adsorbents displayed higher capacity and better solavetivone partitioning capability than the hydrocarbons; however, their selectivity for the sesquiterpene solavetivone was found to be poor in comparison to hexane (Corry et al. 1993). Various hairy root clones were derived by transforming two kinds of *Catharanthus roseus* plants with *Agrobacterium rhizogenes*. Hairy root growth and indole alkaloid production were then investigated to select a high yielding hairy root clone. One of the Amberlite resins tested greatly enhanced the release of catharanthine and ajmalicine from hairy root cultures, with an increase in total production. The researchers introduced a permeabilizing agent (dimethyl sulfoxide) and a fungal elicitor to provide physical and biochemical stress, respectively, together with in situ adsorption with the ultimate aim of enhancing alkaloid production and secretion. The results indicated that in situ adsorption sequentially applied with permeabilization and fungal elicitation had a synergistic effect on the production and secretion of indole alkaloids (Sim et al. 1994). Another compound of interest was digoxin, a cardiac glycoside. Selection of proper resins was attempted in order to use in situ adsorption by biotransformation from digitoxin in plant cell suspension cultures. Among various kinds of resins tested, one of the Amberlite adsorbent was found to be the best for digoxin production in considering adsorption characteristics as well as the effect on cell growth. However, the findings of this study were not in parallel to the rest of the other mentioned studies in this section because of the fact that the presence of resins was suggested to be as short as possible to increase the productivity. Another comment was to prevent the cells from direct contact with resin particles with a suggestion of designing immobilized systems (Hong et al. 1998).

Recent studies are related to continuous removal of value-added compounds, introducing innovative process designs, and testing novel biomaterials as adsorbents. Continuous removal of anthraquinones by Amberlite polymeric adsorbents through in situ adsorption in *Morinda elliptica* cell suspension cultures was studied for product recovery and improvement of the overall titer. Ethanol was the best eluting solvent for effective recovery of anthraquinones from all adsorbents. Pretreatment of the adsorbent with sodium acetate not only enhanced intracellular anthraquinones, but also their release and subsequent recovery from the adsorbent. High amount of adsorbent and longer contact period for the cultures entering stationary growth phase was reported to stimulate anthraquinones release and recovery, but at the expense of cell growth (Chiang and Abdullah 2007). A recent study was conducted to develop an optimal tandem simulated moving bed process containing two four-zone simulated moving bed units in series for separation of paclitaxel, 13-dehydroxybaccatin III and 10-deacetylpaclitaxel. The bed was loaded with a Curosil-PFP resin with an average particle size of 5 μm which was used as the adsorbent. Based on the estimated adsorption isotherm and mass-transfer parameters, the tandem simulated moving bed was optimized to maximize throughput under the constraint on pressure drop while meeting the targeted purities and yields. According to the optimization results, the best strategy leading to the highest throughput was to recover paclitaxel in the first simulated moving bed and then separate the remaining two components in the second (Kang et al. 2010). Another study focused on molecularly imprinted polymers for the selective recovery of nicotine in plant cell cultures. Molecularly imprinted polymers can selectively uptake nicotine from suspension cultures of *Nicotiana tabacum*, and therefore may be useful for improving levels of secondary metabolites in plant cell cultures. Molecularly imprinted polymers and corresponding nonimprinted polymers were introduced aseptically into the suspension cultures of *N. tabacum* permeabilized by Tween 80, the nicotine contents of polymers were determined by gas chromatography. The results demonstrated that molecularly imprinted polymers were able to uptake 50–70 % of released nicotine in suspension cultures of *N. tabacum*, indicating the potential use of molecularly imprinted polymers (Abdelkar et al. 2010). Activated charcoal was reported to be often used in tissue culture to improve cell growth and development. It plays a critical role in micropropagation, orchid seed germination, somatic embryogenesis, anther culture, synthetic seed production, protoplast culture, rooting, stem elongation, and bulb formation. The promotary effects of activated charcoal on morphogenesis may be mainly due to its irreversible adsorption of inhibitory compounds in the culture medium and substancially decreasing the toxic metabolites, phenolic exudation, and brown exudate accumulation. In addition to this activated charcoal was involved in a number of stimulatory and inhibitory activities including the release of substances naturally present in activated charcoal which promote growth, alteration and darkening of culture media, and adsorption of vitamins, metal ions and plant growth regulators, including abscisic acid, and gaseous ethylene. The effect of activated charcoal on growth regulator uptake is still unclear but some researchers believe that certain adsorbed products, such as nutrients and growth regulators may be gradually released to become available to plants (Thomas 2008).

1.3.4.4 Chromatographic Separation

The other separation procedure involving adsorption of the active compounds on the surface of solid particles is chromatography. The basis of chromatography is the selective retardation which is referred as differential migration of solute molecules during passage through a bed of resin particles. The solutes migrate at different speeds depending on their relative affinities for the resin particles as solvent flows through the column. Consequently, if the affinities of the compounds of interest are different from each other, they are separated and can be collected at the end of the column at different times. Chromatography is a high-resolution technique and therefore suitable for recovery of high-purity therapeutics and pharmaceuticals. Chromatographic methods available for purification of proteins, peptides, alkaloids, vitamins, steroids, and many other biological materials include liquid chromatography which is applied both as a laboratory method for sample analysis and as a preparative technique for large-scale purification of value added compounds; adsorption chromatography, partition chromatography, ion exchange chromatography, gel chromatography, and affinity chromatography. These methods differ in the principal mechanism by which molecules are retarded in the chromatography column.

In the early 1990s, a research study was conducted to increase the separation efficiency of β-carotene from cultured carrot cells by avoiding the saponification process and improving the adsorption chromatography and liquid–liquid extraction. The designed process markedly reduced the operation time and yielded highly pure β-carotene which contained no other lipids (Ogawa et al. 1991). On the other hand, the majority of the studies in regards to chromatographic methods are related to purification of enzymes. For instance, a study was conducted to purify an endo-l, 4-beta-glucanase to apparent homogeneity from the culture medium of *Populus alba* L. cells by sequential anion-exchange, hydrophobic, and gel-filtration chromatography. The enzyme specifically cleaved the l, 4-beta-glucosyl linkages of carboxymethylcellulose, swollen cellulose, lichenan, and xyloglucan, although the last was hydrolyzed more slowly than the other tested substrates. The activity of the endo-l, 4-beta-glucanase increased up to the early stages of the mid-logarithmic phase of growth and then decreased rapidly, suggesting that the beta-glucanase was induced before cell development (Nakamura and Hayashi 1993). Another study was the purification of the cell wall associated lanatoside 15′-O-acetylesterase from suspension-cultured *Digitalis lanata* Ehrh. (Scrophulariaceae) cells by cell wall preparation, ion exchange chromatography, and gel filtration (Sutor and Kreis 1996). With an attempt to design a new process for the recovery of raucaffricine-O-beta-D-glucosidase from plant cell suspension cultures of Rauwolfia, a five-step procedure using anion exchange chromatography, chromatography on hydroxylapatite, gel filtration, and FPLC chromatography was developed (Warzecha et al. 1999).

In the more recent studies, chromatographic techniques are related to purification of high value proteins. Human lactoferrin, an iron-binding glycoprotein that has been considered to play many biological roles in the human, including the stimulation of the immune system, antimicrobial, and anti-inflammatory effects, and regulation of iron absorption was produced by transgenic Siberian ginseng (*Acanthopanax senticosus*) cell cultures. Recombinant human lactoferrin was purified from suspension cells of Siberian ginseng by ammonium sulfate precipitation, cation-exchange, and gel filtration chromatography (Jo et al. 2006). The beta-glucosidase derived from plant cell culture of *Pichia pastoris* was recovered by expanded bed adsorption chromatography. Higher bead density made it possible to operate at a higher feedstock concentration and higher flow velocity. The high binding capacity was argued to be caused by the more complex interaction of multimodal ligand(Charoenrat et al. 2006). In another study, the production and purification of the M protein in the methalotrophic yeast, *Pichia pastoris* were reported. M protein was purified to near homogeneity from whole cell lysates using cation exchange, immobilized metal ion affinity chromatography, and gel filtration (Schmidt et al. 2007). The successful purification of such proteins opens the way for biochemical and structural analysis of this class of important plant proteins.

1.3.5 Sequence of Process Steps

In spite of the efforts for increasing productivity in bioreactors, the efficient recovery and purification of target compounds from the broth leaving the bioreactor are still of prime importance. In plant cell and tissue cultures, the target compounds are either intracellular and retain inside the cells or extracellular and are secreted into the medium. A block diagram illustrating the process steps from the broth leaving the bioreactor till a raw extract or purified compound is obtained can be seen in Fig. 1.5.

The downstream processing for plant cell and tissue cultures begins with filtration or centrifugation of the broth leaving the bioreactor, in terms of intracellular compounds, the filtrate is collected and various cell disruption techniques are applied, whereas, the disrupted cells need to be separated by removing the cell debris. Subsequently, leaching is applied in order to capture the value-added compounds. The advantages of intracellular retention for downstream processing lie in the smaller volume of the harvested plant cells. In terms of extracellular compounds, the retentate or the liquid phase is concentrated by various membrane techniques depending on the molecular weight of the compounds of interest. After concentration, liquid–liquid extraction is applied. Alternatively, supercritical fluid extraction can be conducted for both intracellular and extracellular compounds circumventing cell disruption step and providing the possibility of solvent-free green processing. Depending on the end product requirement, the purification steps that follow are generally vacuum evaporation, spray drying, adsorption, and several chromatography options including adsorption chromatography, affinity chromatography, ion exchange chromatography, and size-exclusion chromatography which result in pure compound.

1.4 State of the Art for Downstream Processing of Plant Cell Cultures

Although commercial systems for cultivating plant cell cultures were established more than 25 years ago, there has been little progress toward wide application, and only a limited number of compounds such as shikonin, ginsenosides, taxol, and berberine are produced at large scale today (Yesil-Celiktas et al. 2010).

The majority of the innovations are related to developing new methods, optimizing process parameters, and designing novel bioreactors. However, fewer patents

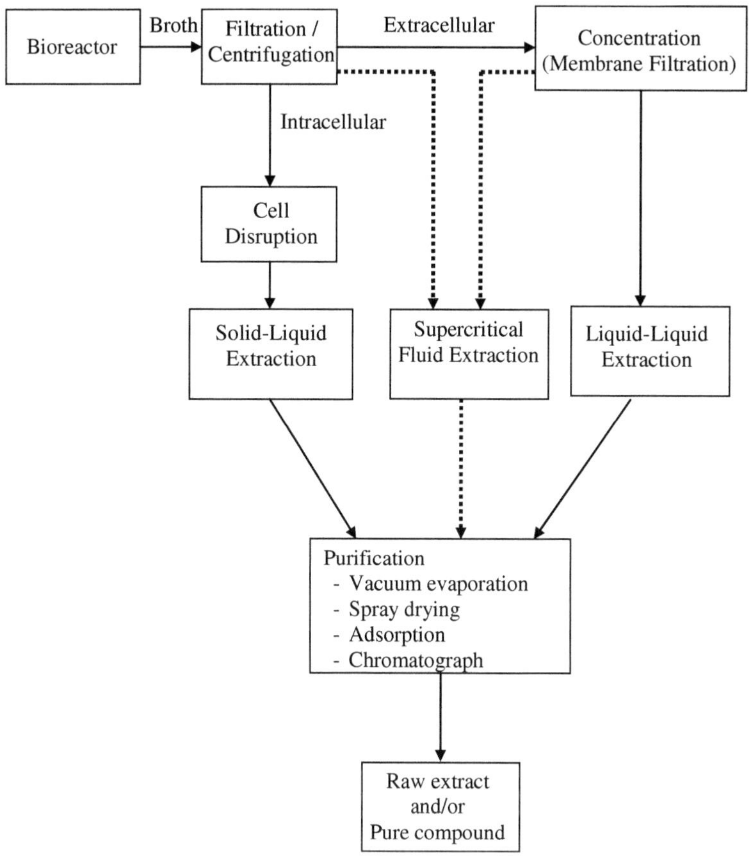

Fig. 1.5 A block diagram illustrating the process steps from the broth leaving the bioreactor till a raw extract or purified compound are obtained

are available in regard to the innovative approaches in downstream processing. Most of the patent applicants were either from universities or research institutes, on the other hand, a few companies were identified from Germany, France, and Japan. This raises the question whether the companies are not eager to file patents regarding the innovations in downstream processes accounting as much as 80 % of the total production costs (Hellwig et al. 2004) and rather keep them as trade secrets which can be in the form of a practice, process, design, or compilation of information providing economic advantage over competitors.

An earlier Japanese patent deals with the extraction of compounds produced in plant cell (Tanaka and Jajima 1985). Treatment of plant cell walls by aqueous extraction to yield a water-soluble fraction and an extraction residue consisting of edible fibres (Thibault et al. 1991) and extraction of α-linolenic acid, tocopherol, or sitosterols especially from Resedaceae family (Masse 1991) were patented as well.

Another invention was a US patent providing cyclodextrin derivatives that were substituted with groups bearing charges in aqueous solutions in their salt forms and their use, also in combinations with other cyclodextrins, as useful components of plant cell and tissue growth media. Addition of charged cyclodextrins to plant growth media were used for isolation of taxol and other bioactive taxanes which were secreted into extracellular media during the growth of *Taxus* cell cultures. These secreted compounds contain moieties such as side chain phenyl rings of the taxol molecule. Compounds of such structure are known to form strong complexes with cyclodextrins. When charged cyclodextrins were present in growth media, they formed complexes with the hydrophobic secreted compounds, particularly when cyclodextrins were present in large excess over the secreted compounds. Upon separation of cells from the growth medium, size exclusion chromatography was suggested for the isolation of formed complexes (Alexey 2004).

The Korean patents identified focus on two pharmaceutically important compounds, interleukin-12 protein and humanization fraction antibody. A production method of biologically active interleukin-12 protein using plant cell suspension culture was provided, thereby cheaply and safely producing a large quantity of biologically active interleukin-12 protein from *Nicotiana tabacum* (Kwon and Yang 2004). A method for preparing a humanization fraction antibody through a suspension culture of transformed plant cells was provided to improve clinical pharmacodynamics and to diagnose and treat cancer. The antibody fragment was isolated from the suspension culture (Yang 2011).

The scope of another invention was producing a stilbene plant secondary metabolite such as resveratrol, comprising, cultivation by suspension culture, including an amount of adsorbent and one or more elicitor agents suitable to increase production of the stilbene within the suspension culture and recovery of the stilbene from the suspension culture by means of methanol extraction (Zhang and Franco 2005). This study was a result of industry–university collaboration in Australia. Procyanidins have received increased attention due to their anticarcinogenic and antioxidant properties. Production and extraction of procyanidins from plant cell cultures were patented as well by Diana Plant Sciences Inc. in US. In particular, the provided methods are harvesting the cells by means of centrifugation and filtration, harvesting cell biomass in an appropriate solvent for extraction of polyphenol rich fraction, isolating a procyanidin rich fraction using liquid–liquid extraction and chromatography and finally drying or concentrating the fraction of interest (Venkatramesh et al. 2010).

A very recent patent filed by Uhde High Pressure Technologies GMBH in Germany can be a good example for supercritical fluid treatment in plant cell and tissue culture. The invention describes a cell lysis method for suspended starting materials by combination of pressure exertion, spraying, decompression, and subsequent selective extraction and separation of valuable cellular materials. A cell extract was produced in a cell lysis unit and a gas was led through the cell extract in an extraction stage and the gas loaded was separated from the valuable cellular materials in a separation stage while reducing the pressure. The starting material suspension and the solvent were brought to a pressure of 100–2,500 bar and mixed to give a solvent mixture. The solvent mixture was sprayed into a container via at least one nozzle which was at a pressure of 100–2,500 bar and a temperature of 10–90 °C, while the container had a lower pressure (Dierkes et al. 2011) just like the principle of supercritical micronization processes.

1.5 Future Challenges

Availability and future use of existing natural resources have become imperative in modern societies subsequent to global concerns for environmental protection, public health, and energy conservation. Furthermore, the overutilization of some value added plant species has led to ecological problems and severe constraints of any further harvesting. Hence, there is an increasing emphasis on economically viable and eco-friendly technologies. Further, significant progress has been made in recent years toward the development of low-cost, highly efficient, and safe bioreactor configurations which were supported by patent applications (Wang et al. 1999; Chen 2005; Yatian 2008). There have also been substantial advances in downstream processing which have been thoroughly reviewed in this chapter. However, the reflection of these advances on industrial-scale applications cannot be fully elucidated due to the confidentiality policies of the companies active in this field.

From the cost point of view, downstream processing in plant cell and tissue cultures may account for as much as 80 % of the total production costs (Hellwig et al. 2004). The recovery and purification of secondary metabolites for therapeutic applications are the most challenging but, on the other hand, costly processes in which high ratios of purities are required and techniques complying with the tight regulatory standards have to be fulfilled. Therefore, efforts to develop new and improved downstream processing methods can significantly reduce the overall costs associated with plant cell and tissue culture production systems (Nikolov and Woodard 2004). With the development of more integrated processes with less impact on environment, the expertise and the know-how would lead to a wide-scale commercialization of plant cell cultures and contribute to the preservation of global biodiversity.

References

Abdelkar MSA, Lockwood B, Sansongsak P (2010) Uptake of nicotine from suspension culture of *Nicotiana tabacum* by molecularly imprinted polymers. J Pharm Pharmacol 62:633–637

Alami I, Mari S, Clerivet A (1998) A glycoprotein from *Ceratocystis fimbriata* f. sp. *platani* triggers phytoalexin synthesis in *Platanus* x *acerifolia* cell-suspension cultures. Phytochemistry 48:771–776

Alexey E (2004) Charged cyclodextrin derivatives and their use in plant cell and tissue culture growth media. AU 2003298824 A1 20040623

Alpay E, Demircioglu M (2006) Mass transfer and mass transfer operations. Ege University publication, Izmir

Castor TP, Chikarmane HM, Hong GT, Shallice C (1998) Methods for fraction of biological derived materials. USP, 5854064

Chan DJC, Ahmad AL, Chan LK (2010) Improvement of biomass and juvenile hormone III (JH III) production from *Cyperus aromaticus* cell suspension culture via in situ membrane filtration technology. Acta Physiol Plant 32:1153–1159

Charoenrat T, Ketudat-Cairns M, Jahic M, Enfors SO, Veide A (2006) Recovery of recombinant beta-glucosidase by expanded bed adsorption from *Pichia pastoris* high-cell-density culture broth. J Biotechnol 122:86–98

Chaterjee A, Shukla S, Mishra P, Rastogi A, Singh SP (2010) Prospects of in vitro production of thebaine in opium poppy (*Papaver somniferum* L.). Ind Crops Prod 32:668–670

Chen SQ (2005) Plant cell tissue simple-pouring bioreactor culture system CN 1710058 A 20051221

Chiang L, Abdullah MA (2007) Enhanced anthraquinones production from adsorbent-treated *Morinda elliptica* cell suspension cultures in production medium strategy. Process Biochem 42:757–763

Choi JW (1992) In situ berberine separation with immobilized adsorbent in cell-suspension cultures of *Thalictrum rugosum*. Korean J Chem Eng 9:128–134

Corry JP, Reed WL, Curtis WR (1993) Enhanced recovery of solavetivone from *Agrobacterium* transformed root cultures of hyoscyamus-muticus using integrated product extraction. Biotechnol Bioeng 42:503–508

Curino A, Milanesi L, Benassati S, Skliar M, Boland R (2001) Effect of culture conditions on the synthesis of vitamin D-3 metabolites in *Solanum glaucophyllum* grown in vitro. Phytochemistry 58:81–89

Darani KK, Farahani EV (2005) Application of supercritical fluid extraction in biotechnology. Crit Rev Biotechnol 25:231–242

Dierkes H, Steinhagen V, Bork M, Lutge C, Knez Z (2011) Cell lysis of plant or animal starting materials by a combination of a spray method and decompression for the selective extraction and seperation of valuable intracellular materials EP 2315825 A1 20110504

Dobbs JM, Wong JM, Lahiere RJ, Johnston KP (1987) Modification of supercritical fluid phase-behavior using polar cosolvents. Ind Eng Chem Res 26:56–65

Donnez D, Kim KH, Antoine S, Conreux A, De Luca V, Jeandet P (2011) Bioproduction of resveratrol and viniferins by an elicited grapevine cell culture in a 2 L stirred bioreactor. Process Biochem 46:1056–1062

Gallois R, Prevot JC, Clement A, Jacob JL (1997) Purification and characterization of phosphoribosylpyrophosphate synthetase from rubber tree latex. Plant Physiol 115:847–852

Ghosh A, Ganapathi TR, Nath P, Bapat VA (2009) Establishment of embryogenic cell suspension cultures and *Agrobacterium*-mediated transformation in an important Cavendish banana cv. Robusta (AAA) Plant Cell Tiss Org 97:131–139

Gichner T, Stavreva DA, Cerovskan N, Wagner ED, Plewa MJ (1995) Metabolic-Activation of m-phenylenediamine to products mutagenic in *Salmonella typhimurium* by medium isolated fiom tobacco Suspension Cell-Cultures. Mutat Res-Fund Mol M 331:127–132

Gromova M, Roby C (2010) Toward *Arabidopsis thaliana* hydrophilic metabolome: assessment of extraction methods and quantitative 1H NMR. Physiol Plantarum 140:111–127

Gurel A (1989) The Effects of Plant Growth Regulators on The Production of Dihaploids of Tobacco and Datura through Callus Culture, PhD thesis, Ege University Graduate School of Natural and Applied Sciences, p.101, Bornova

Harashima H, Kato K, Shinmya A, Sekine M (2007) Auxin is required for the assembly of A-type cyclin-dependent kinase complexes in tobacco cell suspension culture. J Plant Physiol 164:1103–1112

Hartonen K, Parshntsev J, Sandberg K, Bergelin E, Nisula L, Riekkola M-L (2007) Isolation of flavonoids from aspen knotwood by pressurized hot water extraction and comparison with other extraction techniques. Talanta 74:32–38

Hellwig S, Drossard J, Twyman RM, Fischer R (2004) Plant cell cultures for the production of recombinant proteins. Nat Biotechnol 22:1415–1422
Hong HJ, Lee JE, Ahn JE, Kim DI (1998) Enhanced production of digoxin by digitoxin biotransformation using in situ adsorption in *Digitalis lanata* cell cultures. J Microbiol Biotechnol 8:478–483
Hwang SY, Lee EK, Terashima M (2005) Chromatography based purification of rhGM-CSF from recombinant rice cell culture. Chem Eng Technol 28:1421–1425
Jo SH, Kwon SY, Park DS, Yang KS, Kim JW, Lee KT, Kwak SS, Lee HS (2006) High-yield production of functional human lactoferrin in transgenic cell cultures of Siberian ginseng (*Acanthopanax senticosus*) Biotechnol. Bioprocess Eng 11:442–448
Kaiser CS, Römpp H, Schmidt PC (2001) Pharmaceutical Applications of Supercritical Carbon Dioxide. Pharmazie 56:907–926
Kang SH, Kim JH, Mun S (2010) Optimal design of a tandem simulated moving bed process for separation of paclitaxel, 13-dehydroxybaccatin III, and 10- deacetylpaclitaxel. Process Biochem 45:1468–1476
Kim TH, Shin JH, Baek HH, Lee HJ (2001) Volatile flavour compounds in suspension culture of *Agastache rugosa* Kuntze (Korean mint). J Sci Food Agric 81:569–575
Kurosumi A, Kobayashi F, Mtui G, Nakamura Y (2006) Development of optimal culture method of *Sparassis crispa* and a new extraction method of antineoplastic constituent. Biochem Eng J 30:109–113
Kwon TH, Yang MS (2004) Production method of a large amount of biological active interleukin-12 protein using plant cell suspention culture cheaply and safely. KR 20040097575 A 20041118
Machado MW, Neto CS, Salgado J, Zaffari G, Barison A, Campos FR, de Corilo YE, Eberlin MN, Biavatti MW (2010) Search for alkaloids on callus culture of *Passiflora alata*. Brazillian Arch Biol Technol 53:901–910
Masse JP (1991) Therapeutic composition containing alpha-linolenic acid and a compound capable of promoting the passage of the acid through the cell membrane, plant extract comprising the acid and the compound, and process for the preparation of the extract, US 5002767 A 19910326
McHugh MA, Krukonis VJ (1994) Supercritical fluid extraction. Principles and practice, 2nd edn. Butterworth-Heinemann, Stoneham
Nakamura S, Hayashi T (1993) Purification and properties of an extracellular endo-1, 4-beta-glucanase from suspension-cultured poplar cells. Plant Cell Physiol 34:1009–1013
Nikolov ZL, Woodard SL (2004) Downstream processing of recombinant proteins from transgenic feedstock. Curr Opin Biotechnol 15:479–486
Ogawa K, Matsushita T, Funatsu K (1991) Establishment of a simple separation process for highly pure beta-carotene from cultured carrot cells. Kagaku Kogaku Ronbunshu 17:437–441
Prakash G, Srivastava AK (2007) Azadirachtin production in stirred tank reactors by *Azadirachta indica* suspension culture. Process Biochem 42:93–97
Pyo SH, Song BK, Han BH, Kim JH (2004) A large-scale purification of paclitaxel from cell cultures of Taxus chinensis Process Biochem 39:1985–1991
Rivzi SSH, Benado AL, Zollweg JA, Daniels JA (1986) Supercritical fluid extraction: Fundamental principles and modelling methods. Food Technol 7:55–65
Salgin U, Ustun AS, Mehmetoglu U, Calimli A (2005) Supercritical CO_2 extraction of accumulated capsidiol from biotic elicitor-activated *Capsicum annuum* L fruit tissues. J Chem Technol Biotechnol 80:124–132
Sandell J (1996) Balancing market and technology requirements in early stage product development. Dissertation, Department of Industrial Marketing and Economics, Chalmers University of Technology, Goteborg
Schmidt SA, Williams SJ, Wang CIA, Sornaraj P, James B, Kobe B, Dodds PN, Ellis JG, Anderson PA (2007) Purification of the M flax-rust resistance protein expressed in *Pichia pastoris*. Plant J 50:1107–1117
Seader JD, Henly EJ (1998) Separation process principles. Wiley, USA

Sim SJ, Chang HN, Liu JR, Jung KH (1994) Production and secretion of indole alkaloids in hairy root cultures of *Catharanthus roseus*—effects of in situ adsorption, fungal elicitation and permeabilization. J Ferment Bioeng 78:229–234

Soletto D, Binaghi L, Lodi A, Carvalho JCM, Converti A (2005) Batch and fed-batch cultivations of *Spirulina platensis* using ammonium sulphate and urea as nitrogen sources. Aquaculture 243:217–224

Son SH, Choi SM, Lee YH, Choi KB, Yun SR, Kim JK, Park HJ, Kwon OW, Noh EW, Seon JH, Park YG (2000) Large-scale growth and taxane production in cell cultures of *Taxus cuspidata* (Japanese yew) using a novel bioreactor. Plant Cell Rep 19:628–633

Su WW, Lei F, Kao NP (1995) High density cultivation of *Anchusa officinalis* in a stirred-tank bioreactor with in situ filtration. Appl Microbiol Biotechnol 44:293–299

Sutor R, Kreis W (1996) Partial purification and characterization of the cell-wall-associated lanatoside 15′-O-acetylesterase from *Digitalis lanata* suspension cultures. Plant Physiol Biochem 34:763–770

Takahashi T, Warren GS (1994) Studies on an inhibitor of TMV viral-antigen accumulation released by cultured-cells of *Nicotiana-tabacum*. Plant Sci 100:83–96

Tanaka H, Jajima M (1985) Extraction of material produced in plant cell. JP 60180594 A 19850914

Tapia JC, Garcia R, Escamilla EM, Calva G, Rocha JA (1993) Capsaicin recovery from a cell-culture broth. Ind Eng Chem Res 32:2242–2246

Taylor LT (1996) Supercritical fluid extraction. Wiley, USA

Thibault JF, Della VG, Ralet MC (1991) Process for treating products rich in plant cell walls, in particular rich in pectins, said pectins, the extraction residues from the process, their use and compositions containing them. EP 0432210 A1 19910619

Thomas TD (2008) The role of activated charcoal in plant tissue culture. Biotechnol Adv 26:618–631

Thorpe TA (2007) History of plant tissue culture. Mol Biotechnol 37:69–180

Tikhomiroff C, Allais S, Klvana M, Hisiger S, Jolicoeur M (2002) Continuous selective extraction of secondary metabolites from *Catharanthus roseus* hairy roots with silicon oil in a two-liquid-phase bioreactor. Biotechnol Prog 18:1003–1009

Venkatramesh M, Wagner D, Lall S, Lejard F (2010) Production and extraction of procyanidins from plant cell cultures. WO 2010114567 A1 20101007

Wang Y, Zhao B, Ouyang F (1999) Airlift cyclic immersion lighting plant cell tissue organ culture method and culture reactor CN 1063224 C 20010314

Wang CG, Wu JY, Mei XG (2001) Enhanced taxol production and release in *Taxus chinensis* cell suspension cultures with selected organic solvents and sucrose feeding. Biotechnol Prog 17:89–94

Warzecha H, Obitz P, Stockigt J (1999) Purification, partial amino acid sequence and structure of the product of raucaffricine-O-beta-D-glucosidase from plant cell cultures of *Rauwolfia serpentine*. Phytochemistry 50:1099–1109

Williams RD, Chauret N, Bedard C, Archambault J (1992) Effect of polymeric adsorbents on the production of sanguinarine by *Papaver somniferum* cell-cultures. Biotechnol Bioeng 40:971–977

Winkelnkemper T, Schuldt S, Schembecker G (2011) Systematic downstream process development for purification of baccatin III with key performance indicators. Sep Purif Technol 77:355–366

Xu XH, Zhang W, Cao XP, Xue S (2011) Abietane diterpenoids synthesized by suspension-cultured cells of *Cephalotaxus fortunei*. Phytochem Lett 4:52–55

Yang MS (2011) Method of producing humanized fragment antibody via transgenic plant cell suspention culture. KR 20100125870 A 20101201

Yatian D (2008) Curtain-shape anchoring culture bioreactor and method for culturing plant cell, tissue or organ using the same CN 101100641 A 20080109

Yesil-Celiktas O, Nartop P, Gurel A, Bedir E, Vardar Sukan F (2007) Determination of phenolic content and antioxidant activity of extracts obtained from *Rosmarinus officinalis'* calli. J Plant Physiol 164:1536–1542

Yesil-Celiktas O, Gurel A, Vardar-Sukan F (2010) Large scale cultivation of plant cell and tissue culture in bioreactors. Transworld Research Network, Kerala

Zhang W, Franco CMM (2005) Production of plant secondary metabolites using adsorption and elicitation in cell suspention culture. WO 2005012507 A1 20050210

Zhang CH, Xu HB (2001) Improved paclitaxel production by in situ extraction and elicitation in cell suspension cultures of *Taxus chinensis*. Biotecnol Lett 23:189–193

Chapter 2
Agrobacterium rhizogenes-Mediated Transformation in Medicinal Plants: Prospects and Challenges

Dipasree Roychowdhury, Anrini Majumder and Sumita Jha

2.1 Introduction

Root cultures have been studied since the early days of tissue culture research, but have created little interest because of their slow growth rate, although root and shoot organ cultures have been used for studies of alkaloids (Hashimoto et al. 1986; Hirata et al. 1990; Jha et al. 1991; Baíza et al. 1999a; Khanam et al. 2001; Ghosh et al. 2002; Ghosh and Jha; 2005), coumarins (Panichayupakarananta et al. 1998), saponins (Kusakari et al. 2000; Kim et al. 2005a), phenolic acids (Karam et al. 2003), essential oils (Olszowska et al. 1996), terpenes (Pannuri et al. 1993), glycosides (Swanson et al. 1992), steroidal lactones (Ray and Jha 2001), etc.

However, in all but a few species, they are difficult to culture and the auxin concentrations optimal for growth may reduce productivity (Siah and Doran 1991; Bourgaud et al. 2001). Few studies on secondary metabolite production with root and shoot organ cultures in bioreactors have been reported (Kevers et al. 1999; Choi et al. 2000; Bondarev et al. 2002; Piatczak et al. 2005; Kim et al. 2005a).

Although, root and shoot organ cultures are genetically and biosynthetically more stable than cell cultures, interest in root and shoot organ cultures as a source of secondary metabolites has been limited (Flores and Curtis 1992; Flores and Medina-Bolivar 1995).

D. Roychowdhury · A. Majumder · S. Jha (✉)
Department of Botany, Centre of Advanced Study,
University of Calcutta, University College of Science,
35 Ballygunge Circular Road, Kolkata 700019, India
e-mail: sjbot@caluniv.ac.in; sumitajha.cu@gmail.com

S. Chandra et al. (eds.), *Biotechnology for Medicinal Plants*,
DOI: 10.1007/978-3-642-29974-2_2,

2.2 Ri-Transformed Root Cultures

Agrobacterium rhizogenes infects higher plants to produce so-called "hairy roots" with altered phenotype from the wound sites. The transformed roots can be excised to establish axenic root cultures and indefinitely propagated in growth regulator free medium. The root exhibit fast, plagiotropic growth characterized by profuse lateral branching and rapid root tip elongation (Tepfer and Tempé 1981; Chilton et al. 1982; Tepfer 1984). Putatively transformed root lines can be easily screened with these morphological markers. Foreign genes can also be introduced into transformed roots by using binary vectors (Hamill et al. 1988). The rapid biomass accumulation in transformed root cultures is comparable, if not superior, to unorganized cell cultures and the fastest biomass doubling time is 1-day in *Datura stramonium* (Maldonado-Mendoza et al. 1993). The advantage of this transformation approach is that each primary root produced at the infection site is the result of a single transformation event—a clone (Chilton et al. 1982). However, somaclonal variations in transformed root cultures are also known (Sevón et al. 1998; Wilhelmson et al. 2005). Transformed root cultures have been established in several species, including many medicinal plants (Tepfer 1989; Sevón and Oksman-Caldentey 2002).

Transformed clones vary in morphology (Amselem and Tepfer 1992), growth, and metabolite productivity (Aoki et al. 1997; Batra et al. 2004). This is attributed to the nature, site, and number of T-DNA integration into the host genome (Ambros et al. 1986a, b; Jouanin et al. 1987; Amselem and Tepfer 1992). Therefore, clone selection is critical for metabolite productivity through transformed root cultures (Mano et al. 1986). Transformed root clones are genetically and biosynthetically stable for long periods. Growth and alkaloid production were stable over a period of 5 years in transformed roots of *D. stramonium* (Maldonado-Mendoza et al. 1993). Ri-transformed roots also exhibit a high degree of cytogenetic stability (Baíza et al. 1999b). In *Beta vulgaris* and *Nicotiana rustica* (Benson and Hamill 1991), growth, secondary metabolite production, and T-DNA structure in several transformed root lines were unchanged after cryopreservation. However, progressive loss of growth rate under conditions that favor the production of tropane alkaloids was reported in *Duboisia myoporoides* (Yukimune et al. 1994).

Plant roots can synthesize, store, and secrete a vast array of compounds and transformed root cultures have a wide range of biosynthetic capacities (Flores et al. 1999). They can produce the full range of secondary products characteristics of roots of the parent plants (Parr and Hamill 1987; Zárate 1999) as well as novel compounds (Fukui et al. 1998, 1999). The secondary metabolite levels are often comparable to or greater than that of intact plants (Sevón and Oksman-Caldentey 2002). Unlike cell cultures, actively growing transformed roots can continuously produce secondary metabolites (Holmes et al. 1997). Transformed root cultures are reported to synthesize secondary products, including alkaloid, in numerous medicinal plant species (Table 2.1). Many of these have already been reviewed

Table 2.1 Secondary metabolite production in Ri-transformed roots

Plant species	*A. rhizogenes* strain used for transformation	Secondary metabolite	Reference
Aconitum heterophyllum	LBA 9402, LBA 9360, A4	Heteratisine, atisine, hetidine	Giri et al. (1997)
Agastache rugosa	R1000	Rosmarinic acid	Lee et al. (2008)
Ajuga multiflora	A4	20-hydroxyecdysone	Kim et al. (2005b)
Ambrosia maritima	ATCC15834	Thiarubrine A, thiarubrine A epoxide, thiarubrine A diol	Zid and Orihara (2005)
Ammi majus	A4, LBA9402	Umbelliferone	Królicka et al. (2001)
Artemisia annua	–	Artemisinin	Weathers et al. (2005)
	LBA 9402, 8196, A4 1601	–	Mukherjee et al. (1995)
A. dubia, A. indica	LBA 9402 and 8196	Artemisinin	Mannan et al. (2008)
Astragalus mongholicus	LBA 9402, ATCC 15834, R 1601 and TR 105	Astragaloside I, astragaloside II, astragaloside III	Ionkova et al. (1997)
Atropa acuminata	LBA 9402	Atropine, scopolamine	Banerjee et al. (2008)
Atropa belladonna	15834	Atropine, scopolamine	Kamada et al. (1986)
Atropa belladonna	MAFF 03-01724, ATCC 15834	Hyoscyamine, 6β-hydroxyhyoscyamine, scopolamine, littorine	Jaziri et al. (1994)
Atropa belladonna	AR15834	Scopolamine, hyoscyamine	Chashmi et al. (2010)
Azadirachta indica	LBA9402	Azadirachtin, nimbin, salannin, 3-acetyl-1-tigloylazadirachtinin, 3-tigloylazadirachtol	Allan et al. (2002)
Brugmansia candida	LBA 9402	Scopolamine, hyoscyamine	Spollansky et al. (2000)
Brugmansia candida	LBA 9402	Cadaverine, putrescine, spermidine, spermine	Carrizo et al. (2001)
Brugmansia suaveolens	15834, TR 105	Tropine, pseudotropine, scopoline, scopine, aposcopolamine, hyoscyamine	Zayed and Wink (2004)
Calystegia sepium	15834	Calystegine	Scholl et al. (2001)
Camptotheca acuminata	ATCC 15834, R-1000	Camptothecin, 10-hydroxycamptothecin	Lorence et al. (2004)

(continued)

Table 2.1 (continued)

Plant species	*A. rhizogenes* strain used for transformation	Secondary metabolite	Reference
Catharanthus roseus	15834, A2, A2-83, A47-83, R1000	Yohimbine, ajmalicine, tetrahydroalstonine, tabersonine, horhammericine, lochnericine, venalstonine, vindolinine, 19-epi-vindolinine, catharanthine, pericalline, O-acetylvallesamine	Toivonen et al. (1989)
	15834	Catharanthusopimaranoside A, catharanthusopimaranoside B	Chung et al. (2008)
	R1000	Vincristin, vinblastin	Zargar et al. (2010)
C. roseus var. Prabal	A4	Serpentine, ajmalicine	Batra et al. (2004)
Centaurium erythraea, *C. pulchellum*	A4M70GUS	Xanthone	Janković et al. (2002)
Cephaelis ipecacuanha	ATCC 15834	Cephaeline, emetine	Yoshimatsu et al. (2003)
Cichorium intybus	LMG 150	Esculin, esculetin	Bais et al. (1999)
Cinchona ledgerina	LBA 9402	Quinine, cinchonidine, quinidine, quinamine	Hamill et al. (1989)
Cinchona officinalis	LBA 9402 harbouring a binary vector	Tryptamine, strictosidine, quinidine,quinine, cinchonine, cinchonidine	Geerlings et al. (1999)
Coleus blumei	A4	Rosmarinic acid	Bauer et al. (2009)
Coleus forskohlii	MAFF 03-01724	Rosmarinic acid	Li et al. (2005)
	–	Forskolin	Sasaki et al. (1998)
Datura stramonium	LBA 9402	Hyoscyamine, apohyoscyamine,	Robins et al. (1991)
	1855, AR-10, TR-105, ATCC15834, A4, A41027, ATCC13333	Hyoscyamine, scopolamine	Maldonado-Mendoza et al. (1993)
	ATCC 15834	Hygrine, tropinone, apoatropine	Berkov et al. (2003)
Duboisia leichhardtii	15834, A4	Scopolamine	Mano et al. (1989)
Glycyrrhiza glabra	LBR56	Glycyrrhizic acid	Tenea et al. (2008)

(continued)

Table 2.1 (continued)

Plant species	*A. rhizogenes* strain used for transformation	Secondary metabolite	Reference
Glycyrrhiza pallidiflora	pRi 15834	Licoagroisoflavone, licoagroside C, calycosin, erythrinin, isoliquiritigenin, echinatin, maackiain, trifolirhizin, ononin	Li et al. (2001)
Glycyrrhiza uralensis	ACCC 10060 harbouring vector p130/35S-GuSQS1	Glycyrrhizin	Lu et al. (2008)
	pRi A4 harboring the binary vector pCHI	Flavonoids	Zhang et al. (2009)
	pRi A4	Licochalcone A	Zhang et al. (2011)
Gmelina arborea	ATCC 15834	Verbascoside	Dhakulkar et al. (2005)
Gynostemma pentaphyllum	ATCC 15834	Gypenoside	Chang et al. (2005)
Harpagophytum procumbens	ATCC 15834	Harpagoside, gallic acid	Georgiev et al. (2006)
	A4	Harpagoside, verbascoside, isoverbascoside	Grabkowska et al. (2010)
Hyoscyamus albus	LBA9402, A4	Atropine	Zehra et al. (1999)
Hyoscyamus muticus	LBA9402, A4	Scopolamine, hyoscyamine	Zolala et al. (2007)
Linum tauricum ssp. *tauricum*	TR 105, ATCC 15834	4′-demethyl-6-methoxypodophylotoxin, 6-methoxypodophyllotoxin	Ionkova and Fuss (2009)
Nicotiana tabacum cv. Xanthi	LBA9402, LBA9402 pLAL21	Hyoscyamine, scopolamine, nicotine, nornicotine, anabasine, anatabine, anatalline	Häkkinen et al. (2005)
Ocimum basilicum	ATCC 15834, MAFF 03-01724	Rosmarinic acid, lithospermic acid, lithospermic acid B	Tada et al. (1996)
	ATCC 15834	Rosmarinic acid	Bais et al. (2002a)
Ophiorriza pumila	15834	Camptothecin	Saito et al. (2001)
Panax ginseng	A4	Ginsenoside	Bulgakov et al. (1998)

(continued)

Table 2.1 (continued)

Plant species	*A. rhizogenes* strain used for transformation	Secondary metabolite	Reference
Papaver bracteatum	R15834 harbouring the binary vector pBI121	Morphine, noscapine, sanguinarine	Rostampour et al. (2009)
Papaver somniferum var. *album*	LBA 9402	Morphine, codeine, sanguinarine	Le Flem-Bonhomme et al. (2004)
Picrorhiza kurroa	LBA 9402	Kutkoside, picroside I	Verma et al. (2007)
Plantago lanceolata	LBA 9402	Verbascoside, plantamoside	Fons et al. (1999)
Plumbago indica	ATCC 15834	Plumbagin	Gangopadhyay et al. (2011)
Plumbago zeylanica	A4	Plumbagin	Verma et al. (2002)
Przewalskia tangutica	A4	Hyoscyamine, scopolamine	Lan and Quan (2010)
Pueraria phaseoloides	ATCC15834, harboring agropine type plasmid pRiA4b	Puerarin	Shi and Kintzios (2003)
Rauvolfia micrantha	ATCC 15834	Ajmalicine, ajmaline	Sudha et al. (2003)
Ruta graveolens	LBA 9402	Pinnarin, rutacultin, xanthotoxin, bergapten, isopimpinelin, osthole, osthenol, dictamnine, skimmianine, kokusaginine, rybalinine and an isomer of rybalinine	Sidwa-Gorycka et al. (2009)
Salvia broussonetti	ATCC-15834	Brussonol, iguestol	Fraga et al. (2005)
Salvia miltiorrhiza	ATCC 15834	Tanshinones	Chen et al. (2001)
Salvia sclarea	LBA 9402 carrying pRi 1855 plasmid	Salvipisone, aethiopinone, ferruginol	Kuźma et al. (2006)
Saussurea involucrata	LBA 9402, R1000, R1601	Syringin	Fu et al. (2005)
Saussurea medusa	R1601	Jaceosidin	Zhao et al. (2004)
Scopolia parviflora	KCTC 2703 harbouring pBE*pmt* plasmid	Scopolamine, hyoscyamine	Lee et al. (2005)

(continued)

Table 2.1 (continued)

Plant species	*A. rhizogenes* strain used for transformation	Secondary metabolite	Reference
Scutellaria baicalensis	A4	Baicalin, wogonoside, baicalein, wogonin	Kovács et al. (2004)
	ATCC 15834	Baicalin	Hwang (2006)
	A4GUS	Baicalin, wogonin	Tiwari et al. (2008)
Taxus x *media* var. *Hicksii*	LBA 9402	Paclitaxel, 10-deacetylbaccatin III	Furmanowa and Syklowska-Baranek (2000)
Tylophora indica	A4	Tylophorine	Chaudhuri et al. (2005)
Valeriana wallichii	LBA 9402, A4	Isovaleroxyhydroxy didrovaltrate, didrovaltrate, acevaltrate, homodidrovaltrate	Banerjee et al. (1998)
Withania somnifera	LBA 9402, A4	Withasteroids	Ray et al. (1996), Bandyopadhyay et al. (2007)

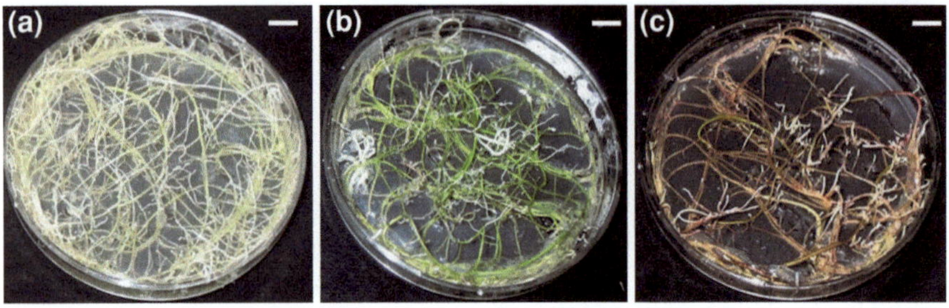

Fig. 2.1 Ri-transformed roots of *Tylophora indica* showing pigmentation **a** white root (bar = 9 mm), **b** green root (bar = 9 mm), and **c** red root (bar = 9 mm)

(Verpoorte et al. 1991; Rhodes et al. 1997; Mukundan et al. 1998; Shanks and Morgan 1999; Sevón and Oksman-Caldentey 2002).

Secretion is a fundamental function of plant cells and it is especially well-developed in plant roots (Roschina and Roschina 1993). Their ability to secrete plethora of compounds into the rhizosphere is a remarkable physiological feature and up to nearly 21 % of all carbon fixed by photosynthesis can be transferred to the rhizosphere in the form of root exudates (Marschner 1995) and contains both low molecular weight (e.g. secondary metabolites) and high molecular weights (e.g. proteins) compounds. Secondary metabolites that are produced in the roots that are transported and stored in other parts of the plants may be released in the culture medium from excised roots (Rhodes et al. 1986). Several secondary products, including a few alkaloids produced in transformed root culture of a number of plant species are released into the culture medium. In transformed root cultures the released secondary metabolites can be adsorbed and removed by a variety of high affinity polymeric resins (Freeman et al. 1993). In addition to a reduction in the production cost, this operation is known to stimulate the productivity of a number of compounds by transformed root cultures (Muranaka et al. 1993a; Holmes et al. 1997; Saito et al. 2001).

Ri-transformed root cultures of *Tylophora indica* shows variation in pigmentation when cultured under light (unpublished data). While, the hairy root cultures of *T. indica* remains white when cultured in dark, on exposure to light, they become green to red depending on the light intensity (Fig. 2.1). Hairy root turning green on exposure to light is also reported in other plants like in *Solanum khasianum* (Jacob and Malpathak 2005), red, and green hairy root lines in addition to white Ri-transformed root cultures are also reported by Yang and Choi (2000) in *Panax ginseng*.

2.3 Ri-Transformed Plants

Regeneration of whole viable plants from hairy root cultures, established from transformation with *A. rhizogenes*, has been reported in a number of plant species. Such transgenic plants frequently show a very characteristic phenotype which differ from their normal counterparts, such as, wrinkled leaf, shortened internodes, decreased apical dominance, altered flower morphology, increase in number of branches, reduced pollen and seed production, and abundant production of highly branched plagiotropic roots. All these altered phenotypic characters the so-called "Hairy Root Syndrome" is due to the combined expression of *rolA*, *rolB*, and *rolC* genes. *rolA* gene is associated with shortening of internodes, wrinkling of leaves, etc., whereas *rolB* gene causes reduced length of stamens, protruding stigmas, and increased adventitious roots on stems. *rolC* gene is responsible for reduced apical dominance, internodes shortening, and increased branching (Nilsson and Olsson 1997; Tepfer 1984). The hairy root phenotype was first described by Ackerman (1977) in Tobacco regenerants. In addition to the above-mentioned changes, biennial species frequently becomes annuals on transformation and regeneration with *A. rhizogenes* (Tepfer 1984; Sun et al. 1991; Kamada et al. 1992). Regeneration of transgenic plants from hairy roots can be either spontaneous or can be induced with the help of plant growth regulators.

2.3.1 Spontaneous Plant Regeneration from Ri-Transformed Root Cultures

One of the various advantages of using *A. rhizogenes*-mediated transformation system is direct regeneration of transgenic plants from root cultures as it avoids the problems due to somaclonal variations. Spontaneous and direct development of adventitious shoot buds from older regions of transformed hairy roots in hormone-free media without any callus formation is reported in a number of plant species like *Armoracia lapathifolia* (Noda et al. 1987), *Taraxacum platycarpum* (Lee et al. 2004), *Centaurium erythraea* (Subotić et al. 2003), *Hypericum perforatum* (Vinterhalter et al. 2006), *T. indica* (Chaudhuri et al. 2006), *Bacopa monnieri* (Majumdar et al. 2011), *Atropa belladonna* (Jaziri et al. 1994), *Plumbago indica* (Gangopadhyay et al. 2010), *Brassica oleracea* var. Botrytis (David and Tempé 1988), *B. oleracea* var. sabauda, *B. oleracea* var. capitata (Christey et al. 1997; Sretenović-Rajičić et al. 2006), *Populus tremula* (Tzfira et al. 1996), *Lotus corniculatus* (Petit et al. 1987), *Blackstonia perforiata* (Bijelović et al. 2004), *Pelargonium graveolens* cv. Hemanti (Saxena et al. 2007), *Ajuga reptans* var. atropurpurea (Tanaka and Matsumoto 1993), etc. These adventitious shoots when excised and cultured on hormone-free basal media regenerated into whole plants. However, culture conditions and time required for regeneration varied from plant to plant.

Shoot regeneration from transformed roots can be light dependent or independent. In *A. lapathifolia* (Noda et al. 1987), roots maintained in dark showed induction of shoot buds on transfer to light throughout the root except the root tips, but no adventitious shoot bud formation took place in those kept in dark. Non-transformed roots rarely developed adventitious shoot buds on transfer to light. LBA 9402 transformed roots of *P. tremula* (Tzfira et al. 1996) and A4 transformed roots of *B. monnieri* (Majumdar et al. 2011) showed spontaneous shoot bud regeneration when cultured under 16/8 h (light/dark) photoperiod. Interestingly, LBA 9402 transformed roots of *B. monnieri* did not show any regeneration but spontaneously dedifferentiated into callus. In *L. corniculatus* (Petit et al. 1987) and *P. indica* (Gangopadhyay et al. 2010), spontaneous shoot organogenesis is reported, when transformed roots were transferred to continuous light from dark. The hairy roots of *P. indica* did not regenerate in dark even after application of exogenous hormones to the media. Contrastingly, in *T. indica* (Chaudhuri et al. 2006), 17 % of transformed root clones are reported to regenerate shoots directly on hormone-free MS media in light independent manner. Subotić et al. (2003) also reported development of adventitious shoot primordial on older regions of hairy root cultures of *C. erythraea* both under light and in dark. In *Amoracia rusticana*, effect of light on shoot regeneration from transformed root cultures have been studied in details by Saitou et al. (1992). In dark conditions, shoot formation was rarely observed in the hairy roots, but the longer the culture period in the light, higher the frequency of shoot formation, and number of shoots per explant were noted. When cultured under light conditions (16 h light/8 h dark) for the first 4 weeks, no shoot bud regeneration was observed from the hairy roots, but when they were precultured in darkness for 2 weeks and then transferred to light, shoot formation was observed within 1 week of transfer. Shoot regeneration also varied with exposure time of hairy roots to light. A short exposure (1.68 h) to a high light intensity (78 μmol/m^2 s) showed less shoot formation than a long exposure (168 h) to lower light intensity (0.78 μmol/m^2 s). Thus, for adventitious shoot formation from transformed roots of *A. rusticana*, duration of light exposure is more important than the light intensity. The authors also showed that shoot formation in the hairy roots can be induced by white light and red light when precultured for 12 weeks in dark but not by the far red light. Far red light irradiation after red light irradiation partially inhibited shoot formation. Red light is also reported to stimulate shoot formation in *Pseudotsuga menziesi*, Petunia, and Apple (Kadkade and Jopson 1978; Economou and Read 1986; Predieri and Malavasi 1989). Shoot formation in tobacco calli is reported to be stimulated by blue light and inhibited by red light (Seibert et al. 1975; Weis and Jaffe 1969). Hence, effect of light on adventitious shoot organogenesis in hairy roots appeared to be plant specific. Time requirement for regeneration of shoot buds from the Ri-transformed roots also varied from plant to plant. In *A. lapathifolia* (Noda et al. 1987) and *B. monnieri* (Majumdar et al. 2011), shoot bud development was observed within 7 and 10 days of transfer to light, respectively. In others like, *L. corniculatus* (Petit et al. 1987), *P. tremula* (Tzfira et al. 1996), and *P. indica* (Gangopadhyay et al. 2010), shoot bud regeneration took place within 3–4 weeks.

Again, in *T. indica* (Chaudhuri et al. 2006) and *A. belladona* (Jaziri et al. 1994) shoot buds were seen only when transformed roots were kept for more than 8 weeks without subculture.

Regeneration potential of hairy roots also varied with the concentration of sucrose used in the media. When hairy roots of *A. belladona* (Jaziri et al. 1994) were cultured on half strength MS media with 1.5 % sucrose, spontaneous shoot bud regeneration took place. For rooting and further development, these shoot buds were transferred to the same media with 3 % sucrose. Hairy roots of *H. perforatum* (Vinterhalter et al. 2006) is reported to have high potential for spontaneous shoot regeneration which increased from first subculture (40.5 %) to second subculture (62 %) in hormone-free basal media. However, the regeneration frequency varied with the concentration of the sucrose used in the media. Highest shoot bud regeneration took place on hormone-free medium with 1 and 2 % sucrose with 50.3 and 48.8 shoot buds per culture, respectively. Higher concentration of sucrose at and above 4 % have got an adverse effect on both root growth and shoot differentiation, making them necrotic with characteristic brown color. Thus, low level of sucrose seems to favor regeneration of shoot buds from hairy roots in *A. belladona* and *H. perforatum.*

In addition to direct shoot bud regeneration from transformed root cultures, spontaneous callus formation is reported in many plant species like in *T. indica* (Chaudhuri et al. 2006), *Carica papaya* (Cabrera-Ponce et al. 1996), *Catharanthus roseus* (Brillanceau et al. 1989), *Solanum nigrum* (Wei et al. 1986), etc. Shoots regenerated from these calli either in hormone-free medium or with application of hormones. In *T. indica* (Chaudhuri et al. 2006), 44 % of the root clones spontaneously produced yellow friable callus from older regions of root, when kept for more than 8 weeks without subculture. Regeneration occurred via somatic embryogenesis from these calli within 3 weeks on hormone free MS medium (Fig. 2.2). Spontaneous callus development is also reported in many hairy root lines of *C. roseus* (Brillanceau et al. 1989), one of which evolved rapidly on hormone-free media with spontaneous shoot bud formation.

2.3.2 *Plant Regeneration from Ri-Transformed Root Cultures Induced by Plant Growth Regulators*

Induction of shoot bud directly from the transformed roots in presence of plant growth regulators are reported in many plants like *A. belladonna* (Aoki et al. 1997), *A. reptans* var. *atropurpurea* (Uozumi et al. 1996), *T. platycarpum* (Lee et al. 2004), *Plumbago rosea* L. (Satheeshkumar et al. 2009), *Linum usitatissimum* (Zhan et al. 1988), etc. To induce shoot formation, in most of the cases the cytokinin BA (6-Benzyladenine) is used either alone or in combination with some auxin. Concentration of hormones used for effective regeneration varied from plant to plant along with other conditions like, size of transformed root used as explant, concentration of sucrose used, photoperiod at which cultures were maintained, media used for rooting, etc.

Fig. 2.2 Regeneration of Ri-transformed plant from hairy roots of *Tylophora indica*. **a** Spontaneous induction of callus in transformed roots (bar = 3.3 mm), **b** developing somatic embryos from the embryogenic callus (bar = 5 mm), **c** a single somatic embryo (bar = 1 mm), **d** germinated somatic embryo developing into plantlet (bar = 1.4 mm), **e** Ri-transformed plantlet developed from somatic embryo (bar = 2.5 mm), and **f** Ri-transformed plant transferred to potted soil (bar = 25 mm)

Shoot buds regenerated from hairy roots of *A. reptans* var. *atropurpurea* (Uozumi et al. 1996), *T. platycarpum* (Lee et al. 2004), *Aesculus hippocastanum* (Zdravković-Korać et al. 2004), and *P. rosea* (Satheeshkumar et al. 2009), when cultured on media with BA as the sole hormone supplement. Whereas, in *A. reptans* var. *atropurpurea* (Uozumi et al. 1996) high concentration of BA (10 mg/l) supplementation in media exhibited highest number of plantlet formation in hairy roots. In *T. platycarpum* (Lee et al. 2004) 100 % shoot bud regeneration was noted when the hairy roots were cultured on media supplemented with 1 mg/l BA. In *P. rosea* (Satheeshkumar et al. 2009) when concentration of BA used was within 0.5–2 mg/l shoot formation took place without callusing. But with increase of BA concentration from 2.5 mg/l onwards, callusing increased with less number of shoot bud production. Best shoot bud regeneration frequency was noted on media supplemented with 2 mg/l BA. Root explants cultured in medium containing BA with auxin did not differentiate into shoot buds but proliferated into callus.

As mentioned above, in some species, induction of shoot buds from transformed root cultures was also noted when auxin was used in combination with BA. In *A. belladonna* (Aoki et al. 1997), *Cichorium intybus* cv. Lucknow local (Bais et al. 2001), and Mexican lime (Pérez-Molphe-Balch and Ochoa-Alejo 1998), shoot buds regenerated when root segments were cultured on medium supplemented with low concentration of NAA along with BA. Whereas, in apple (Lambert and Tepfer 1992), low concentration of IBA is reported to be used along with high concentration of BA for development of adventitious shoot buds from the transformed roots. Transformed roots of *L. usitatissimum* (Zhan et al. 1988) were cultured on MS medium containing NAA, BA, and adenine, for successful shoot bud regeneration. Hence, for efficient regeneration of shoot buds from hairy roots, type, combination, and concentration of hormones to be used varies from plant to plant.

Hormone supplementation in media increases the regeneration frequency in some plants when compared to that on hormone-free media. For example, in hairy roots of *T. platycarpum* (Lee et al. 2004), the regeneration frequency increased from 88.5 ± 9.8 % on hormone-free media to 100 % on BA supplemented media. Similarly, in hairy roots of *A. hippocastanum* (Zdravković-Korać et al. 2004), regeneration frequency was significantly increased on media supplemented with BA compared to that in phytohormone-free media. Contrastingly, in some other plants, like in *B. oleracea* var. sabauda and *B. oleracea* var. capitata (Sretenović-Rajičić et al. 2006), average number of shoots regenerated from transformed roots were lower on hormone supplemented media compared to hormone-free media. Thus, use of phytohormone does not always increase the regeneration frequency.

Hairy roots of different species were maintained under different photoperiod for efficient regeneration of shoot buds from the roots. Ri-transformed roots of *A. belladonna* (Aoki et al. 1997), *A. hippocastanum* (Zdravkovic-Korac et al. 2004), and *L. usitatissimum* (Zhan et al. 1988) were maintained under 16/8 h photoperiod, while hairy root explants of *A. reptans* var. *atropurpurea* (Uozumi et al. 1996) and *P. rosea* (Satheeshkumar et al. 2009) were cultured under 14/10 h (light/dark) and 12 h photoperiod, respectively, for shoot bud induction. Continuous exposure to light was given to transformed root explants of Mexican lime (Pérez-Molphe-Balch and Ochoa-Alejo 1998). Thus, like in case of spontaneous regeneration, effect of light on shoot bud regeneration in presence of phytohormones also varies from species to species.

Apart from use of plant growth regulators to induce shoot from transformed roots, some other agents are also reported to induce shoot regeneration either alone or in addition to exogenous hormones. In *Antirrhinum majus*, Hoshino and Mii (1998) studied effect of phosphinothricin-based herbicide, Bialaphos (a tripeptide antibiotic produced by *Streptomyces hygroscopicus*), and plant growth regulators on regeneration from transformed roots. The various concentrations of phytohormones, NAA in combination with BA or TDZ, tried by them did not result in any improvement in shoot bud induction compared to that in hormone-free media. But bialaphos, although toxic at and above 0.9 mg/l, when added at or below 0.5 mg/l, enhanced shoot regeneration. Fifty-six percent of hairy roots regenerated shoots when cultured on half strength MS with 0.5 mg/l bialaphos after 3 months.

Shoot organogenesis from transformed root cultures induced by phytohormones via callus formation is also reported in many plant species like, *Stylosanthes humilis* cv. Paterson (Manners and Way 1989), *Medicago arborea* L. (Damiani and Arcioni 1991), *Glycine argyrea* (Kumar et al. 1991), *Lotus japonicus* (Stiller et al. 1997), *Alhagi pseudoalhagi* (Wang et al. 2001), *C. roseus* (Choi et al. 2004), *Crotalaria juncea* (Ohara et al. 2000), cultivars of *B. oleracea*, and *Brassica campestris* (Christey et al. 1997), etc. Concentration and combinations of plant growth regulators used to induce callus formation and for regeneration of shoot buds from callus varied from plant to plant.

In *S. humilis* cv. Paterson (Manners and Way 1989), *G. argyrea* (Kumar et al. 1991), *C. roseus* (Choi et al. 2004), *C. juncea* (Ohara et al. 2000), *L. japonicus* (Stiller et al. 1997) and different cultivars of *B. oleracea* and *B. campestris* (Christey et al. 1997), callus induction from Ri-transformed roots and shoot organogenesis from those calli took place in the same hormone supplemented media. In some of these plants, the development and elongation of the regenerated shoot buds required transferring the regenerating callus or regenerated shoot buds to a different medium. When hairy root segments of *S. humilis* cv. Paterson (Manners and Way 1989) were cultured on MS with 2 mg/l BA, callus induction took place and within 2–3 weeks shoot regeneration occurred on the same media. When less than 2 mg/l BA was used, callus formation took place without any shoot bud regeneration, while shoot regeneration frequency from callus was very low when 4 mg/l BA was used. In *G. argyrea* (Kumar et al. 1991), when Ri-transformed root segments were cultured on B5 medium with BA and IBA (Indole-3-butyric acid), green nodular callus was induced within 20 days. These nodular calli produced shoots within 40–50 days. Shoot elongation was noted when regenerating callus was transferred to B5 medium with IBA and reduced level of BA. When hairy root segments of *C. roseus* (Choi et al. 2004) were cultured on MS supplemented with BA and NAA (α-Naphthaleneacetic acid), callus induction was noted after 2 weeks, from which shoot buds regenerated with 80 % frequency on the same medium. Transformed root fragments of *C. juncea* (Ohara et al. 2000) when cultured on solid B5 medium supplemented with 3 mg/l BA, callus induction followed by shoot bud regeneration was observed. Regeneration frequency which was 30 % in 3 mg/l BA supplemented media reduced to 14 % when concentration of BA increased to 5 mg/l. The induced shoot buds showed severe hyperhydricity, thus mostly failed to develop further on 0.8 % agar and 0.2 % gelrite solidified media. Elevated concentrations of gelling agent (1.2 % agar) was found to be effective in lowering the hyperhydricity and in promoting further development of the shoot buds. In *L. japonicus* (Stiller et al. 1997), when hairy root segments were cultured on B5 medium supplemented with BA and NAA, callogenesis followed by development of shoot primordial took place. Shoot elongation of these regenerated shoot buds was achieved when concentration of NAA was reduced to half.

Contrastingly, in *M. arborea* (Damiani and Arcioni 1991), the Ri-transformed callus only proliferated on the callus induction media (B5 media containing 2,4-D and kinetin). Shoot regeneration needed transfer of these calli to a different media, i.e. on hormone-free solid MS media. Similarly, Wang et al. (2001) reported use of two different media, one for callus induction from the hairy root of *A. pseudoalhagi* and other for regeneration from those calli. Hence, for induction of callus from Ri-transformed roots and subsequent shoot bud regeneration, choice and concentration of exogenous hormone supplementation varies with species.

Induction of embryogenic callus from hairy roots using phytohormones and regeneration of transgenic plants from such calli is reported in many plant species like in *Cucumis sativus* L. (Trulson et al. 1986), *P. ginseng* (Yang and Choi 2000), *Prunus avium* x *P. pseudocerasus* (Gutièrrez-Pesce et al. 1998), etc.

Yang and Choi (2000) cultured green, white, and red root lines of *P. ginseng* on MS medium supplemented with 1 mg/l 2,4-D. Low frequency of embryogenic callus formation was noted until 1 month which increased with duration of the culture. Among the three root lines, best induction of embryogenic callus was noted in the red line. Efficient production of somatic embryos from this embryogenic callus was achieved by transferring them to MS medium containing 0.5 mg/l 2,4-D, which matured into cotyledonary stage after 2 months. Germination of these cotyledonary somatic embryos were only found when transferred to MS medium containing 10 mg/l GA_3 within 15 days, which continued to grow on half strength MS medium without any growth regulators.

2.4 Production of Secondary Metabolites in Ri-Transformed Plants

While extensive study have been done on secondary metabolite production in hairy root cultures obtained through transformation with *A. rhizogenes*, only a few reports are available regarding analysis of secondary metabolites in roots and shoots of Ri-transformed plants (Table 2.2). Secondary metabolite productions in Ri-transformed plants are at levels, comparable to or even greater than that in non-transformed plant in many cases, whereas in some plants reduction of specific secondary metabolite is also reported. In A4 transformed shoots of *T. indica*, tylophorine content was 20–60 % higher than that in the control (Chaudhuri et al. 2006). Similarly, in A4 transformed plants of *B. monnieri*, the content of four bacopa saponins (bacopasaponin D, bacopasaponin F, bacopaside II, and bacopaside V) were up to five times higher than non-transformed plants of same age (Majumdar et al. 2011). Ri-transformed plants of *P. indica* are also reported to have an increased plumbagin content compared to non-transformed plants (Gangopadhyay et al. 2010). On the contrary, Ri-transformed plants of *Hyoscyamus muticus* showed reduced alkaloid production (Sevón et al. 1997) and same was in case of transgenic plants of *D. myoporoides* × *D. leichhardtii* for scopolamine and hyoscyamine (Celma et al. 2001).

2.5 Large-Scale Production of Secondary Metabolites by *A. rhizogenes* Mediated Transformed Roots in Bioreactors

Hairy roots appear to be potential systems for culture in bioreactors in large scale because of their organized nature, fast growth rate, and stability in metabolite production. As hairy roots grow continuously (Jeong et al. 2002) bioreactors used

Table 2.2 Secondary metabolites from transgenic plants obtained through transformation with *Agrobacterium rhizogenes*

Plant species	Family	*A. rhizogenes* strain used for transformation	Secondary metabolite	Reference
Ajuga reptans var. *atropurpurea*	Lamiaceae	*A. rhizogenes* MAFF03-01724	20-hydroxyecdysone (20-HE)	Tanaka and Matsumoto (1993)
Armoracia lapathifolia	Brassicaceae	–	Isoperoxidase	Saitou et al. (1991)
Atropa belladonna	Solanaceae	*A. rhizogenes* strain 15834	Hyoscyamine and other tropane alkaloids	Aoki et al. (1997)
Bacopa monnieri	Scrophulariaceae	*A. rhizogenes* strain A4	Bacopasaponin D, Bacopasaponin F, Bacopaside II, Bacopaside V, Bacoside A3 and Bacopasaponin C	Majumdar et al. (2011)
Centaurium erythraea	Gentianaceae	*A. rhizogenes* strain LBA 9402	Secoiridoids glucosides	Piatczak et al. (2006)
Convolvulus arvensis	Convolvulaceae	*A. rhizogenes* with synthetic *crypt* gene	Calystegine	Chaudhuri et al. (2009)
Duboisia myoporoides X *D. leichhardtii*	Solanaceae	*A. rhizogenes* strain A4	Scopolamine	Celma et al. (2001)
Hyoscyamus muticus	Solanaceae	*A. rhizogenes* stain LBA9402	Hyoscyamine, scopolamine, Calystegins	Sevón et al. (1997)
Papaver somniferum	Papaveraceae	*A. rhizogenes* strain MAFF 03-01724	Morphinan alkaloid	Yoshimatsu and Shimomura (1992)
Pelargonium graveolens cv. Hemanti	Geraniaceae	*A. rhizogenes* strains A4 and LBA9402	Essential Oil, Linalool, Rose oxides(cis + trans), Isomenthone, Citronellol, Geraniol, 6-Caryophyllene, Guaia-6,9-diene, 10-epi-Y-Eudesmol, Citronellyl esters, Geranyl esters	Saxena et al. (2007)
Pimpinella anisum	Umbelliferae	–	Total phenolics	Andarwulan and Shetty (1999)
Plumbago indica	Plumbaginaceae	*A. rhizogenes* strain ATCC 15834	Plumbagin	Gangopadhyay et al. (2010)
Scoparia dulcis	Scrophulariaceae	*A. rhizogenes* strain 15834 with *bar* gene	Scopadulcic acid B.	Yamazaki et al. (1996)
Solanum khasianum	Solanaceae	*A. rhizogeness* strains A4 and LBA 9402	Solasodine	Jacob and Malpathak (2005)
Tylophora indica	Asclepiadaceae	*A. rhizogenes* strain A4	Tylophorine	Chaudhuri et al. (2006)
Vinca minor	Apocynaceae	*A. rhizogenes* strain DC-AR2	Vincamine	Tanaka et al. (1995)

for the cultivation of hairy roots are complex and quite different from the conventional ones utilized for the culture of plant cell suspensions (Mishra and Ranjan 2008). Reactors with unique configurations are required due to the organized and entangled nature of these roots. The structured as well as delicate hairy roots distribute unevenly throughout the bioreactor (Jeong et al. 2002) and form continuous root clumps composed of interconnected primary and lateral roots which hinder the percolation of oxygen into the roots (Kino-Oka et al. 1999; Bordonaro and Curtis 2000). Root hairs also limit fluid flow and the availability of oxygen (Bordonaro and Curtis 2000). Limitations of nutrient and oxygen mass transfer, which increase with increase in root biomass, are the main drawbacks associated with culture of hairy roots in bioreactors, leading to cell death and necrosis at the core of the biomass (Jeong et al. 2002; Suresh et al. 2005). This in turn causes reduction in growth and metabolite production (Suresh et al. 2005). Transfer of oxygen from the air bubbles to the hairy roots via the medium and from one cell to another within the hairy roots is complicated (Neelwarne and Thimmaraju 2009). Mass transfer can be enhanced by vigorous mixing, but this might lead to reduction in root viability due to increase in hydrodynamic shear stress (Hitaka et al. 1997). Large-scale culture of hairy roots is difficult as nutrients need to be provided simultaneously from the gas and liquid phases (Jeong et al. 2002). Several factors such as growth characteristics, morphological changes of hairy roots during proliferation, nutrient requirements, availability and utilization rates, medium composition, mass transfer, mechanical properties, methods for providing a support matrix, protection from shear damage, inoculum density and even distribution of the inoculum, and chances of flow restriction caused by the highly tangled root masses should be considered while designing a bioreactor for hairy roots (Taya et al. 1989; Yu and Doran 1994; Mishra and Ranjan 2008). Because of continuous proliferation and repeated branching, the rheological properties of the hairy roots change continuously (Neelwarne and Thimmaraju 2009). Thus, it is generally difficult to select a single type of bioreactor for the cultivation of hairy roots as the rheological properties of the roots vary not only from species to species but also within clones of a single species (Mishra and Ranjan 2008). Productivity in bioreactors also depends on certain other parameters like temperature, pH, composition of gases, selection of hairy root clone, reactor operation, removal of toxic byproducts, etc. (Kim et al. 2002a; Mishra and Ranjan 2008). To a large extent, bioreactor design also depends on the location of the product which may be either intracellular or extracellular (Mishra and Ranjan 2008).

2.5.1 Types of Bioreactors Used for the Cultivation of Hairy Roots

Reactors used for the cultivation of hairy roots can be of different types based on the continuous phase viz. liquid phase, gas phase, or the hybrid ones which are a combination of both.

2.5.1.1 Liquid Phase Bioreactors

In liquid phase reactors, also known as submerged reactors, roots remain submerged in the medium and air is bubbled through the culture medium to supply oxygen (Kim et al. 2002a, b). Stirred tank, bubble column, air lift, liquid impelled loop, and submerged connective flow reactors (Wilson et al. 1987; Taya et al. 1989; Buitelaar et al. 1991; Tescione et al. 1997; Carvalho and Curtis 1998) are examples of liquid phase reactors.

Stirred Tank Bioreactors

Among the liquid phase reactors, stirred tank reactor is mechanically agitated (Mishra and Ranjan 2008); aeration and medium currency are regulated by mortar-derived impeller or turbine blades (Choi et al. 2006). Compressed air is sparged into the reactor in the form of bubbles to supply oxygen for the aerobic processes (Mishra and Ranjan 2008). The impeller region has an aeration device which produces a well dispersed gas phase in the continuous liquid phase (Mishra and Ranjan 2008). However, the impeller rotation damaged the roots by shearing, which resulted in callus formation and poor biomass production (Hilton et al. 1988; Wilson 1997). The problem was alleviated by isolating roots from the impeller by using a steel cage or nylon mesh to fix the roots (Kondo et al. 1989; Hilton and Rhodes 1990) or by developing modified stirred tank reactors having flat blade turbines (Mishra and Ranjan 2008). Cardillo et al. (2010) used a 1.5 l modified stirred tank reactor equipped with a plastic mesh to culture hairy roots of *Brugmansia candida*. These cultures produced an increased biomass (49.33 g FW, GI of 3.93), corresponding to a 36 % higher yield than that obtained in Erlenmeyer flasks; alkaloid yields in the reactor were 2.6- to 3-folds higher than the cultures in Erlenmeyer flasks with a fivefold increase in scopolamine yield. The specific productivities for both the systems were similar except a 31 % increase in anisodamine production in the bioreactor. 2.8–4 times higher volumetric productivities were obtained for the processes carried out in the bioreactor when compared with the Erlenmeyer processes. Choi et al. (2006) reported that compared to other types of reactors, temperature, pH, amount of dissolved oxygen, and nutrient concentration can be controlled in a better way within a stirred tank bioreactor. Growth and hyoscyamine production in transformed root cultures of *D. stramonium* was studied in a modified 14 l stirred tank reactor (Hilton and Rhodes 1990). The effects of batch and continuous modes of operation, three different temperatures and half and full strength Gamborg's B5 salts were studied. When cultured at half strength B5 medium, the dry matter content and hyoscyamine levels of the roots were higher than those grown on full strength B5 salts. The amounts of hyoscyamine were similar at both 25 and 30 °C but about 40 % lower at 35 °C. However, highest production rate of 8.2 mg/l/day hyoscyamine was obtained when the roots were grown at 30 °C. Hyoscyamine released into the culture medium was low in continuous fermentation at 25 °C but up to sevenfolds higher when the fermentors were operated at 30 or 35 °C.

Reduced productivities, often recorded during scale-up cultures, might also result from handling problems during inoculation and uneven distribution of the roots in the bioreactor (Hilton and Rhodes 1990; Kwok and Doran 1995; Woo et al. 1996). Lee et al. (1999) cultivated randomly cut hairy roots of *A. belladonna* in 3 l and 30 l modified stirred reactors. For a good distribution, roots were inoculated and immobilized on a stainless steel mesh in the reactors that separated the roots from the stirrer. After a period of 1 month 1490 mg tropane alkaloids were produced by the roots cultured in the 30 l reactor; 5.4 mg/gDW atropine could be detected in the roots which was equivalent to the amount found in plants grown in the field for 12 months and contained considerable amounts of other alkaloids like 1.6 mg/gDW 6-β-hydroxyhyoscyamine, 0.9 mg/gDW scopolamine, and 2.0 mg/gDW littorine. The study demonstrated that the use of this type of modified stirred bioreactor would provide sufficient supply of oxygen and nutrition for the growth of roots and alkaloid production. In another study, *A. belladonna* hairy roots were cultured in a bioreactor using porous polypropylene membrane tubing as a supplementary aeration device and an emulsion of Fluorinert^{TM} FC-43 perfluorocarbon was added to the medium (Kanokwaree and Doran 1998). The treatments were applied to improve oxygen supply to the roots. Combination of air sparging and membrane tubing aeration in a gas driven bioreactor supported 32–65 % higher biomass levels than sparging only of oxygen-enriched air at the same total gas flow rate. However, growth was not improved by the addition of perfluorocarbon to the medium in the sparged stirred tank reactor. The study demonstrated the need for site directed aeration of hairy root cultures for supplying oxygen into the zones of highest root biomass. Large-scale culture of *Ophiorrhiza pumila* in a 3 l modified stirred tank reactor fitted with a stainless steel net (to minimize physiological stress of the hairy roots by agitation) decreased the biomass yield compared to shake flasks; while a biomass of 162 ± 20 gFW/l was achieved in a 100 ml shake flask, only 87.2 ± 12.4 gFW/l was achieved in the aerated 3 l reactor (Hiroshi et al. 2002). Camptothecin yields in the reactor (8.7 ± 1.3 mg/l) were, however, almost equivalent to that of the roots cultured in shake flasks (8.9 ± 1.4 mg/l).

Nuutila et al. (1994) demonstrated that growth and alkaloid content in hairy roots of *C. roseus* was affected by different shear levels. For culturing the roots a stirred tank reactor with a metal mesh used to isolate the roots from the impeller system was used. Although the impeller was used to avoid shear stress, root growth was severely inhibited. Best growth and alkaloid production was obtained in an air sprayed bioreactor with no other mixing.

Mehrotra et al. (2008) cultured a fast growing hairy root clone G6 of *Glycyrrhiza glabra* in a 5 l bench top air sparged and mechanically agitated bioreactor, provided with a nylon mesh septum containing 4 l modified NB (Nitsch and Nitsch 1969) medium. After 30 days of incubation, the root biomass harvested (310 g) represented 20 times increase over initial inoculum (16 g).

Bubble Column Bioreactors

In bubble column reactors roots are submerged in the medium and an air distributor situated at the bottom of the column generates an upflow of air bubbles leading to liquid mixing (Mishra and Ranjan 2008). Compared to other stirred types of reactors this reactor is advantageous for the culture of organized structures like hairy roots as the bubbles create less shearing stress. Kwok and Doran (1995) improved supply of oxygen to the roots in a 2.5 l bubble column reactor by introducing gas at multiple points of the reactor, divided into three segments with wire mesh, each segment containing a sparger. After 43 days of culture 9.9 gDW/l of *A. belladonna* hairy roots were harvested. However, bulk mixing was very poor in the reactor. McKelvey et al. (1993) reported that in hairy root cultures of *H. muticus*, the yields of tissue mass in submerged air sparged reactors was 31 % of that accumulated in shake flask controls. The authors suggested that liquid phase channeling and stagnation leading to impaired oxygen transfer were the probable causes resulting in poor growth of the roots. When *Lobelia inflata* hairy roots were cultivated in a cylindrical 10 l bubble column bioreactor, a 3.5 times increase in biomass was noted at the end of the culture period (Bálványos et al. 2004). High levels of polyacetylene (36.5 mg/g lobetyolin and 15.9 mg/g lobetyolinin) were detected by HPLC.

Jeong et al. (2002) used two types of bioreactors viz. an air bubble four stage column bioreactor and a modified stirred bioreactor, both fitted with stainless steel meshes, for the culture of *P. ginseng* hairy roots. After 32 days of culture in the four stage bubble column reactor, each stage of the column was closely packed with hairy roots and a 36.3-fold increase in biomass was noted. The crude saponin content was 20 % on dry weight basis, which was similar to flask cultures. A 52-fold increase from the initial inoculum was observed in the modified stirred bioreactor after 42 days. In both the types of reactors growth was about three times as high as in flask cultures.

A prototype basket bubble bioreactor was used for the coculture of hairy roots and shoots of *Genista tinctoria* (Shinde et al. 2008). Large amounts of isoliquiritigenin were produced by the hairy roots.

Connective Flow Bioreactor

Carvalho and Curtis (1998) developed a connective flow reactor which consisted of a 14 l stirred tank, equipped with agitation and temperature control, a peristaltic pump and a tubular reactor. Sparging and agitation in the fermentor transferred oxygen into the medium and a peristaltic pump recirculated the medium between the stirred tank and the tubular reactor. 556 $\pm$ 4 gFW/l of *H. muticus* hairy roots, were produced after a period of 30 days whereas a bubble column reactor produced only 328 $\pm$ 5 gFW/l hairy roots. Although better than a bubble column reactor, it was suggested that a connective flow reactor may not be suitable from a realistic point of view as a pressure is required to circulate the culture medium at a high velocity to overcome the flow resistance of the root bed (Carvalho and Curtis 1998). However, reactors similar to connective flow reactors have been proved to be effective as research tools. Williams and Doran (1999) minimized the liquid solid hydrodynamic boundary layer at the root surface and determined the critical oxygen level of hairy roots of *A. belladonna* by using a packed bed recirculation reactor.

Air Lift Bioreactors

Air lift reactors are basically bubble column reactors containing a draught tube (Mishra and Ranjan 2008). In these reactors air is supplied through a sparger ring into the bottom of a central draught tube that controls the circulation of air and the medium (Mishra and Ranjan 2008). The draught tube also prevents bubble coalescence by causing the bubbles to move in one direction and the shear stress is distributed equally throughout the reactor. Thus cells can grow in a more stable physical environment. Shear stress generated in air lift reactors is lower than that generated in bubble column reactors, giving the air lift reactors an advantage over the bubble column reactors (Al-Masry 1999). Although reports indicate the use of conventional air lift reactors for the scale-up culture of hairy roots (Buitelaar et al. 1991; Oka et al. 1992; Uozumi et al. 1995), optimum biomass could not be yielded mainly because of two factors—uneven distribution of roots at certain regions of the reactor and excessive channelling of gas phase, blocking liquid flow due to clumping of roots (Kim and Yoo 1993). However, immobilization of roots has been reported to increase biomass of hairy roots in air lift reactors (Taya et al. 1989). In another study, after 39 days of culture in a 5 l airlift bioreactor, growth of hairy roots of *P. ginseng* was about 55-fold of the initial inoculum whereas a 38-fold increase in biomass was noted after 40 days of culture in a 19 l airlift bioreactor (Jeong et al. 2003). When cultured in a 30 l airlift bioreactor for 20 days, the dry weight of hairy roots (11.5 g/l dry weight) and astragaloside IV yields (1.4 mg/g) from hairy root cultures of *Astragalus membranaceus* were higher than a 10 l bioreactor (dry weight 9.4 g/l and astragaloside IV 0.9 mg/g) (Du et al. 2003). Yields from the 30 l bioreactor were almost similar to cultures from 250 ml and 1 l flasks. Growth of hairy roots and the production of scopolamine in hairy root cultures of *Datura metel* were improved following treatment with permeabilizing agent Tween 20 in a 4 l airlift bioreactor with root anchorage (Cusido et al. 1999). After 4 weeks of culture, biomass yield was 2.3 and 0.84 mg/l/day scopolamine was produced. 30 g FW of *Duboisia leichhardtii* hairy root clone DL47-1 was inoculated into 3 l airlift bioreactor packed with Amberlite XAD-2 (Muranaka et al. 1993b). Biomass of the roots increased eightfolds during the culture period and 0.5 g/l scopolamine was obtained after 11 weeks. When a polyurethane foam was used in the vessel, the root tissue spread uniformly and grew well, showing a 14-fold increase in biomass after 12 weeks, with a 0.8 g/l yield of scopolamine. In the same study, a two-stage culture system, the first stage for the growth of hairy roots, and the second one for scopolamine release was carried out in a 1 l turbine blade reactor fitted with a stainless steel mesh as a support for the uniform growth of the roots and also packed with Amberlite XAD-2 column. The root biomass increased 12-folds during 4 weeks of culture in the growth medium; after culturing in the medium for scopolamine release for 11 weeks 1.3 g/l scopolamine was recovered.

Tikhomiroff et al. (2002) reported the use of a two-liquid phase bioreactor for the extraction of indole alkaloids from *C. roseus* hairy roots with silicon oil. The roots were inoculated and immobilized in a stainless steel screen mesh box placed 1 cm from the bottom of the reactor in the aqueous phase. The second phase was DC200 silicon oil. Although growth of the hairy roots was not affected by silicon oil, the specific yields of tabersonine and löchnericine were improved by 100–400 and 14–200 %, respectively. When jasmonic acid was used as an elicitor, 10–55 % tabersonine and 20–65 % löchnericine were produced. The alkaloids accumulated in the silicon oil phase and were never detected in the culture medium.

Liu et al. (1998) investigated the effect of four different culture systems—250 ml Erlenmeyer flasks, bubble column reactor, modified bubble column reactor, and modified inner loop air lift reactor, for the production of artemisinin. The modified inner loop air lift bioreactor was designed to improve the transfer of nutrient medium and supply of oxygen and to increase growth homogeneity. When compared it was found that growth rate and artemisinin content (26.8 g/l and 536 mg/l, respectively) in the hairy roots cultivated in the modified air lift reactor was higher than the other types of reactors. The hairy roots grew more homogenously in the modified inner loop air lift reactor, proving that optimization of design of air lift reactor can promote growth and artemisinin production. In another study, *Pueraria phaseoloides* hairy roots were cultured in 2.5 l airlift bioreactors for 3 weeks (Kintzios et al. 2004). 5,570 μg/gDW puerarin, corresponding to 200 times as much as in 250 ml flask cultures was produced after incubation for 3 weeks. Puerarin was also released into the culture medium at concentrations higher than that found in the hairy roots themselves. Sharp and Doran (1990) cultured root tips of *A. belladonna* in hormone-free MS medium containing 3 % sucrose in two 2.5 l airlift bioreactors. A 33-fold increase in biomass yield was noted over a 26 day culture period and concentration of atropine in the dried roots was 0.37 %. In another study, Caspeta et al. (2005a) studied the growth of *Solanum chrysotrichum* hairy roots in shake flasks, a glass-draught internal-loop 2-l basic design air lift reactor (BDR) and a novel modified mesh-draught with wire-helix 2 l reactor (MR). Growth patterns were different in each of them with specific growth rates being 0.08, 0.067, and 0.112 per day for shake flasks, BDR, and MR, respectively. After 42 days of culture, tissue density in the MR was almost the same as that found in the shake flasks and twice as that obtained in the BDR. Cultures were scaled-up into 10 l MR level. Results indicated the overall tissue density reduced slightly compared to that of 2 l reactor. *S. chrysotrichum* hairy roots were also cultivated in 250 ml flasks and 2 l modified draught-tube internal-loop airlift reactors for the production of five antifungal saponins (SC-2 to SC-6) (Caspeta et al. 2005b). In the 2 l reactor, yield of SC-2 was 0.7 % on dry weight basis, which were sixfolds greater than that found in plant leaves. SC-4 was recovered both from the flasks and the reactor while SC-5 and SC-6 were only detected from the culture medium of roots grown in the 2 l reactor.

2.5.1.2 Gas Phase Bioreactors

Trickle bed, droplet phase, liquid dispersed, and nutrient mist reactors are examples of gas phase reactors (Taya et al. 1989; Flores and Curtis 1992; Wilson 1997; Williams and Doran 2000; Woo et al. 1996; Liu et al. 1999). Hairy roots are intermittently exposed to air or other gaseous mixtures and the nutrient liquid in these reactors. Liquid nutrients are either sprayed onto the roots or the roots get nutrient media as droplets, the size of which varies considerably (Kim et al. 2002a, b). The droplets are 0.5–30 μm in mist reactors using ultrasonic transducers (Weathers et al. 1999) whereas they are of much larger sizes for trickle bed or other gas phase reactors using spray nozzles (Wilson 1997). Any oxygen deficiency in the dense root clumps can be eliminated using these reactors; stress caused by shearing is also minimized (Kim et al. 2002b). However, excessive delivery of the medium can cause liquid phase channeling and medium retention. Oxygen can be completely depleted in the submerged regions of the root clumps and nutrient concentrations can be different from that in the bulk medium (Singh and Curtis 1994; Williams and Doran 2000). Since gas is the continuous phase in these reactors, roots need to be immobilized. Horizontal sheets of mesh, vertical structures, packing rings made of nylon mesh, or intalox metal process packing have been used variously for the purpose of immobilization (Woo et al. 1996; Liu et al. 1999; Chatterjee et al. 1997; Williams and Doran 2000).

Nutrient Mist Bioreactors

In nutrient mist reactors roots are dispersed in an air phase by immobilization on a mesh support and a mist phase, consisting of liquid medium is introduced into the reactor (Mishra and Ranjan 2008). The hairy roots are continuously bathed in nutrient mist, thereby replenishing the nutrients rapidly, and removing toxic byproducts (Dilorio et al. 1992). Compared to other reactors, these reactors have certain advantages viz. these are easy to operate and scale up, gas composition can be closely controlled, and oxygen is not a limiting factor and provides an environment of low shearing forces and pressure drops (Weathers et al. 1999; Mishra and Ranjan 2008). As the nutrient mist can be dispersed homogenously within the culture chamber, mechanical agitation is not required, and this reduces root damage caused by shearing (Dilorio et al. 1992). Certain parameters need to be taken into account while designing a nutrient mist reactor. These are composition of the mist, availability of the nutrients, recycling of medium, feeding schedules, and a feed rate sufficient for the growth of the roots (Dilorio et al. 1992).

Whitney (1992) used various types of bioreactors like nutrient mist reactor, trickle bed reactor, stirred tank reactor, and an air lift reactor for the culture of hairy roots of *Nicotiana tabacum* and *D. stramonium* and found that cultivation in nutrient mist reactors resulted in a higher growth rate and yield of tobacco compared to the other reactors used. Kim et al. (2001) found that the amount of artemisinin was significantly higher (2.64 μg/g dry weight) when *Artemisia annua* hairy root clone YUT16 was cultured in a nutrient mist bioreactor, compared to a bubble column reactor (0.98 μg/g dry weight). On the contrary, overall biomass concentration was higher in a bubble column reactor (15.3 gDW/l) than in a nutrient mist reactor (14.4 gDW/l) (Kim et al. 2002b). Also, the average growth rate was higher in the bubble column reactor than in the nutrient mist reactor. *A. annua* hairy roots have also been cultivated in three different types of nutrient mist bioreactors, each of 2.3 l working volume, fitted with three stainless steel meshes, namely, a nutrient mist bioreactor, an inner-loop nutrient mist bioreactor and a modified inner-loop nutrient mist bioreactor (Liu et al. 1999). After a culture period of 25 days, growth indices in the three types of bioreactors were 42, 61, and 68, respectively. 13.6 gDW/l medium biomass yield was achieved in the modified inner-loop nutrient mist bioreactor. The development and growth kinetics of single hairy roots of *A. annua* were characterized in nutrient mist reactor and shake flasks (Wyslouzil et al. 2000). The effects of mist duty cycle, medium formulation, gas composition on growth kinetics, and morphology of the roots were also studied. Dilorio et al. (1992) reconfigured a nutrient mist bioreactor for batch operation to homogenously deliver nutrient mist to the hairy root cultures of *Carthamus tinctorius*. Misting cycle, inoculum size, batch or continuous operation, and sucrose concentration were varied to obtain maximum growth over a period of 1 week. Williams and Doran (2000) scaled-up *A. belladonna* hairy roots in a liquid dispersed bioreactor, where the liquid medium was sprayed onto the roots growing on a stainless steel support.

Bioreactors of various configurations viz. bubble column reactor, nutrient sprinkle reactor, and an acoustic mist bioreactor have been used to study the growth of *C. intybus* hairy roots (Bais et al. 2002b). Among all the types of bioreactors used, roots grown in acoustic mist bioreactor had greater final biomass (295 gFW/l, 29.8 gDW/l); this reactor also showed better specific growth rate of 0.075/day that was very close to that of shake flasks (0.086/day). Also, accumulation of esculin was maximum in the acoustic mist bioreactor (18.5 g/l) which was 1.4 times greater than nutrient sprinkle and bubble column reactors.

In an interesting study, Souret et al. (2003) examined the effect of three different culture systems like shake flasks, a mist reactor and a bubble column reactor on the expression levels of four key terpenoid biosynthetic genes, 3-hydroxy-3-methylglutaryl coenzyme A reductase (HMGR), 1-deoxy-D-xylulose-5-phosphate synthase (DXS), 1-deoxy-D-xylulose-5-phosphate reductoisomerase (DXR), and farnesyl diphosphate synthase (FPS) in hairy root cultures of *A. annua*. All the genes showed temporal regulations when the roots were cultured in shake flasks and the production of artemisinin could only be correlated with the expression of FPS. The expression of the genes in the reactors was equivalent to or greater than that of the hairy roots grown in the shake flasks. However, the expression level within the six different zones of each reactor could not be correlated with their respective oxygen levels, light and root packing density. The authors also observed that the genes were expressed unusually; the position of the roots in the reactors affected the transcriptional regulation of all the genes under study. It was thus concluded that to characterize gene activity in a whole reactor, analysis of a single reactor sample could be misleading. Also it was found that the terpenoid gene expression in hairy roots of *A. annua* cultured in a bubble column reactor was different when compared with a mist bioreactor.

In another study, Flores and Curtis (1992) showed that when grown in a trickle bed reactor, production of solavetivone by hairy root cultures of *H. muticus* was 3.5 times higher when compared with roots grown in a submerged reactor.

2.5.1.3 Hybrid Bioreactors

Hybrid reactors represent the third type of reactor used for the cultivation of hairy roots. In these reactors, roots are grown in a gas phase after being exposed to liquid phase (Mishra and Ranjan 2008). A major drawback of gas phase reactors is that manual loading is required to uniformly distribute roots in the growth chamber (Kim et al. 2002a; Mishra and Ranjan 2008). Ramakrishnan et al. (1994) used a hybrid reactor to solve the problem by initially running the reactor as a bubble column to suspend, distribute, and attach the roots to the packing rings in the reactor. After 2 weeks the reactor was switched to a trickle bed operation, thus exposing the dense root clumps to a gas environment. Roots were also pumped as slurry into the reactor after being chopped in a blender, thereby eliminating the manual labor of cutting and inoculation. Using these processes a 14 l reactor was run for 4 weeks and at harvest 20 gDW/l packing density of *H. muticus* was obtained (Ramakrishnan et al. 1994).

Datura stramonium hairy roots were cultured in a 500 l hybrid reactor that used a droplet phase for 40 days after an initial submerged culture for 21 days (Wilson 1997). At the end, 39.8 kgFW biomass was harvested, yielding a packing density of 79.6 gFW/l. *D. stramonium* hairy roots have also been cultivated in 8.2 l root tube bioreactor; good growth and productivities have been recorded (Greens and Thomas 1996).

2.5.1.4 Rotating Drum and Ebb-and-Flow Bioreactors

Cycles of liquid and gas phases are alternated in rotating drum reactors and ebb-and-flow reactors (Kim et al. 2002a). Rotating drum reactors are also mechanically agitated (Mishra and Ranjan 2008). Kondo et al. (1989) reported that as the roots rotated during drum rotation they did not attach well to the vessel wall; the roots were lifted above the medium and then dropped back. As a result of these repeated drops, cells were damaged resulting in low biomass yield. This effect was overcome by laying polyurethane foam on the inner wall of the drum, which acted as a support for the hairy roots. The authors also demonstrated that hairy roots of *Daucus carota* grew well in turbine blade reactors (0.63 g/l/day) yielding 10 g/l dry mass after 30 days whereas only 4 gDW/l biomass was obtained when the roots were cultured in shake flasks. They concluded that a stirred tank reactor fitted with a turbine blade or an immobilized rotating drum reactor with a high volumetric mass transfer coefficient is more conducive for the culture of hairy roots. Repeated ebbing and flowing or periodic rising and falling of the liquid medium is the main characteristic feature of an ebb-and-flow bioreactor. *H. muticus* hairy roots were also cultivated in a 2.5 l bench top ebb-and-flow reactor for a period of 18 days (Cuello et al. 2003). A productivity of 0.481 gDW/l/day was noted demonstrating a 50-fold scale up compared to cultures in 50 ml shake flasks.

2.5.1.5 Wave Bioreactors

Wave bioreactors are another type of novel reactors. Stress levels are significantly reduced as these reactors are based on wave-induced agitation (Mishra and Ranjan 2008). Palazon et al. (2003) showed that both the accumulation of biomass and production of ginsenosides by *P. ginseng* were higher in a 2 l wave bioreactor than in shake flasks.

2.6 Use of Ri-Transformed Roots in Research and Practical Application

Apart from the production of secondary metabolites, Ri-transformed root cultures have also been used in other research and practical applications like phytoremediation, biotransformation, and plant environment interactions.

Phytoremediation refers to the ability of plants to uptake chemicals from polluted soil, water or air, and thus clean up environmental pollution. Ri-transformed root cultures are excellent tools for phytoremediation. The roots provide a large surface area of contact between the contaminant and the tissue as they grow fast and are highly branched. As mentioned earlier, the roots can be scaled up in bioreactors and the biomass can be used for clean up of the environment. Also, hairy root exudates contain enzymes and metal chelating compounds which can be used to detoxify or sequester harmful complexes (Gujarathi and Linden 2005; Doty 2008), thereby recycling roots for future use. These roots are also potential tools to better understand the enzymatic machinery involved in the bioconversion of toxic pollutants to non-toxic metabolites (Macek et al. 2000) and the mechanism involved in metal tolerance and hyperaccumulation. As shoots are absent, these root cultures help in assessing the mechanisms present only in the roots for remediation of contaminants without the effects of translocation. Also, foreign genes introduced in the hairy roots can be expressed for a long period of time and the resultant functional proteins can be used to metabolise environmental contaminants (Banerjee et al. 2002). There are several reports indicating the use of Ri-transformed root cultures in the uptake and degradation of various environmental pollutants like phenols, chlorophenols, polychlorinated biphenyls, heavy metals, etc. (Macková et al. 1997; Agostini et al. 2003; Coniglio et al. 2008; Vinterhalter et al. 2008).

Biotransformation is the process whereby a substance is changed from one chemical to another (transformed) by a chemical reaction within the body of a living organism. Plant cell cultures have the ability to specifically convert exogenously administered organic compounds into useful analogues. The main problem with use of cell suspension culture is the phenomenon of somaclonal variation, which might lead to unstable biochemical behavior. This problem can be overcome with the use of plant root and shoot cultures. The reactions involved in biotransformation include oxidation, reduction, esterification, methylation, isomerization, hydroxylation, and glycosylation. Glycosylation only takes place readily in plant cells but laboriously in microorganisms. Ri-transformed root cultures of many plants are reported to be useful in biotransformation. For example, the hairy roots of *Coleus forskohlii* biotransformed methanol to *β*-D-glucopyranosides, ethanol to *β*-D-*ribo*-hex-3-ulopyranosides, and 2-propanol to its *β*-D-glucopyranoside (Li et al. 2003). *B. candida* hairy roots bioconverted hydroquinone into arbutin (Casas et al. 1998). Yan et al. (2007) reported use of Ri-transformed root cultures of *Polygonum multiflorum* for the biotransformation of 4-hydroxybenzen derivatives (1,4-benzenediol, 4-hydroxybenzaldehyde, 4-hydroxybenzyl alcohol, and 4-hydroxybenzoic acid) to their corresponding glucosides. The conversion of 4-hydroxybenzoic acid into its α-D-glucopyranoside was for the first time reported in a plant biotransformation system. In addition, these hairy root cultures were able to reduce the 4-hydroxybenzaldehyde to its corresponding alcohol. Thus, biotransformation using Ri-transformed root cultures has got a good prospective to generate novel products or to produce known products more efficiently.

Ri-transformed root cultures have also been used to study plant environment interactions. Isoflavone reductase (IFR) and (+) 6a-hydroxymaackiain 3-*O*-methyltransferase (HMM) are enzymes apparently involved in the synthesis of pisatin, an isoflavonoid phytoalexin synthesized by *Pisum sativum*. Wu and VanEtten (2004) produced transgenic pea hairy root cultures using sense and antisense-oriented cDNAs of *Ifr* and *Hmm* fused to the 35 s CaMV promoter and *A. rhizogenes* R1000. The virulence of *Nectria haematococca* (a pea pathogenic fungus) on the transgenic roots that produced the minimum amount of pisatin was studied in order to assess the effectiveness of pisatin in disease resistance. It was found that control hairy roots were less susceptible to isolates of *N. haematococca* that are either virulent or nonvirulent on wild-type pea plants than hairy roots expressing antisense *Hmm*. It was demonstrated for the first time that transgenic plant tissue with reduced ability to produce phytoalexin can be generated and such tissues are more susceptible to fungal infection. Interactions between roots and rhizobia (Quandt et al. 1993), mycorrhizal fungi (Mugnier 1997) and nematodes (Narayanan et al. 1999) have also been elucidated using hairy roots.

2.7 Conclusion

Since the first reports of establishment of Ri-transformed hairy root cultures, researchers have attempted to utilize these systems for the production of stable transgenic plants, analysis of genes, production of secondary metabolites and engineering of the biosynthetic pathways, production of therapeutically recombinant proteins, trapping of biomolecules released into the culture medium, for elucidating molecular aspects of biological processes, etc. Compared to plant cell suspension cultures, hairy root cultures appear to be potential systems for continuous production of valuable secondary metabolites because of their fast growth rates, ease of maintenance, genetic and biosynthetic stability, and ability to synthesize a vast array of compounds. Till date hairy root cultures have been established from several plant species, many of which are endangered and pharmaceutically important. Although these cultures have shown tremendous potential for the production of several important phytochemicals, their culture in large scale is still very challenging. As the culture environments in bioreactors and shake flasks are totally different, the results obtained from the studies on improvement in yields from shake flasks may not be directly applicable to bioreactors. Development of more effective and economic scale-up culture systems is required so that bioreactors can be used successfully for the large-scale production of secondary metabolites. Plants regenerated from hairy roots following transformation with *A. rhizogenes* have got a lot of use in different fields. Higher levels of some target metabolite production in transformed plants are very important as a tool for improvement of secondary metabolite production in medicinal plants. Apart from pharmaceutical use, other importance of hairy root regenerants include, use of transgenic plants in micropropagation of plants that are difficult to multiply for example in *A. majus* (Hoshino and Mii 1998), in Mexican lime (Pérez-Molphe-Balch and Ochoa-Alejo 1998), in cherry (Gutièrrez-Pesce et al. 1998), etc. Morphological characters such as adventitious shoot formation, reduced apical dominance, altered leaf and flower morphology have ornamental value in some plants while proved effective in plant breeding programmes in some others (Giovanni et al. 1997; Pellegrineschi et al. 1994; Handa et al. 1995). The promising developments and applications of hairy root cultures indicate that in the near future these cultures will provide researchers with powerful tools for further biotechnological research.

References

Ackerman C (1977) Pflanzen aus *Agrobacteriium rhizogenes*- tumoren an *Nicotiana tabactum*. Plant Sci Lett 8:23–30

Agostini E, Coniglio MS, Milrad SR, Tigier HA, Giulietti AM (2003) Phytoremediation of 2,4-dichlorophenol by *Brassica napus* hairy root cultures. Biotechnol Appl Biochem 37:139–144

Allan EJ, Eeswara JP, Jarvis AP, Mordue (Luntz) AJ, Morgan ED, Stuchbury T (2002) Induction of hairy root cultures of *Azadirachta indica* A. Juss. and their production of azadirachtin and other important insect bioactive metabolites. Plant Cell Rep 21:374–379

Al-Masry WA (1999) Effect of scale-up on average shear rates for aerated non-Newtonian liquids in external loop airlift reactors. Biotechnol Bioeng 62:494–498

Ambros PF, Matzke AJM, Matzke MA (1986a) Localization of *Agrobacterium rhizogenes* T-DNA in plant chromosomes by in situ hybridization. Eur Mol Biol Organ J 5:2073–2077

Ambros PF, Matzke MA, Matzke AJM (1986b) Detection of a 17 kb unique sequence (T-DNA) in plant chromosomes by in situ hybridization. Chromosoma 94:11–18

Amselem J, Tepfer M (1992) Molecular basis of novel root phenotypes induced by *Agrobacterium rhizogenes* A4 on cucumber. Plant Mol Biol 19:421–432

Andarwulan N, Shetty K (1999) Phenolic content in differentiated tissue cultures of untransformed and Agrobacterium-transformed roots of Anise (*Pimpinella anisum* L.). J Agric Food Chem 47:1776–1780

Aoki T, Matsumoto H, Asako Y, Matsunaga Y, Shimomura K (1997) Variation of alkaloid productivity among several clones of hairy roots and regenerated plants of *Atropa belladonna* transformed with *Agrobacterium rhizogenes* 15834. Plant Cell Rep 16:282–286

Bais HP, Sudha G, Ravishankar GA (1999) Putrescine influences growth and production of coumarins in hairy root cultures of Witloof Chicory (*Cichorium intybus* L. cv. Lucknow Local). J Plant Growth Regul 18:159–165

Bais HP, Venkatesh RT, Chandrashekar A, Ravishankar GA (2001) *Agrobacterium rhizogenes*-mediated transformation of Witloof chicory—in vitro shoot regeneration and induction of flowering. Curr Sci 80(1):83–87

Bais HP, Walker TS, Schweizer HP, Vivanco JM (2002a) Root specific elicitation and antimicrobial activity of rosmarinic acid in hairy root cultures of *Ocimum basilicum*. Plant Physiol Biochem 40:983–995

Bais HP, Suresh B, Raghavarao KSMS, Ravishankar GA (2002b) Performance of hairy root cultures of *Cichorium intybus* L. in bioreactors of different configurations. In Vitro Cell Dev Biol Plant 38:573–580

Baíza AM, Quiroz A, Ruíz JA, Maldonado-Mendoza I (1999a) Growth patterns and alkaloid accumulation in hairy root and untransformed root cultures of *Datura stramonium*. Plant Cell Tiss Org 54:123–130

Baíza AM, Quiroz-Moreno A, Ruíz JA, Loyola-Vargas VM (1999b) Genetic stability of hairy root cultures of *Datura stramonium*. Plant Cell Tiss Org 59:9–17

Bálványos I, Kursinszki L, Bányai P, Szöke É (2004) Analysis of polyacetylenes by HPLC in hairy root cultures of *Lobelia inflata* cultivated in bioreactor. Chromatographia 60:S235–S238

Bandyopadhyay M, Jha S, Tepfer D (2007) Changes in morphological phenotypes and withanolide composition of Ri-ransformed roots of *Withania somnifera*. Plant Cell Rep 26:599–609

Banerjee S, Rahman L, Uniyal GC, Ahuja PS (1998) Enhanced production of valepotriates by *Agrobacterium rhizogenes* induced hairy root cultures of *Valeriana wallichii* DC. Plant Sci 131:203–208

Banerjee S, Shang TQ, Wilson AM, Moore AL, Strand SE, Gordon MP, Doty Sl (2002) Expression of functional mammalian P450 2E1 in hairy root cultures. Biotechnol Bioeng 77(4):462–466

Banerjee S, Madhusudanan KP, Chattopadhyay SK, Rahman LU, Khanuja SPS (2008) Expression of tropane alkaloids in the hairy root culture of *Atropa acuminata* substantiated by DART mass spectrometric technique. Biomed Chromatogr 22:830–834

Batra J, Dutta A, Singh D, Kumar S, Sen J (2004) Growth and terpenoid indole alkaloid production in *Catharanthus roseus* hairy root clones in relation to left- and right-termini-linked Ri T-DNA gene integration. Plant Cell Rep 23:148–154

Bauer N, Kiseljak D, Jelaska S (2009) The effect of yeast extract and methyl jasmonate on rosmarinic acid accumulation in *Coleus blumei* hairy roots. Biol Plantarum 53(4):650–656

Benson EE, Hamill JD (1991) Cryopreservation and post freeze molecular and biosynthetic stability in transformed roots of *Beta vulgaris* and *Nicotiana rustica*. Plant Cell Tiss Org 24:163–172

Berkov S, Pavlov A, Kovatcheva P, Stanimirova P, Philipov S (2003) Alkaloid spectrum in diploid and tetraploid hairy root cultures of *Datura stramonium*. Z Naturforsch 58c:42–46

Bijelović A, Rosić N, Miljuš-Djukić J, Ninković S, Grubišić D (2004) In vitro regeneration and transformation of *Blackstonia perfoliata*. Biol Plantarum 48(3):333–338

Bondarev N, Reshetnyak O, Nosov A (2002) Features of development of *Stevia rebaudiana* shoots cultivated in the roller bioreactor and their production of steviol glycosides. Planta Med 68:759–762

Bordonaro JL, Curtis WR (2000) Inhibitory role of root hairs on transport within root culture bioreactors. Biotechnol Bioeng 70:176–186

Bourgaud F, Gravot A, Milesi S, Gontier E (2001) Production of plant secondary metabolites: a historical perspective. Plant Sci 161:839–851

Brillanceau MH, David C, Tempé J (1989) Genetic transformation of *Catharanthus roseus* G. Don by *Agrobacterium rhizogenes*. Plant Cell Rep 8:63–66

Buitelaar R, Langenhoff AAM, Heidstra R, Tramper J (1991) Growth and thiophene production by hairy root cultures of *Tagetes patula* in various two-liquid-phase bioreactors. Enzyme Microb Tech 13:487–494

Bulgakov VP, Khodakovskaya MV, Labetskaya NV, Chernoded GK, Zhuravlev YN (1998) The impact of plant *rolC* oncogene on ginsenoside production by ginseng hairy root cultures. Phytochemistry 49(7):1929–1934

Cabrera-Ponce J, Vegas-Garcia A, Herrera-Estrella L (1996) Regeneration of transgenic papaya plants via somatic embryogenesis induced by *Agrobacterium rhizogenes*. In Vitro Cell Dev Biol Plant 32(2):86–90

Cardillo AB, Otálvaro AM, Busto VD, Talou JR, Velásquez LME, Giulietti AM (2010) Scopolamine, anisodamine and hyoscyamine production by *Brugmansia candida* hairy root cultures in bioreactors. Process Biochem 45:1577–1581

Carrizo CN, Pitta-Alvarez SI, Kogan MJ, Giulietti AM, Tomaro ML (2001) Occurrence of cadaverine in hairy roots of *Brugmansia candida*. Phytochemistry 57:759–763

Carvalho EB, Curtis WR (1998) Characterization of fluid-flow resistance in root cultures with a convective flow tubular bioreactor. Biotechnol Bioeng 60:375–384

Casas DA, Pitta-Alvarez SI, Giulietti AM (1998) Biotransformation of hydroquinone by hairy roots of *Brugmansia candida* and effect of sugars and free-radical scavengers. Appl Biochem Biotechnol 69:127–136

Caspeta L, Quintero R, Villarreal ML (2005a) Novel airlift reactor fitting for hairy root cultures: developmental and performance studies. Biotechnol Prog 21:735–740

Caspeta L, Nieto I, Zamilpa A, Alvarez L, Quintero R, Villarreal ML (2005b) *Solanum chrysotrichum* hairy root cultures: characterization, scale-up and production of five antifungal saponins for human use. Planta Med 71:1084–1087

Celma CR, Palazón J, Cusidó RM, Piñol MT, Keil M (2001) Decreased scopolamine yield in field-grown *Duboisia* plants regenerated from hairy roots. Planta Med 67(3):249–253

Chang CK, Chang KS, Lin YC, Liu SY, Chen CY (2005) Hairy root cultures of *Gynostemma pentaphyllum* (Thunb.) Makino: a promising approach for the production of gypenosides as an alternative of ginseng saponins. Biotechnol Lett 27:1165–1169

Chashmi NA, Sharifi M, Karimi F, Rahnama H (2010) Differential production of tropane alkaloids in hairy roots and in vitro cultured two accessions of *Atropa belladonna* L. under nitrate treatments. Z Naturforsch 65c:373–379

Chatterjee C, Correll MJ, Weathers PJ, Wyslouzil BE, Walcerz DB (1997) Simplified acoustic window mist bioreactor. Biotechnol Tech 11:155–158

Chaudhuri KN, Ghosh B, Tepfer D, Jha S (2005) Genetic transformation of *Tylophora indica* with *Agrobacterium rhizogenes* A4: growth and tylophorine productivity in different transformed root clones. Plant Cell Rep 24:25–35

Chaudhuri KN, Ghosh B, Tepfer D, Jha S (2006) Spontaneous plant regeneration in transformed roots and calli from *Tylophora indica*: changes in morphological phenotype and tylophorine accumulation associated with transformation by *Agrobacterium rhizogenes*. Plant Cell Rep 25:1059–1066

Chaudhuri KN, Das S, Bandyopadhyay M, Zalar A, Kollmann A, Jha S, Tepfer D (2009) Transgenic mimicry of pathogen attack stimulates growth and secondary metabolite accumulation. Transgenic Res 18:121–134

Chen H, Chen F, Chiu FCK, Lo CMY (2001) The effect of yeast elicitor on the growth and secondary metabolism of hairy root cultures of *Salvia miltiorrhiza*. Enzyme Microb Technol 28:100–105

Chilton MD, Tepfer DA, Petit A, David C, Casse-Delbart F, Tempé J (1982) *Agrobacterium rhizogenes* inserts T-DNA into the genomes of the host-plant root cells. Nature 295:432–434

Choi SM, Son SH, Yun SR, Kwon OW, Seon JH, Paek KY (2000) Pilot scale culture of adventitious roots of ginseng in a bioreactor system. Plant Cell Tiss Org 62:187–193

Choi PS, Kim YD, Choi KM, Chung HJ, Choi DW, Liu JR (2004) Plant regeneration from hairy-root cultures transformed by infection with *Agrobacterium rhizogenes* in *Catharanthus roseus*. Plant Cell Rep 22:828–831

Choi Y, Kim Y, Paek K (2006) Types and designs of bioreactors for hairy culture. In: Duttagupta S, Ibaraki Y (eds) Plant tissue culture engineering. Springer, Dordrecht, pp 161–172

Christey MC, Sinclair BK, Braun RH, Wyke L (1997) Regeneration of transgenic vegetable brassicas (*Brassica oleracea* and *B. campestris*) via Ri-mediated transformation. Plant Cell Rep 16:587–593

Chung IM, Ali M, Chun SC, San KY, Peebles CAM, Ahmad A (2008) New catharanthusopimaranoside A and B from hairy root cultures of *Catharanthus roseus*. Chem Nat Compd 44(4):458–462

Coniglio MS, Busto VD, González PS, Medina MI, Milrad S, Agostini E (2008) Application of *Brassica napus* hairy root cultures for phenol removal from aqueous solutions. Chemosphere 72:1035–1042

Cuello JL, Walker PN, Curtis WR (2003) Design of ebb-and-flow bioreactor (EFBR) for immobilized "hairy root" cultures Part II: Growth studies and model verifications. Trans ASAE 46(5):1469–1476

Cusido RM, Palazon J, Pinol MT, Bonfill M, Morales C (1999) *Datura metel*: in vitro production of tropane alkaloids. Planta Med 65:144–148

Damiani F, Arcioni S (1991) Transformation of *Medicago arborea* L. with an *Agrobacterium rhizogenes* binary vector carrying the hygromycin resistance gene. Plant Cell Rep 10:300–303

David C, Tempé J (1988) Genetic transformation of cauliflower (*Brassica oleracea L.* var. *botrytis*) by *Agrobacterium rhizogenes*. Plant Cell Rep 7:88–91

Dhakulkar S, Ganapathi TR, Bhargava S, Bapat VA (2005) Induction of hairy roots in *Gmelina arborea* Roxb. and production of verbascoside in hairy roots. Plant Sci 169:812–818

Dilorio AA, Cheetham RD, Pamela J, Weathers PJ (1992) Growth of transformed roots in a nutrient mist bioreactor: reactor performance and evaluation. Appl Microbiol Biotechnol 37:457–462

Doty SL (2008) Enhancing phytoremediation through the use of transgenics and endophytes. New Phytol 179:318–333

Du M, Wu XJ, Ding J, Hu ZB, White KN, Branford-White CJ (2003) Astragaloside IV and polysaccharide production by hairy roots of *Astragalus membranaceus* in bioreactors. Biotechnol Lett 25:1853–1856

Economou AS, Read PE (1986) Light quality control of shoot and root formation in vitro in petunia cultures (abstract No. 750). Hortscience 21:S-761

Flem-Bonhomme VL, Laurain-Mattar D, Fliniaux MA (2004) Hairy root induction of *Papaver somniferum* var. *album*, a difficult-to-transform plant, by *A. rhizogenes* LBA 9402. Planta 218:890–893

Flores HE, Curtis WR (1992) Approaches to understanding and manipulating the biosynthetic potential of plant roots. Ann N Y Acad Sci 665:188–209

Flores HE, Medina-Bolivar F (1995) Root culture and plant natural products: "unearthing" the hidden half of plant metabolism. Plant Tissue Cult Biotech 1:59–74

Flores HE, Vivanco JM, Loyola-Vargas VM (1999) "Radicle" biochemistry: the biology of root-specific metabolism. Trend Plant Sci 4:220–226

Fons F, Tousch D, Rapior S, Gueiffier A, Roussel JL, Gargadennec A, Andary C (1999) Phenolic profiles of untransformed and hairy root cultures of *Plantago lanceolata*. Plant Physiol Biochem 37(4):291–296

Fraga BM, Diäaz CE, Guadaño A, Lez-Coloma AG (2005) Diterpenes from *Salvia broussonetii* transformed roots and their insecticidal activity. J Agric Food Chem 53:5200–5206
Freeman A, Woodley JM, Lilly MD (1993) In situ product removal as a tool for bioprocessing. Nat Biotechnol 11:1007–1012
Fu CX, Zhao DX, Xue XF, Jin ZP, Ma FS (2005) Transformation of *Saussurea involucrata* by *Agrobacterium rhizogenes*: Hairy root induction and syringin production. Process Biochem 40:3789–3794
Fukui H, Feroj-Hasan AFM, Ueoka T, Kyo M (1998) Formation and secretion of a new brown benzoquinone by hairy root cultures of *Lithospermum erythrorhizon*. Phytochemistry 47:1037–1039
Fukui H, Feroj-Hasan AFM, Kyo M (1999) Formation and secretion of a unique quinone by hairy root cultures of *Lithospermum erythrorhizon*. Phytochemistry 51:511–515
Furmanowa M, Syklowska-Baranek K (2000) Hairy root cultures of *Taxus* x *media* var. *Hicksii* Rehd. as a new source of paclitaxel and 10-deacetylbaccatin III. Biotechnol Lett 22:683–686
Gangopadhyay M, Chakraborty D, Bhattacharyya S, Bhattacharya S (2010) Regeneration of transformed plants from hairy roots of *Plumbago indica*. Plant Cell Tiss Org 102:109–114
Gangopadhyay M, Dewanjee S, Chakraborty D, Bhattacharya S (2011) Role of exogenous phytohormones on growth and plumbagin accumulation in *Plumbago indica* hairy roots and conservation of elite root clones via synthetic seeds. Ind Crop Prod 33:445–450
Geerlings A, Hallard D, Martinez Caballero A, Lopes Cardoso I, van der Heijden R, Verpoorte R (1999) Alkaloid production by a *Cinchona officinalis* 'Ledgeriana' hairy root culture containing constitutive expression constructs of tryptophan decarboxylase and strictosidine synthase cDNAs from *Catharanthus roseus*. Plant Cell Rep 19:191–196
Georgiev M, Heinrich M, Kerns G, Bley T, Pavlov A (2006) Production of iridoids and phenolics by transformed *Harpagophytum procumbens* root cultures. Eng Life Sci 6(6):593–596
Ghosh S, Jha S (2005) Colchicine accumulation in in vitro cultures of *Gloriosa superba*. In: D' Souza L, Anuradha M, Nivas S, Hegde S, Rajendra K (eds) Biotechnology for a better future. SAC Publications, Mangalore, pp 92–98
Ghosh B, Mukherjee S, Jha TB, Jha S (2002) Enhanced colchicine production in root cultures of *Gloriosa superb* by direct and indirect precursors of the biosynthetic pathway. Biotechnol Lett 24:231–234
Giovanni A, Pecchioni N, Rabaglio M, Allavena A (1997) Characterization of ornamental *Datura* plants transformed by *Agrobacterium rhizogenes*. In Vitro Cell Dev Biol Plant 33:101–106
Giri A, Banerjee S, Ahuja PS, Giri CC (1997) Production of hairy roots in *Aconitum heterophyllum* Wall. using *Agrobacterium rhizogenes*. In Vitro Cell Dev Biol Plant 33:280–284
Grąbkowska R, Królicka A, Mielicki W, Wielanek M, Wysokińska H (2010) Genetic transformation of *Harpagophytum procumbens* by *Agrobacterium rhizogenes*: iridoid and phenylethanoid glycoside accumulation in hairy root cultures. Acta Physiol Plant 32:665–673
Greens KD, Thomas NH (1996) An integrated "root tube" bioreactor/separator for transformed root cultures. J Ferment Bioeng 81:453–457
Gujarathi NP, Linden JC (2005) Oxytetracycline inactivation by putative reactive oxygen species released to nutrient medium of *Helianthus annuus* hairy root cultures. Biotechnol Bioeng 92(4):393–402
Gutièrrez-Pesce P, Taylor K, Muleo R, Rugini E (1998) Somatic embryogenesis and shoot regeneration from transgenic roots of the cherry rootstock Colt (*Prunus avium* × *P. pseudocerasus*) mediated by pRi 1855 T-DNA of *Agrobacterium rhizogenes*. Plant Cell Rep 17:574–580
Häkkinen ST, Moyano E, Cusidó RM, Palazón J, Piñol MT, Oksman-Caldentey KM (2005) Enhanced secretion of tropane alkaloids in *Nicotiana tabacum* hairy roots expressing heterologous hyoscyamine-6β-hydroxylase. J Exp Bot 56(420):2611–2618
Hamill JD, Evans D, Robins RJ, Rhodes MJC, Prescott A, Martin C (1988) Foreign gene insertion into hairy roots with binary vectors and *Agrobacterium rhizogenes*: potential for genetic manipulation of plant secondary metabolism in culture. In: Robins RJ, Rhodes MJC (eds) Manipulating secondary metabolism in culture. Cambridge University Press, Cambridge, pp 145–153

Hamill JD, Robins RJ, Rhodes MJ (1989) Alkaloid production by transformed root cultures of *Cinchona ledgeriana*. Planta Med 55(4):354–357

Handa T, Sujimura T, Kato E, Kamada H, Takayanagi K (1995) Genetic transformation of *Eustoma grandiflorum* with rol genes. Acta Hort 392:209–218

Hashimoto T, Yukimune Y, Yamada Y (1986) Tropane alkaloid production in *Hyoscyamus* root cultures. J Plant Physiol 124:61–76

Hilton MG, Rhodes MJC (1990) Growth and hyoscyamine production of hairy root cultures of *Datura stramonium* in a modified stirred tank reactor. Appl Microbiol Biotechnol 33:132–138

Hilton MG, Wilson PDG, Robins RJ, Rhodes MJC (1988) Transformed root cultures—fermentation aspects. In: Robins RJ, Rhodes MJC (eds) Manipulating secondary metabolism in culture. Cambridge University Press, Cambridge, pp 239–245

Hirata K, Horiuchi M, Ando T, Miyamoto K, Miura Y (1990) Vindoline and catharanthine production in multiple shoot culture of *Catharanthus roseus*. J Ferment Bioeng 70:193–195

Hiroshi S, Yamakawa T, Yamazaki M, Aimi N, Saito K (2002) Bioreactor production of camptothecin by hairy root cultures of *Ophiorrhiza pumila*. Biotechnol Lett 24:359–363

Hitaka Y, Kino-Oka M, Taya M, Tone S (1997) Effect of liquid flow on culture of red beet hairy roots in single column reactor. J Chem Eng Jpn 30:1070–1075

Holmes P, Si Li, Green KD, Ford-Lloyd BV, Thomas NH (1997) Drip tube technology for continuous culture of hairy roots with integrated alkaloid extraction. In: Doran PM (ed) Hairy roots: culture and application. Harwood Academic Publisher, Amsterdam, pp 201–208

Hoshino Y, Mii M (1998) Bialaphos stimulates shoot regeneration from hairy roots of snapdragon (*Antirrhinum majus* L.) transformed by *Agrobacterium rhizogenes*. Plant Cell Rep 17:256–261

Hwang SJ (2006) Baicalin production in transformed hairy root clones of *Scutellaria baicalensis*. Biotechnol Bioprocess Eng 11:105–109

Ionkova I, Fuss E (2009) Influence of different strains of *Agrobacterium rhizogenes* on induction of hairy roots and lignan production in *Linum tauricum* ssp. *tauricum*. Pharmacognosy Mag 5:14–18

Ionkova I, Kartnig T, Alfermann W (1997) Cycloartane saponin production in hairy root cultures of *Astragalus mongholicus*. Phytochemistry 45(8):1597–1600

Jacob A, Malpathak N (2005) Plantlet regeneration enhances Solasodine productivity in hairy root cultures of *Solanum khasianum* Clarke. In Vitro Cell Dev Biol Plant 41:291–295

Janković T, Krstić D, Savikin-Fodulović K, Menković N, Grubisić D (2002) Xanthones and secoiridoids from hairy root cultures of *Centaurium erythraea* and *C. pulchellum*. Planta Med 68(10):944–946

Jaziri M, Yoshimatsu K, Homes J, Shimomura K (1994) Traits of transgenic *Atropa belladonna* doubly transformed with different *Agrobacterium rhizogenes* strains. Plant Cell Tiss Org Cult 38:257–262

Jeong G, Park D, Hwang B, Park K, Kim S, Woo J (2002) Studies on mass production of transformed *Panax ginseng* hairy roots in bioreactor. Appl Biochem Biotech 98–100:1115–1127

Jeong G, Park D, Hwang B, Woo J (2003) Comparison of growth characteristics of *Panax ginseng* hairy roots in various bioreactors. Appl Biochem Biotech 105–108:493–503

Jha S, Sahu NP, Sen J, Jha TB, Mahato SB (1991) Production of Emetine and Cephaeline from cell suspension and excised root cultures of *Cephaelis ipecacuanha*. Phytochemistry 30:3999–4004

Jouanin L, Guerche D, Pamboukdjian N, Tourneur C, Casse-Delbart F, Tourneur J (1987) Structure of T-DNA in plants regenerated from roots transformed by *Agrobacterium rhizogenes* strain A4. Mol Gen Genet 206:387–392

Kadkade PG, Jopson H (1978) Influence of light quality on organogenesis from the embryo-derived callus of Douglas fir (*Pseudotsuga menziesii*). Plant Sci Lett 13:67–73

Kamada H, Okamura N, Satake M, Harada H, Shimomura K (1986) Alkaloid production by hairy root cultures in *Atropa belladonna*. Plant Cell Rep 5:239–242

Kamada H, Saitou T, Harada H (1992) No requirement of vernalization for flower formation in Ri-transformed *Cichorium* plants. Plant Tiss Cult Lett 9:206–208

Kanokwaree K, Doran PM (1998) Application of membrane tubing aeration and perfluorocarbon to improve oxygen delivery to hairy root cultures. Biotechnol Prog 14:479–486

Karam NS, Jawad FM, Arikat NA, Shib RA (2003) Growth and rosmarinic acid accumulation in callus, cell suspension, and root cultures of wild *Salvia fruiticosa*. Plant Cell Tiss Org 73:117–121

Kevers C, Jacques P, Thonart P, Gaspar T (1999) In vitro root cultures of *Panax ginseng* and *P. quinquefolium*. Plant Growth Regul 27:173–179

Khanam N, Khoo C, Close R, Khan AG (2001) Tropane alkaloid production by shoot cultures of *Duboisia myoporoides* R. Br Phytochem 56:59–65

Kim YH, Yoo YJ (1993) Development of a bioreactor for high density culture of hairy roots. Biotechnol Lett 7:859–862

Kim Y, Wyslouzil BE, Weathers PJ (2001) A comparative study of mist and bubble column reactors in the in vitro production of artemisinin. Plant Cell Rep 20:451–455

Kim Y, Wyslouzil BE, Weathers PJ (2002a) Secondary metabolism of hairy root cultures in bioreactors. In Vitro Cell Dev Biol Plant 38:1–10

Kim YJ, Weathers PJ, Wyslouzil BE (2002b) Growth of *Artemisia annua* hairy roots in liquid- and gas-phase reactors. Biotechnol Bioeng 80(4):454–464

Kim YS, Hahn EJ, Murthy HH, Paek KY (2005a) Adventitious root growth and ginsenoside accumulation in *Panax ginseng* cultures as affected by methyl jasmonate. Biotechnol Lett 26:1619–1622

Kim OT, Manickavasagm M, Kim YJ, Jin MR, Kim KS, Seong NS, Hwang B (2005b) Genetic transformation of *Ajuga multiflora* Bunge with *Agrobacterium rhizogenes* and 20-Hydroxyecdysone production in hairy roots. J Plant Biol 48(2):258–262

Kino-Oka MK, Hitaka Y, Taya M, Tone S (1999) High-density culture of red beet hairy roots by considering medium flow condition in a bioreactor. Chem Eng Sci 54:3179–3186

Kintzios S, Makri O, Pistola E, Matakiadis T, Shi HP, Economou A (2004) Scale-up production of puerarin from hairy roots of *Pueraria phaseoloides* in an airlift bioreactor. Biotechnol Lett 26:1057–1059

Kondo O, Honda H, Taya M, Kobayashi T (1989) Comparison of growth properties of carrot hairy root in various bioreactors. Appl Microbiol Biotechnol 32:291–294

Kovács G, Kuzovkina IN, Szőke É, Kursinszki L (2004) HPLC determination of flavonoids in hairy-root cultures of *Scutellaria baicalensis* Georgi. Chromatographia 60:S81–S85

Królicka A, Staniszewska I, Bielawski K, Maliński E, Szafranek J, Łojkowska E (2001) Establishment of hairy root cultures of *Ammi majus*. Plant Sci 160:259–264

Kumar V, Jones B, Davey MR (1991) Transformation by *Agrobacterium rhizogenes* and regeneration of transgenic shoots of the wild soybean *Glycine argyrea*. Plant Cell Rep 10:135–138

Kusakari K, Yokoyama M, Inomata S (2000) Enhanced production of saikosaponins by root culture of *Bupleurum falcatum* L. using two-step control of sugar concentration. Plant Cell Rep 19:1115–1120

Kuźma Ł, Skrzypek Z, Wysokińska H (2006) Diterpenoids and triterpenoids in hairy roots of *Salvia sclarea*. Plant Cell Tiss Org 84:171–179

Kwok KH, Doran PM (1995) Kinetic and stoichiometric analysis of hairy roots in a segmented bubble column reactor. Biotechnol Prog 11:429–435

Lambert C, Tepfer D (1992) Use of *Agrobacterium rhizogenes* to create transgenic apple trees having an altered organogenic response to hormones. Theor Appl Genet 85:105–109

Lan X, Quan H (2010) Hairy root culture of *Przewalskia tangutica* for enhanced production of pharmaceutical tropane alkaloids. J Med Plants Res 4(14):1477–1481

Lee KT, Suzuki T, Yamakawa T, Kodama T, Igarashi Y, Shimomura K (1999) Production of tropane alkaloids by transformed root cultures of *Atropa belladonna* in stirred bioreactors with a stainless steel net. Plant Cell Rep 18:567–571

Lee MH, Yoon ES, Jeong JH, Choi YE (2004) *Agrobacterium rhizogenes*-mediated transformation of *Taraxacum platycarpum* and changes of morphological characters. Plant Cell Rep 22:822–827

Lee OS, Kang YM, Jung HY, Min JY, Kang SM, Karigar CS, Prasad DT, Bahk JD, Choi MS (2005) Enhanced production of tropane alkaloids in *Scopolia parviflora* by introducing the pmt (putrescine n-methyltransferase) gene. In Vitro Cell Dev Biol Pant 41:167–172

Lee SY, Xu H, Kim YK, Park SU (2008) Rosmarinic acid production in hairy root cultures of *Agastache rugosa* Kuntze. World J Microbiol Biotechnol 24:969–972

Li W, Asada Y, Koike K, Hirotani M, Rui H, Yoshikawa T, Nikaido T (2001) Flavonoids from *Glycyrrhiza pallidiflora* hairy root cultures. Phytochemistry 58:595–598

Li W, Koike K, Asada Y, Yoshikawa T, Nikaido T (2003) Biotransformation of low-molecular-weight alcohols by *Coleus forskohlii* hairy root cultures. Carbohyd Res 338:729–731

Li W, Koike K, Asada Y, Yoshikawa T, Nikaido T (2005) Rosmarinic acid production by *Coleus forskohlii* hairy root cultures. Plant Cell Tiss Org 80:151–155

Liu CZ, Wang YC, Ouyang F, Ye HC, Li GF (1998) Production of artemisinin by hairy root cultures of *Artemisia annua* L in bioreactor. Biotechnol Lett 20(3):265–268

Liu CZ, Wang YC, Zhao B, Guo C, Ouyang F, Ye HC, Li GF (1999) Development of a nutrient mist bioreactor for growth of hairy roots. In Vitro Cell Dev Biol Plant 35:271–274

Lorence A, Medina-Bolivar F, Nessler CL (2004) Camptothecin and 10-hydroxycamptothecin from *Camptotheca acuminata* hairy roots. Plant Cell Rep 22:437–441

Lu HY, Liu JM, Zhang HC, Yin T, Gao SL (2008) Ri-mediated transformation of *Glycyrrhiza uralensis* with a squalene synthase gene (gusqs1) for production of glycyrrhizin. Plant Mol Biol Rep 26:1–11

Macek T, Macková M, Káš J (2000) Exploitation of plants for the removal of organics in environmental remediation. Biotechnol Adv 18:23–34

Macková M, Macek T, Kučerová P, Burkhard J, Pazlarová J, Demnerová K (1997) Degradation of polychlorinated biphenyls by hairy root culture of *Solanum nigrum*. Biotechnol Lett 19(8):787–790

Majumdar S, Garai S, Jha S (2011) Genetic transformation of *Bacopa monnieri* by wild type strains of *Agrobacterium rhizogenes* stimulates production of bacopa saponins in transformed calli and plants. Plant Cell Rep 30(5):941–954

Maldonado-Mendoza IE, Ayora-Talavera T, Loyola-Vargas VM (1993) Establishment of hairy root cultures of *Datura stramonium* characterization and stability of tropane alkaloid production during long periods of subculturing. Plant Cell Tiss Org 33:321–329

Mannan A, Shaheen N, Arshad W, Qureshi RA, Zia M, Mirza B (2008) Hairy roots induction and artemisinin analysis in *Artemisia dubia* and *Artemisia indica*. Afr J Biotechnol 7(18): 3288–3292

Manners JM, Way H (1989) Efficient transformation with regeneration of the tropical pasture legume *Stylosanthes humilis* using *Agrobacterium rhizogenes* and a Ti plasmid binary vector system. Plant Cell Rep 8:341–345

Mano Y, Nabeshima S, Matsui C, Ohkawa H (1986) Production of tropane alkaloids by hairy root cultures of *Scopolia japonica*. Agr Biol Chem 50:2715–2722

Mano Y, Ohkawa H, Yamada Y (1989) Production of tropane alkaloids by hairy root cultures of *Duboisia leichhardtii* transformed by *Agrobacterium rhizogenes*. Plant Sci 59(2):191–201

Marschner H (1995) Mineral nutrition of higher plants. Academic press, London

McKelvey SA, Gehrig JA, Hollar KA, Curtis WR (1993) Growth of plant root cultures in liquid- and gas-dispersed reactor environments. Biotechnol Prog 9:317–322

Mehrotra S, Kukreja AK, Khanuja SPS, Mishra BN (2008) Genetic transformation studies and scale up of hairy root culture of *Glycyrrhiza glabra* in bioreactor. Electron J Biotechnol 11(2):1–7

Mishra BN, Ranjan R (2008) Growth of hairy-root cultures in various bioreactors for the production of secondary metabolites. Biotechnol Appl Biochem 49:1–10

Mugnier J (1997) Mycorrhizal interactions and the effects of fungicides, nematicides and herbicides on hairy root cultures. In: Doran PM (ed) Hairy roots: culture and applications. Harwood Academic Publishers, Amsterdam, pp 123–132

Mukherjee S, Ghosh B, Jha TB, Jha S (1995) Genetic transformation of *Artemisia annua* by *Agrobacterium rhizogenes*. Indian J Exp Biol 33:868–871

Mukundan U, Bhide V, Singh G, Curtis WR (1998) pH-mediated release of betalains from transformed root cultures of *Beta vulgaris* L. Appl Microbiol Biotechnol 50:241–245

Muranaka T, Kazuoka T, Ohkawa H, Yamada Y (1993a) Characteristics of scopolamine-releasing hairy root clones of *Duboisia leichhardtii*. Biosci Biotech Bioch 57:1398–1399

Muranaka T, Ohkawa H, Yamada Y (1993b) Continuous production of scopolamine by a culture of *Duboisia leichhardtii* hairy root clone in a bioreactor system. Appl Microbiol Biotechnol 40:219–223

Narayanan RA, Atz R, Denny R, Young ND, Somers DA (1999) Expression of soybean cyst nematode resistance in transgenic hairy roots of soybean. Crop Sci 39:1680–1686

Neelwarne B, Thimmaraju R (2009) Bioreactor for cultivation of red beet hairy roots and in situ recovery of primary and secondary metabolites. Eng Life Sci 3:227–238

Nilsson O, Olsson O (1997) Getting to the root: the role of the *Agrobacterium rhizogenes rol* genes in the formation of hairy roots. Physiol Plantarum 100:463–473

Nitsch JP, Nitsch C (1969) Haploid plants from pollen grains. Science 163:85–87

Noda T, Tanaka N, Mano Y, Nabeshima S, Ohkawa H, Matsui C (1987) Regeneration of horseradish hairy roots incited by *Agrobacterium rhizogenes* infection. Plant Cell Rep 6:283–286

Nuutila AM, Toivonen L, Kauppinen V (1994) Bioreactor studies on hairy root cultures of *Catharanthus roseus:* comparison of three bioreactor types. Biotechnol Tech 8:61–66

Ohara A, Akasaka Y, Daimon H, Mii M (2000) Plant regeneration from hairy roots induced by infection with *Agrobacterium rhizogenes* in *Crotalaria juncea* L. Plant Cell Rep 19:563–568

Oka KM, Hongo Y, Taya M, Tone S (1992) Culture of red beet hairy roots in a column-type reactor. J Chem Eng Jpn 25:490–495

Olszowska O, Alfermann AW, Furmanowa M (1996) Eugenol from normal and transformed root cultures of *Coluria geoides*. Plant Cell Tiss Org 45:273–276

Palaz6n J, Mallol A, Eibl R, Lettenbaucer C, Cusidó RM, Piñol MT (2003) Growth and ginsenoside production in hairy root cultures of *Panax ginseng* using a novel bioreactor. Planta Med 69(4):344–349

Panichayupakarananta P, Noguchi H, De Eknamkul W (1998) A new biscoumarin from *Impatiens balsamina* root cultures. Planta Med 64:774

Pannuri S, Reddy GR, McNeill D, Curtis WR (1993) Interpreting the role of phosphorous and growth rate in enhanced fungal induction of sesquiterpenes from *Hyoscyamus muticus* Root cultures. Appl Microbiol Biotechnol 38:550–555

Parr AJ, Hamill JD (1987) Relationship between *Agrobacterium rhizogenes* transformed hairy roots and intact, uninfected *Nicotiana* plants. Phytochemistry 26:3241–3245

Pellegrineschi A, Damon JP, Valtorta N, Paillard N, Tepfer D (1994) Improvement of ornamental characters and fragrance production in lemon scented geranium through genetic transformation by *Agrobacterium rhizogenes.* Nat Biotechnol 12:64–68

Pérez-Molphe-Balch E, Ochoa-Alejo N (1998) Regeneration of transgenic plants of Mexican lime from *Agrobacterium rhizogenes*-transformed tissues. Plant Cell Rep 17:591–596

Petit A, Stougaard J, Kiihle A, Marcker KA, Tempé J (1987) Transformation and regeneration of the legume *Lotus corniculatus*: a system for molecular studies of symbiotic nitrogen fixation. Mol Gen Genet 207:245–250

Piatczak E, Chmiel A, Wysokinska H (2005) Mist trickling bioreactor for *Centaurium erythraea* Rafn.: growth of shoots and production of secoiridiodes. Biotechnol Lett 27:721–724

Piatczak E, Krolicka A, Wysokinska H (2006) Genetic transformation of *Centaurium erythraea* Rafn by *Agrobacterium rhizogenes* and the production of secoiridoids. Plant Cell Rep 25:1308–1315

Predieri S, Malavasi FFF (1989) High frequency shoot regeneration from leaves of the apple rootstock M26 (*Malus pumila* Mill.). Plant Cell Tiss Org 17:133–142

Quandt HJ, PuÈhler A, Broer I (1993) Transgenic root nodules of *Vicia hirsuta*: a fast and efficient system for the study of gene expression in indeterminate-type nodules. Mol Plant Microb Interact 6:699–706

Ramakrishnan D, Salim J, Curtis WR (1994) Inoculation and tissue distribution in pilot-scale plant root culture bioreactors. Biotechnol Tech 8:639–644

Ray S, Jha S (2001) Production of withaferin A in shoot cultures of *Withania somnifera*. Planta Med 67:432–436

Ray S, Ghosh B, Sen S, Jha S (1996) Withanolide production by root cultures of *Withania somnifera* transformed with *Agrobacterium rhizogenes.* Planta Med 62(6):571–573

Rhodes MJC, Hilton M, Parr AJ, Hamill JD, Robins RJ (1986) Nicotine production by "hairy root" cultures of *Nicotiana rustica*: fermentation and product recovery. Biotechnol Lett 8:415–420

Rhodes MJC, Parr AJ, Walton NJ (1997) Studies of secondary metabolic pathways in transformed roots. In: Doran PM (ed) Hairy roots: culture and applications. Harwood Academic Publishers, Amsterdam, pp 99–110

Robins RJ, Parr AJ, Bent EG, Rhodes MJC (1991) Studies on the biosynthesis of tropane alkaloids in *Datura stramonium* L. transformed root cultures 1. The kinetics of alkaloid production and the influence of feeding intermediate metabolites. Planta 183:185–195

Roschina VV, Roschina VD (1993) The excretory function of higher plants. Springer, Berlin 311

Rostampour S, Sohi HH, Jourabchi E, Ansari E (2009) Influence of *Agrobacterium rhizogenes* on induction of hairy roots and benzylisoquinoline alkaloids production in Persian poppy (*Papaver bracteatum* Lindl.): preliminary report. World J Microbiol Biotechnol 25:1807–1814

Saito K, Sudo H, Yamazaki M, Nakamura MK, Kitajima M, Takayama H, Aimi N (2001) Feasible production of camptothecin by hairy root cultures of *Ophiorrhiza pumila*. Plant Cell Rep 20:267–271

Saitou T, Kamada H, Harada H (1991) Isoperoxidase in hairy roots and regenerated plants of horseradish (*Armoracia lapathifolia*). Plant Sci 75:195–201

Saitou T, Kamada H, Harada H (1992) Light requirement for shoot regeneration in horseradish hairy roots. Plant Physiol 99:1336–1341

Sasaki K, Udagava A, Ishimaru H, Hayashi T, Alfermann AW, Nakanishi F, Shimomura K (1998) High forskolin production in hairy roots of *Coleus forskohlii*. Plant Cell Rep 17:457–459

Satheeshkumar K, Jose B, Soniya EV, Seeni S (2009) Isolation of morphovariants through plant regeneration in *Agrobacterium rhizogenes* induced hairy root cultures of *Plumbago rosea* L. Indian J Biotechnol 8(4):435–441

Saxena G, Banerjee S, Rahman L, Verma PC, Mallavarapu GR, Kumar S (2007) Rose-scented geranium (*Pelargonium* sp.) generated by *Agrobacterium rhizogenes* mediated Ri-insertion for improved essential oil quality. Plant Cell Tiss Org 90:215–223

Scholl Y, Höke D, Dräger B (2001) Calystegines in *Calystegia sepium* derive from the tropane alkaloid pathway. Phytochemistry 58:883–889

Seibert M, Wetherbee PJ, Job DD (1975) The effects of light intensity and spectral quality on growth and shoot initiation in tobacco callus. Plant Physiol 56:130–139

Sevón N, Oksman-Caldentey KM (2002) *Agrobacterium rhizogenes*-mediated transformation: root cultures as a source of alkaloids. Planta Med 68:859–868

Sevón N, Dräger B, Hiltunen R, Oksman-Caldentey KM (1997) Characterization of transgenic plants derived from hairy roots of *Hyoscyamus muticus*. Plant Cell Rep 16:605–611

Sevón N, Hiltunen R, Oksman-Caldentey KM (1998) Somaclonal variation in transformed roots and protoplast-derived hairy root clones of *Hyoscyamus muticus*. Planta Med 64:37–41

Shanks JV, Morgan J (1999) Plant 'hairy root' culture. Curr Opin Biotechnol 10:151–155

Sharp JM, Doran PM (1990) Characteristics of growth and tropane alkaloid synthesis in *Atropa belladonna* roots transformed by *Agrobacterium rhizogenes*. J Biotechnol 16:171–186

Shi HP, Kintzios S (2003) Genetic transformation of *Pueraria phaseoloides* with *Agrobacterium rhizogenes* and puerarin production in hairy roots. Plant Cell Rep 21:1103–1107

Shinde AN, Malpathak N, Fulzele DP (2008) Production of phytoestrogens by plant cell and tissue cultures: recent scenario and exciting prospects. Pharmacognosy Rev 2:43–53

Siah CL, Doran PM (1991) Enhanced codeine and morphine production in suspended *Papaver somniferum* cultures after removal of exogenous hormones. Plant Cell Rep 10:349–353

Sidwa-Gorycka M, Krolicka A, Orlita A, Malinski E, Golebiowski M, Kumirska J, Chromik A, Biskup E, Stepnowski P, Lojkowska E (2009) Genetic transformation of *Ruta graveolens* L. by *Agrobacterium rhizogenes*: hairy root cultures a promising approach for production of coumarins and furanocoumarins. Plant Cell Tiss Org 97:59–69

Singh G, Curtis WR (1994) Reactor design for plant root culture. In: Shargool PD, Ngo TT (eds) Biotechnological applications plant cultures. CRC series of current topics in plant molecular biology. CRC Press, Boca Raton, pp 185–206

Souret FF, Kim Y, Wyslouzil BE, Wobbe KK, Weathers PJ (2003) Scale-up of *Artemisia annua* L. hairy root cultures produces complex patterns of terpenoid gene expression. Biotechnol Bioeng 83(6):653–667

Spollansky TC, Pitta-Alvarez SI, Giulietti AM (2000) Effect of jasmonic acid and aluminum on production of tropane alkaloids in hairy root cultures of *Brugmansia candida*. Electron J Biotechnol 3(1):72–75

Sretenović-Rajičić T, Ninković S, Miljuš-Đukić J, Vinterhalter B, Vinterhalter D (2006) *Agrobacterium rhizogenes*-mediated transformation of *Brassica oleracea* var. sabauda and *B. oleracea* var. capitata. Biol Plantarum 50(4):525–530

Stiller J, Martirani L, Tuppale S, Chian RJ, Chiurazzi M, Gresshoff PM (1997) High frequency transformation and regeneration of transgenic plants in the model legume *Lotus japonicus*. J Exp Bot 48(312):1357–1365

Subotić A, Budimir S, Grubišić D, Momčilović I (2003) Direct regeneration of shoots from hairy root cultures of *Centaurium erythraea* inoculated with *Agrobacterium rhizogenes*. Biol Plantarum 47(4):617–619

Sudha CG, Reddy BO, Ravishankar GA, Seeni S (2003) Production of ajmalicine and ajmaline in hairy root cultures of *Rauvolfia micrantha* Hook f., a rare and endemic medicinal plant. Biotechnol Lett 25:631–636

Sun LY, Touraud G, Charbonnier C, Tepfer D (1991) Modification of phenotype in Belgian endive (*Cichorium intybus*) through genetic transformation by *Agrobacterium rhizogenes*: conversion from biennial to annual flowering. Transgenic Res 1:14–22

Suresh B, Bais HP, Raghavarao KSMS, Ravishankar GA, Ghildyal NP (2005) Comparative evaluation of bioreactor design using *Tagetes patula* L. hairy roots as a model system. Process Biochem 40:1509–1515

Swanson SM, Mahady GB, Beecher CWW (1992) Stevioside biosynthesis by callus, root, shoot and rooted-shoot cultures in vitro. Plant Cell Tiss Org 28:151–157

Tada H, Murakami Y, Omoto T, Shimomura K, Ishimaru K (1996) Rosmarinic acid and related phenolics in hairy root cultures of *Ocimum basilicum*. Phytochemistry 42(2):431–434

Tanaka N, Matsumoto T (1993) Regenerants from *Ajuga* hairy roots with high productivity of 20-hydroxyecdysone. Plant Cell Rep 13:87–90

Tanaka N, Takao M, Matsumoto T (1995) Vincamine production in multiple shoot culture derived from hairy roots of *Vinca minor*. Plant Cell Tiss Org 41:61–64

Taya M, Yoyama A, Kondo O, Kobayashi T, Matsui C (1989) Growth characteristics of plant hairy roots and their cultures in bioreactors. J Chem Eng Jpn 22:84–88

Tenea GN, Calin A, Gavrila L, Cucu N (2008) Manipulation of root biomass and biosynthetic potential of *Glycyrrhiza glabra* L. plants by *Agrobacterium rhizogenes* mediated transformation. Roumanian Biotechnol Lett 13(5):3922–3932

Tepfer D (1984) Genetic transformation of several species of higher plants by *Agrobacterium rhizogenes*: phenotypic consequences and sexual transmission of the transformed genotype and phenotype. Cell 37:959–967

Tepfer D (1989) Ri T-DNA from *Agrobacterium rhizogenes:* a source of genes having applications in rhizosphere biology and plant development, ecology and evolution. In: Kosuge T, Nester E (eds) Plant-microbe interactions, vol 3. McGraw Hill, New York, pp 294–342

Tepfer D, Tempé J (1981) Production of d'agropine par des racines transformes sous I'action d'*Agrobacterium rhizogenes* souche A4. CR Acad Sci 292:153–156

Tescione LD, Ramakrishnan D, Curtis WR (1997) The role of liquid mixing and gas phase dispersion in a submerged, sparged root reactor. Enzyme Microb Technol 20:207–213

Tikhomiroff C, Allais S, Klvana M, Hisiger S, Jolicoeur M (2002) Continuous selective extraction of secondary metabolites from *Catharanthus roseus* hairy roots with silicon oil in a two-liquid-phase bioreactor. Biotechnol Prog 18:1003–1009

Tiwari RK, Trivedi M, Guang ZC, Guo GQ, Zheng GC (2008) *Agrobacterium rhizogenes* mediated transformation of *Scutellaria baicalensis* and production of flavonoids in hairy roots. Biol Plantarum 52(1):26–35

Toivonen L, Balsevich J, Kurz WGW (1989) Indole alkaloid production by hairy root cultures of *Catharanthus roseus*. Plant Cell Tiss Org 18:79–93

Trulson AJ, Simpson RB, Shahin EA (1986) Transformation of cucumber (*Cucumis sativus* L.) plants with *Agrobacterium rhizogenes*. Theor Appl Genet 73:11–15

Tzfira T, Ben-Meir H, Vainstein A, Altman A (1996) Highly efficient transformation and regeneration of aspen plants through shoot-bud formation in root culture. Plant Cell Rep 15:566–571

Uozumi N, Makino S, Kobayashi T (1995) 20-Hydroxyecdysone production in *Ajuga* hairy root controlling intracellular phosphate based on kinetic model. J Ferment Bioeng 80:362–368

Uozumi N, Ohtake Y, Nakashimada Y, Morikawa Y, Tanaka N, Kobayashi T (1996) Efficient regeneration from gus-transformed *Ajuga* hairy root. J Ferment Bioeng 81(5):374–378

Verma PC, Singh D, Rahman LU, Gupta MM, Banerjee S (2002) In vitro-studies in *Plumbago zeylanica*: rapid micropropagation and establishment of higher plumbagin yielding hairy root cultures. J Plant Physiol 159:547–552

Verma PC, Rahman LU, Negi AS, Jain DC, Khanuja SPS, Banerjee S (2007) *Agrobacterium rhizogenes*-mediated transformation of *Picrorhiza kurroa* Royle ex Benth.: establishment and selection of superior hairy root clone. Plant Biotechnol Rep 1:169–174

Verpoorte R, van der Heijden R, van Gulik WM, ten Hoopen HJG (1991) Plant biotechnology for the production of alkaloids: present status and prospects. In: Brossi A (ed) The Alkaloids 40. Academic Press, San Diego, pp 1–187

Vinterhalter B, Ninković S, Cingel A, Vinterhalter D (2006) Shoot and root culture of *Hypericum perforatum* L. transformed with *Agrobacterium rhizogenes* A4M70GUS. Biol Plantarum 50 (4):767–770

Vinterhalter B, Savić J, Platiša J, Raspor M, Ninković S, Mitić N, Vinterhalter D (2008) Nickel tolerance and hyperaccumulation in shoot cultures regenerated from hairy root cultures of *Alyssum murale* Waldst et Kit. Plant Cell Tiss Org 94:299–303

Wang YM, Wang JB, Luo D, Jia JF (2001) Regeneration of plants from callus cultures of roots induced by *Agrobacterium rhizogenes* on *Alhagi pseudoalhagi*. Cell Res 11(4):279–284

Weathers PJ, Wyslouzil BE, Wobbe KK, Kim YJ, Yigit E (1999) The biological response of hairy roots to O_2 levels in bioreactors. In Vitro Cell Dev Biol Plant 35:286–289

Weathers PJ, Bunk G, McCoy MC (2005) The effect of phytohormones on growth and artemisinin production in *Artemisia annua* hairy roots. In Vitro Cell Dev Biol Plant 41:47–53

Wei ZM, Kamada H, Harada H (1986) Transformation of *Solanum nigrum* L. protoplasts by *Agrobacterium rhizogenes*. Plant Cell Rep 5:93–96

Weis JS, Jaffe MJ (1969) Photoenhancement by blue light of organogenesis in tobacco pith cultures. Physiol Plant 22:171–176

Whitney PJ (1992) Novel bioreactors for the growth of roots transformed by *Agrobacterium rhizogenes*. Enzyme Microb Technol 14(1):13–17

Wilhelmson A, Kallio PT, Oksman-Caldentey KM, Nuutila AM (2005) Expression of *Vitreoscilla* hemoglobin enhances growth of *Hyoscyamus muticus* hairy root cultures. Planta Med 71:48–53

Williams GRC, Doran PM (1999) Investigation of liquid–solid hydrodynamic boundary layers and oxygen requirements in hairy root cultures. Biotechnol Bioeng 64:729–740

Williams GRC, Doran PM (2000) Hairy root culture in a liquid-dispersed bioreactor: characterization of spatial heterogeneity. Biotechnol Prog 16:391–401

Wilson PDG (1997) The pilot-scale cultivation of transformed roots. In: Doran PM (ed) Hairy roots: culture and applications. Harwood Academic, Amsterdam, pp 179–190

Wilson PDG, Hilton MG, Robins RJ, Rhodes MJC (1987) Fermentation studies of transformed root cultures. In: Moody GW, Baker PB (eds) International conference on bioreactors and biotransformation. Elsevier, London, pp 38–51

Woo SH, Park JM, Yang J (1996) Root culture using a mist culture system and estimation of scale-up feasibility. J Chem Tech Biotechnol 66:355–362

Wu Q, VanEtten HD (2004) Introduction of plant and fungal genes into pea (*Pisum sativum* L.) hairy roots reduces their ability to produce pisatin and affects their response to a fungal pathogen. Mol Plant Microb Interact 17(7):798–804

Wyslouzil BE, Waterbury RG, Weathers PJ (2000) The growth of single roots of *Artemisia annua* in nutrient mist reactors. Biotechnol Bioeng 70(2):143–150

Yamazaki M, Son L, Hayashi T, Morita N, Asamizu T, Mourakoshi T, Saito K (1996) Transgenic fertile *Scoparia dulcis* L., a folk medicinal plant, conferred with a herbicide-resistant trait using an Ri binary vector. Plant Cell Rep 15:317–321

Yan CY, Yu RM, Zhang Z, Kong LY (2007) Biotransformation of 4-hydroxybenzen derivatives by hairy root cultures of *Polygonum multiflorum* Thunb. J Integr Plant Biol 49(2):207–212

Yang DC, Choi YE (2000) Production of transgenic plants via *Agrobacterium rhizogenes* mediated transformation of *Panax ginseng*. Plant Cell Rep 19:491–496

Yoshimatsu K, Shimomura K (1992) Transformation of opium poppy (*Papaver somniferum* L.) with *Agrobacterium rhizogenes* MAFF 03-01724. Plant Cell Rep 11:132–136

Yoshimatsu K, Shimomura K, Yamazaki M, Saito K, Kiuchi F (2003) Transformation of ipecac (*Cephaelis ipecacuanha*) with *Agrobacterium rhizogenes*. Planta Med 69(11):1018–1023

Yu S, Doran PM (1994) Oxygen requirements and mass transfer in hairy root culture. Biotechnol Bioeng 44:880–887

Yukimune Y, Yara Y, Yamada Y (1994) Tropane alkaloid production in root cultures *Duboisia myoporoides* obtained by repeated selection. Biosci Biotech Bioch 58:1443–1446

Zárate R (1999) Tropane alkaloid production by *Agrobacterium rhizogenes* transformed hairy root cultures of *Atropa baetica* Wilk. (Solanaceae). Plant Cell Rep 18:418–423

Zargar M, Farahani F, Nabavi T (2010) Hairy roots production of transgenic *Catharanthus roseus* L. plants with *Agrobacterium rhizogenes* under in vitro conditions. J Med Plants Res 4(21):2199–2203

Zayed R, Wink M (2004) Induction of tropane alkaloid formation in transformed root cultures of *Brugmansia suaveolens* (Solanaceae). Z Naturforsch 59c:863–867

Zdravković-Korać S, Muhovski Y, Druart P, Ćalić D, Radojević L (2004) *Agrobacterium rhizogenes*-mediated DNA transfer to *Aesculus hippocastanum* L. and the regeneration of transformed plants. Plant Cell Rep 22:698–704

Zehra M, Banerjee S, Sharma S, Kumar S (1999) Influence of *Agrobacterium rhizogenes* strains on biomass and alkaloid productivity in hairy root lines of *Hyoscyamus muticus* and *H. albus*. Planta Med 65(1):60–63

Zhan X, Jones DA, Kerr A (1988) Regeneration of flax plants transformed by *Agrobacterium rhizogenes*. Plant Mol Biol 11:551–559

Zhang HC, Liu JM, Lu HY, Gao SL (2009) Enhanced flavonoid production in hairy root cultures of *Glycyrrhiza uralensis* Fisch by combining the over-expression of chalcone isomerase gene with the elicitation treatment. Plant Cell Rep 28:1205–1213

Zhang HC, Liu JM, Chen HM, Gao CC, Lu HY, Zhou H, Li Y, Gao SL (2011) Up-regulation of licochalcone A biosynthesis and secretion by tween 80 in hairy root cultures of *Glycyrrhiza uralensis* Fisch. Mol Biotechnol 47:50–56

Zhao D, Fu C, Chen Y, Ma F (2004) Transformation of *Saussurea medusa* for hairy roots and jaceosidin production. Plant Cell Rep 23:468–474

Zid SA, Orihara Y (2005) Polyacetylenes accumulation in *Ambrosia maritima* hairy root and cell cultures after elicitation with methyl jasmonate. Plant Cell Tiss Org 81:65–75

Zolala J, Farsi M, Gordan HR, Mahmoodnia M (2007) Producing a high scopolamine hairy root clone in *Hyoscyamus muticus* through transformation by *Agrobacterium rhizogenes*. J Agric Sci Technol 9:327–339

Chapter 3
Scutellaria: Biotechnology, Phytochemistry and Its Potential as a Commercial Medicinal Crop

N. Joshee, A. Tascan, F. Medina-Bolivar, P. Parajuli, A. M. Rimando, D. A. Shannon and J. W. Adelberg

3.1 Introduction

Medicinal plants play a considerable role, not only as traditional medicines used across many cultures, but also as trade commodities which meet the demand of often distant markets. Recent trends in medicinal plants studies show that there has been an appreciable increase in research activity in the area of bioactivity of natural products. For example, between 1981 and 2002, 5% of the 1,031 new chemical entities approved as drugs by the U. S. Food and Drug Administration were natural products, and another 23% were natural product-derived molecules (Clardy and Walsh 2004). As many as 84% of pediatric oncology patients, 50% of breast cancer patients and 37% of prostate cancer patients use complementary and alternative medicine (CAM), including predominantly herbal approaches

N. Joshee (✉)
Agricultural Research Station, Fort Valley State University,
1005 State University Dr, Fort Valley, GA 31030, USA
e-mail: josheen@fvsu.edu

A. Tascan · J. W. Adelberg
Department of Horticulture, Clemson University, Clemson, SC, USA

F. Medina-Bolivar
Department of Biological Sciences, Arkansas Bioscience Institute,
Arkansas State University, Jonesboro, AR, USA

P. Parajuli
Department of Neurosurgery, Karmanos Cancer Institute,
Wayne State University, Detroit, MI, USA

A. M. Rimando
Natural Products Utilization Research Unit, USDA-ARS, Oxford, MS, USA

D. A. Shannon
Auburn University, Auburn, AL, USA

S. Chandra et al. (eds.), *Biotechnology for Medicinal Plants*,
DOI: 10.1007/978-3-642-29974-2_3,

(Richardson 2001). One in six American men will develop prostate cancer and the disease claims >30,000 lives yearly. To reduce this tremendous health burden, new approaches have been directed toward extremes of the disease spectrum centering on strategies for prostate cancer prevention and for treating advanced androgen-independent cancers (Bonham et al. 2005). According to Srivastava et al. (1996), in 1990 alone, traditional Chinese medicine (TCM) practitioners used 700,000 tons of plant material and 80% of it was collected from the wild. The concept of multitargeted therapy exists in traditional medical practice that employs multicomponent extracts of natural products (Chavan et al. 2006). Botanical derivatives usually contain several classes of compounds, which simultaneously act on multiple targets. There has been a paradigm shift from single-target drugs to multi-target drugs during the past decade (Wermuth 2004). Multi-target approaches are often directed toward activation of defense, protective, and repair mechanisms of the body rather than focusing only on destruction of the damage-causing agent. This may be achieved by the use of a combination of drugs as in AIDS, cancer, atherosclerosis, or use of single compounds that affect many targets simultaneously (Csermely et al. 2005; Cappuzzo et al. 2005). Further, there is a renewed interest in exploring plants as a novel means by which large quantities of vaccine and therapeutic proteins can be produced in a safe and cost-effective manner (Hefferon 2010). The value of medicinal plants is in their unique phytochemicals commonly known as secondary metabolites. Secondary metabolites mediate many vital functions in plant survival against herbivory, predators, environmental stress, and competition. Very little is known about the genetics of plant secondary metabolism, as the genes of most pathways have not been identified and little is understood of their regulation and functions.

Scutellaria is a perennial, herbaceous genus in the Lamiaceae family with 350–360 species (Bailey and Bailey 1978; Paton 1990; Cole et al. 2007) and new species are being added to the expanding list continuously (Turner 2011). It is the second largest genus in the family after *Salvia*. The skullcap (also spelled scullcap) is a North American perennial plant that grows in wet places in Canada and the northern and eastern U.S. Shang et al. (2010) recently reviewed the ethnopharmacology and the biological activities in the genus *Scutellaria*. According to their report more than 295 compounds have been isolated from genus *Scutellaria* and the majority of the compounds fall under flavonoids and diterpenes. Modern pharmacology research has confirmed that the extracts or monomeric compounds of the genus *Scutellaria* possess antitumor, hepatoprotective, antioxidant, anti-inflammatory, anticonvulsant, antibacterial, and antiviral effects (Shang et al. 2010). Many species of *Scutellaria* have been extensively used as medicinal plants, culinary herbs, and garden flowers (Nishikawa et al. 1999a).

The Convention on Biological Diversity (CBD), an international agreement adopted in 1992 and ratified by over 190 countries, promotes the conservation and sustainable use of biodiversity. It also aims to ensure the fair and equitable sharing of benefits arising from the utilization of biodiversity. As the majority of the medicinal plants are still wildcrafted, their sustainable use and conservation is the most critical issue requiring immediate attention.

3.2 Distribution and Botany

The generic name *Scutellaria* is derived from the Latin word *scutella* (little dish), describing the lid of the calyx (Fig. 3.1a, b). *Scutellaria* species are annuals, herbaceous perennials, and shrubs that grow under full sunlight, in well-drained soils with limited nutrient status. The fibrous root system which is a short creeping stolons, supports a branching stem growing from 1 to 3 feet tall, with opposite, ovate, and serrate leaves. Leaves are 0.5–2.5 inches long with scalloped or toothed margins. Leaves, stem, calyx, and corolla surface are full of many different types of trichomes (Fig. 3.1b, d). The blue to lavender flowers are in racemes and grow from the leaf axils. Figure 3.2 shows the range of flower color in *Scutellaria* genus. Flowers of *Scutellaria* species are hooded, tube shaped, and two lipped with the upper lip being the hood and the lower lip having two shallow lobes. Flowers range a great deal in size (Fig. 3.1c). Most of the species bloom from May to August and attract pollinators. Chromosome numbers in various *Scutellaria* species have been studied and reported as $2n = 24–34$ (Paton 1990); however, there are many exceptions to it.

3.2.1 Planting

Skullcap can be grown through direct seeding, transplanting, or by dividing the roots. The preferred method is to germinate seeds indoors. Seeds can be sown in early spring after danger of frost is over. *Scutellaria* species are usually pest and disease-free. Plants growing in high humidity conditions usually succumb to an early seedling damping off or powdery mildew at a later stage. Skullcap seeds require stratification and light to germinate. Seeds are planted in flats covered with a thin layer of prepared soil mix. Soil mix is moistened and kept at 5–10 °C for 7 days. After the required stratification period, flats are transferred to the greenhouse for germination. Emerging plants are spaced eight to twelve inches apart in rows one and a half to three feet apart or in three-foot wide beds. Once established, plants quickly fill the bed. It is important to keep the beds free of weeds to promote good growth. These plants do well under conditions of partial shade to full sun, limited feeding, and well-drained soil (Similicn 2009).

3.3 *Scutellaria* in Alternative Medicine System

In traditional medical systems, scullcap is used in the treatment of a wide range of nervous system conditions, including epilepsy, insomnia, hysteria, anxiety, delerium tremens, and withdrawal from barbiturates and tranquilizers.

Fig. 3.1 **a** The name skullcap originates from the structure of the calyx; **b** scanning electron micrograph (SEM) of the calyx in *S. ocmulgee*; **c** variation in flower size, *S. drummondii* vs *S. montana* from our germplasm collection at FVSU; **d** SEM of lower leaf surface of *S. ocmulgee* showing trichomes (*arrows*) and stomata; **e** shoot bud induction in *S. ocmulgee*; **f** comparison of biomass growth in *S. lateriflora* using MS and B5 liquid medium; **g** use of *S. lateriflora* shoot tip for encapsulation (60 days at 4 °C); **h** regeneration of encapsulated shoot in MS medium; **i** *S. ocmulgee* as an ornamental plant

Scutellaria baicalensis is one of the most studied species under this genus due to its extensive use in TCM and Japanese Kampo medicine systems (Cole et al. 2007). Most of the studies done on skullcap have examined Chinese skullcap. This species is native to China and parts of Russia and has been used in TCM to treat allergies, infections, inflammation, cancer, and headaches.

Fig. 3.2 Potentially ornamental species in the genus *Scutellaria*. **a** *S. suffrutescens*, **b** *S. angulosa*, **c** *S. baicalensis*, **d** *S. montana*, **e** *S. scandens*, and **f** *S. ocmulgee*

3.3.1 Commonly Used Medicinal Scutellaria *Species*

3.3.1.1 *Scutellaria baicalensis*

Scutellaria baicalensis Georgi has been used in China and Japan as a medicinal plant for over 2,000 years (Zobayed et al. 2004). It is officially listed in the Japanese Pharmacopeia JPXIII and Chinese Pharmacopeia, and is one of the most widely used crude drugs for the treatment of bronchitis, hepatitis, diarrhea, and tumors. Wogon, Scutellariae Radix, is the dry root of *S. baicalensis* which is collected in spring and fall, and is a very old and well-known drug in TCM (Tang and Eisenbrand 1992). Chinese physicians use the root of *S. baicalensis* as an antibacterial, diuretic, antispasmodic, and promoter of bile flow.

Recommended oral dosage of *S. baicalensis* is from 5 to 15 g of dried root per day (Loon 1997). *S. baicalensis* can be taken as decoction (hot water extract of dried root) or as commercial extract (in tablets, granules, spray-dried extract, as an alcohol extract and by injection) (Hart 2002; Wichtl et al. 1994). Chemical comparison (baicalin, baicalein, and wogonin) of one gram dried sample each of traditional decoction (as crude drug) and commercial extract of *S. baicalensis* employing HPLC analyses indicated that the traditional decoction had less bioactive compounds than the commercial preparations. The reason might be that the preparation process differed for traditional decoction and commercial drug (Lai et al. 2001; Wichtl et al. 1994). The traditional decoction was prepared by boiling the plant root in water a couple of times. In contrast there are three ways to prepare commercial extract; infusion (boiled water poured over the prescribed amount of drug, then covered), decoction (prescribed amount of drug placed in cold water then boiled, after a while strained), and maceration (required amount of cold water poured over drug, and allowed to stand in room temperature for several hours). During boiling process, some of the secondary products may be reduced, or storage conditions may also have a detrimental effect on secondary metabolites.

3.3.1.2 *Scutellaria barbata*

Scutellaria barbata D. Don is a perennial herb which is native to Korea and China. This plant often grows in wet meadows and nearby pools and brooks. This herb is known as Ban-Zhi-Lian (BZL) in TCM and has been used to cure pain and swelling of throat, edema, and hemorrhoids. It is known as Banjiryun in traditional Korean medicine and has been used as an anti-inflammatory, antitumor, and diuretic agent. It is specifically used for the treatment of liver diseases such as hepatitis and hepatocellular carcinoma. Extracts of *S. barbata* have shown growth inhibitory effects in vivo in a number of cancers. The herb has been used in the treatment of digestive system cancers, hepatoma, lung cancer, breast cancer, and chorioepithelioma. In particular, 62% of patients suffering from hepatoma were completely cured when treated with *S. barbata* (Qian 1987). Isolation of the antitumor compound pheophorbide and its potential in the induction of apoptosis in hepatoma cells validates its antitumor property (Chan et al. 2006).

3.3.1.3 *Scutellaria lateriflora*

Scutellaria lateriflora has been used as a medicinal plant to help promote menstrual cycle. It is one of the most commonly grown and marketed species (Wills and Stuart 2004). Aerial part of three to 4-year-old plants is used for medicinal purposes (Hart 2002). For children, it can be taken as herbal tea, for adults it can be taken as dried herb, tea, fluid extract, and tincture (Hart 2002). A natural essential oil is present in the aerial part of *S. lateriflora*. Sesquiterpenes make up 78.3% of the oil, with τ-cadinene (27%), calamenane (15.2%), β-elemene (9.2%), α-cubebene (4.2%), and α-humulene (4.2%) present in high concentration. Non-terpenoid constituents are also present in low concentration in *S. lateriflora* (Yaghmai 1998).

3.3.1.4 Other *Scutellaria* Species Used as Medicinal Herb

Many *Scutellaria* species, including *Scutellaria iyoensis*, *Scutellaria ventenatii*, *Scutellaria incana*, *Scutellaria orientalis*, *Scutellaria taurica*, *Scutellaria pontica*,

Scutellaria galericulata, and *Scutellaria alpina* have flavonoid accumulation mostly in root rather than in stem and leaf tissues of the plant (Nishikawa et al. 1999a). In Nepal leaves of *Scutellaria discolor* are used as a folk remedy for cold, cut, and insect stings (Sinha et al. 1999). Many *Scutellaria* species with folk medicinal use reported from Turkey are *S. galericulata, Scutellaria hastifolia, Scutellaria altissima, Scutellaria tournefortii, Scutellaria albida, Scutellaria velenovsky, Scutellaria rubicunda, Scutellaria megalaspis, Scutellaria glaphyrostachys, Scutellaria salvifolia, Scutellaria diffusa, S. orientalis, Scutellaria tomentosa*, and several subspecies (Davis 1982).

3.4 Field Studies

During 2001, about 70% of the world market demand for scullcap was supplied by sources outside of North America (Tascan et al. 2010). Cultivation of medicinal plants seems to be a suitable practice to meet current and future demands for large volume production of plant-based drug and herbal remedies. Additionally, it will relieve harvest pressure on wild populations, provide supplemental income to growers.

3.4.1 Herb Processing

Above ground parts are gathered during the summer as flowers bloom and are dried and stored for later use as herbs. Once flowering begins the plant is cut with shears or a mower. A light cutting during the first year is possible, followed by two cuttings each consecutive year. While harvesting skullcap, the freshly cut herb is kept in the shade until harvesting is complete or taken immediately to the drying area. Skullcap can dry in 3–5 days with frequent turning. The fresh herb is approximately 80% water. A warm location with adequate airflow is needed for drying. Drying is carried out at 35–38 °C, turning the herb often to allow for aeration and to prevent molds from developing. For a good quality product, full color of the herb must be retained after drying. Packaged herb is stored in a cool, dry, and dark place.

According to Tascan et al. (2007), especially in plants where roots are the desirable tissue, it is possible that during harvesting and processing some of the thinner roots and outer layer may be lost. This may lead to batch to batch variation in the composition of bioactive compounds. For this reason scaled up tissue culture would be a recommendable method for growing *Scutellaria* species in order to get clean preparations including entire root fractions.

3.4.2 Harvesting

American Skullcap is cut when it begins to flower or in the late flowering period when seed pods are present (Crop Development Branch, Saskatchewan Agriculture and Food 2005). *Scutellaria* blooms from May to August in southeastern USA (Joshee et al. 2007). As the plants are cut, the tops should be piled thinly in a shaded location. According to research in Australia, compaction of product can reduce the flavonoid levels (Wills and Stuart 2004). Baikal Skullcap is harvested in spring or fall with roots being dug from three or 4-year-old plants. The aboveground parts of this species are discarded and roots are saved and dried for further use and storage.

3.4.3 Post Harvest

Good care is needed in storage. Soon after harvesting, American skullcap needs special care for minimal physical damage because fresh plant is still metabolically active and such damage could result in enzymatic or chemical changes to the flavonoids (Wills and Stuart 2004). High temperature (40–70 °C) drying is considered feasible as it is economically advantageous, time-saving, and does not cause significant changes in chemical composition (Wills and Stuart 2004). Loss of flavonoids is not directly related to temperature but dried ground skullcap stored at any temperature up to 30 °C loses about 0.1% of flavonoid per day and greater losses could occur if the product reabsorbs moisture (Wills and Stuart 2004). Harvested material should not be allowed to heat up after harvesting (Crop Development Branch, Saskatchewan Agriculture and Food 2005). Little information is available on storage conditions.

3.4.4 Field Experiments

There is very little data on skullcap production. The amount of Skullcap genus harvested and sold in world market in 2001 was 15,910 kg which is 2.5 times from approximately 6,364 kg in 1997 (Greenfield and Davis 2004). Growing requirements of *S. lateriflora* reflect its classification by the USDA/Natural Resources Conservation Service as a facultative wetland plant in Southeastern United States, indicating that it is primarily found in wetlands (NRCS 2010). Under natural conditions, it is most often found in wooded areas (Awad et al. 2003; Foster and Duke 2000). The plant has low tolerance for drought, such that irrigation should be provided during the summer months when the plant is grown in the open field, especially if no shade is provided. An experiment was conducted in Shorter, Alabama on a Marvyn loamy sand (fine-loamy, kaolinitic, Thermic Typic Kanhapludults) to test the effects of irrigation, shade, and nutrients on yield and flavonoid content of *S. lateriflora* (Similien 2009). Aboveground plant parts were harvested twice during each of the growing seasons in 2007 and 2008. Without irrigation, the plant stands significantly declined in full sun in the second year, whereas plant density increased in plots having shade and irrigation. The highest shoot and flavonoid yields were obtained with a combination of shade, irrigation, and nutrients, supplied as manure or inorganic fertilizer. Higher concentrations of the flavonoid, baicalin, were observed in full sun than in shade, but shade had little effect on concentrations of baicalein, wogonin, and chrysin. According to Similien (2009), yield of American skullcap can be increased from 283 kg per hectare to 1,280 kg per hectare with proper treatments like integrating shade with manures.

The primary constraint that has been encountered in production of *S. lateriflora* besides drought stress is powdery mildew (*Golovinomyces cichoracearum*, syn. *Erysiphe cichoracearum*) which appears during periods of high humidity, particularly in mid- to late-summer. It is readily controlled with neem oil-based products, or with potassium bicarbonate and horticultural oil. More than one application may be required to control the disease. Powdery mildew is a much greater problem under shade than in full sun. In our field experiments during 2009, following unusually wet weather, plants at one site died off due to a soil-borne disease, probably *Pythium*, but were unaffected at a better-drained site. Weed control is important. Since no herbicides are labeled for *Scutellaria*, hand weeding or mulch is required to control weeds between plants. The advantage of *S. lateriflora* is that it can be managed as a forage crop with harvest of the stems and leaves at least twice a summer without the need for replanting before each harvest. The need for irrigation and shade, together with control of powdery mildew under shade add to the cost of production.

Baikal skullcap grown in Mississippi provided 128 kg/ha of baicalin and 2.32 kg/ha of baicalein and 184 to 1,173 kg/ha root yield (Zheljazkov et al. 2007). Baikal skullcap (*S. baicalensis*) may be less difficult and costly to produce than American skullcap (*S. lateriflora*) because it does not require shade and requires little or no irrigation.

3.5 Micropropagation and Molecular Studies

With progressive depletion of native populations, there is an urgent need for conservation and domestication of these plants. The challenge is to produce plants economically and market their products as demanded by the industry and consumers. Currently, many pharmaceutical and nutraceutical companies market products based on traditional medicine and scientific research. However, various serious problems may arise with the adaptation of traditional medicines. Misidentification of a Chinese medicinal plant resulted in the loss of renal function through irreversible interstitial fibrosis in more than 100 patients (Betz 1998). Potential problems with medicinal plant preparations are contamination with insects, bacteria, fungi, and environmental pollutants, seasonal variability in bioactive compounds, degradation of active ingredients in processing and storage of plant materials, and the lack of understanding of the unique physiology of medicinal plants (Li et al. 2000a, b). Correct identification is a critical step at the beginning of an extensive process of quality assurance and is of importance for the characterization of genetic diversity, phylogeny, and phylogeography as well as the protection of endangered species. Micropropagation is one of the plant tissue culture technologies for producing a large number of genetically superior and pathogen-free transplants in a limited time and space. Genomic fingerprinting can further be used to differentiate between individuals, species, and populations and is useful for determining homogeneity of the samples and presence of adulterants (Sucher and Carles 2008).

3.5.1 Tissue Culture of Scutellaria *Species*

Tissue culture is widely used for rapid multiplication, genotype modification, germplasm preservation, cell culture for secondary metabolites, and other scientific studies (Ravindran and Babu 2005). In vitro plant propagation techniques help to standardize production of biochemically and medicinally active secondary products, and select clones of superior individual genotypes (Li et al. 2000b). *Scutellaria* has attracted the attention of scientists to produce superior plants for medicinal purpose. Successful micropropagation of *Scutellaria* has been reported in recent publications (Li et al. 2000b; Joshee et al. 2007; Tascan 2007; Tascan et al. 2007; Sinha et al. 1999; Stojakowska et al. 1999). The in vitro propagation systems developed so far will be helpful in rapidly providing large quantities of uniform, pest- and disease-free plants to isolate and investigate the efficacy of medicinal components. In the last few years discoveries of cell cultures capable of producing specific medicinal compounds at a rate similar or superior to that of intact plants have substantially advanced. It has been demonstrated that the biosynthetic activity of cultured cells can be enhanced by regulating environmental factors, as well as by artificially selecting the induction of variant clones. It will assist future commercial production of standard, consistent, and safe herbal preparations. Alan et al. (2007) combined flow cytometry analysis with morphological and chemical profiling to assess the genetic stability and bioactive compound diversity in a *S. baicalensis* germplasm collection that was clonally maintained in vitro for over 6 years. The analysis showed no significant variation in the nuclear DNA contents and ploidy levels in the long-term in vitro-maintained germplasm lines. The high level of genetic stability observed in in vitro maintained *S. baicalensis* lines opens up a variety of opportunities for other species of *Scutellaria* in the area of long-term aseptic preservation and dissemination of well-characterized germplasm lines.

3.5.2 Organogenesis

One of the earliest records of micropropagation in the *Scutellaria* genus has been reported in *Scutellaria costaricana* using liquid culture (Uhring 1982). Rapid multiplication and callus induction studies in *S. discolor* revealed that cotyledons with shoot tip were the most responsive explants (Sinha et al. 1999). Maximum multiple shoot production was observed in the treatment containing 1 mg/l BAP and 0.1 mg/l NAA using MS (Murashige and Skoog 1962) medium. From one explant, in 10 weeks, 15–20 shoots could be obtained. Microshoots (six to eight leaves) produced roots in 2–3 weeks when placed in non-sterile sand. Three auxins (NAA, 2,4-D and IAA) at various concentrations were also used individually for callus induction and 2,4-D was found to be the most suitable (Sinha et al. 1999). Stojakowska et al. (1999) found that most efficient production of shoot buds was accomplished on MS basal medium supplemented with 0.5 μM naphtaleneacetic acid (NAA) and 2.5 μM kinetin. Thidiazuron (TDZ: [N-phenyl-N′-(1,2,3-thidiazol-5-ylurea)]), a synthetic cytokinin, has been reported to induce somatic embryogenesis as well as shoot regeneration in some plants (viz., peanut, tobacco, geranium, St. John's Wort). Using intact seedlings, leaves, and petiole explants of *S. baicalensis* Li et al. (2000b) observed the effectiveness of TDZ for induction of de novo shoot organogenesis. Later, a two-stage procedure was established to increase the multiplication rate. First, thidiazuron (0.1–0.5 μM) was used to induce organogenic callus, and shoots were then regenerated on medium containing 2.2 μM BA. This procedure resulted in over 20 shoots per explant. Figure 3.1e exhibits a typical shoot bud induction response in *Scutellaria ocmulgee* (Joshee et al. unpublished results).

3.5.3 Liquid Culture

Bioreactors provide highly controlled environments for the rapid multiplication of cells, tissues, and plantlets. Growing tissues in this manner allows production of secondary metabolites, until the media is depleted of nutrients or some other culture variable is changed. Six different culture systems using semi-solid and liquid medium were analyzed to study in vitro multiplication of *S. baicalensis* using a fast-growing germplasm line HQ 111 (Zobayed et al. 2004). The culture systems employed were balloon type bubble bioreactor (BB bioreactor), temporary immersion bioreactor system with liquid medium (TI bioreactor), airlift bioreactor (AL bioreactor), silicone tubing bioreactor (ST bioreactor), large vessel with gelled medium under forced ventilation (LVF bioreactor), magenta vessels with gelled medium (MV system; control), and standard closed system (natural ventilation) in order to obtain biomass production, plant growth, and production of bioactive metabolites. The greatest accumulation of biomass occurred in a temporary immersion bioreactor with liquid media, and the most effective system for the production of specific phytochemicals was a large vessel under forced aeration on gelled medium. This work showed the potential of the application of biotechnology and analytical chemistry in the production of chemically optimized medicinal plant tissues. According to Zobayed et al. (2004) flavonoid concentration in bioreactor raised plants was close to that found in the field grown plants, whereas LVF bioreactor system produced the highest concentration of baicalin

(5.5%) and baicalein (1.7%). In our studies using liquid culture medium, we observed that *Scutellaria* species vary a great deal in their response. American skullcap tolerate liquid culture conditions very well and in a comparative study on the effect of medium on growth and biomass accumulation revealed that MS medium is superior to Gamborg's B5 (Gamborg et al. 1968) (Fig. 3.1f).

In vitro liquid culture of *Scutellaria* species (*S. lateriflora*, *S. costaricana* and *S. baicalensis*) was investigated using MS medium supplemented with BA (5 μM), NAA (0.025 μM), and sucrose (30 g l^{-1} with Agar, liquid stationary (LS), liquid agitated (LA), and liquid culture with floating paper (FP) cultured for 4–8 weeks. LA and LS plantlets were more hyperhydric and had less multiplication compared to agar and FP plantlets. Biomass was higher in liquid culture except *S. baicalensis* plantlets in LS. Since LA and LS gave higher biomass, our further research is focused on a polyester fiber matrix in liquid to help prevent hyperhydricity and provide greater biomass. Agar, LS and a fiber matrix infused with liquid that kept plantlets in the gaseous phase were compared. The matrix was a polyester fiber covered with seed germination paper. Agar and matrix grown plantlets were less hyperhydric and had higher MR compared to LS plantlets. Matrix prevented hyperhydricity and provided greater percent dry weight with relatively higher multiplication ratios (Tascan et al. 2010). *S. costaricana* plantlets in vitro showed high levels of the three flavonoids compared with *S. baicalensis* and *S. lateriflora*. Growing non-hyperhydric tissues on fiber-supported paper, in vitro, allowed the clonal propagation of *Scutellaria* species to proceed in a simple, controlled environment.

3.5.4 Tissue/Cell Culture and Secondary Metabolite Biosynthesis

The possible use of plant cell cultures for the specific biotransformation of natural compounds has been demonstrated (Berlin 1988; Verpoorte et al. 2000; Vanisree et al. 2004; Sarin 2005). Apart from micropropagation, callus, and cell culture techniques have been used for commercial propagation because of their potentially high capacity for multiplication. One of the methods employed to enhance production of secondary metabolites is the use of elicitors like salicylic acid and jasmonic acid. Suspension cultures have been scaled up to large bioreactors to study secondary metabolites leached into the medium. Exposure to biotic elicitors or to stress agents (abiotic elicitors) frequently induces the synthesis of secondary metabolites in plants. When cell suspension cultures of *S. baicalensis* were treated with 50 μg/l of yeast elicitor preparation, both oleanolic acid and ursolic acid transiently increased in the culture medium. The maximal triterpenoid concentration following methyl jasmonate application was 2.1 and 13.7 mg/l after 20 and 35 h, respectively. Elicitor treatment also doubled phospholipase A_2 activity and the simultaneous treatment of aristolochic acid, a phospholipase A_2 inhibitor, inhibited triterpenoids accumulation as well as phospholipase A_2 activity (Yoon et al. 2000).

Production of flavonoids has been reported from tissue cultures of a number of plant species, such as *Cicer arietinum*, *Crotolaria juncea*, *Emblica officinalis*, and *Tylophora indica* (Sarin 2005). Relatively few secondary metabolic pathways have been successfully manipulated in medicinal plants. This is in part because of the enormous complexity of secondary metabolism, which is regulated at different levels. The highly compartmentalized nature of enzymes, substrate precursors, and metabolic intermediates also contribute to these complexities (Gomez-Galera et al. 2007). Organ and callus culture techniques have been used to study phenolic compound biosynthesis in *Scutellaria.* Shoot (from commercial seeds) and callus (from leaf segments) cultures of nine species of *Scutellaria* (*S. iyoensis, S. ventenatii, S. lateriflora, S. orientalis, S. taurica, S. pontica, S. galericulata, S. alpina*) were studied for the distribution and concentration of acteoside, baicalin, wogonin 7-glucuronide, baicalein, wogonin, skullcapflavone I, and skullcapflavone II. Shoot and callus cultures were established in MS medium supplemented with different combinations of 2,4-D (0.1, 1.0 mg/l), NAA (0.5, 2.0 mg/l), IAA (3.0 mg/l), and BA (0.0, 0.1 mg/l) for over 6 months in dark as well as light (Nishikawa et al. 1999a). As reported by Nishikawa et al. 1999a, phytochemical analysis at the end of 6 months of culture revealed that *S. lateriflora* root cultures had the highest wogonin concentration (3.02% DW) which is four times higher than that of cultured roots of *S. baicalensis* (0.78% DW). Most effective plant growth regulators for higher fresh weight/biomass were BA (0.1 mg/l) and 2,4-D (0.1 mg/l). Among the explants studied, root was found to have higher phenolic accumulation in all species. However, *S. pontica* and *S. taurica* gave the highest amount of baicalin in shoot cultures. The reason that baicalin was used as a reference was that calluses from all species produce baicalin as a major phenolic. For all species callus produced wogonin and wogonoin 7-*O*-glucuronide, and *S. ventenatii* produced the highest concentration of wogonin in callus culture (Nishikawa et al. 1999a). Successful production of baicalin has also been recorded in the callus cultures of *S. baicalensis* (Shin and Lee 1995).

3.5.5 *Molecular Analysis of* Scutellaria *Germplasm*

Genetic polymorphism in many plant genera has been documented by use of various fingerprinting procedures. Distinction between dried aerial parts that are used as botanicals on the basis of their morphology alone is extremely difficult. Successful use of randomly amplified polymorphic DNA (RAPD) technique has been reported to differentiate three species of medicinal *Scutellaria* (Hosokawa et al. 2000). The dendrogram analysis exhibited that the similarity among three species (*S. galericulata*, *S. lateriflora*, and *S. baicalensis*) of *Scutellaria* ranged from 61 to 77%. Authentications have been attempted on *Radix Scutellariae* and its adulterants by chemical and microscopic methods (American Herbal Pharmacopoeia 2009). Microscopic analysis depends too much on the availability of experts, and is time consuming and laborious. Therefore, an accurate, sensitive, and simple method is urgently needed. Recently, Guo et al. (2011), reported successful use of 'DNA barcode' technology using specific gene sequences to identify *Radix Scutellariae* (Huang Qin) from other *Scutellaria* species as adulterants. According to their research, the *psb*A-*trn*H intergeneric spacer from the chloroplast genome can be used as effective tool as 'DNA barcode' for species identification.

3.6 Genetic Transformation Studies on *Scutellaria*

3.6.1 Transgenic Studies

Medicinal properties of *Scutellaria* species have been attributed to many different flavonoids distributed in above ground plant parts and roots. Flavonoids are low molecular weight compounds and their biosynthesis is among the best described secondary metabolic pathways, and in many plants, genes encoding flavonoids biosynthetic enzymes have been cloned and characterized. Our transformation models will help to elucidate the possible role of transcription factors (e.g. MYB) in the expression of a subset of genes involved in the biosynthesis of flavonoids. Research on hairy root cultures is aimed to understand which factors contribute to the production of specific flavonoids taking advantage of the controlled conditions of this biological system.

The ability of manipulating flavonoid biosynthesis in plant species is gaining importance due to the emergence of economically important new uses in the areas of food and feed quality and nutraceuticals, and rapid advances in technology has facilitated achieving these goals. Metabolic engineering will assist to improve the plant cell factory for the production of desired flavonoid (Jedinak et al. 2004). Introduction of target genes into plants for the purpose of metabolic engineering can be achieved through genetic transformation, either via *Agrobacterium tumefaciens* or *Agrobacterium rhizogenes*.

We have established a successful protocol for the transformation of *S. ocmulgee* via *A. tumefaciens* (EHA105) harboring plasmid pq35SGR following the method of Li et al. 2004. The *Agrobacterium* culture was grown overnight in 20 ml YEP medium containing 20 mg l^{-1} rifampicin and 50 mg l^{-1} kanamycin on a rotary shaker (200 rpm) at 28 °C. The pq35GR vector consists of the cauliflower mosaic virus (CaMV) 35S promoter-derived bi-directional promoters containing two divergently arranged enhancer repeats, a fusion between the nptII and GUS genes, and the EGFP gene (Joshee et al. 2010).

3.6.2 Hairy Root Culture

Agrobacterium rhizogenes is the causal organism for hairy root disease in the above ground parts of many plants. There have been efforts to establish a transformed hairy root culture and optimum culture conditions for flavonoid production using *Scutellaria* explants (Nishikawa and Ishimaru 1997). Successful establishment of *S. baicalensis* hairy roots has been attained with *A. rhizogenes* strain A13. Six hairy root clones were chosen for further study of flavonoid content. One hairy root clone (Clone D), registered higher amounts of glucuronide-type flavonoids (baicalin). Using the hairy root culture system, Hirotani (1999), isolated a new flavone glucoside, 15 known flavonoids, and five known phenylethanoids. Transformed root culture of *S. baicalensis* induced by *A. rhizogenes* (ATCC 15834) produced both non-glucuronidated-type flavonoids (baicalein, wogonin) and glucuronidated type flavonoids (baicalin). *A. rhizogenes* strain LBA 9402 induced hairy root cultures in the explants of *S. bacailensis.* Flavonoid accumulation was optimized to produce up to 7% baicalein, 1.9% wogonin, and 1.3% oroxylin on dry weight basis in this system (Stojakovska and Malarz 2000).

In the Medina-Bolivar laboratory (Arkansas State University) ongoing research on hairy roots is directed to several medicinally active plants including *S. lateriflora*. Several hairy root cultures of this species were successfully established by inoculation of stem explants with *A. rhizogenes* ATCC 15834 (Joshee et al. 2010). PCR analyses confirmed the presence of *rol* and *aux* genes in these hairy root lines. The hairy roots showed profusion of root hairs and plagiotropic growth on semi-solid medium. A modified MS medium supplemented with IBA was effective for sustainable growth of the hairy roots in liquid medium. Chemical characterization of flavonoids was conducted at different stages of growth of the hairy root cultures. The known anticancer flavonoids wogonin, bacalein, and baicalin were identified in the root tissue. Among these, wogonin was present in higher amount. All these compounds accumulated in the root tissue and were not detected in the culture medium.

Treatment of hairy root cultures with abiotic and biotic elicitors have been shown to induce the production phenolic and other secondary "specialized" metabolites in hairy root cultures. Previous work in the Medina-Bolivar laboratory has shown that hairy root cultures of peanut treated with distinct elicitors induce the biosynthesis of the resveratrol and its analogs (Medina-Bolivar et al. 2007, 2010; Condori et al. 2010). The majority of these compounds were found in the medium facilitating their purification by centrifugal partition chromatography (Abbott et al. 2010) and further elucidation of biological activity (Brents et al. 2011). Similar aims which include the use of elicitors such as methyl jasmonate and cyclodextrin to induce the biosynthesis of flavonoids and other potential bioactive compounds followed by novel purification approaches are currently being studied with *S. lateriflora* hairy roots (Medina-Bolivar et al. unpublished).

3.7 Bioactive Phytochemicals and Recent Medical Studies

Scutellaria species have been used in many biomedical studies and their anticancer, antiviral, anti-inflammatory, antibacterial, antiallergenic, and antioxidant properties have been summarized in a recent review article (Li and Khan 2006).

More than 50 phenolic compounds and essential oils (monoterpenoids, sesquiterpenoids, phenylpropane derivatives, or iridoid glycosides) have been reported from many *Scutellaria* species studied so far (Li and Khan 2006). Flavonoids are low molecular weight compounds and their biosynthesis is among the best described secondary metabolic pathways (Winkel-Shirley 1998). Metabolomic analysis of *S. baicalensis* resulted in over 2,400 different compounds including 781 with possible therapeutic importance and novel amino derivatives of baicalin and wogonin (Murch et al. 2004). *Scutellaria*-based formulations have been used to establish medical/scientific efficacy of skullcaps employing in vitro cell culture systems (Chen 2001; Sato et al. 2000; Shao et al. 1999; Li et al. 1993). With an initial focus in the post-genomic era on tracking gene expression changes for target identification, microarray applications quickly expanded spanning the entire drug discovery pipeline (Gerhold et al. 2002; Koppal 2004). Several studies have indicated that extracts of *S. barbata* have growth inhibitory effects on a number of human cancers. However, the mechanism underlying the antitumor activity was unclear. Complementary DNA microarray analysis showed that 16 genes, involved in DNA damage, cell cycle control, nucleic acid binding, and protein phosphorylation, underwent more than fivefold change. The data indicated that these processes are involved in *S. barbata*-mediated killing of cancer cells (Yin et al. 2004).

3.7.1 Phytochemicals in Genus Scutellaria

Flavones are the most commonly reported bioactive phytochemicals in *Scutellaria*, present in the plant either as glucuronides (such as baicalin and wogonin-7-*O*-glucuronide) or aglycone (baicalein and wogonin, respectively) (Nishikawa et al. 1999b).

Scutellaria baicalensis has been the most extensively studied species and its root is known to contain a number of flavone derivatives having different methoxylation and hydroxylation pattern on the aromatic rings. Chemical constituents of other *Scutellaria* species, including *Scutellaria rivularis* Wall. (Chou 1978; Tomimori et al. 1984, 1990), *S. discolor* (Tomimori et al. 1985, 1986), *S. indica* (Chou and Lee 1986; Miyaichi et al. 1987, 1989), and *Scutellaria scandens* (Miyaichi et al. 1988a, b) have also been investigated. Wogonin was the first flavone isolated from the roots of *S. baicalensis* and Hattori (1930) determined its chemical structure. However, wogonin is present in only small amounts in the root; the predominant flavone glycoside is baicalin (Shibata et al. 1923). Acid hydrolysis of baicalin yielded glucuronic acid and a flavone aglycone named baicalein.

Other related bioactive compounds isolated from *Scutellaria* are some 50 kinds of flavonoids, including apigenin, chrysin, luteolin, neobaicalein, oroxylin a, skullcapflavone I, scutellarein, and isoscutellarein. Monoterpenes of the ridoid type as well as, neoclerodane diterpene derivatives have also been reported in

Scutellaria species (Tang and Eisenbrand 1992; Miyaichi and Tomimori 1994, 1995; Zhang et al. 1994; Ishimaru et al. 1995; Zhou et al. 1997; Loon 1997).

HPLC analysis revealed different flavone distribution in different plant parts of *Scutellaria baicalensis* based on percent dry weight. Baicalin was uniformly distributed in all parts of the root ($\cong$10% DW), levels are reduced in shoot ($\cong$5% DW). Leaves and flowers only had traces of baicalin. Baicalein and wogonin are mostly found in thinner feeder roots at the distal root tips and adventitious roots proximal to the crown (Tani et al. 1985). A metabolic analysis of *S. baicalensis* showed the plant contains more than 2,000 compound and 781 are putatively medicinal (Murch et al. 2004).

3.7.2 Immune Modulatory and Anti-inflammatory Activities

Scientific studies on the extracts or isolated active components from *Scutellaria* have been mostly limited to their anti-inflammatory activities (Lin and Shieh 1996; Nijveldt et al. 2001; Huang et al. 2006). An extract of *S. rivularis* has shown anti-inflammatory and hepatoprotective activity in a test animal (Ching and Den 1996). The anti-inflammatory activity of formulations having skullcap has been associated with capability to inhibit the enzyme cyclooxygenase-2 (COX-2). The group of pharmaceutical drugs known as nonsteroidal anti-inflammatories (NSAID) directly inhibits COX-2. Unfortunately, most of these also inhibit the highly protective and beneficial enzyme, COX-1. Chinese skullcap is acknowledged to selectively inhibit the enzyme COX-2 (Chi et al. 2001). *Scutellaria* extracts or flavonoids also exert anti-inflammatory activity by inhibiting secretion of inflammatory mediators and chemoattractants such as leukotriene C4 (Butenko et al. 1993) and MCP-1 (Chang et al. 2001). Baicalin inhibits superantigen-induced inflammatory cytokines and chemokines (Krakauer et al. 2001). Baicalin may also exhibit anti-inflammatory activity by binding to chemokines (Li et al. 2000a). Involvement of NF-κB in the inhibition of interleukin-12 production from mouse macrophages by baicalein has also been reported (Kang et al. 2003). *S. baicalensis* and some of its constituents are also suggested to enhance TGF-β1 gene expression in RAW 264.7 murine macrophage cell line (Chuang et al. 2005). Apigenin and chrysin suppresses IgE and Th2 cytokines in OVA-immunized BALB/c mice (Yano et al. 2007).

Some studies have shown that flavonoids, including apigenin and luteolin inhibit functions of dendritic cells via inhibition of MAPK and NF-kB signaling (Kim and Jobin 2005; Yoon et al. 2006). Interestingly, wogonin has been shown to remove immunosuppression without affecting inflammatory response to glucocorticoids (Enomoto et al. 2007). Wogonin has also been shown to enhance tumor necrosis factor activity in RAW 264.7 murine macrophage cell line (Chiu et al. 2002). In our own studies, leaf extracts of various *Scutellaria* species and the flavonoids baicalein and wogonin did not affect maturation and antigen-presenting functions of dendritic cells in vitro (Parajuli et al. 2009). On the other hand, a leaf extract of *S. ocmulgee* and the flavonoid wogonin reduced glioma-mediated immune suppression and enhanced the activation of T cells via inhibition of TGF-β1 activity (Dandawate et al. 2011).

3.7.3 Antibacterial and Antiviral Activities

Crude extracts prepared from *S. barbata* having apigenin and luteolin as active constituents were found to be selectively toxic to *Staphylococcus aureus*, including both methicillin-resistant and sensitive strains (Sato et al. 2000). Flavonoids from the roots of *S. baicalensis* have been reported to have inhibitory effect on human immunodeficiency virus (HIV-1) (Li et al. 1993), human T-cell leukemia virus type I (HTLV-I) (Baylor et al. 1992), and mouse skin tumor promotion (Konoshima et al. 1992). There is already a patent for a flavone as sialidase inhibitor of influenza virus for therapeutic use from the leaves of *S. baicalensis* (Akishiro et al. 1992).

3.7.4 Antioxidant Activities

It has been shown that baicalein and baicalin have antioxidant property (Heo et al. 2004). In addition, baicalein has direct scavenging effects on superoxide radicals (Shao et al. 1999). A number of studies have demonstrated that reactive oxygen species (ROS), including superoxide, hydrogen peroxide, and hydroxyl radicals, contribute to myocardial ischemia–reperfusion injury (Halpern et al. 1995). A similar role for ROS has been demonstrated in the contractile dysfunction and cell death observed during both simulated ischemia–reperfusion and mitochondrial inhibition in a perfused cardiomyocyte model. In vitro studies revealed that baicalein; a major flavone component of *S. baicalensis* extract can directly scavenge superoxide, hydrogen peroxide, and hydroxyl radicals (Shieh et al. 2000; Shao et al. 1999) protecting cells from lethal oxidant damage.

3.7.5 Preclinical and Clinical Studies on Antitumor Activities

Scutellaria is one of the most widely used anticancer medicines in both Eastern and Western cultures. Among various species of *Scutellaria*, antitumor activities of *S. baicalensis* has been most commonly studied (Ikemoto et al. 2000; Hsieh et al. 2002; Zhang et al. 2003; Bonham et al. 2005; Ma et al. 2005; Scheck et al. 2006; Himeji et al. 2007; Kumagai et al. 2007; Ye et al. 2007). Extract of *S. baicalensis* also comprises one of the eight constituents of the herbal mixture PC-SPES, which is widely used by prostate cancer patients as a supplementary treatment (Hsieh et al. 2002). *S. barbata* has also been studied to some extent with respect to their antiproliferative, apoptosis-inducing, and COX-2/PGE-2 inhibiting activities (Yin et al. 2004; Dai et al. 2006; Kim et al. 2007; Yu et al. 2007). A recent study has illustrated anti-proliferative effects of an ethanolic extract from *S. baicalensis* root extract in drug-resistant gliomas when used alone and in conjunction with 1,3-bis(2-chloroethyl)-1-nitrosourea (BCNU) (Scheck et al. 2006). A phase I clinical trial has been concluded using BZL101, an aqueous extract from *S. barbata*, in patients with advanced breast cancer with encouraging outcome (Rugo et al. 2007). The clinical study consisted of 21 patients, with a mean number of prior cancer treatments of 3.9. Notably, adverse events (AEs) of category III or IV were not reported at all. AEs reported were of levels I and II, consisting of nausea, diarrhea, headache, flatulence, vomiting, constipation, and fatigue. The researchers reported that the likely causes for the symptoms were the bitter and nausea-inducing taste of the extract, the raw amount of plant fiber in the extract, and the underlying symptoms of a heavily pretreated patient population.

This clinical study was continued in a phase 1B trial with a few modifications (Perez et al. 2010), the principal change being modification of the orally administered extract from a bitter liquid form to a freeze-dried powder. The maximum dose administered was 40 g/day. The most common negative side effect was diarrhea, which was only minor and did not reach grade 3 or 4 severity like other chemotherapeutics like capecitabine (Saif et al. 2010). Moreover, the authors speculate that further reduction in the fiber mass of the extract would ameliorate this problem (Perez et al. 2010).

Studies in our laboratory have demonstrated tumor-specific and dose-dependent antiproliferative and pro-apoptotic activities of leaf, stem, and root extracts obtained from 13 *Scutellaria* species against malignant breast cancer, glioma, and prostate cancer cells in vitro as well as in vivo (Parajuli et al. 2009, 2011). These antitumor activities were associated with a significant inhibition of Akt/PKB signaling activity in the tumor cells following treatment with *Scutellaria* extracts (Parajuli et al. 2011). Further, comparisons of the cellular effects induced by the entire extract versus the four-compound combination produced comparable cell cycle changes, and levels of growth inhibition. Individual compounds exhibited antiandrogenic activities with reduced expression of the androgen receptor and androgen-regulated genes. Most of the anticancer activities of *S. baicalensis* can be recapitulated with four purified constituents (baicalein, wogonin, neobaicalein, and skullcapflavone) that function in part through inhibition of the androgen receptor signaling pathway (Bonham et al. 2005).

Among the flavonoids, antiproliferative and pro-apoptotic activities of baicalein and baicalin have been studied against various cancer cells (Kyo et al. 1998; Chan et al. 2000; Hsu et al. 2001; Nijveldt et al. 2001; Ma et al. 2005; Shieh et al. 2006). Baicalein has also shown a favorable effect in cisplatin-induced cell death of human glioma cells (Lee et al. 2005). There have also been a few reports on anticancer activities of apigenin (Way et al. 2004; Brusselmans et al. 2005; Ujiki et al. 2006) and wogonin (Ikemoto et al. 2000; Chang et al. 2002; Himeji et al. 2007; Yu et al. 2007). We have compared the antitumor activities of six predominant *Scutellaria* flavonoids, namely, apigenin, baicalein, baicalin, chrysin, scutellarein, and wogonin against breast cancer, glioma, and prostate cancer cells in vitro. All these flavonoids significantly inhibited the proliferation, induced G1/G2 arrest, and enhanced apoptosis in the three cell lines tested (Parajuli et al. 2009). Similar to the results of *Scutellaria* extracts, as mentioned above, antitumor activities of all six flavonoids also involved significant inhibition of Akt/PKB, GSK-3β, and NF-κB signaling activity in the tumor cells (Parajuli et al. 2011).

3.7.6 Other Bio-medical Activities

It has been shown that baicalin may modulate the lipid metabolism, lipogenic, and lipolytic pathways of adipose cells (Eun et al. 1994; Chung et al. 1995) suggesting a potential application of *Scutellaria* or its constituent flavonoids in the control of obesity and related health issues.

3.8 Endangered and Threatened Species

To achieve at least the taxon level survival in ex situ collections for endemic and critically endangered species, often the only recourse is to collect material from remaining individuals (Maunder et al. 2000). Tissue culture approaches to conservation are appealing (Bunn et al. 2011) for species where, (a) source explants are in limited abundance, (b) inbreeding has resulted in lowered seed yields (Hendrix and Kyhl 2000), (c) species possess complex, often unresolved seed dormancy mechanisms (Merritt and Dixon 2003; Merritt et al. 2007), and (d) seed availability is hampered by low and erratic seed set and viability due to environmental stresses, such as drought, predation, or disease.

3.8.1 Conservation Studies on Threatened and Endangered Species

Demand for a wide variety of wild species is increasing with growth in human needs, numbers, and commercial trade. With the increased realization that some wild species are being over-exploited, a number of agencies are recommending that wild species be brought into cultivation systems. Cultivation can also have conservation impacts; however, these need to be better understood. Medicinal plant production through cultivation, for example, can reduce the extent to which wild populations are harvested, but it also may lead to environmental degradation and loss of genetic diversity as well as loss of incentives to conserve wild populations. Sustainable harvesting needs to be recognized as the most important conservation strategy for most wild-harvested species and their habitats, given their current and potential contributions to local economies and their greater value to harvesters over the long term. The basic idea is that non-destructive harvests and local benefits will maintain population, species, and ecosystem diversity.

We have been conducting research on germplasm collection and medicinal properties of *Scutellaria* species (Joshee et al. 2002) that are commercially in demand (*S. baicalensis*, *S. barbata*, and *S. lateriflora*), or have been found therapeutically promising (*S. ocmulgee*), and the conservation of ones that are threatened (*Scutellaria montana* and *S. ocmulgee*). Preliminary studies conducted on the alginate bead encapsulation in our laboratory seems promising as the shoot tip and nodal explants were encapsulated in 3% sodium alginate and kept at 4–6 °C for 60 days (Fig. 3.1g). These explants showed good regeneration potential when they were taken out from cold treatment and transferred to regeneration medium (Fig. 3.1h).

3.8.2 Scutellaria montana *Chapm*

Flowers bloom during May and June, blue and white in color, and 26–35 mm long on a 10–15 cm raceme. Plants grow up to 20–60 cm height. The leaf is 50–80 mm long by 30–40 mm wide on petioles up to 10–15 mm long. The skullcap's habitat is dry soil on rocky slopes in undisturbed mature oak and hickory woodlands where trees range from 70 to over 200 years old. This is an endangered species reported from the border areas of Georgia and Tennessee. Major factors contributing to the loss of existing population are exotic weeds, loss of habitat for agriculture and development, quarrying, and pine plantation. It is an endangered species at both Federal and State levels (Patrick et al. 1995). Constant monitoring has helped *S. montana* population increase and recently it has been moved to threatened category.

3.8.3 Scutellaria ocmulgee *Small*

This species has violet blue flowers that bloom from April to August and are up to 23 mm long. Plants are 60 cm tall with stems clothed throughout with spreading capitate glands and curled hairs in the upper parts. The leaves are hairy, round to ovate ranging from 50 to 65 mm in diameter and petioles are 20–25 mm long. This endemic species to Georgia has a threatened status at present and predominantly reported from the banks of river Ocmulgee in Bibb and adjoining counties in central Georgia (Patrick et al. 1995).

3.8.4 Germplasm Collection at Fort Valley State University

Our research at Fort Valley State University (FVSU) focuses on the *Scutellaria* species found in and around Georgia. Some of these species are becoming rare or threatened because of population pressure, environmental pollution, and destruction of their natural habitat. Our future studies include growing these species at FVSU, developing in vitro systems (for rapid multiplication and secondary metabolite biosynthesis), and conservation of vulnerable candidates. We have 19 *Scutellaria* species in our collection representing North America, Central America, Europe, and Asia.

3.9 Economic Potential

3.9.1 Scutellaria-*Based Products in the Market*

It is evident that skullcap has been used as a potent medicinal herb in many traditional healing systems. Many herbal formulations containing *Scutellaria* species are available in the world market; for example, PC SPES (BotanicLab, Brea, California), Zyflamend PM, Zyflamend Creme (New Chapter), and Migra-Profen (Gaia Herbs). In most of the formulations *S. baicalensis* and *S. lateriflora* have been used.

3.9.2 Potential for Developing as a Premium Medicinal Crop

Skullcap requires areas of constant moisture, such as moist thickets or marshlands, to grow. It prefers a fertile soil, is hardy to zone four, and grows well in full sun or partial shade. When growing in a hot, dry area, shade, and moisture must be provided. Once harvesting begins, fertilizing with high nitrogen compost is suggested to gather more biomass. Proper agricultural practices may yield up to 2,000 lbs/acre of dried material (Similien 2009).

The amount of skullcap harvested and sold in the world market increased 2.5 times from approximately 6,364 kg in 1997 to 15,910 kg in 2001 (Greenfield and Davis 2004). It has been estimated that the annual consumption of skullcap increased by 23% between 2000 and 2001. The dollar value of the 2001 skullcap harvest was between \$185,000 and \$195,000, that was 3.5 times higher than that of 1997 (Greenfield and Davis 2004). Skullcap is currently recommended as an alternative sedative treatment to kava which could increase the demand for this herb. The price of skullcap has steadily increased during the past 5 years and is currently around \$8.80–17.60/kg. In Canada organically grown skullcaps fetch premium price of \$17.60–33.00/kg (Porter 2006). The demand for skullcap in the world market is predicted to grow at an annual rate of 20–30% (Greenfield and Davis 2004).

We have 19 species in our germplasm collection and *S. ocmulgee*, *Scutellaria suffratescens*, *S. scandens*, and *Scutellaria angulosa* stand out as ornamental plant (Fig. 3.2). These species have a long blooming period (2–4 months) and bear plenty of flowers. *S. suffratescens* (Fig. 3.2a), also known as pink Texas skullcap, is quite hardy withstanding a wide temperature range typical of the central Georgia conditions blooming for almost 7 months (May to October). Another outstanding candidate in our experience is *S. ocmulgee* (Figs. 3.1i and 3.2f) that is restricted to Georgia in its geographical distribution and highly medicinal. Adopting this species as an ornamental will aid to its conservation.

3.10 Conclusion

Natural plant products have been invaluable as tools for deciphering the logic of biosynthesis and as platforms for developing frontline drugs (Newman et al. 2000). The CBD thus acknowledges the need to place the conservation of biodiversity in the context of sustainable development. To be successful, it is clear that the conservation and sustainable use of biodiversity need to make sense as a solution and integral part of socioeconomic needs and strategies. The vast majority of current drug discovery carried out by the pharmaceutical industry relies on molecular approaches, involving defined molecular targets. These molecular approaches in turn are dominated by high-throughput screening. On the other hand studies indicate that even though the establishment of transgenic cell lines and hairy root cultures are valuable approaches to study the biosynthesis of the biologically active compounds in *Scutellaria*, they have inherent problems of high cost of bioreactors for large-scale production. The pharmaceutical companies that have the best and most diverse chemical collections will ultimately dominate the industry. Broadly speaking, such chemically diverse collections can only come from two main sources of large numbers of compounds: combinatorial chemistry and natural products. Although combinatorial chemistry can provide large numbers of compounds for high-throughput screening they tend to have limited structural diversity. In contrast, natural products provide a wealth of small molecules with drug-like properties and with incredible structural diversity.

Acknowledgments This review article is a collective effort of scientists, students, and technicians that were part of *Scutellaria* research from many institutions during past several years. This is still an ongoing project. We are thankful to USDA for the three Capacity building grants (NJ, one as Co-PI and two as PI), and to Wayne State University for WSU-FMRE (PP as PI). We are thankful to Vicki Owen, Research Assistant, and FVSU for maintaining and taking care of *Scutellaria* germplasm.

References

Abbott J, Medina-Bolivar F, Martin E, Engelberth AS, Villagarcia H, Clausen EC, Carrier DC (2010) Purification of resveratrol, arachidin-1 and arachidin-3 from hairy root culture of peanut (*Arachis hypogaea*) and determination of their antioxidant activity and cytotoxicity. Biotechnol Prog 26(5):1344–1351

Akishiro Y, Takayuki N, Takeshi T (1992) Flavones as inhibitors for therapeutic use. Japan Kokkai Koho JP 04, 18, 019 (Cl. A61 K31/35) (Patent)

Alan AR, Zeng H, Assani AS, Wendy L, McRae HE, Murch SJ, Saxena PK (2007) Assessment of genetic stability of the germplasm lines of medicinal plant *Scutellaria baicalensis* Georgi (Huang-qin) in long-term, in vitro maintained cultures. Plant Cell Rep 26:1345–1355

American Herbal Pharmacopoeia and Therapeutic Compendium (2009) Skullcap aerial parts. In: Upton R (ed) *Scutellaria lateriflora* L. Standard of analysis quality control, and therapeutics, Scotts Valley, USA, pp 1–40

Awad R, Arnason T, Trudeau V, Bergeron C, Budzinski JW, Foster BC, Merali Z (2003) Phytochemical and biological analysis of Skullcap (*S. lateriflora* L.): a medicinal plant with anxiolytic properties. Phytomedicine 10:640–649

Bailey LH, Bailey EZ (1978) Hortus third: a concise dictionary of plants cultivated in the United States and Canada. Macmillan Publishing Co., Inc., New York

Baylor NW, Fu T, Yan Y-D, Ruscetti FW (1992) Inhibition of human T cell leukemia virus by the plant flavonoid baicalin (7-glucuronic acid 5,6-dihydroxyflavone). J Infect Dis 165:433–437

Berlin J (1988) Formation of secondary metabolites in cultured plant cells and its impact on pharmacy. In: Bajaj YPS (ed) Biotechnology in agriculture and forestry. Medicinal and aromatic plants I, vol 4. Springer, Berlin

Betz W (1998) Epidemic of renal failure due to herbals. Sci Rev Altern Med 2:12–13

Bonham M, Posakony J, Coleman I, Montgomery B, Simon J, Nelson PS (2005) Characterization of chemical constituents in *Scutellaria baicalensis* with antiandrogenic and growth-inhibitory activities toward prostate carcinoma. Clin Cancer Res 11:3905–3914

Brents LK, Medina-Bolivar F, Seely KA, Nair V, Bratton SM, Ñopo-Olazabal L, Patel RY, Liu H, Doerksen RJ, Prather PL, Radominska-Pandya A (2011) Natural prenylated resveratrol analogs, arachidin-1 and -3, demonstrate improved glucuronidation profile and have affinity for cannabinoid receptors. Xenobiotica 42(2):139–156

Brusselmans K, Vrolix R, Verhoeven G, Swinnen JV (2005) Induction of cancer cell apoptosis by flavonoids is associated with their ability to inhibit fatty acid synthase activity. J Biol Chem 280:5636–5645

Bunn E, Turner SR, Dixon KW (2011) Biotechnology for saving rare and threatened flora in a biodiversity hotspot. In Vitro Cell Dev Biol Plant 47:188–200

Butenko IG, Gladtchenko SV, Galushko SV (1993) Anti-inflammatory properties and inhibition of leukotriene C4 biosynthesis in vitro by flavonoid baicalein from *Scutellaria baicalensis* Georgi roots. Agents Actions 39:49–51

Cappuzzo F, Finocchiaro G, Trisoloini R, Toschi L, Bartolini S, Metro G et al (2005) Perspectives on salvage therapy for non-small-cell lung cancer. Oncology 19:989–995

Chan FL, Choi HL, Chen ZY, Chan PS, Huang Y (2000) Induction of apoptosis in prostate cancer cell lines by a flavonoid, baicalin. Cancer Lett 160:219–228

Chan JY, Tang PM, Hon P, Au SW, Sui SK, Waye MM, Kong S, Mak TC, Fung K (2006) Pheophorbide a, a major antitumor component purified from *Scutellaria barbata*, induces apoptosis in human hepatocellular carcinoma cells. Planta Med 72:28–33

Chang YL, Shen JJ, Wung BS, Cheng JJ, Wang DL (2001) Chinese herbal remedy wogonin inhibits monocyte chemotactic protein-1 gene expression in human endothelial cells. Mol Pharmacol 60:507–513

Chang WH, Chen CH, Lu FJ (2002) Different effects of baicalein, baicalin and wogonin on mitochondrial function, glutathione content and cell cycle progression in human hepatoma cell lines. Planta Med 68:128–132

Chavan P, Joshi K, Patwardhan B (2006) DNA microarrays in herbal drug research. Review eCAM 3:447–457

Chen S (2001) In vitro mechanism of PC SPES. Urology 58:28–35

Chi YS, Cheon BS, Kim HP (2001) Effect of wogonin, a plant flavone from *Scutellaria radix*, on the suppression of cyclooxygenase-2 and the induction of inducible nitric oxide synthase in lipopolysaccharide-treated RAW 264.7 cells. Biochem Pharmacol 61:1195–1203

Ching LC, Den ES (1996) The anti-inflammatory activity of *Scutellaria rivularis* extracts and its active components, baicalin, baicalein, and wogonin. Am J Chin Med 24:31–36

Chiu JH, Lay IS, Su MY, Chiu HL, Chiu AC, Lui WY, Wu CW (2002) Tumor necrosis factor-producing activity of wogonin in RAW 264.7 murine macrophage cell line. Planta Med 68:1036–1039

Chou C-J (1978) Rivularin, a new flavone from *Scutellaria rivularis*. J Taiwan Pharm Assoc 30:36–43

Chou C-J, Lee S-Y (1986) Studies on the constituents of *Scutellaria indica* root (I). J Taiwan Pharm Assoc 38:107–118

Chuang H-N, Wang J-Y, Chiu J-H, Tsai T-H, Yeh S-F, Shu LF, Lui W-Y, Wu C-W (2005) Enhancing effects of *Scutellaria baicalensis* and some of its constituents on TGF-β1 gene expression in RAW 264.7 murine macrophage cell line. Planta Med 71(5):440–445

Chung CP, Park JB, Bae KH (1995) Pharmacological effects of methanolic extracts from the root of *Scutellaria baicalensis* and its flavonoids on human gingival fibroblast. Planta Med 61:150–153

Clardy J, Walsh C (2004) Lessons from natural molecules. Nature 432:829–837

Cole IB, Saxena PK, Murch SJ (2007) Medicinal biotechnology in the genus *Scutellaria*. In Vitro Cell Dev Biol Plant 43:318–327

Condori J, Sivakumar S, Hubstenberger J, Dolan MC, Sobolev VS, Medina-Bolivar F (2010) Induced biosynthesis of resveratrol and the prenylated stilbenoids arachidin-1 and arachidin-3 in hairy root cultures of peanut: Effects of culture medium and growth stage. Plant Physiol Biochem 48:310–318

Crop Development Branch, Saskatchewan Agriculture and Food (2005) Crops Overview. http://www.agriculture.gov.sk.ca/Default.aspx?DN=a6cd8eb3-e7fd-4b57-9d21-5d83a74c75bf. Verified 2 Dec 2008

Csermely P, Agoston V, Pongor S (2005) The efficiency of multi-target drugs: the network approach might help drug design. Trends Pharmacol Sci 26:178–182

Dai SJ, Tao JY, Liu K, Jiang YT, Shen L (2006) Neo-Clerodane diterpenoids from *Scutellaria barbata* with cytotoxic activities. Phytochemistry 67:1326–1330

Dandawate S, Williams L, Joshee N, Rimando AM, Mittal S, Thakur A, Lum L, Parajuli P (2011) *Scutellaria* extract and wogonin inhibit tumor-mediated induction of T_{reg} cells via inhibition of TGF-beta1 activity. Cancer Immunol Immunother. doi:10.1007/s00262-011-1130-3

Davis PH (1982) Flora of Turkey, vol 7. University Press, Edinburgh, pp 78–99

Enomoto R, Suzuki C, Koshiba C, Nishino T, Nakayama M, Hirano H, Yokoi T, Lee E (2007) Wogonin prevents immunosuppressive action but not anti-inflammatory effect induced by glucocorticoid. Ann N Y Acad Sci 1095:412–417

Eun JS, Suh ES, So JN, Oh SH (1994) Effect of baicalin on the differentiation of 3T3-Li cells. Yakahak Hoeji 38:48

Foster S, Duke JA (2000) Medicinal plants and herbs of Eastern and Central North America, 2nd edn. Houston Mifflin Co, New York 211

Gamborg OL, Miller R, Ojima AK (1968) Nutrient requirements of suspension cultures of soybean root cells. Exp Cell Res 50:151–158

Gerhold DL, Jensen RV, Gullans SR (2002) Better therapeutics through microarrays. Nat Genet 32:547–551

Gomez-Galera S, Pelacho AM, Gene A, Capell T, Christou P (2007) The genetic manipulation of medicinal and aromatic plants. Plant Cell Rep 26:1689–1715

Greenfield J, Davis JM (2004) Skullcap (*Scutellaria laterifolia* L.). Medicinal herb production guide. http://www.naturalmedicinesofnc.org. Accessed July 12, 2006

Guo X, Wang X, Su W, Zhang G, Zhou R (2011) DNA barcodes for discriminating the medicinal plant *Scutellaria baicalensis* (Lamiaceae) and its adulterants. Biol Pharm Bull 34:1198–1203

Halpern HJ, Yu C, Barth E, Peric M, Rosen GM (1995) In situ detection by spin trapping of hydroxyl radical markers produced from ionizing radiation in the tumor of the living mouse. Proc Natl Acad Sci USA 95:796–800

Hart J (2002) Skullcap alternative medicine. A.D.A.M., Inc. www.healthandage.com

Hattori S (1930) Spectrography of the flavone series. III. The constitution of wogonin. Acta Phytochim 5:99–116

Hefferon KL (2010) Biopharmaceutical in plants: toward the next century of medicine. CRC Press, Boca Raton, p 206

Hendrix SD, Kyhl JF (2000) Population size and reproduction in *Phlox pilosa*. Conserv Biol 14:304–313

Heo HJ, Kim D, Choi SJ, Shin DH, Lee CY (2004) Potent inhibitory effect of flavonoids in *Scutellaria baicalensis* on amyloid β protein-induced neurotoxicity. J Agric Food Chem 52:4128–4132

Himeji M, Ohtsuki T, Fukazawa H, Tanaka M, Yazaki S, Ui S, Nishio K, Yamamoto H, Tasaka K, Mimura A (2007) Difference of growth-inhibitory effect of *Scutellaria baicalensis*-producing flavonoid wogonin among human cancer cells and normal diploid cell. Cancer Lett 245:269–274

Hirotani M (1999) Genetic transformation of *Scutellaria baicalensis*. In: Bajaj YPS (ed) Biotechnology in agriculture and forestry. Transgenic medicinal plants, vol 45. Springer, New York, pp 271–283

Hosokawa K, Minami M, Kawahara K, Nakamura I, Shibata T (2000) Discrimination among three species of medicinal *Scutellaria* plants using RAPD markers. Planta Med 66:270–272

Hsieh TC, Lu X, Chea J, Wu JM (2002) Prevention and management of prostate cancer using PC-SES: a scientific perspective. J Nutr 132:3513–3517

Hsu SL, Hsieh YC, Hsieh WC, Chou CJ (2001) Baicalein induces a dual growth arrest by modulating multiple cell cycle regulatory molecules. Eur J Pharmacol 425:165–171

Huang WH, Lee AR, Yang CH (2006) Antioxidative and anti-inflammatory activities of polyhydroxyflavonoids of *Scutellaria baicalensis* GEORGI. Biosci Biotechnol Biochem 70:2371–2380

Ikemoto S, Sugimura K, Yoshida N, Yasumoto R, Wada S, Yamamoto K, Kishimoto T (2000) Antitumor effects of *Scutellariae radix* and its components baicalein, baicalin, and wogonin on bladder cancer cell lines. Urology 55:951–955

Ishimaru K, Nishikawa K, Omoto T, Asai I, Yoshihira K, Shimomura K (1995) Two flavone 2′-glucosides from *Scutellaria baicalensis*. Phytochemistry 40:279–281

Jedinak A, Farago J, Psenakova I, Maliar T (2004) Approaches to flavonoids production in plant tissue cultures. Biologia Bratislava 59:697–710

Joshee N, Patrick TS, Mentreddy SR, Yadav AK (2002) Skullcap: potential medicinal crop. In: Janick J, Whipkey A (eds) Trends in new crops and new uses. ASHS Press, Alexandria

Joshee N, Mentreddy SR, Yadav AK (2007) Mycorrhizal fungi and growth and development of micropropagated *Scutellaria integrifolia* plants. Ind Crops Prod 25:169–177

Joshee N, Parajuli P, Medina-Bolivar F, Rimando AM, Yadav AK (2010) *Scutellaria* biotechnology: achievements and future prospects. Bull UASVM Horticulture 67:24–32. Print ISSN:1843-5254; Electronic ISSN:1843-5394

Kang BY, Chung SW, Kim SH, Cho D, Kim TS (2003) Involvement of nuclear factor-kappaB in the inhibition of interleukin-12 production from mouse macrophages by baicalein, a flavonoid in *Scutellaria baicalensis*. Planta Med 69:687–691

Kim JS, Jobin C (2005) The flavonoid luteolin prevents lipopolysaccharide-induced NF-kappaB signalling and gene expression by blocking IkappaB kinase activity in intestinal epithelial cells and bone-marrow derived dendritic cells. Immunology 115:375–387

Kim EK, Kwon KB, Han MJ, Song MY, Lee JH, Ko YS, Shin BC, Yu J, Lee YR, Ryu DG, Park JW, Park BH (2007) Induction of G1 arrest and apoptosis by *Scutellaria barbata* in the human promyelocytic leukemia HL-60 cell line. Int J Mol Med 20:123–128

Konoshima T, Kokumai M, Kozuka M, Iinuma M, Mizuno M, Tanaka T, Tokuda H, Nishino H, Iwashima A (1992) Studies on inhibitors of skin tumor promotion. XI. Inhibitory effects of flavonoids from *Scutellaria baicalensis* on Epstein-Barr virus activation and their anti-tumor-promoting activities. Chem Pharm Bull 40:531–533

Koppal T (2004) Microarrays: migrating from discovery to diagnostics. Drug Discov Dev 7:30–34

Krakauer T, Li BQ, Young HA (2001) The flavonoid baicalin inhibits superantigen-induced inflammatory cytokines and chemokines. FEBS Lett 500:52–55

Kumagai T, Muller CI, Desmond JC, Imai Y, Heber D, Koeffler HP (2007) *Scutellaria baicalensis*, a herbal medicine: anti-proliferative and apoptotic activity against acute lymphocytic leukemia, lymphoma and myeloma cell lines. Leukemia Res 31:523–530

Kyo R, Nakahata N, Sakakibara I, Kubo M, Ohizumi Y (1998) Baicalin and baicalein, constituents of an important medicinal plant, inhibit intracellular Ca^{2+} elevation by reducing phospholipase C activity in C6 rat glioma cells. J Pharm Pharmacol 50:1179–1182

Lai MY, Chen CC, Hsiu SL, Chao PDL (2001) Analysis and comparison of baicalin, baicalein, and wogonin contents in traditional decoctions and commercial extracts of *Scutellaria radix*. J Food Drug Anal 9:145–149

Lee SW, Song GS, Kwon CH, Kim YK (2005) Beneficial effect of flavonoid baicalein in cisplatin-induced cell death of human glioma cells. Neurosci Lett 382:71–75

Li J, Khan I (2006) The advanced bioactivity studies of *Scutellaria baicalensis* Georgi and its phenolic compounds. Acta Hort (ISHS) 720:157–169

Li B-Q, Fu T, Yan Y-D, Baylor NW, Ruscetti FW, Kung HF (1993) Inhibition of HIV infection by baicalin—a flavonoid compound purified from Chinese herbal medicine. Cell Mol Biol Res 39:119–124

Li B-Q, Fu T, Gong WH, Dunlop N, Kung H, Yan Y, Kang J, Wang JM (2000a) The flavonoid baicalin exhibits anti-inflammatory activity by binding to chemokines. Immunopharmacology 49:295–306

Li H, Murch SJ, Saxena PK (2000b) Thidiazuron-Induced de novo shoot organogenesis on seedlings, etiolated hypocotyls and stem segments of Huanq-qin. Plant Cell Tissue Organ Cult 62:169–173

Li ZJ, Jayasankar S, Gray DJ (2004) Bi-directional duplex promoters with duplicated enhancers significantly increase transgene expression in grape and tobacco. Transgenic Res 13:143–154

Loon IM (1997) The golden root: clinical application of *Scutellaria baicalensis* Georgi flavonoids as modulators of the inflammatory response. Altern Med Rev 3:472–480

Ma Z, Otsuyama K, Liu S, Abroun S, Ishikawa H, Tsuyama N, Obata M, Li FJ, Zheng X, Maki Y, Miyamoto K, Kawano MM (2005) Baicalein, a component of *Scutellaria radix* from Huang-Lian-Jie-Du-Tang (HLJDT), leads to suppression of proliferation and induction of apoptosis in human myeloma cells. Blood 105:3312–3318

Maunder M, Culham A, Alden B, Zizka G, Orliac C, Lobin W, Bordeu A, Ramirez JM, Glissman-Gough S (2000) Conservation of the Toromiro tree: case study in the management of a plant extinct in the wild. Conserv Biol 14:1341–1350

Medina-Bolivar F, Condori J, Rimando A, Hubstenberger J, Shelton K, Bennett S, Dolan M (2007) Production and secretion of resveratrol in hairy root cultures of peanut. Phytochemistry 68:1992–2003

Medina-Bolivar F, Dolan M, Bennett S, Condori J, Hubstenberger J (2010) Production of stilbenes in hairy roots. US Patent N. 7666677

Merritt DJ, Dixon KW (2003) Seed storage characteristics and dormancy of Australian indigenous plant species. In: Smith RD, Dickie JB, Linington SH, Pritchard HW, Probert RJ (eds) Seed conservation: turning science into practice. Royal Botanic Gardens Kew, London

Merritt DJ, Turner SR, Clarke S, Dixon KW (2007) Seed dormancy and germination stimulation syndromes for Australian temperate species. Aust J Bot 55:336–344

Miyaichi Y, Tomimori T (1994) Studies on the constituents of *Scutellaria* species XVI. On the phenol glycosides of the root of *Scutellaria baicalensis* Georgi. Nat Med 48:215–218

Miyaichi Y, Tomimori T (1995) Studies on the constituents of *Scutellaria* species XVII. On the phenol glycosides of the root of *Scutellaria baicalensis* Georgi. Nat Med 49:350–353

Miyaichi Y, Imoto Y, Tomimori T, Lin C-C (1987) Studies on the constituents of *Scutellaria* species XI. On the flavonoid constituents of the aerial parts of *Scutellaria indica* L. Chem Pharm Bull 35:3720–3725

Miyaichi Y, Imoto Y, Kizu H, Tomimori T (1988a) Studies on the Nepalese crude drugs. X. On the flavonoid constituents of the root of *Scutellaria scadens* Buch.-Ham. D. Don. Chem Pharm Bull 36:2371–2376

Miyaichi Y, Imoto Y, Kizu H, Tomimori T (1988b) Studies on the Nepalese crude drugs. X. On the flavonoid and stilbene constituents of the leaves of *Scutellaria scadens* Buch.-Ham. D. Don. Shoyakugaku Zasshi 42:204–207

Miyaichi Y, Imoto Y, Tomimori T, Lin C-C (1989) Studies on the constituents of *Scutellaria* species IX. On the flavonoid constituents of the root *Scutellaria indica* L. Chem Pharm Bull 37:794–797

Murashige T, Skoog F (1962) A revised medium for rapid growth and bio assays with tobacco tissue cultures. Physiol Plantarum 15:473–492

Murch SJ, Rupasinghe HPV, Goodenowe D, Saxena PK (2004) A metabolomic analysis of medicinal diversity in Huang-qin (*Scutellaria baicalensis* Georgi.) genotypes discovery of novel compounds. Plant Cell Rep 23:419–425

Newman DJ, Cragg GM, Snader KM (2000) The influence of natural products upon drug discovery. Nat Prod Rep 17:215–234

Nijveldt RJ, van Nood E, van Hoorn DE, Boelens PG, van Norren K, van Leeuwen PA (2001) Flavonoids: a review of probable mechanisms of action and potential applications. Am J Clin Nutr 74:418–425

Nishikawa K, Ishimaru K (1997) Flavonoids in root cultures of *Scutellaria baicalensis*. J Plant Physiol 151:633–636

Nishikawa K, Furukawa H, Fujioka T, Fuji H, Mihashi K, Shimomura K, Ishmaru K (1999a) Phenolics in tissue cultures of *Scutellaria*. Nat Med 53:209–213

Nishikawa K, Furukawa H, Fujioka T, Fuji H, Mihashi K, Shimomura K, Ishimaru K (1999b) Flavone production in transformed root cultures of *Scutellaria baicalensis* Georgi. Phytochemistry 52:885–890

NRCS (2010) *Scutellaria lateriflora* L. Blue skullcap. United States Department of Agriculture/ Natural Resources Conservation Service. http://plants.usda.gov/java/profile?symbol=SCLA2. Accessed March 24, 2010

Parajuli P, Joshee N, Rimando A, Mittal S, Yadav AK (2009) In vitro anti-tumor mechanisms of various *Scutellaria* extracts and constituent flavonoids. Planta Med 75:41–48

Parajuli P, Joshee N, Chinni SR, Rimando AM, Mittal S, Sethi S, Yadav AK (2011) Delayed growth of glioma by *Scutellaria* flavonoids involve inhibition of Akt, GSK-3 and NF-kB signaling. J Neurooncol 101:15–24

Paton A (1990) A global taxonomic investigation of *Scutellaria* (Labiateae). Kew Bull 45:399–450

Patrick TS, Allison JR, Krakow GA (1995) Protected plants of Georgia. Georgia Department of Natural Resources, Atlanta, pp 171–174

Perez AT, Arun B, Tripathy D, Tagliaferri MA, Shaw HS, Kimmick GG, Cohen I, Shtivelman E, Caygill KA, Grady D, Schactman M, Shapiro CL (2010) A phase 1B dose escalation trial of *Scutellaria barbata* (BZL101) for patients with metastatic breast cancer. Breast Cancer Res Treat 120:111–118

Porter B (2006) Skullcap production in Saskatchewan. http://www.agr.gov.sk.ca/docs/crops/ horticulture/skullcap.asp. Accessed July 12, 2006

Qian B (1987) Clinical effect of anticancer Chinese medicine. Shanghai Translation Publishing House, Shanghai, pp 6–7

Ravindran PN, Babu NK (2005) Ginger, the genus Zingiber, medicinal and aromatic plants-industrial profiles. CRC Press, Boca Raton

Richardson MA (2001) Biopharmacologic and herbal therapies for cancer: research update from NCCAM. J Nutr 131:3037–3040

Rugo H, Shtivelman E, Perez A, Vogel C, Franco S, Tan Chiu E, Melisko M, Tagliaferri M, Cohen I, Shoemaker M, Tran Z, Tripathy D (2007) Phase I trial and antitumor effects of BZL101 for patients with advanced breast cancer. Breast Cancer Res Treat 105:17–28

Saif MW, Lansigan F, Ruta S, Lamb L, Mezes M, Elligers K, Grant N, Jiang ZL, Liu SH, Cheng YC (2010) Phase I study of the botanical formulation PHY906 with capecitabine in advanced pancreatic and other gastrointestinal malignancies. Phytomedicine 17:161–169

Sarin R (2005) Useful metabolites from plant tissue cultures. Biotechnology 4:79–93

Sato Y, Suzuki S, Nishikawa T, Kihara M, Shibata H, Higuti T (2000) Phytochemical flavones isolated from *Scutellaria barbata* and antibacterial activity against methicillin-resistant *Staphylococcus aureus*. J Ethnopharmacol 72:483–488

Scheck AC, Perry K, Hank NC, Clark WD (2006) Anticancer activity of extracts derived from the mature roots of *Scutellaria baicalensis* on human malignant brain tumor cells. BMC Complement Altern Med 6:27

Shang X, He X, He X, Li M, Zhang R, Fan P, Zhang Q, Jia Z (2010) The genus *Scutellaria* an ethnopharmacological and phytochemical review. J Ethnopharmacol 128:279–313

Shao Z-H, Li C-Q, Vanden Hoek TL, Becker LB, Schumacker PT, Wu JA, Attele AS, Yuan C-S (1999) Extract from *Scutellaria baicalensis* Georgi attenuates oxidant stress in cardiomyocytes. J Mol Cell Cardiol 31:1885–1895

Shibata K, Iwata S, Nakamura M (1923) Baicalin, a new flavone-glucuronic acid compound from the roots of *Scutellaria baicalensis*. Acta Phytochim 1:105–139

Shieh DE, Liu LT, Lin CC (2000) Antioxidant and free radical scavenging effects of baicalein, baicalin and wogonin. Anticancer Res 20:2861–2865

Shieh DE, Cheng HY, Yen MH, Chiang LC, Lin CC (2006) Baicalin-induced apoptosis is mediated by Bcl-2-dependent, but not p53-dependent, pathway in human leukemia cell lines. Am J Chin Med 34:245–261

Shin SW, Lee HK (1995) Production of baicalin by cell cultures of *Scutellaria baicalensis*. Korean J Pharmacog 26:159–163

Similien A (2009) Effect of shade, irrigation and nutrients on dry matter yield and flavonoid content in American skullcap. M.S. thesis, Department of Agronomy and Soils, Auburn University

Sinha S, Pokhrel S, Vaidya BN, Joshee N (1999) In vitro micropropagation and callus induction in *Scutellaria discolor* COLEBR.—A medicinally important plant of Nepal. Indian J Plant Genet Resour 12:219–223

Srivastava J, Lambert J, Vietmeyer N (1996) Medicinal plants: an expanding role in development. World Bank technical paper 320. The World Bank, Washington

Stojakovska A, Malarz J (2000) Flavonoid production in transformed root cultures of *Scutellaria baicalensis*. Plant Physiol 156:121–125

Stojakowska A, Malarz J, Kohlmuenzer S (1999) Micropropagation of *Scutellaria baicalensis* Georgi. Acta Soc Bot Pol 68:103–107

Sucher NJ, Carles MC (2008) Genome based approaches to the authentication of medicinal plants. Planta Med 74:603–623

Tang W, Eisenbrand G (1992) Chinese drugs of plant origin. Springer, Berlin

Tani T, Katsuki T, Kubo M, Arichi S (1985) Histochemistry. VII. Flavones in *Scutellaria Radix*. Chem Pharm Bull 33:4894–4900

Tascan A (2007) Liquid culture systems for *Scutellaria* species in vitro. M.S. thesis. Clemson University, Clemson, SC

Tascan A, Adelberg JW, Joshee N, Yadav AK, Tascan M (2007) Liquid culture systems for *Scutellaria* spp. Acta Hort 756:163–170

Tascan A, Adelberg JW, Rimando AM, Tascan M, Joshee N, Yadav AK (2010) Hyperhydricity and flavonoid content of *Scutellaria* species in vitro on polyester-supported liquid culture systems. HortScience 45(10):1723–1728

Tomimori T, Miyaichi Y, Imoto Y, Kizu H, Namba T (1984) Studies on the constituents of *Scutellaria* species V. On the flavanoid constituents of Ban Zhi Lian, the whole herb of *Scutellaria rivularis* Wall (1). Shoyakugaku Zasshi 38:249–252

Tomimori T, Miyaichi Y, Imoto Y, Kizu H, Namba T (1985) Studies on Nepalese crude drugs. V. On the flavonoid constituents of the root of *Scutellaria discolor* Colebr. Chem Pharm Bull 33:4457–4463

Tomimori T, Miyaichi Y, Imoto Y, Kizu H (1986) Studies on the constituents of Scutellaria species. VIII. On the flavonoid constituents of Ban Zhi Lian, the whole herb of *Scutellaria rivularis* Wall (2). Shoyakugaku Zasshi 40:432–433

Tomimori T, Imoto Y, Miyaichi Y (1990) Studies on the constituents of *Scutellaria* species XIII. On the flavonoid constituents of the root of *Scutellaria rivularis* Wall. Chem Pharm Bull 38:3488–3490

Turner BL (2011) A new species of *scutellaria* (Lamiaceae) from Oaxaca, Mexico. Phytologia 93:241–244

Uhring J (1982) In vitro propagation of *Scutellaria costaricana*. HortScience 17:533

Ujiki MB, Ding XZ, Salabat MR, Bentrem DJ, Golkar L, Milam B, Talamonti MS, Bell RH Jr, Iwamura T, Adrian TE (2006) Apigenin inhibits pancreatic cancer cell proliferation through G2/M cell cycle arrest. Mol Cancer 5:76

Vanisree M, Lee C-Y, Lo S-F, Nalawade SM, Lin CY, Tsay H-S (2004) Studies on the production of some important secondary metabolites from medicinal plants by plant tissue cultures. Bot Bull Acad Sin 45:1–22

Verpoorte R, van der Heijden R, Memelink J (2000) Engineering the plant cell factory for secondary metabolite production. Transgenic Res 9:323–343

Way TD, Kao MC, Lin JK (2004) Apigenin induces apoptosis through proteasomal degradation of HER2/neu in HER2/neu-overexpressing breast cancer cells via the phosphatidyl-inositol 3-kinase/Akt-dependent pathway. J Biol Chem 279:4479–4489

Wermuth CG (2004) Multitargeted drugs: the end of the 'one-target-one-disease' philosophy? Drug Discov Today 9:826–827

Wichtl M, Franz C, Czygan FZ, Frohne D, Holtzel C, Nagell A, Pfander HJ, Willuhn G, Buff W (1994) Herbal drugs and phytopharmaceuticals. CRC press, Boca Raton (translated by Bisset NG)

Wills RBH, Stuart DL (2004) Generation of high quality Australian skullcap products. A report for the Rural Industries Research and Development Corporation, RIRDC publication no. 04/020, Australian Government. ISBN:0642587302, ISSN:1440-6845

Winkel-Shirley B (1998) Flavonoids in seeds and grains: physiological function, agronomic importance and the genetics of biosynthesis. Seed Sci Res 8:415–422

Yaghmai MS (1998) Volatile constituents of *Scutellaria lateriflora* L. Flavour Fragrance J 3: 27–31

Yano S, Umeda D, Yamashita T, Ninomiya Y, Sumida M, Fujimura Y, Yamada K, Tachibana H (2007) Dietary flavones suppresses IgE and Th2 cytokines in OVA-immunized BALB/c mice. Eur J Nutr 46:257–263

Ye F, Jiang S, Volshonok H, Wu J, Zhang DY (2007) Molecular mechanism of anti-prostate cancer activity of *Scutellaria baicalensis* extract. Nutr Cancer 57:100–110

Yin X, Zhou J, Jie C, Xing D, Zhang Y (2004) Anticancer activity and mechanism of *Scutellaria barbata* extract on human lung cancer cell line A549. Life Sci 75:2233–2244

Yoon HJ, Kim HK, Ma C-J, Huh H (2000) Induced accumulation of triterpenoids in *Scutellaria baicalensis* suspension cultures using a yeast elicitor. Biotech Lett 22:1071–1075

Yoon MS, Lee JS, Choi BM, Jeong YI, Lee CM, Park JH, Moon Y, Sung SC, Lee SK, Chang YH, Chung H, Park YM (2006) Apigenin inhibits immuno-stimulatory function of dendritic cells: implication of immunotherapeutic adjuvant. Mol Pharmacol 70:1033–1044

Yu J, Liu H, Lei J, Tan W, Hu X, Zou G (2007) Antitumor activity of chloroform fraction of *Scutellaria barbata* and its active constituents. Phytother Res 21:817–822

Zhang Y-Y, Guo Y-Z, Onda M, Hashimoto K, Ikeya Y, Okada M, Maruno M (1994) Four flavonoids from *Scutellaria baicalensis*. Phytochemistry 35:511–514

Zhang DY, Wu J, Ye F, Xue L, Jiang S, Yi J, Zhang W, Wei H, Sung M, Wang W, Li X (2003) Inhibition of cancer cell proliferation and prostaglandin E2 synthesis by *Scutellaria baicalensis*. Cancer Res 63:4037–4043

Zheljazkov VD, Cantrell CL, Ebelhar M, Coker C, Evans WB (2007) Quality assessment and yield of Baikal skullcap (*Scutellaria baicalensis*) grown at multiple locations across Mississippi. HortScience 42:1183–1187

Zhou Y, Hirotani M, Yoshikawa T, Furuya T (1997) Flavonoids and phenylethanoids from hairy root cultures of *Scutellaria baicalensis*. Phytochemistry 44:83–87

Zobayed SMA, Murch SJ, Rupasinghe HPV, de Boer JG, Glickman BW, Saxena PK (2004) Optimized system for biomass production, chemical characterization and evaluation of chemo-preventive properties of *Scutellaria baicalensis* Georgi. Plant Sci 167:439–446

Chapter 4
Microbial Endophytes: Progress and Challenges

Dnyaneshwar Rathod, Mudasir Dar, Aniket Gade, Ravi B. Shrivastava, Mahendra Rai and Ajit Varma

4.1 Introduction

Endophytes are microbes (fungi or bacteria) that live within the plant tissues without causing any noticeable symptoms of disease (Tejesvi et al. 2007). Endophytes invade the tissues of living plants and reside between living plant cells (Vanessa and Christopher 2004). Some form a mutually beneficial relationship (symbiosis) with the host plants while others are opportunistic pathogens. Petrini et al. (1992) reported that there may be more than one type of mycoendophytes found within a single plant. For example, 13 taxa of mycoendophyte were isolated from the leaf, stem, and root tissues of *Catharanthus roseus* (Kharwar et al. 2008).

It is reported that fungal metabolites are not only indispensable for medicine but also important for plant protection. The demand for new highly effective agricultural agents to control farm pests and pathogens is huge, and partly arises from the removal of synthetic compounds from the market because of their toxicity toward the environment. Shipunov et al. (2008) tested the co-introduction and host-jumping hypotheses in plant communities by comparing endophytes isolated from the invasive spotted knapweed (*Centaurea stoebe*, Asteraceae) in its

D. Rathod · M. Dar · A. Gade · M. Rai
Department of Biotechnology Sant Gadge Baba, Amravati University,
Amravati 444602, India

R. B. Shrivastava
Defense Institute of High Altitude Research (DIHAR), Defense Research and Development Organization, Ministry of Defense, Government of India, Leh-Ladakh, India

A. Varma (✉)
Amity Institute of Microbial Technology (AIMT), Amity University,
Noida 201303, Uttar Pradesh, India
e-mail: ajitvarma@aihmr.amity.edu

S. Chandra et al. (eds.), *Biotechnology for Medicinal Plants*,
DOI: 10.1007/978-3-642-29974-2_4, © Springer-Verlag Berlin Heidelberg 2013

native and invaded ranges. Shipunov et al. (2008) reported that endophytes can affect the competitiveness of *C. stoebe*. Both co-introduction and host-jumping of endophytes align with hypotheses of plant invasion that are based on enhanced competitiveness. Kharwar et al. (2008) reported 183 mycoendophytes representing 13 fungal taxa isolated from leaf, stem, and root tissues of *C. roseus* from two sites representing two different ecosystems in North India. The leaf tissues showed more diversity of endophytes such as *Drechslera*, *Curvularia*, *Bipolaris*, *Alternaria,* and *Aspergillus* sp. Wei et al. (2009) studied the colonization frequencies of endophytic *Pestalotiopsis* species diverse with host plants, ages, tissues, and sites. Ya-li et al. (2010) reported 49 endophytic fungi which were recovered from *Saussurea involucrata* and identified using morphological and molecular techniques. Among these fungi *Cylindrocarpon* sp. was the dominant species followed by *Phoma* sp. and *Fusarium* species. Li and Shun (2009) reported the recovery of 300 isolates in which 172 isolates were from *Dracaena cambodiana* and 128 from *Aquilaria sinensis.*

4.2 Fungal Endophyte Diversity and Bioactive Compounds

The maximum biological diversity in terrestrial ecosystems is in tropical and temperate rainforests. Interestingly, they also have the greatest number of mycoendophytes. These ecosystems cover only 1.44 % of the land's surface, yet they harbor more than 60 % of the world's terrestrial biodiversity (Strobel and Daisy 2003). Hazalin et al. (2009) isolated 300 endophytes from various parts of plants collected from the National Park, Penang in Malaysia. Some of these endophytes demonstrated cytotoxic activity against the murine leukemic P388 cell line and 1.7 % against a human chronic myeloid leukemic cell line K562 (Hazalin et al. 2009). Survey of the literature provides evidence of increasing research on endophytes and their secondary metabolites. Studies on medicinal plants used by indigenous communities to treat various diseases have resulted in a wealth of scientific discoveries.

Fungal endophytes are being increasingly accepted as an ecological group of microorganisms that may provide sources for new secondary metabolites with useful biological activities. An array of active principles has been isolated and characterized from endophytes and many of these have diverse bioactivities (anticancerous, antioxidants, antifungal, antibacterial, antiviral, anti-insecticidal, and immune suppressants). Isolation of endophytic fungi from coffee plants (*Coffea arabica* and *C. robusta*) was shown to have antimicrobial activity against various human pathogenic bacteria (Sette et al. 2006). At present, there is an urgent need to search for endophytic metabolite that can be developed as safe, effective antifungal agents that are nonpetrochemical, eco-friendly, and easily obtained (Liu et al. 2006). Endophyte *Taxomyces andreanae,* which produces taxol in vitro, was isolated from *Taxus* sp. (Stierle et al. 1993). Vennila et al. (2010) studied the effect of taxol extracted from the endophytic fungus *Pestalotiopsis pauciseta* recovered from *Tabebuia pentaphylla* Hems. *T. pentaphylla* (family Bignoniaceae) is distributed in Northern Mexico, Southern Florida, and Cuba.

Zhou et al. (2010) summarized the recent advances in taxol-producing endophytic fungi all over the world. Kajula et al. (2010) studied the extracellular siderophore production as well as production of antibacterial and antioxidant compounds by endophytic fungi of Scots pine (*Pinus sylvestris* L.) and Labrador tea (*Rhododendron tomentosum* Harmaja). The pinolenic acid contained in pine nut oil helps curb appetite. It is used as a pain reliever in arthritis, aches, and sore muscles. Yang et al. (1994) reported that the phenol and phenolic acids detected in culture medium of endophytes often have remarkable biological activities. 2-Hydroxy-6-methylbenzoic acid was isolated from endophytic *Phoma* species which showed noteworthy antibacterial activity. *Phoma medicaginis* exists as a prolonged asymptomatic infection of its host plant (*Medicago* species). Suryanarayanan et al. (2005) studied the cactus *Cylindropuntia fulgida* for its endophytic diversity. The authors reported 900 endophytic isolates belonging to 22 fungal species from 21 cactus sp.

Karsten et al. (2007) reported herbicidal and algaecidal activity in ethyl acetate extract of an endophytic *Phoma* sp. isolated from *Fagonia cretica*, as ethnomedicine, *F. cretica* is used against fever, thirst, vomiting, dysentery, asthma, urinary discharges, liver trouble, dropsy, delirium, typhoid, toothache, stomach troubles, and skin diseases. Randa et al. (2010) isolated a *Botryosphaeria rhodina* from the stem of the medicinal plant *Bidens pilosa* (Asteraceae) that showed anti-inflammatory, antiseptic, and antifungal effects. Borges and Pupo (2006) reported two novel hexahydroanthraquinone derivatives, dendryol E and F isolated from *Phoma sorghina*, which was found as endophyte in association with the medicinal plant *Tithonia diversifolia.*

Schwarz et al. (2004) optimized the culture conditions of *Phoma* species and reported highest nematicidal activity in yeast malt glucose medium. Phomodione, 2,6-diacetyl-7-hydroxy-4a,9-dimethoxy-8,9b-dimethyl-4a.9b-dihydrodibenzo furan-1,3, an usnic acid derivative, was isolated from culture broth of a *Phoma* species which was an endophyte in the Guinea plant (*Saurauia scaberrinae*). Smith et al. (2008) provided direct evidence from bioassays of endophytes isolated from tropical plants and bioinformatic analyses that will give a novel chemistry of potential value. Raviraja (2005) studied 18 species of mycoendophytes, isolated from bark, stem, and leaf segments of five medicinal plant species growing within the Kudremukh range in the Western Ghats of India. The most common endophytic fungi were *Curvularia clavata*, *C. lunata*, *C. pallescens,* and *Fusarium oxysporum.* The greatest species richness and frequency was found in the leaf segments, rather than the stem and bark segments of the host plant species. Thus, if endophytes could produce the same rare and important bioactive compounds as their host plants, this would not only reduce the need to harvest slow-growing and possibly rare plants but also help to preserve the world's ever-diminishing biodiversity.

Large number of eco-friendly drugs produced by fungal endophytes is large as compared to endophytic bacteria and actinomycetes. Natural products from fungal endophytes can be grouped into several categories, including alkaloids, steroids, terpenoids, isocoumarins, quinones, phenylpropanoids and lignans, phenol and phenolic acids, aliphatic metabolites, lactones, etc. Zhen-Liang et al. (2011) reported novel bioactive compounds with spiro-5, 6-lactone ring skeleton isolated from endophyte *Massrison* sp. which is recovered from wild *Rehmannia glutinosa*. Further, they performed the antifungal and cytotoxic activities of the compounds. Debbab et al. (2009) reported secondary metabolites in *Chaetomium* sp. isolated from *Salvia officinalis* and evaluated their cytotoxic activity. Rukachaisirikul et al. (2007) reported endophytic *Phomopsis* species which produces secondary metabolites such as phomoenamide, phomonitroester, and deacetylphomoxanthone, and showed antibacterial activity against *Mycobacterium tuberculosis*. Tuberculosis (TB) infections are continuously increasing. Gordien et al. (2010) studied extracts from Scottish plants, lichens, and mycoendophyte which were screened for activity against *Mycobacterium aurum* and *M. tuberculosis*. The greatest activity against *M. aurum* was shown by extracts of *Juniperus communis* roots of the lichen *Cladonia arbuscula* and of a mycoendophyte isolated from *Vaccinium myrtillus* (Gordien et al. 2010). It is obvious that mycoendophytes serve as a source of potentially useful medicinal compounds. For example, 3-Nitropropionic acid was isolated from *Phomopsis* species which inhibited *Mycobacterium tuberculosis* (Copp and Pearce 2007). Guo et al. (2008) studied the new antimicrobial metabolites isolated and extracted from the culture of *Colletotrichum* species from *Artemisia annua,* which is a traditional Chinese herb. It is well recognized for its synthesis of artemisinin (Antimalarial drug).

4.3 Endophytic Actinomycetes

The actinomycetes that reside in the tissues of living plants and do not visibly harm the plants are known as endophytic actinomycetes (Strobel et al. 2004). Endophytic actinomycetes exist in the inner tissues of living plants. They have been isolated from the stem, leaf, fruit, root, and interior of other parts of many plants species (Okazaki et al. 1995; de Araújo et al. 2000; Coombs and Franco 2003). These organisms have attracted taxonomists, agronomists, chemists, ecologists, and evolutionary biologists. It has been indicated by numerous studies that these prolific actinomycetes appear to have a capacity to produce an impressive and important array of secondary metabolites. These metabolites exhibit a wide range of biologically active compounds, such as antibiotics, antitumor, anti-infectives, and other important chemical agents, such as plant growth promoters and enzymes. Similarly, actinomycetes also contribute to their host by promoting growth and enhancing their ability of withstanding in unfavorable environmental conditions (Qin et al. 2011a, b). Moreover, it represents under explored reservoir of novel species of potential interest in the discovery of novel lead compounds and for exploitation in pharmaceutical, agriculture, and industry (Qin et al. 2009a, b).

The actinomycetes accounts for a high proportion of soil microbial diversity and contains the most economically significant prokaryotes, producing about half of the known bioactive compounds (Lazzarini et al. 2000), including antibiotics (Bérdy 2005), antitumor agents (Cragg et al. 2005), immunosuppressive agents (Mann 2001), and enzymes (Oldfield, 1998). Endophytic actinomycetes have attracted attention in recent years, with increasing reports of isolates from a range of plant types, including medicinal plants (Taechowisan et al. 2003; Zin et al. 2007) and crop plants (cereals, such as wheat and rice, as well as potatoes, carrots, tomatoes, and citrus) (Nejad and Johnson 2000; Araújo 2002; Coombs and Franco 2003; Surette et al. 2003; Sessitsch et al. 2004; Tian et al. 2007). Endophytic actinomycetes promote the growth of host plants as well as to reduce disease caused by plant pathogens through various mechanisms, including the production of secondary metabolites, which are used in direct antagonism against plant diseases and pests (Cao et al. 2004, 2005; Castillo et al. 2007), also changes in host physiology (Igarashi et al. 2002), and the induction of systemic acquired resistance in plants (Conn et al. 2008). Similarly, other metabolites have antibiotic activity, suggesting that these organisms can be an important source for bioprospecting. Day-by-day the new species of endophytic actinomycetes have been increasingly reported. Thus, these endophytes are expected to be potential sources of new species with new bioactive agents (Gu et al. 2006; Duangmal 2008).

Higher plants contain a promising niche inside their tissues. The studies have demonstrated that some actinomycetes can form important associations with plants and colonize themselves in their inner tissues. For example, *Frankia* species and *Streptomyces scabies* important actinomycetes can penetrate their respective hosts and establish a relation as either endophyte or pathogen (Benson and Silvester 1993; Doumbou et al. 1998). The actinomycetes that live within the tissue of living plants and do not cause any symptom or do not visibly harm the plants are known as endophytic actinomycetes (Stone et al. 2000). It is suggested that there are approximately 300,000 plant species on the Earth, and each plant is considered to be the host for one or more type of microbial endophytes (Strobel and Daisy 2003), which creates an enormous biodiversity of endophytic microorganisms. But, a few of these endophytic actinomycetes associated with plants have been studied, indicating the opportunity to find new species and new related natural products among plants living in different niches and ecosystems. Recent studies revealed that large group of endophytic actinomycetes species and their diverse compounds with different biological functions are present (Araujo et al. 2002; Coombs and Franco 2003; Ryan et al. 2008; Bascom-Slack et al. 2009). But in some cases, actinomycetes act as biological agents (Cao et al. 2005) that enhance the plant growth promotion activity (Igarashi et al. 2002) and promote plant development in adverse environmental conditions (Hasegawa et al. 2006). These organisms are relatively unstudied and are potential sources of novel natural products used in medicine, agriculture, and industry (Strobel et al. 2004). Continuously, new findings in the recent study of the endophytic actinomycetes have an enormous biotechnological potential in the areas of natural products discovery and related applications.

Moreover, the natural products play highly significant role in drug discovery and development process. Cragg (2005) reported that 28 % of the new chemicals were discovered and 42 % of these chemicals were natural products and their derivatives, including anticancerous drugs. More than 22,000 biologically active metabolites were recovered from microorganisms, out off these 45 % compounds were recovered from actinomycetes, such as *Streptomyces* which is an excellent producer (Bérdy 2005). The application of actinomycetes for health and well-being of people throughout the world is a revolution in medicine (Demain and Sanchez 2009).

4.3.1 Diversity of Endophytic Actinomycetes in Medicinal Plants

Medicinal plants play a very important role in medicine to cure different types of diseases. In 1886, the genus *Frankia* was isolated from non-legume root nodules, which indicate the association between actinomycetes and host (Okazaki 2003). In the last two decades, endophytic actinomycetes have been isolated from all examined vascular plants, ranging from woody trees to herbaceous plants. The symbiotic colonization of terrestrial plants by actinomycetes is ubiquitous and common in nature. Therefore, the endophytic actinobacteria are important components of microbial biodiversity because of the frequent plant species. The plants which maintain the ethnobotanic values it is considered as promising sources of endophytes which produce novel bioactive compounds.

Endophytic actinomycetes also reported from crop plants, such as wheat, rice, potato, carrots, tomato, and citrus (Nejad and Johnson 2000; Araujo et al. 2002; Coombs and Franco 2003; Surette et al. 2003; Sessitsch et al. 2004; Tian et al. 2007; Velazquez et al. 2008), woody tree species (Taechowisan et al. 2003; Zin et al. 2007; Yuan et al. 2008; Zhao et al. 2010a, b, c), mosses, and ferns (Janso and Carter 2010). This study reported that the greatest diversity of endophytes is probable to occur in the tropical and temperate regions (Strobel and Daisy 2003). The occurrence of actinomycetes in roots compared to other tissues is common. Taechowisan et al. (2003) screened 5,400 roots, leaves, and stem tissues from 36 species of plants and recovered 212 isolates from roots, 97 from leaves, and 21 from stems. Furthermore, Verma et al. (2009a, b) recovered more than double the number of all isolates from roots than from stems or leaves from 20 different Indian lilac trees and *Azadirachta indica*. Root plays a vital role for water and nutrient uptake for plants, and therefore it is an ideal substrate for actinomycete colonization (Janso and Carter 2010).

Endophytic actinomycetes diversity and antibiotic activity was reported recovered from medicinal plants of tropical regions (Li et al. 2008a, b; Bascom-Slack et al. 2009; Qin et al. 2009a). Qin et al. (2009b) reported 2,174 endophytic actinomycetes isolated from different medicinal plants of tropical rainforest of China. Among these only from one kind of plant, *Maytenus austroyunnanensis*, one new genus, and seven new species of endophytes were isolated. The richness of actinomycetes and their diversity from tropical rainforest are much higher compared to biogeographical regions. Similarly, tropical rainforest has a promising potential as a source of actinomycetes and new endophytes. It is reported that the endophytic actinomycetes are diverse and the diversity may vary between different host plant regions and different plant species.

Taechowisan et al. (2003) isolated 330 endophytes from different medicinal plants belonging to four different genera (*Streptomyces, Microbispora, Nocardia* and *Micromonospora*). Also El-Shatoury et al. (2009) reported endophytic actinomycetes associated with the medicinal plant, *Achillea fragrantissima* (Forssk) Sch. Bip. (Compositae), of Saint Katherine, South Sinai, Egypt. The diversity of endophytic actinomycetes showed no significant differences between different locations of the studied regions. Twenty-five isolates were identified and categorized into seven genera of actinomycetes (viz. *Streptomyces*, *Nocardioides*, *Kitasatosporia*, *Nocardia*, *Kibdelosporangium*, *Pseudonocardia,* and *Promicromonospora).* In addition to these one unidentified isolate was also obtained. Genera *Nocardia* and *Streptomyces* were the most recorded in *A. fragrantissima* internal tissues. An endophytic actinomycete was recovered from four medicinal plants such as *Artemisia herba-alba, Echinops spnosus, Mentha longifolia,* and *Ballota undulata* belonging to different genera. Among these the *Streptomyces* species was dominant as compared to other (El-Shatoury et al. 2006). An endophytic actinomycete novel species *Pseudonocardia endophytica* sp. nov. was isolated from the inner tissues of the traditional Chinese medicinal plant *Lobelia clavata* by Chen et al. (2009). Wang et al. (2008) reported the diversity of uncultured microbes associated with medicinal plant *Mallotus nudiflorus* and found that actinomycetes were the most dominant microbial group, covering about 37.7 % of whole endophytes isolated. *Glycomyces endophyticus* sp. nov. is an endophytic actinomycete, designated strain YIM 56134T, isolated from the root of a Chinese medicinal plant, *Carex baccans* Nees, collected from Yunnan, South West China. This medicinal plant harbors other endophytic actinomycetes in their internal tissues (Qin et al. 2008).

The diversity of endophytic actinomycetes associated with tropical, native plants is essentially unexplored. Janso and Carter (2010) isolated 123 endophytic actinomycetes from tropical plants collected from several locations in Papua New Guinea, Mborokua Island, and Solomon Islands. These isolates were found to be prevalent in roots but uncommon in leaves. Characterization of 105 isolates of unique strains by 16S rRNA gene sequence analysis revealed that 17 different genera were represented and rare genera also, such as *Sphaerisporangium* and *Planotetraspora*, which have never been previously reported to be endophytic. Phylogenetic analyses grouped many of the strains into clades distinct from known genera within Thermomonosporaceae and Micromonosporaceae, indicating that they may be unique genera. Although *Streptomyces* was the most frequently isolated genus, the Streptosporangiaceae was also well-represented family, which consisted of five different genera: *Sphaerisporangium, Planotetraspora, Nonomuraea, Streptosporangium,* and *Microbispora. Planotetraspora* and *Sphaerisporangium* were common among rare genera of actinomycetes and were consistently represented in plants from each sampling location.

Also *Microbispora*, which belongs to Streptosporangiaceae, is common in a variety of plants. They gave new report on the occurrence of *Planotetraspora* and *Sphaerisporangium* in plants as endophytes. Also three *Kibdelosporangium* strains, from the roots of a *Heterospathe* sp. and *Pandanus* sp., are the first report of endophytes isolated from these genera (Janso and Carter 2010).

Zhu et al. (2009) isolated more than 160 endophytic actinomycetes from multiple tissues of both wild and cultured specimens of the native, tropical tree *Trewia nudiflora*. They reported the presence of actinomycetes from each tissue, provided relatively little information on the biodiversity of these isolated actinomycetes, and characterized only *Streptomyces* sp. Castillo et al. (2007) isolated endophytic actinomycetes from plant species growing in Southern Patagonia, but they only recovered *Streptomyces* sp. from the stems. *Pseudonocardia sichuanensis* sp. nov., a novel endophytic actinomycetes has been isolated from the root of *Jatropha curcas* L. (Qin et al. 2011a, b). Three novel endophytic *Streptomyces* have been isolated from plants with ethnobotanical properties on the Malay Peninsula including *Thottea grandiflora* (Aristolochiaceae), *Polyalthia* sp. (Annonaceae), and *Mapania* sp. (Cyperaceae) (Zhao et al. 2007; Zin et al. 2007). Traditional Chinese medicinal plants are large sources of biologically active compounds, providing raw material for agriculture, pharmaceutical, cosmetics, and fragrance industries. The endophytes of these medicinal plants take part in biochemical pathways and produce analogous or novel bioactive compounds. In a study, 560 endophytic actinomycetes were isolated from 26 medicinal plant species in Panxi plateau. According to the phyllogenetic analysis, seven isolates were *Streptomyces* sp., while the remainder belonged to genera *Micromonospora, Oerskovia, Nonomuraea, Promicromonospora,* and *Rhodococcus* (Zhao et al. 2011).

4.3.2 Bioactive Compounds from Actinomycetes

Endophytic Actinomycetes of medicinal plants participate in metabolic pathways of medicinal plants and produce analogous or novel bioactive compounds, for example taxol (Strobel et al. 1999). Endophytic actinomycetes are considered as rich and potential sources of novel bioactive compounds and various bioactive compounds are continuously isolated from them until now (Taechowisan et al. 2005; Castillo et al. 2006; Igarashi et al. 2007a, b). Endophytic actinomycetes associated with medicinal plants could be a rich source of functional metabolites (Strobel et al. 2004). Natural products from endophytes is evident from the number of review articles that have appeared in the recent literatures (Hasegawa et al. 2006; Zhang et al. 2006; Gunatilaka 2006; Guo et al. 2008; Staniek et al. 2008; Ryan et al. 2008; Verma et al. 2009a, b). Study of the endophytic actinomycetes has resulted in the identification of many new natural products with wide range of biological activities.

Endophytic actinomycetes, mainly from medicinal plants, show the ability to inhibit or kill a wide variety of harmful microorganisms such as pathogenic bacteria, fungi, and viruses. Thus, there is great application value to develop antimicrobial drugs from endophytic actinobacteria. Till now, number of new antibiotics have been isolated, such as kakadumycins (Castillo et al. 2003), celastramycins A-B (Pullen et al. 2002), munumbicins A-D (Castillo et al. 2002), and demethylnovobiocins (Igarashi 2004). This was observed from the culture broth of some endophytic *Streptomyces sp.* a simple compound 6-prenylindole was isolated. This chemical exhibited a significant antifungal activity against plant pathogen *Fusarium oxysporum*. The chemical was originally isolated from the plant liverwort (*Hepaticae*). This compound is isolated from both plant and endophyte (Igarashi 2004). From the strain *Streptomyces* sp. TP-A0456 two novel compounds cedarmycins A and B were isolated, which were also isolated from a twig of cedar. The Cedarmycins A showed antifungal activity against *Candida glabrata* in vitro (Igarashi 2004). Another endophytic *Streptomyces* sp. Tc022 isolated from *Alpinia galanga* strongly inhibited the fungi *Colletotrichum musae* and *Candida albicans*. The metabolites extracted from *Streptomyces* sp. Tc 022 in the broth culture medium showed the presence of a major component actinomycin D, which displayed strong antifungal activity (Taechowisan et al. 2006). Two new chromophoric peptides antibiotics designated as munumbicins E-4 and E-5 were isolated from endophytic *Streptomyces* NRRL 30562 which produces broadspectrum antibiotics such as munumbicins A-D. These compounds showed broadspectrum activity against both Gram-positive and Gram-negative bacteria (Castillo et al. 2006). Recently, a new antifungal compound saadamycin was isolated from endophytic *Streptomyces* sp. Hedaya48, it confers potent antifungal activity against clinical fungi such as dermophytes, etc. (El-Gendy and EL-Bondkly 2010). These studies suggest that actinomycetes possess strong microbial compounds, which are utilized as drugs and can be modified to combat the future challenges by use of biotechnological approaches.

Nowadays, there has been an increasing interest in discovering antitumor agents from endophytes. Endophytic actinomycetes live in close association with their host plants and close relationship with them, then it becomes a real possibility that genes involved in natural products biosynthesis could be exchanged via horizontal gene transfer between endophytic actinomycetes and plants, resulting in synthesis of plant-derived compounds by a microbial endophyte such as the paclitaxel-producing *Kitasatospora* sp. isolated from *Taxus baccata* in Italy (Caruso et al. 2000; Janso and Carter 2010). This is the first report on synthesis of taxol from endophytic actinomycetes. Also, maytansinoids (19-membered macrocyclic lactams ansamycin antibiotics), a potent antitumor agent related to antitumor chemical compounds that were originally isolated from members of the higher plant families Celastraceae, Rhamnaceae, and Euphorbiaceae (Kupchan et al. 1972; Powel et al. 1982) and some mosses (Suwanborirux et al. 1990), were also produced by plant-associated actinomycete *Actinosynnema pretiosum* (Higashide et al. 1977). One novel chlorine-containing ansamycin, namely naphthomycin K, was obtained from the endophytic strain of *Streptomyces* sp. CS was isolated from *Maytenus hookeri* a medicinal plant producing maytansinoids. This compound showed cytotoxic activity against P388 and A-549 cell lines (Lu and Shen 2007). It was quite interesting that the ansacarbamitocins, a new family of maytansinoids, were isolated from an actinomycete strain *Amycolatopsis* CP2808 that belongs to the family Pseudonocardiaceae; also, the ansamitocin-producing strain *A. pretiosum* belongs to the same family (Snipes et al. 2007). The 24-demethylbafilomycin C1, newly added member of the afilomycin subfamily, and two more new bafilomycin derivatives were isolated from the strain *Streptomyces* sp. (Lu and Shen 2003, 2004).

In the subsequent research in the field of antitumor by researchers all over found new bioactive chemical compounds including five 16-membered macrolides, belonging to the bafilomycin subfamily were isolated from *Streptomyces* sp. CS. was isolated from *M. hookeri* and showed cytotoxic activity against MDA-MB-435 cell line in vitro (Li et al. 2010). Another strain Streptomyces sp. ls9131, which was also isolated from *M. hookeri*, produced novel macrolides. This compound dimeric dinactin had strong antineoplastic activity and antibacterial activity (Zhao et al. 2005). Two novel compounds anthraquinones and lupinacidins A and B were isolated from the fermented broth of newly identified endophytic actinomycetes, *Micromonospora lupine*. The lupinacidins showed considerable inhibitory effects on the invasion of murine colon 26-L5 carcinoma cells (Igarashi et al. 2007a, b). Kim et al. (2006) isolated two new 6-alkylsalicylic acids, salaceyins A and B from another strain *Streptomyces laceyi* MS53. Salaceyins A and B showed cell toxicity against human breast cancer cell line SKBR3. Pterocidin is a new antitumor cytotoxic chemical compound, isolated from the endophytic *Streptomyces hygroscopicus* TP-A0451 exhibited cytotoxicity against cancer cell lines (Igarashi et al. 2006). Two compounds 5, 7-dimethoxy-4-phenylcoumarin and 5, 7-dimethoxy-4-p-methoxyl-phenylcoumarin isolated from endophytic Streptomyces aureofaciens CMUAc130 were originally produced by numerous species of plants. These bioactive compounds showed high and strong antifungal and antitumor activity (Taechowisan et al. 2005, 2007a). These findings support the application of them as anti-inflammatory agent (Taechowisan et al. 2007b). From the endophytic strain *Streptomyces* sp. SUC1, four novel secondary metabolites, lansai A to D were isolated. Tuntiwachwuttikul et al. (2008) reported lansai B which showed weak activity against the BC cell line, lansai C also showed significant anti-inflammatory activity in LPS-induced RAW 264.7 cells. In a word, endophytic actinobacteria still remain a relatively untapped source of novel natural products, presumed to push forward the frontiers of drug discovery against deadly cancers.

Endophytic actinomycetes also produced various other different bioactive compounds with a large magnitude of functioning. An endophytic *Streptomyces* sp. isolated from a Chinese traditional medicine plant *Cistanche deserticola* produced tyrosol. This compound acts as a new possible ligand for GPR12 (Lin et al. 2008). Coronamycin is a complex of novel peptide antibiotics extracted from verticillate *Streptomyces* sp. isolated from an *Epiphytic vine*, *Monstera* sp., which shows activity against pythiaceous fungi and the human fungal pathogen *Cryptococcus neoformans*. It also shows activity against the malarial parasite *Plasmodium falciparum*. It also showed antifungal activity against human fungal pathogen *Cryptococcus neoformans* (Ezra et al. 2004). Endophytic actinomycetes associated with higher plants, particularly with medicinal plants (used traditional medicines), are harboring bioactive compounds with medicinal properties; it is supported by the examples of taxol and maytansinoids discussed above. Endophytic actinomycetes as a key player of biocontrol agents have been partly discussed (Hasegawa et al. 2006). Studies are continuously conducted to unravel the mechanisms of action of these endophytes for production of bioactive compounds, such as antibiotics and cell wall degrading enzyme (El-Tarabily and Sivasithamparam 2006).

4.4 Bacterial Endophytes

The survival of endophytic bacteria inside different plant tissues is a well-predictable phenomenon (Sharrock et al. 1991; Fisher et al. 1992; McInroy and Kloepper 1995a, b). In general, endophytic bacteria are described as bacteria that are able to colonize the living plant tissues without harming a plant (Kado 1992). Endophytic bacteria isolated from the internal plant tissue of healthy plants comprises over 129 species representing over 54 genera, with *Pseudomonas, Bacillus, Enterobacter,* and *Agrobacterium* being the most commonly isolated bacterial genera (Mundt and Hinkle 1976; Gardener et al. 1982; Sturz 1995; McInroy and Kloepper 1995a; Hallmann et al. 1997a, b; Mahafee and Kloepper 1997).

Endophytic bacteria are usually isolated from the rhizosphere (roots), seeds, or plant material (leaf, branch, and other parts of plant) (Hallmann et al. 1997a, b). A division of rhizobacteria may enter the interior of the root by hydrolyzing cell wall cellulose, through tumors, along water flow, through wounds, or through lateral branching sites (Hallmann et al. 1997a, b; Siciliano et al. 1998). Evidences of plant-associated microbes has been discovered in the fossilized tissues of stems and leaves and these endophytic relationships may have begun from the time that higher plants first have appeared on the Earth, hundreds of millions of years ago to establish these types of relationships (Taylor and Taylor 2000). As a result of these associations between bacteria and plants, there is possibility to happen that some of these endophytic bacteria modified their genetic systems allowing for the transfer of information between themselves and the higher plant and vice versa (Stierle et al. 1993). The endophytic bacteria are known to contribute to the host plants growth and development by producing plant growth regulators. The bacterial endophytes are also known to increase host plants resistance to plant pathogens and parasites to promote biological nitrogen fixation and to produce antibiotics. The major group of endophytic bacteria comes under the category of actinomycetes, which are mostly studied and established group of endophytic bacteria. In the next section we will be confined to the endophytic bacteria other than Endophytic actinomycetes, which have already been discussed above.

4.4.1 Diversity of Endophytic Bacteria in Medicinal Plants

Various researchers reported endophytic microbes from various plants either medicinally useful or not. Nearly 300,000 plant species that exists on the earth and each individual plant is host to one or more endophytes. Only a few these plants have ever been completely studied for their endophytic nature. Consequently, the opportunity to find new and interesting microorganism among myriads of plants in different ecosystems is great (Strobel and Daisy 2003). Endophytic bacteria are commonly associated with medicinal plants.

Based on the visible morphological differences totally 20 bacterial isolates were isolated from *Andrographis paniculata* plant samples in India. Among all these 20 endophytic bacterial isolates, 11 were Gram-positive cocci and nine Gram-positive rods. Total seven isolates showed positive result for endospore staining and they were suspected as *Bacillus* species. The other isolates could not be identified (Arunachalam and Gayathri 2010). Cho et al. (2002) reported a strain of Bacillus CY22 from the roots of Balloon flower *Platycodon grandiflorum* in Republic of Korea. This was the endophytic bacterium capable of secreting various biologically active compounds showing activity against various pathogenic bacteria and fungi. A total of 98 isolates of bacterial endophytes were obtained from 27 medicinal plants by Berde et al. (2010). Among these isolates 58.43 % were Gram-positive rods, 19.10 % were Gram-negative rods, 10 % Gram-positive, and 10 % Gram-negative cocci. Bhore et al. (2010) screened three bacterial endophytes *Pseudomonas resinovorans, Paenibacillus polymaxa*, and *Acenitobacter calcoaceticus* from the leaves of Sambung Nyawa [*Gynura procumbens* (Lour.) Merr.].

The species *Jacaranda decurrens* Cham. is a shrub present in diverse physiognomies of the Brazilian Cerrado, popularly described as a medicinal plant that acts in gynecological infections, Giardia and amoeba's diseases, and as an anti-syphilitic. A total of 18 isolates of endophytic bacteria were isolated from *Jacaranda decurrens* Cham. with predomination of the bacteria of the genera *Bacillus* (39 %), *Pseudomonas* (27.6 %), *Corynebacterium* (16.7 %), *Actinomyces* (11.1 %) and *Staphylococcus* (5.6 %) (Carrim et al. 2006). In a study, Laorpaksa et al. (2008) screened 106 isolates of endophytic bacteria from 35 Thai medicinal plants. These endophytic bacteria belong to the diverse genera of bacteria, most of these endophytic bacteria showed antimicrobial properties against bacterial and fungal human pathogens.

The diversity of endophytic bacteria associated with ginseng (*Panax ginseng* C.A. Meyer) plants of varying age levels in Korea was investigated. A total of 51 colonies of endophytic bacteria were isolated from the interior of ginseng stems. These endophyte communities recovered from ginseng plant based on the results of 16S rDNA analysis, *Bacillus* and *Staphylococcus,* were dominant in age group 1–4-year-old plants. Other bacteria such as *Pseudomonas, Agrobacterium,* and *Stenotrophomonas, Mycobacterium Paenibacillus,* and *Staphylococcus* also being predominant genera (Vendan et al. 2010).

Echinacea is one of the most popular herbal supplements and the popularity of this remedy is due to its use for the treatment of infections and its immune-stimulatory and anti-inflammatory effects (Giles et al. 2000). Thirty-nine isolates of endophytic bacteria were isolated and selected based on colony morphology from four different micropropagated species of *Echinacea* (20 isolates from *E. pallida* var. *angustifolia*, four from *E. pallida* var. pallida, seven from

E. purpurea, and eight isolates from *E. pallida* var. *tenneseensis*). These isolates mainly include *Bacillus pumilus* (from *E. purpurea*), *Acinetobacter lwofii* and *Stenotrophomonas maltophila* (from *E. pallida* var. *angustifolia*), *Bacillus pumilus* (from *E. pallida* var. *pallida*), *Pseudomonas stutzeri,* and *Wautersia paucula* (from *E. pallida* var. *tenneseensis*). *W. paucula* was previously known as *Ralstonia paculua* (Lata et al. 2006).

4.4.2 Chemical Diversity of Endophytic Bacteria

The need for new and useful compounds to provide resistance and relief in all aspects of human conditions is continuously over growing. Natural products have been the traditional pathfinder compounds offering an untold diversity of chemical structures (Strobel and Daisy 2003). The word *Bioprospecting* is today most commonly used to express the collection and screening of biological material for commercial purposes (Synnes 2007). Molecules derived from natural products, particularly those products by plant and microbes, have an excellent record of providing novel chemical structural compounds for development of new pharmaceutical products. It was not until Pasteur discovered that fermentation is caused by living cells that people seriously began to investigate microorganisms as a source of high value metabolites. Since then people have been engaged in the discovery and application of microbial metabolites with the activity against both plant and human pathogens. Further, most of the discovery of plethora of microbes for application that span a broad spectrum of utility in medicine, agriculture, and industry is now practical (Balagurunathan and Radhakrishnan 2007).

Bacterial endophytes are mostly responsible for the production of the chemical metabolites which are used by the plants for their growth and development, against pathogens, etc. Apart from these functional metabolites which are utilized by plants, these endophytic bacteria also produce chemicals which are having pharmaceutical functions such as antibacterial, antifungal, anticancer, and many more biological activities. Bhore et al. (2010) screened cytokinin and cytokinin-like compounds from the endophytic bacteria (*Pseudomonas resinovorans, Paenibacillus polymaxa,* and *Acenitobacter calcoaceticus*) isolated from the medicinal plant *Gynura procumbens* (Lour.) Merr. resulted in growth and development of the host plant. In a study, Carrim et al. (2006) reported proteolytic, amilolytic, lipolytic, and esterasic activity showing enzymes from isolates of endophytic bacteria isolated from the *Jacaranda decurrens* Cham. a medicinal plant. All these enzymes are of biotechnological and pharmaceutical interest used for various purposes. The amylolytic activity was shown by isolates *B. coagulans* CAR3, *B. licheniformis* CAR4, *B. megaterium* CAR5, *Act. pyogenes* CAR1, and *P. stutzeri* CAR8. Esterases and lipases were produced by *Coryne. renale* CAR 7, *Bacillus, Ps. stutzeri,* and *Corynebacterium.* The presence of proteolytic activity was observed in *B. circulans* CAR 2, *B. coagulans*

CAR3, *B. licheniformis* CAR4, *B. megaterium* CAR5, *Coryne. renale* CAR7, and *Corynebacterium.*

In another study, Vendan et al. (2010) studied 18 isolates of endophytic bacteria isolated from Ginseng (*Panax ginseng* C.A. Meyer) medicinal plants. Among these 18 isolates 14 isolates of endophytic bacteria were producing Indole Acetic Acid in nutrient broth. The ability to produce IAA is considered for the development of plant growth. Also, IAA-producing endophytic bacterial screening was studied for the isolates from micropropagated plantlets of *Echinacea.* These bacteria include *Acinetobacter, Bacillus, Pseudomonas, Wautersia (Ralstonia),* and *Stenotrophomonas. Pseudomonas stutzeri* P3 strain which have been extensively exploited for the production of IAA plant growth regulator (Lata et al. 2006).

4.5 Conclusion

The need for new bioactive compounds to overcome the growing problems of drug resistance in microorganisms and the appearance of new diseases is of increasing importance. The capability of fungi to produce bioactive metabolites has encouraged researchers to isolate and screen fungi from diverse habitat and environments to search for novel bioactive metabolites. Therefore, endophytes have been proved as an outstanding source of both novel and bioactive natural products, which have an enormous potential for the development of new drugs and agricultural products. Consequently, endophytes are known to be a rich and reliable source of biologically active compounds with potential benefits in medicinal, industrial, and agricultural application. In contrast, the development of techniques such as combinatorial chemistry and equipment such as peptide synthesizers gives rise to exciting opportunities and expectations for the synthesis of biological active compounds. Additionally, as most endophyte research has been conducted on cultivated species, the diversity of studied endophytes has been further restricted by limitations in the ability to cultivate endophytes in the laboratory.

In spite of the increased number of reports in the last decade, endophytes are still a relatively poorly investigated group of microorganisms. Therefore, the research focusing on endophytes is a promising field in the chemistry and biological properties of natural products. To overcome the infectious disease, there is need for a variety of novel antimicrobial compounds of biological origin. The endophytes hold enormous potential as sources of antimicrobials. These endophytes may open new vistas for the development of new drugs and agricultural products. The multi-drug resistance problem in microbes underscores the need for further research on novel metabolites obtained from endophytes.

Acknowledgments The corresponding author is thankful to DRDO and DBT, New Delhi for partial financial assistance. Mahendra Rai and Aniket Gade thank Department of Environment, Government of India, New Delhi for financial assistance.

References

Araujo WL, Marcon J, Maccheroni W Jr, Van Elsas JD, Van Vuurde JW, Azevedo JL (2002) Diversity of endophytic bacterial populations and their interaction with *Xylella fastidiosa* in citrus plants. Appl Environ Microbiol 68:4906–4914

Arunachalam C, Gayathri P (2010) Studies on bioprospecting of endophytic bacteria from the medicinal plant of *Andrographis paniculata* for their antimicrobial activity and antibiotic susceptibility pattern. Int J Curr Pharm Res 2(4):63–68

Balagurunathan R, Radhakrishnan M (2007) Exploiting the less explored microbiol endophytes. Adv Biotech 6(2):20–23

Bascom-Slack CA, Ma C, Moore E, Babbs B, Fenn K, Greene JS, Hann BD, Keehner J, Kelley-Swift EG, Kembaiyan V, Lee SJ, Li P, Light DY, Lin EH, Schorn MA, Vekhter D, Boulanger LA, Hess WM, Vargas PN, Strobel GA, Strobel SA (2009) Multiple novel biologically active endophytic actinomycetes isolated from upper Amazonian rainforests. Microb Ecol 58(2):374–383

Be′rdy J (2005) Bioactive microbial metabolites. J Antibiot 58:1–26

Benson DR, Silvester WB (1993) Biology of Frankia strain actinomycetes symbionts of actinorrhizal plants. Microbiol Rev 57:293–319

Berde CV, Bhosale PP, Chaphalkar SR (2010) Plasmid of endophytic bacteria as vectors for transformation in plants. Int J Interg Biol 9(3):113–118

Bhore SJ, Nithya R, Loh CY (2010) Screening of endophytic bacteria isolated from leaves of Sambung Nyawa [*Gynura procumbens* (Lour)] for cytokinin-like compounds. Bioinformation 5(5):191–197

Borges WS, Pupo MT (2006) Novel anthraquinone derivatives produced by *Phoma sorghina* an endophyte found in association with the medicinal llant *Tithonia diversifolia* (Asteraceae). J Braz Chem Soc 17:929–934

Cao LX, Qiu ZQ, You JL, Tan HM, Zhou S (2004) Isolation and characterization of endophytic *Streptomyces* strains from surface-sterilized tomato (*Lycopersicon esculentum*) roots. Lett Appl Microbiol 39:425–430

Cao LX, Qiu ZQ, You JL, Tan HM, Zhou S (2005) Isolation and characterization of endophytic streptomycete antagonists of fusarium wilt pathogen from surface-sterilized banana roots. FEMS Microbiol Lett 247:147–152

Carrim AJI, Barbosa EC, Vieira JDG (2006) Enzymatic activity of endophytic bacterial isolates of *Jacaranda decurrens* cham (Carobinha-do-campo). Braz Arch Biol Technol 49(3):353–359

Caruso M, Colombo AL, Crespi-Perellino N, Fedeli L, Malyszko J, Pavesi A, Quaroni S, Saracchi M, Ventrella G (2000) Studies on a strain of Kitasatospora sp paclitaxel producer. Ann Microbiol 50:89–102

Castillo UF, Strobel GA, Ford EJ, Hess WM, Porter H, Jensen JB, Albert H, Robison R, Condron MAM, Teplow DB, Stevens D, Yaver D (2002) Munumbicins wide-spectrum antibiotics produced by Streptomyces NRRL 30562 endophytic on *Kennedia nigriscans*. Microbiology 148:2675–2685

Castillo P, Navas-Cortés JA, Gomar-Tinoco D, Di Vito M, Jiménez Díaz RM (2003) Interactions between *Meloidogyne artiellia*, the cereal and legume root-knot nematode, and *Fusarium oxysporum* f. sp. ciceris race 5 in chickpea. Phytopathology 93(12):1513–1523

Castillo UF, Strobel GA, Mullenberg K, Condron MM, Teplow DB, Folgiano V, Gallo M, Ferracane R, Mannina L, Viel S, Codde M, Robison R, Porter H, Jense J (2006) Munumbicins E-4 and E-5: novel broad-spectrum antibiotics from Streptomyces NRRL 3052. FEMS Microbiol Lett 255:296–300

Castillo UFL, Browne G, Strobel WM, Hess S, Ezra G, Pacheco Ezra D (2007) Biologically active endophytic streptomycetes from *Nothofagus* spp and other plants in Patagonia. Microb Ecol 53:12–19

Chen HH, Qin S, Li J, Zhang YQ, Xu LH, Jiang CL, Kim CJ, Li WJ (2009) *Pseudonocardia endophytica* sp nov isolated from the pharmaceutical plant *Lobelia clavata*. Int J Syst Evol Microbiol 59:559–563

Cho SJ, Park SR, Kim MK, Lim WJ, Ryu SK, An CL, Hong SY, Lee YH, Jeong SG, Cho YU, Yun HD (2002) Endophytic *Bacillus* sp. isolated from the interior of balloon flower root. Biosci Biotechnol Biochem 66:1270–1275

Conn VM, Walker AR, Franco CMM (2008) Endophytic actinobacteria induce defense pathways in *Arabidopsis thaliana*. Mol Plant-Microbe Interact 21:208–218

Coombs JT, Franco CMM (2003) Isolation and identification of actinobacteria isolated from surface-sterilized wheat roots. Appl Environ Microbiol 69:5603–5608

Copp BR, Pearce AN (2007) Natural product growth inhibitors of *Mycobacterium tuberculosis*. Nat Prod Rep 2:278–297

Cragg GM, Kingston DGI, Newman DJ (2005) Anticancer agents from natural products. Brunner-Routledge Psychology Press, Taylor and Francis Group, Boca Raton, pp 47–70

de Araújo JM, da Silva AC, Azevedo JL (2000) Isolation of endophytic actinomycetes from roots and leaves of maize (Zea ays L). Braz Arch Biol Technol 43:1516–1591

Debbab A, Aly AH, Edrada-Ebel R, Mueller WEG, Mosaddak M, Hakikj A, Ebel R, Proksch P (2009) Bioactive secondary metabolites from the endophytic fungus *Chaetomium* sp isolated from *Salvia officinalis* growing in Morocco. Biotech Agro Soc Environ 13(2):229–234

Demain AL, Sanchez S (2009) Microbial drug discovery: 80 years of progress. J Antibiot 62:5–16

Doumbou CL, Akimov V, Beaulieu C (1998) Selection and characterization of microorganisms utilizing Thaxtomin A a phytotoxin produced by *Streptomyces scabies*. Appl Environ Microbiol 64:4313–4316

Duangmal K, Thamchaipenet A, Ara I, Matsumoto A, Takahashi Y (2008) *Kineococcus gynurae* sp nov isolated from a Thai medicinal plant. Int J Syst Evol Microbiol 58:2439–2442

El-Gendy MMA, El-Bondkly AMA (2010) Production and genetic improvement of a novel antimycotic agent saadamycin against dermatophytes and other clinical fungi from endophytic *Streptomyces* sp Hedaya 48. J Ind Microbiol Biotechnol 37(8):831–841

El-Shatoury S, Abdullah H, El-Kraly O, El-Kazzaz W, Dewedar A (2006) Bioactivities of endophytic actinomycetes from selected medicinal plants in the World Heritage Site of Saint Katherine Egypt. Int J Bot 2(3):307–312

El-Shatoury S, El-Kraly O, El-Kazzaz W, Dewedar A (2009) Antimicrobial activities of actinomycetes inhabiting *Achillea Fragrantissima* (Family: Compositae). Egyptian J Nat Tox 6(2):1–15

El-Tarabily KA, Sivasithamparam K (2006) Potential of yeasts as biocontrol agents of soil-borne fungal plant pathogens and as plant growth promoters. Mycoscience 47:25–35

Ezra D, Castillo UF, Strobel GA, Hess WM, Porter H, Jensen JB, Condron MAM, Teplow DB, Sears J, Maranta M, Hunter M, Weber B, Yaver D (2004) Coronamycins peptide antibiotics produced by a verticillate *Streptomyces* sp (MSU-2110) endophytic on *Monstera* sp. Microbiology 150:785–793

Fisher PJ, Petrini O, Scott HML (1992) The distribution of some fungal and bacterial endophytesin maize (*Zea mays* L). New Phytol 122:299–305

Gardener JM, Feldman AW, Zablotowicz RM (1982) Identity and behavior of xylem-residing bacteria in rough lemon roots of Florida citrus tree. Appl Environ Microbiol 43:1335–1342

Giles JR, Palat CT, Chien SH, Chang ZG, Kennedy DT (2000) Evaluation of echinacea for treatment of common colds. Pharmacotherapy 20:690–697

Gordien AY, Gray AI, Ingleby K, Franzblau SG, Seidel V (2010) Activity of Scottish plant, lichen and fungal endophyte extracts against *Mycobacterium aurum* and *Mycobacterium tuberculosis*. Phytother Res 24(5):692–698

Gu QHL, Luo W, Zheng ZH, Liu Y, Huang H (2006) *Pseudonocardia oroxyli* sp nov isolated from *Oroxylum indicum* root. Int J Syst Evol Microbiol 56:2193–2197

Gunatilaka AAL (2006) Natural products from plant-associated microorganisms: distribution structural diversity bioactivity and implications of their occurrence. J Nat Prod 69:509–526

Guo B, Wang Y, Sun X, Tang K (2008) Bioactive natural products from endophytes: a review. Appl Biochem Microbiol 44:153–158

Hallmann J, Rodriguez-Kabana R, Kloepper JW, Quadt-Hallmann A, Mahaffee WF (1997a) Bacterial endophytes in agricultural crops. Can J Microbiol 43:895–914

Hallmann QA, Hallmann J, Kloepper JW (1997b) Bacterial endophytes in cotton: location and interaction with other plant associated bacteria. Can J Microbiol 43:254–259

Hasegawa S, Meguro A, Shimizu M, Nishimura T, Kunoh H (2006) Endophytic actinomycetes and their interactions with host plants. Actinomycetologica 20:72–81

Hazalin NAMN, Ramasamy K, Lim SM, Wahab IA, Cole ALJ, Majeed AB (2009) Cytotoxic and antibacterial activities of endophytic fungi isolated from plants at the National Park Pahang Malaysia. Complem Alt Med 9:46

Higashide E, Asai M, Ootsu KTS, Kozay Y, Hasegawa T, Kishi T, Sugino Y, Yoneda M (1977) Ansamitocins a group of novel maytansinoid antibiotics with antitumour properties from *Nocardia*. Nature 270:721–722

Igarashi Y (2004) Screening of novel bioactive compounds from plant-associated actinomycetes. Actinomycetologica 18:63–66

Igarashi Y, Iida T, Yoshida R, Furumai T (2002) Pteridic acids A and B novel plant growth promoters with auxin-like activity from *Streptomyces hygroscopicus* TP-A0451. J Antibiot 55:764–767

Igarashi Y, Miura S, Fujita T, Furumai T (2006) Pterocidin a cytotoxic compound from the endophytic *Streptomyces hygroscopicus*. J Antibiot 59:193–195

Igarashi Y, Rujillo ME, Martínez-Molina E, Miyanaga S, Obata T, Sakurai H, Saiki I, Fujita T, Furumai T (2007a) Antitumor anthraquinones from an endophytic actinomycete *Micromonospora lupini* sp nov. Bioorg Med Chem Lett 17:3702–3705

Igarashi Y, Trujillo ME, Martínez-Molina E, Yanase S, Miyanaga S, Obata T, Sakurai H, Saiki I, Fujita T, Furumai T (2007b) Antitumor anthraquinones from an endophytic actinomycete *Micromonospora lupini* sp. nov. Bioorg Med Chem Lett 17(13):3702–3705

Janso JE, Carter GT (2010) Biosynthetic potential of phylogenetically unique endophytic actinomycetes from tropical plants. Appl Environ Microbiol 76:4377–4386

Kado CI (1992) Plant pathogenic bacteria. In: Balows A, Truper HG, Dworkin M, Harder W, Schleifer KH (eds) Prokaryotes. Springer, New York

Kajula M, Tejesvi MV, Kolehmainen S, Mäkinen A, Hokkanen V, Mattila S, Pirttilä AM (2010) The siderophore ferricrocin produced by specific foliar endophytic fungi in vitro. Fungal Biol 114(2–3):248–254

Karsten K, Umar F, Ulrich F, Barbara S, Siegfried D, Gennaro P, Piero S, Sándor A, Tibor K (2007) Secondary metabolites isolated from an Endophytic *Phoma* sp absolute configuration of tetrahydropyrenophorol using the solid-state TDDFT CD methodology. Eur J Org Chem 19:3206–3211

Kharwar RN, Verma VC, Strobel G, Ezra D (2008) The endophytic fungal complex of *Catharanthus roseus* (L) G Don. Curr Sci 95(2):228–233

Kim N, Shin JC, Kim W, Hwang BY, Kim BS, Hong YS, Lee D (2006) Cytotoxic 6-alkylsalicylic acids from the endophytic *Streptomyces laceyi*. J Antibiot 59:797–800

Kupchan SM, Komoda Y, Court WA, Thomas GJ, Smith RM, Karim A, Gilmore CJ, Haltiwanger RC, Bryan RF (1972) Maytansine a novel antileukaemic ansa macrolide from *Maytenus ovatus*. J Am Chem Soc 94:1355–1356

Laorpaksa A, Jianmongkol S, Pothiwong W (2008) Antimicrobial activity of endophytic bacteria isolated from Thai medicinal plant. Thai J Pharm Sci 32:21–32

Lata H, Li XC, Silva B, Moraes RM, Halda-Alija L (2006) Identification of IAA-producing endophytic bacteria from micropropagated *Echinacea* plants using 16S rRNA sequencing. Plant Cell Tissue Org Cul 85:353–359

Lazzarini A, Cavaletti L, Toppo G, Marinelli F (2000) Rare genera of actinomycetes as potential producers of new antibiotics. Antonie Van Leeuwenhoek 78:399–405

Li JG, Shun XG (2009) Endophytic fungi from *Dracaena cambodiana* and *Aquilaria sinensis* and their antimicrobial activity. Afr J Biotechnol 8(5):731–736

Li J, Zhao GZ, Chen HH, Wang HB, Qin S, Zhu WY, Xu LH, Jiang CL, Li WJ (2008a) Antitumour and antimicrobial activities of endophytic *Streptomycetes* from pharmaceutical plants in rainforest. Lett Appl Microbiol 47:574–580

Li J, Zhao GZ, Chen HH, Qin S, Xu LH, Jiang CL, Li WJ (2008b) Rhodococcus cercidiphylli sp nov a new endophytic actinobacterium isolated from leaf of *Cercidiphyllum japonicum*. Syst Appl Microbiol 31:1108–1113

Li J, Lu CH, Shen YM (2010) Macrolides of the bafilomycin family produced by *Streptomyces* sp CS. J Antibiot. doi:10.1038/ja.2010.95

Lin ZJ, Lu XM, Zhu TJ, Fang YC, Gu QQ, Zhu WM (2008) GPR12 Selections of the metabolites from an endophytic *Streptomyces* sp. asociated with *Cistanches deserticola*. Arch Pharm Res 31:1108–1114

Liu JY, Huang LL, Ye YH, Zou WX, Guo ZJ, Tan RX (2006) Antifungal and new metabolites of *Myrothecium* sp Z16 a fungus associated with white croaker *Argyrosomus argentatus*. J Appl Microbiol 100:195–202

Lu CH, Shen YM (2003) A new macrolide antibiotics with antitumor activity produced by *Streptomyces* sp CS a commensal microbe of Maytenus hookeri. J Antibiot 56:415–418

Lu CH, Shen YM (2004) Two new macrolides produced by *Streptomyces* sp. CS. J Antibiot 57:597–600

Lu CH, Shen YM (2007) A novel ansamycin naphthomycin K from *Streptomyces* sp. J Antibiot 60:649–653

Mahafee WF, Kloepper JW (1997) Changes in the bacterial communities of soil rhizosphere and endorhiza associated with field grown cucumber (*Cucumis sativus* L). Microb Ecol 34:210–223

Mann J (2001) Natural products as immunosuppressive agents. Nat Prod Rep 18:417–430

McInroy JA, Kloepper JW (1995a) Population dynamics of endophytic bacteria in field-grown sweet corn and cotton. Can J Microbiol 41:95–901

McInroy JA, Kloepper JW (1995b) Survey of indigenous bacterial endophytes from cotton and sweet corn. Plant Soil 173:337–342

Mundt JO, Hinkle NF (1976) Bacteria with in ovules and seeds. Appl Environ Microbiol 32:694–698

Nejad P, Johnson PA (2000) Endophytic bacteria induce growth promotion and wilt disease suppression in oilseed rape and tomato. Biol Control 18:208–215

Okazaki T (2003) Studies on actinomycetes isolated from plant leaves In: Kurtböke DI (ed) Selective isolation of rare actinomycetes. Queensland Complete Printing Service, Australia

Okazaki T, Takahashi K, Kizuka M, Enokita R (1995) Studies on actinomycetes isolated from plant leaves. Annu Rev Sankyo Res Lab 47:97–106

Oldfield C, Wood NT, Gilbert SC, Murray FD, Faure FR (1998) Desulphurisation of benzothiophene and dibenzothiophene by actinomycete organisms belonging to the genus *Rhodococcus* and related taxa. Antonie Van Leeuwenhoek 74:119–132

Petrini OTN, Sieber LT, Viret O (1992) Ecology metabolite production and substrate utilization in edophytic fungi. Natl Toxin 1:185–196

Powel RG, Weisleder D, Smith CR, Kozlowski J, Rohwedder WK (1982) Treflorine, trenudine, and Nmethyltrenudone: novel maytansinoids tumour inhibitors containing two fused macrocyclic rings. J Am Chem Soc 104:4929–4934

Pullen C, Schmitz P, Meurer K, Bamberg DD, Lohmann S, Franc SDC, Groth I, Schlegel B, Möllmann U, Gollmick F, Gräfe U, Leistner E (2002) New and bioactive compounds from streptomyces strains residing in the wood of Celastraceae. Planta 216:162–167

Qin S, Wang HB, Chen HH, Zhang YQ, Jiang CL, Xu LH, Li WJ (2008) *Glycomyces* endophytic sp nov an endophytic actinomycete isolated from the root of *Carex baccans* Nees. Int J Syst Evol Microbiol 58:2525–2528

Qin S, Chen HH, Klenk HP, Zhao GZ, Li J, Xu LH, Li WJ (2009a) Glycomyces scopariae sp nov and Glycomyces mayteni sp nov isolated from two medicinal plants in China. Int J Syst Evol Microbiol 59:1023–1027

Qin S, Jie L, Hua-Hong C, Guo-Zhen Z, Wen-Yong Z, Cheng-Lin J, Li-Hua X, Wen-Jun L (2009b) Isolation diversity and antimicrobial activity of rare actinobacteria from medicinal plants of tropical rain forests in xishuangbanna China. App Environ Microbiol 75:6176–6186

Qin S, Xing K, Jiang J, Xu L, Li W (2011a) Biodiversity bioactive natural products and biotechnological potential of plant-associated endophytic actinobacteria. Appl Microbiol Biotechnol 89:457–473

Qin S, Xing K, Fei S, Lin Q, Chen X, Cao C, Sun Y, Wang Y, Li W, Jiang J (2011b) *Pseudonocardia sichuanensis* sp nov a novel endophytic actinomycete isolated from the root of *Jatropha curcas* L. Antonie Van Leeuwenhoek 99:395–401

Randa A, Kirstin S, Hans-Martin D, Isabel S, Christian H (2010) Botryorhodines A–D antifungal and cytotoxic depsidones from Botryosphaeria rhodina an endophyte of the medicinal plant Bidens pilosa. Photochemistry 71(1):110–116

Raviraja NS (2005) Fungal endophytes in five medicinal plant species from Kudremukh Range Western Ghats of India. J Bas Microbiol 45(3):230–235

Rukachaisirikul T, Saekee A, Tharibun C, Watkuolham S, Suksamrarn A (2007) Biological activities of the chemical constituents of *Erythrina stricta* and *Erythrina subumbrans*. Arch Pharm Res 30(11):1398–1403

Ryan RP, Germaine K, Franks A, Ryan DJ, Dowling DN (2008) Bacterial endophytes: recent development and applications. FEMS Micribiol Lett 278:1–9

Schwarz M, Köpcke B, Weber RWS, Sterner O, Anke H (2004) 3-Hydroxypropionic acid as a nematicidal principle in endophytic fungi. Phytochemistry 65:2239–2245

Sessitsch A, Reiter B, Berg G (2004) Endophytic bacterial communities of field-grown potato plants and their plant-growth promoting and antagonistic abilities. Can J Microbiol 50:239–249

Sette LD, Passarini MRZ, Delarmelina C, Salati F, Duarte MCT (2006) Molecular characterization and antimicrobial activity of endophytic fungi from coffee plants. J Microbiol Biotech 22:1185–1195

Sharrock KR, Parkes SL, Jack HK, Rees-George J, Hawatthorne BT (1991) Involvement of bacterial endophytes in storage roots of buttercup squash (*Cucurbita maxima* D hybrid 'Delica'). NSJ Crop Hortic Sci 19:157–165

Shipunov A, Newcombe G, Raghavendra AKH, Anderson CL (2008) Hidden diversity of endophytic fungi in an invasive plant. Am J Bot 95(9):1096–1108

Siciliano SD, Thoreat CM, de Freitas JR, Huci PJ, Germida JJ (1998) Differences in the microbial communities associated with the roots of different cultivars of canola and wheat. Can J Microbiol 44:844–851

Smith SA, Tank DC, Boulanger LA, Bascom-Slack CA, Eisenman K, Kingery D, Babbs B, Fenn K, Greene JS, Hann BD, Keehner J, Kelley-Swift EG, Kembaiyan V, Lee SJ, Li P, Light DY, Lin EH, Ma C, Moore E, Schorn MA, Vekhter D, Nunez PV, Strobel GA, Donoghue MJ, Strobel SA (2008) Bioactive endophytes warrant intensified exploration and conservation. PLoS ONE 3(8):e3052. doi:10.1371/journal.pone.0003052

Snipes CE, Duebelbeis DO, Olson M, Hahn DR, Dent WH, Gilbert JR, Werk TL, Davis GE, Lee-Lu R, Graupner PR (2007) The ansacarbamitocins: polar ansamitocin derivatives. J Nat Prod 70(10):1578–1581

Staniek A, Woerdenbag HJ, Kayser O (2008) Endophytes: exploiting biodiversity for the improvement of natural product-based drug discovery. J Plant Interact 3:75–93

Stierle A, Strobel G, Stierle D (1993) Taxol and taxane production by *Taxomyces andreanae* an endophytic fungus of Pacific yew. Science 260:214–216

Stone JK, Bacon CW, White JF (2000) An overview of endophytic microbes: endophytism defined. In: Bacon CW, White JF (eds) Microbial endophytes. Marcel Dekker Inc, New York, pp 3–29

Strobel G, Daisy B (2003) Bioprospecting for microbial endophytes and their natural products. Microbiol Mol Biol Rev 67:491–502

Strobel GA, Miller RV, Martinez-Miller C, Condron MM, Teplow DB, Hess WM (1999) Cryptocandin a potent antimycotic from the endophytic fungus *Cryptosporiopsis quercina*. Microbiology 145:1919–1926

Strobel G, Daisy B, Castillo U, Harper J (2004) Natural products from endophytic microorganisms. J Nat Prod 67:257–268

Sturz AV (1995) The role of endophytic bacteria during seed piece decayan and potato tuberisation. Plant Soil 175:257–263

Surette MA, Sturz AV, Lada RR, Nowak J (2003) Bacterial endophytes in processing carrots (*Daucus carota* L var sativus): their localization population density biodiversity and their effects on plant growth. Plant Soil 253:381–390

Suryanarayanan TS, Wittlinger SK, Faeth SH (2005) Endophytic fungi associated with cacti in Arizona. Mycol Res 109(5):635–639

Suwanborirux K, Chang CJ, Spjut RW, Cassady JM (1990) Ansamitocin P-3 a maytansinoid from *Claopodium crispifolium* and *Anomodon attenuatus* or associated actinomycetes. Experientia 46:117–120

Synnes M (2007) Bioprospecting of organisms from the deep sea: Scientific and environmental aspects. Clean Tech Environ Policy 9:53–59

Taechowisan T, Peberdy JF, Lumyong S (2003) Isolation of endophytic actinomycetes from selected plants and their antifungal activity. World J Microbiol Biotechnol 19:381–385

Taechowisan T, Lu C, Shen Y, Lumyong S (2005) Secondary metabolites from endophytic *Streptomyces aureofaciens* CMUAc130 and their antifungal activity. Microbiology 151:1691–1695

Taechowisan T, Wanbanjob A, Tuntiwachwuttikul P, Taylor WC (2006) Identification of *Streptomyces* sp. Tc022 an endophyte in *Alpinia galangal* and the isolation of actinomycin D. Ann Microbiol 56(2):113–117

Taechowisan T, Lu CH, Shen YM, Lumyong S (2007a) 4-arylcoumarin inhibits immediate-type allergy. Food Agric Immunol 18:203–211

Taechowisan T, Lu CH, Shen YM, Lumyong S (2007b) Antitumor activity of 4-arylcoumarins from endophytic *Streptomyces aureofaciens* CMUAc130. J Cancer Res Trer 3:86–91

Taylor TN, Taylor EL (2000) The rhynie chert ecosystem: a model for understanding fungal interactions In: Bacon CW, White JF (eds). Microbial endophytes. Marcel Decker Inc, New York, pp 31–48

Tejesvi MV, Nalini MS, Mahesh B, Prakash HS, Kini KR, Shetty HS, Ven S (2007) New hopes from endophytic fungal secondary metabolites. Bol Soc Quím Méx 1(1):19–26

Tian XL, Cao LX, Tan HM, Han WQ, Chen M, Liu YH, Zhou SN (2007) Diversity of cultivated and uncultivated actinobacterial endophytes in the stems and roots of rice. Microb Ecol 53:700–707

Tuntiwachwuttikul P, Taechowisan T, Wanbanjob A, Thadaniti S, Taylor WC (2008) Lansai A–D secondary metabolites from Streptomyces sp SUC1. Tetrahedron 64:7583–7586

Vanessa MC, Christopher MMF (2004) Analysis of the endophytic actinobacterial population in the roots of wheat (*Triticum aestivum* L) by terminal restriction fragment length polymorphism and sequencing of 16S rRNA Clones. Appl Environ Microbiol 70:31787–31794

Velazquez E, Rojas M, Lorite MJ, Rivas R, Zurdo-Pineiro JL, Heydrich M, Bedmar EJ (2008) Genetic diversity of endophytic bacteria which could be find in the apoplastic sap of medullary parenchym of the stem of healthy sugarcane plants. J Basic Microbiol 48:118–124

Vendan RT, Yu YJ, Lee SH, Rhee YH (2010) Diversity of endophytic bacteria in ginseng and their potential for plant growth promotion. J Microbiol 48(5):559–565

Vennila R, Thirunavukkarasu SV, Muthumary J (2010) Evaluation of fungal taxol isolated from an endophytic fungus *Pestalotiopsis pauciset* AVM1 against experimentally induced breast cancer in sprague dawley rats. Res J Pharmacol 4(2):38–44

Verma VC, Gond SK, Kumar A, Mishra A, Kharwar RN, Gange AC (2009a) Endophytic actinomycetes from *Azadirachta indica* A Juss: isolation diversity and anti-microbial activity. Microb Ecol 57:749–756

Verma VC, Kharwar RN, Strobel GA (2009b) Chemical and functional diversity of natural products from plant associated endophytic fungi. Nat Prod Commun 11:1511–1532

Wang HX, Geng ZL, Zeng Y, Shen YM (2008) Enrichment plant microbiota for a metagenomic library construction. Environ Microbiol 10:2684–2691

Wei J, Xu T, Guo L, Liu A, Zhang Y, Pan X (2009) Endophytic *Pestalotiopsis* species associated with plants of *Podocarpaceae Theaceae* and *Taxaceae* in southern China. Fun Div 24:55–74

Ya-li LV, Fu-sheng Z, Juan C, Jin-long C, Yong-mei X, Xiang-dong L, Shun-xing G (2010) Diversity and antimicrobial activity of endophytic fungi associated with the *Alpine* plant *Saussurea involucrate*. Biol Pharm Bull 33(8):1300–1306

Yang X, Strobel G, Stierle A, Hess WM, Lee J, Clardy J (1994) A fugal endophyte-tree relationship: *Phoma* sp in *Taxus wallachiana*. Plant Sci 102:1–9

Yuan HM, Zhang XP, Zhao K, Zhong K, Gu YF, Lindstrom K (2008) Genetic characterization of endophytic actinobacteria isolated from the medicinal plants in Sichuan. Ann Microbiol 58(4):597–604

Zhang HW, Song YC, Tan RX (2006) Biology and chemistry of endophytes. Prod Rep 23:753–771

Zhao PJ, Fan LM, Li GH, Zhu N, Shen YM (2005) Antibacterial and antitumor macrolides from isp ls9131. Arch Pharm Res 28:1228–1232

Zhao GZ, Li J, Huang HY, Zhu WY, Zhao LX, Tang SK, Xu LH, Li WJ (2010a) Pseudonocardia artemisiae sp anovel actinobacterium isolated from surface-sterilized *Artemisia annua* L. Int J Syst Evol Microbiol. doi:10.1099/ijs0021931-0

Zhao J, Mou Y, Shan T, Li Y, Zhou L, Wang M, Wang J (2010b) Antimicrobial metabolites from the endophytic fungus *Pichia guilliermondii* isolated from *Paris polyphylla* var. *yunnanensis*. Molecules 15:7961–7970

Zhao J, Zhou L, Wub J (2010c) Promotion of *Salvia miltiorrhiza* hairy root growth and tanshinone production by polysaccharide–protein fractions of plant growth-promoting rhizobacterium *Bacillus cereus*. Pro Biochem 45:1517–1522

Zhao K, Penttinen P, Guan T, Xiao J, Chen Q, Xu J, Lindstrom K, Zhang L, Zhang X, Strobel GA (2011) The diversity and anti-microbial activity of endophytic actinomycetes isolated from medicinal plants in Panxi Plateau China. Curr Microbiol 62:182–190

Zhen-Liang S, Ming Z, Ji-Fa Z, Jing F (2011) Antifungal and cytotoxic activities of the secondary metabolites from endophytic fungus *Massrison* sp. Phytomedicine 18(15):859–862

Zhou X, Zhu H, Liu L, Lin J, Tang K (2010) A review: recent advances and future prospects of taxol-producing endophytic fungi. Chem Mater Sci 86(6):1707–1717

Zhu N, Zhao P, Shen Y (2009) Selective isolation and ansamycintargeted screenings of commensal actinomycetes from the "maytansinoids-producing" arboreal *Trewia nudiflora*. Curr Microbiol 58:87–94

Zin NM, Sarmin NIM, Ghadin N, Basri DF, Sidik NM, Hess WM, Strobel GA (2007) Bioactive endophytic *Streptomycetes* from the Malay Peninsula. FEMS Microbiol Lett 274:83–88

Chapter 5
The Role of Biotechnology in *Cannabis sativa* Propagation for the Production of Phytocannabinoids

Suman Chandra, Hemant Lata, Ikhlas A. Khan and Mahmoud A. ElSohly

5.1 Introduction

Cannabis sativa L. (Cannabaceae) is one of the oldest plant sources of medicine, food, and textile fiber (Doyle and Spence 1995; Kriese et al. 2004; Zuardi 2006). Hemp cultivation for fiber originated in western Asia and Egypt, and was subsequently introduced to Europe between 2000 and 1000 BC. Cultivation of hemp in Europe became widespread around 500 AC. The crop was brought to South America (Chile) in 1545 and North America (Port Royal, Canada) in 1606 (Small and Marcus 2002). Currently, the cultivation of cannabis, including hemp, is prohibited in the United States under the federal law.

Cannabis has a long history of medicinal use in the Middle East and Asia, dating back to the sixth century BC, while it was introduced in Western Europe as

S. Chandra (✉) · H. Lata · I. A. Khan · M. A. ElSohly
National Center for Natural Product Research, Research Institute of Pharmaceutical Sciences, School of Pharmacy, University of Mississippi, University, MS 38677, USA
e-mail: suman@olemiss.edu

H. Lata
e-mail: hlata@olemiss.edu

I. A. Khan
e-mail: ikhan@olemiss.edu

M. A. ElSohly
e-mail: melsohly@olemiss.edu

I. A. Khan
Department of Pharmacognosy, School of Pharmacy, University of Mississippi, University, MS 38677, USA

M. A. ElSohly
Department of Pharmaceutics, School of Pharmacy, University of Mississippi, University, MS 38677, USA

S. Chandra et al. (eds.), *Biotechnology for Medicinal Plants*,
DOI: 10.1007/978-3-642-29974-2_5,

a medicine in the early nineteenth century to treat epilepsy, tetanus, rheumatism, migraine, asthma, trigeminal neuralgia, fatigue, and insomnia (Doyle and Spence 1995; Zuardi 2006). Cannabis is valued for its hallucinogenic and medicinal properties, having been used to treat a variety of ailments, including pain (Guindon and Hohmann 2009), glaucoma (Jarvinen et al. 2002), nausea (Slatkin 2007), depression (Viveros and Marco 2007), and neuralgia (Liang et al. 2004).

C. sativa contains a unique class of terpenophenolic compounds (the cannabinoids), which accumulates mainly in the glandular trichomes of the plant (Hammond and Mahlberg 1977). Approximately 110 cannabinoids have been isolated from the plant, the major biologically active compound being Δ^9-tetrahydrocannabinol, commonly referred as THC (Mechoulam and Ben-Shabat 1999). Besides its psychoactivity, THC possesses analgesic, antiinflammatory, appetite stimulant, and antiemetic properties making this compound a very promising therapeutic agent, especially for cancer and AIDS patients (Sirikantaramas et al. 2005). The pharmacologic and therapeutic potency of preparations of cannabis and its main active constituent (THC) has been extensively reviewed (Mechoulam 1986; Formukong et al. 1989; Grinspoon and Bakalar 1993; Mattes et al. 1993, 1994; Brenneisen et al. 1996; Pryce and Baker 2005; Abrams et al. 2007).

Because of the high commercial value of THC in the pharmaceutical area, the production cost of THC as the bulk active pharmaceutical becomes an important factor in its development. Since *C. sativa* is a source of THC, efforts to select *Cannabis* varieties with high THC content are warranted. However, due to the allogamous (cross-fertilization) nature of the species, it is very difficult to maintain the chemical profile of selected chemotypes under field conditions. Considering these limitations, indoor cultivation of a screened and selected high yielding cultivars/clones under controlled environmental conditions using biotechnological tools is the most suitable way to maintain the genetic identity and chemical profile of the plant. This chapter describes the role of biotechnology and our efforts to propagate *Cannabis sativa* for the production of phytocannabinoids. This includes, screening of high yielding genotypes based on their chemical profile, propagation of these genotypes using biotechnological tools, comparison of micropropagated plants with the mother plants for consistency of chemical and genetic profiles and the utility of micropropagation in the conservation of elite clones for future use.

5.2 *Cannabis sativa*: A Dioecious Plant

Cannabis is normally a dioecious species, i.e., male and female flowers develop on separate plants. Sex is determined by heteromorphic chromosomes, with males being heterogametic (XY) and females homogametic (XX). Morphologically, it is difficult to identify male and female plants at vegetative stage since sexual dimorphism occurs late in plant development. However, after the onset of flowering, male plants can be differentiated from female plants. The sexual phenotype of cannabis occasionally shows flexibility leading to the differentiation of hermaphrodite flowers or bisexual inflorescences, i.e., a monoecious phenotype. Molecular techniques can be employed to differentiate male from female plants at an early stage (Sakamoto et al. 1995, 2005; Mandolino et al. 1999; Flachowsky et al. 2001; Törjék 2002; Techen et al. 2010).

5.3 The Species Debate

The number of species in the *Cannabis* genus had been a controversial subject, with a number of reports proposing it a polytypic genus (Emboden 1974; Hillig 2004, 2005; Hillig and Mahlberg 2004). However, at present, based on morphological, anatomical, phytochemical, and genetic studies, the family Cannabaceae is currently recognized as containing only one genus, namely *Cannabis*, which includes a single, highly polymorphic species, *Cannabis sativa* L. (Small 1975a, b; Small and Cronquist 1976; Gilmore et al. 2003). Other previously reported species include *Cannabis indica* Lam. and *Cannabis ruderalis* Janisch. Plants considered to have belonged to these species are now recognized as varieties of *C. sativa* L. (var. *indica* and var. *ruderalis*, respectively).

5.3.1 Sativa Versus Indica

The main morphological difference between the *indica* and *sativa* varieties is in their leaves. Leaves of sativa variety appear much smaller and thinner than *indica*. Leaves of *indica* variety have wide leaflets and are deep green, often tinged with purple. At maturity, *indica* leaves turn dark purple. The *indica* variety has shorter and bushier plants (usually under 6 ft tall and rarely over 8 ft tall plants) compared to *sativa*. The plants have short branches laden with thick, dense buds, which mature early, usually at the beginning of September. The *indica* buds also vary in color from dark green to purple, with cooler condition inducing more intense coloration. Plants of *indica* variety flower early as compared to *sativa*. Natural distribution of the *indica* variety is reported around Afghanistan, Pakistan, India, and the surrounding areas.

Plants of the *sativa* variety grow about 6 ft to more than 20 ft high with 8–12 ft being the most common height. *Sativa* leaves are long with thin leaflets, and are light green, especially equatorial varieties, which have less chlorophyll and more yellow pigments in order to protect the plant from intense light. Temperate varieties are darker green as compared to the tropical plants. The leaves of some *sativa* varieties turn yellow and fall off during maturity. The plants of *sativa* variety have long branches, with the lower branches spreading 4 ft or more from the central stalk, forming a conical "Christmas tree" shaped plants. Buds of *sativa* variety are long and thin, far less densely populated than the *indica*, although longer in length (sometimes stretching up to 3 ft or more). The stomas of the flowering bud may be tinged slightly purple in a cool climate, but in a warmer environment will turn dark orange or red. Maturation time varies considerably depending on weather. Some low THC varieties mature in August and September, while equatorial varieties mature from October to December. Buds of *sativa* require intense light to thicken and swell as compared to *indica*. The biomass produced by *sativa* plants is generally categorized as higher in Δ^9-THC and lower in Cannabidiol (CBD) than that from *indica* plants. Plants of *sativa* variety are found all over the world and include most of the drug type equatorial varieties, such as Colombian, Mexican, Nigerian, and South African where marijuana plants can be very potent.

5.4 Chemotaxonomy

Cannabis is considered a chemically complex species based on its numerous natural constituents (ElSohly and Slade 2005). The concentration of Δ^9-THC in the dried inflorescence (leaves and buds) is used to determine its psychoactivity. Quantitative and qualitative analysis of cannabis can be employed to characterize its phenotype and phytocannabinoid profile (De Meijer et al. 2003).

Cannabis can be divided into three phenotypes: phenotype I (drug-type), with Δ^9-THC > 0.5 % and CBD < 0.5 % (Δ^9-THC/CBD $>>$ 1); an intermediate phenotype II (intermediate-type), with CBD as the major cannabinoid but with Δ^9-THC also present at various concentrations (Δ^9-THC/CBD $\sim$ 1); and phenotype III (fiber-type or hemp), with especially low Δ^9-THC content (Δ^9-THC/CBD $<<$ 1). Hemp usually contains nonpsychoactive cannabinoids as major constituents, e.g., CBD or Cannabigerol (CBG) (De Backer et al. 2009; Galal et al. 2009). Although environmental factors play a role in the amount of cannabinoids present in different parts of the plant at different growth stages (Bócsa et al. 1997), the distribution of CBD: Δ^9-THC ratios in most populations are under genetic control (De Meijer et al. 2003).

The phytocannabinoid content of cannabis is determined by the interaction of several genes, cultivation technique, and environmental factors (Chandra et al. 2008, 2010a, b; De Meijer et al. 1992, 2003; Hemphill et al. 1980; Mendoza et al. 2009). Numerous biotic and abiotic factors affect phytocannabinoid production, including the sex and maturity of the plant (Small et al. 1976), light cycle (Valle et al. 1978), temperature (Chandra et al. 2008), fertilization (Bócsa et al. 1997), and light intensity (Chandra et al. 2008). Variations in phytocannabinoid content in different tissues of a specific plant have also been reported (Hemphill et al. 1980) and are quite wide from large leaves to the female flowers bract.

5.5 Metabolites of *Cannabis sativa*

As stated above, the vast number of constituents present in *C. sativa* and their possible interaction with one another makes the chemistry of this species very complex. These compounds represent various chemical classes, e.g., mono- and sesquiterpenes, sugars, hydrocarbons, steroids, flavonoids, nitrogenous compounds, and amino acids, among others, with the most specific class being the C_{21} terpenophenolic cannabinoids. (−)-Δ^9-Trans-(6aR, 10aR)-tetrahydrocannabinol (Δ^9-THC) is the most psychologically active constituent (Mechoulam and Gaoni 1967). The development of synthetic cannabinoids and the discovery of chemically different endogenous cannabinoid receptor ligands (endocannabinoids) have prompted the use of the term "phytocannabinoids" to describe the plant-derived cannabinoids (Pate 1999). The total number of natural compounds identified in *C. sativa* L. was 423 (with 61 phytocannabinoids) in 1980 (Turner et al. 1980), 483 (with 66 phytocannabinoids) in 1995 (Ross and ElSohly 1995), and 489 (with 70 phytocannabinoids) in 2005 (ElSohly and Slade 2005). During the year 2008, 17 new cannabinoids, including esters and other derivatives have been isolated from high potency variety of Cannabis in our laboratory (Ahmed et al. 2008; Radwan et al. 2008). To date, a total number of 537 constituents (including 101 phytocannabinoids) could be accounted for in *C. sativa*, adding 48 new constituents (including 39 new cannabinoids) since the 2005 reviewed by ElSohly and Slade.

5.5.1 Biosynthesis of Phytocannabinoids

The biosynthesis of cannabinoids from *Cannabis sativa* is almost fully understood (Raharjo et al. 2004). Phytocannabinoids such as Δ^9-tetrahydrocannabinol, CBD, and CBC are well-known to be derived from their acid forms (Shoyama et al. 1975; Kajima and Piraux 1982; Fellermeier and Zenk 1998; Fig. 5.1). Two independent pathways, the cytosolic mevalonate and the plastidial methylerythritol phosphate (MEP), are responsible for plant terpenoid biosynthesis, with the latter responsible for the biosynthesis of the cannabinoid terpenoid moiety (Sirikantaramas et al. 2007). Olivetolic acid (OLA) and geranyl diphosphate (GPP) are derived from the polyketide and the deoxyxylulose phosphate (DOXP)/MEP pathways, respectively, followed by condensation under the influence of the prenylase, GPP: olivetolate geranyltransferase, yielding cannabigerolic acid (CBGA). CBGA, in turn, is oxidocyclized by flavin adenine dinucleotide-dependent oxidases, namely, cannabichromenic acid (CBCA) synthase, cannabidiolic acid (CBDA) synthase, and Δ^9-tetrahydrocannabinolic acid (Δ^9-THCA) synthase, producing and CBCA, CBDA, and Δ^9-THCA, respectively (Flores-Sanchez and Verpoorte 2008a, b). The first step in the biosynthesis of cannabinoids, the formation of OLA, has not been fully elucidated; however, it has been proposed that a type III polyketide synthase, olivetol synthase, could be involved (Taura et al. 2009).

5.5.2 The Phytocannabinoids

The chemistry and pharmacognosy of cannabis has been studied for more than a century, with the psychoactivity initially ascribed to (an) alkaloid(s). CBN was the first cannabinoid to be isolated (Wood et al. 1896, 1899) and identified (Adams

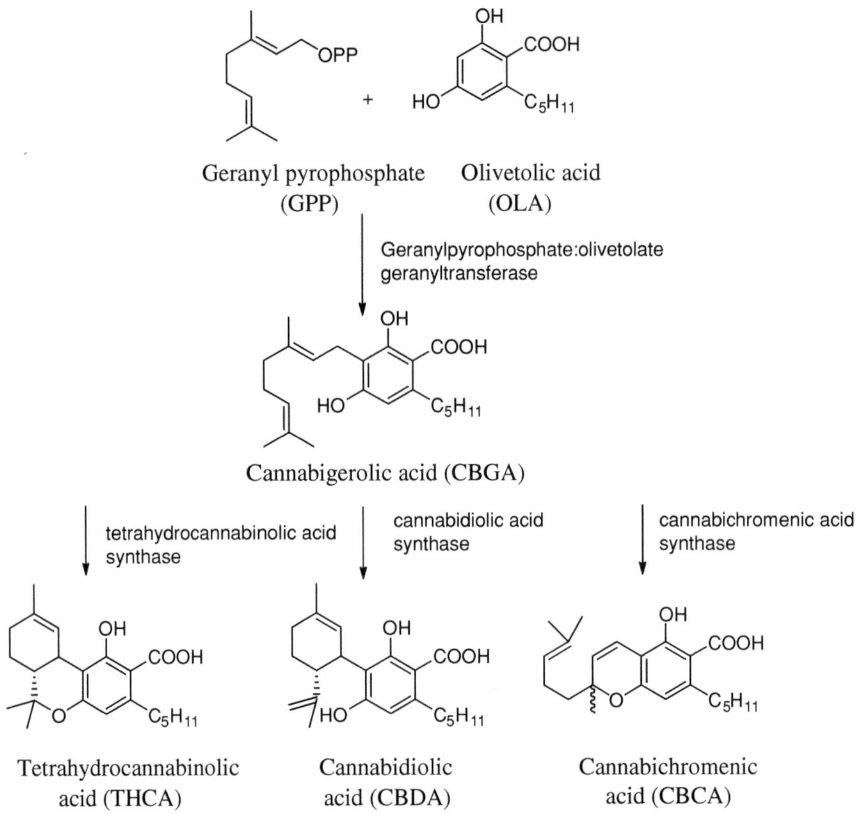

Fig. 5.1 Biosynthetic pathway of tetrahydrocannabinolic acid, cannabidiolic acid, and cannabichromenic acid

et al. 1940a; Cahn 1932; Ghosh et al. 1940) from *Cannabis sativa*. The elucidation of CBN led to speculation that the psychotropically active constituents of cannabis could be THCs.

The nonpsychotropic compound, CBD, was subsequently isolated from Mexican marijuana (Adams et al. 1940b) and the structure determined (Mechoulam and Shvo 1963). Gaoni and Mechoulam, two pioneers of cannabis research, determined the structure of Δ^9-THC after finally succeeding in isolating and purifying the elusive compound (Gaoni and Mechoulam 1964). The structure was elucidated via NMR analysis, derivatization, and partial synthesis from CBD, while the absolute configuration of CBD and Δ^9-THC were established by correlation with known terpenoids (Mechoulam and Gaoni 1967). Δ^9-THC, in contrast to other natural psychoactive compounds, e.g., the alkaloids morphine and cocaine, is highly hydrophobic, a characteristic that probably impeded full understanding of its mode of action until the discovery of the cannabinoid receptors (Devane et al. 1988; Matsuda et al. 1990).

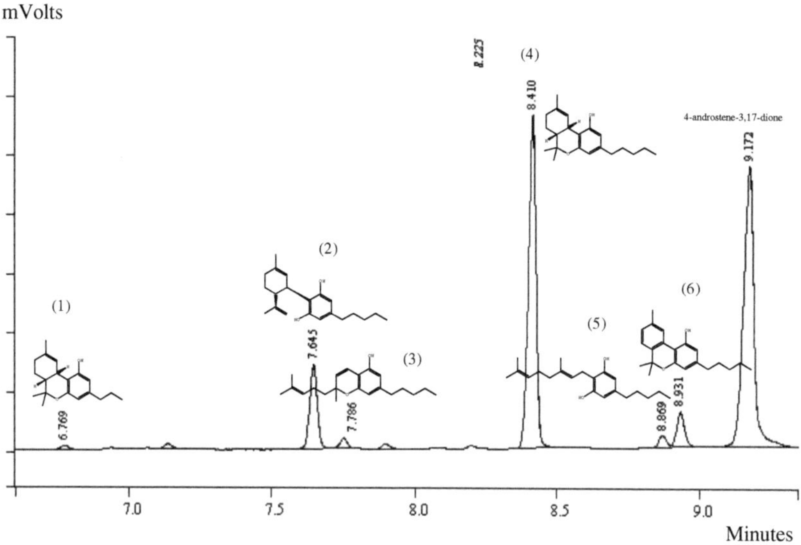

Fig. 5.2 GC-FID analysis of cannabis biomass. Peacks (1) THCV, (2) CBD, (3) CBC, (4) Δ^9-THC, (5) CBG and (6) CBN

Due to the development of synthetic cannabinoids (classical and nonclassical) and the discovery of the endocannabinoid system, which includes cannabinoid metabotropic receptors (CB1 and CB2) and endogenous lipid ligands (endocannabinoids) (Mouslech and Valla 2009), naturally occurring plant cannabinoids have been labeled phytocannabinoids. Exocannabinoid is also used to distinguish plant cannabinoids from animal cannabinoids (endocannabinoids), while classical and non-classical cannabinoids refer to synthetic cannabinoid receptor agonists (Makriyannis et al. 2005), indicating the relative degree of structural similarity with phytocannabinoids.

Phytocannabinoids are terpenoid hydrocarbons, which include a phenolic moiety (terpenophenols) with an n-alkyl side chain at C-3 derived from OLA (Fig. 5.1). A GC-FID and HPLC analysis of cannabis biomass showing major bioactive phytocannabinoids are shown in Figs. 5.2 and 5.3, respectively. Phytocannabinoids can be subdivided into eleven structural groups, with an additional miscellaneous group accounting for compounds that do not conform to any of these groups.

The isolation and structure elucidation of Δ^9-THC in 1964 stimulated intense phytochemical investigations of cannabis, which resulted in the identification of 61 phytocannabinoids by 1980 (Turner et al. 1980). Between 1981 and 2005, only nine new phytocannabinoids were characterized (ElSohly and Slade 2005; Ross and ElSohly 1995), probably due to the focus of cannabis research shifting toward the newly discovered endocannabinoid system. The availability of high-potency cannabis (Δ^9-THC > 10 %, w/w) (Mehmedic et al. 2010) sparked renewed interest in the phytochemistry of cannabis, leading to the discovery of 31 new phytocannabinoids between 2006 and 2010. The 101 known phytocannabinoids can be classified into 12 different groups as shown in Table 5.1.

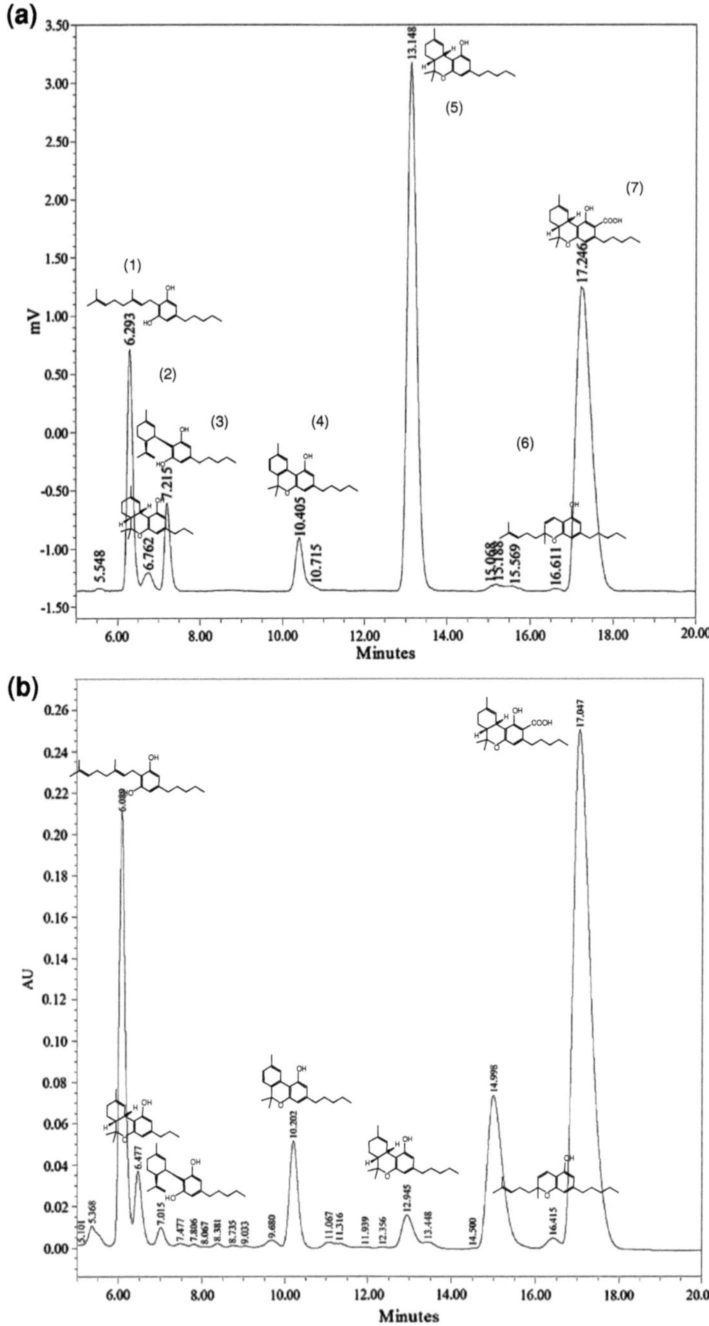

Fig. 5.3 HPLC analysis of cannabis extract using **a** ELS detector and **b** UV (254 nm) detector. Peaks (1) CBG, (2) THCV, (3) CBD, (4) CBN, (5) Δ^9-THC, (6) CBC and (7) Δ^9-THCA

Table 5.1 Total number of known phytocannabinoids in different groups of *Cannabis sativa*

No.	Cannabinoids groups	Number of known compound
1	CBG type	16
2	CBC type	8
3	CBD type	8
4	Δ^9-THC type	17
5	Δ^8-THC type	2
6	CBL type	3
7	CBE type	5
8	CBN type	10
9	CBND type	3
10	CBT type	9
11	Benzoquinone	2
12	Miscellaneous	18
	Total	101

CBG Cannabigerol, *CBC* Cannabichromene, *CBD* Cannabidiol, Δ^9 *-THC* Δ^9 -tetrahydrocannabinol, Δ^8 *-THC* Δ^8 -tetrahydrocannabinol, *CBL* Cannabicyclol, *CBE* Cannabielsoin, *CBN* Cannabinol, *CBND* Cannabinodiol, and *CBT* Cannabitriol

5.6 Selection of Desirable Cannabis Chemotypes

To screen a desirable elite chemotype, plants of different varieties of *Cannabis sativa* can be grown from the seeds in our laboratory. On flowering, male plants are separated from female plants to avoid cross pollination. Vegetative propagation or micropropagation (of nodal segments) is done from few randomly selected growing female plants (identified as mother plants) of each variety at vegetative growing stage. Biomass samples of mature mother plants were analyzed and chemotype of vegetatively grown plants were characterized based on the chemical profile of their respective mother plants.

Considering the complexity of the chemistry of cannabis and the variation in its chemical content, a simple and efficient method for qualitative and quantitative determination of the concentration of the major cannabinoids, namely Δ^9-THC, Tetrahydrocannabivarin (THCV), CBD, CBC, CBG, and CBN was developed using gas chromatography-flame ionization detection (GC-FID) (Ross et al. 1996). Biomass samples were dried at 120 °F and individually manicured by hand. Three 0.1 g samples were each extracted with 3 mL of internal standard/extracting solution (100 mg of 4-androstene-3, 17-dione + 10 mL chloroform + 90 mL methanol) at room temperature for 1 h. The extracts were withdrawn into disposable transfer pipettes through cotton plugs for filtration and are transferred into GC vials, which are then capped and placed on the auto sampler. 1 μL aliquots were injected. Gas chromatography analyses were performed using Varian CP-3380 gas chromatograph equipped with a Varian CP-8400 automatic liquid sampler, a capillary injector and flame ionization detector. The column was a 15 m × 0.25 mm DB-1, 0.25 μ film (J&W Scientific, Inc.). Data are recorded using a Dell Optiplex GX1 computer and Varian Star (version 6.41) workstation software. Helium is used as the carrier gas.

An indicating moisture trap and an indicating oxygen trap located in the helium line from upstream to downstream, respectively, were used. Helium was used as the "make-up" gas at the detector. Hydrogen and compressed air were used as the combustion gases. The instrument parameters used for monitoring samples are: Air—30 psi (400 mL/min); Hydrogen—30 psi (30 mL/min); column head pressure—14 psi (1.0 mL/min); split flow rate—50 mL/min; split ratio—50:1; septum purge flow rate—5 mL/min; make up gas pressure—20 psi (20 mL/min); injector temp—240 °C; detector temp—260 °C; initial oven temp—170 °C; initial temperature hold time—1 min; temperature rate—10 °C/min; final oven temperature—250 °C, and final temperature hold time—3 min. Currently, this method is being used by our group to analyze confiscated marijuana samples submitted by the US Drug Enforcement Administration (DEA) and other law enforcement agencies through a contract with the National Institute on Drug Abuse (NIDA) (Potency monitoring program at the University of Mississippi, ElSohly et al. 2000; Mehmedic et al. 2010).

5.7 Cultivation/Propagation of *Cannabis sativa*

Cannabis is an annual plant that can successfully be grown under indoor and outdoor conditions; however, both techniques have advantages and disadvantages. Under outdoor conditions, the life cycle of the plant completes in 5–7 months depending on the time of plantation and the variety. With indoor growing, flowering, and maturation can be triggered by regulation of the photoperiod.

5.7.1 Outdoor Cultivation

Worldwide cultivation of cannabis continues to be predominantly outdoors, with the plants generally, but not always, grown from seeds. Outdoor cultivation normally starts late March to early April (temperature dependent) and lasts until November to early December (variety dependent) in the Northern Hemisphere. Starting from seeds, plants may be raised in small biodegradable pots and the selected healthy seedlings or seeds may be directly transplanted in the field.

Male flowers, and subsequently female flowers, appear within 2–3 months (middle of July) of seeding/transplanting. Male plants are generally culled for several reasons: (1) male plants contain less Δ^9-THC compared to female plants, (2) to avoid pollination within a variety which produces seeds in mature crop, resulting in less biomass, and consequently lower Δ^9-THC content compared to seedless (sinsemilla) mature plants, (3) to avoid cross pollination among different varieties.

The Δ^9-THC content of the plant increases with maturation, reaching a maximum at the full budding stage, followed by a constant level before the onset of senescence. The maturity of the crop is determined through visual inspection and chemical analysis, e.g., GC-FID, of samples collected at different growth stages. Mature upper buds are harvested first, while the lower branches are given more time to be developed. Outdoor cultivated plants have more biomass compared to indoor grown plants; however, due to the allogamous (cross-fertilization) nature of cannabis, it is difficult to maintain a constant chemical profile under outdoor conditions if grown from seeds.

5.7.2 Indoor Cultivation

Indoor cultivation of cannabis under controlled environmental conditions allows for complete control of the plant life cycle, resulting in management of the quality and quantity of biomass. This is important when producing cannabis with a specific chemical profile for pharmaceutical use.

5.7.2.1 Environmental Parameters

The following environmental parameters are critical for indoor production:

Photosynthetic Photon Flux Density and Photoperiod

Cannabis requires high photosynthetic photon flux density (PPFD) for photosynthesis and growth (Chandra et al. 2008). Different light sources can be used for indoor propagation, namely, fluorescent light bulbs for juvenile cuttings, and metal halide (MH) and/or high pressure sodium (HPS) bulbs for established plants. Separate ballasts are required to regulate MH and HPS bulbs. MH and HPS bulbs should be 3–4 ft from the plants to avoid overexposure. 12 and 18 h photoperiods are optimum for initiation of flowering and vegetative growth, respectively.

Humidity

Humidity plays a crucial role in plant growth, starting at seed germination or vegetative propagation/reproduction to budding and harvesting. Juvenile plants require high humidity (ca. 75 %), vegetative cuttings require a regular water spray on the leaves to maintain a high humidity in its microclimate until the plants are well rooted, while the active vegetative and flowering stages require 55–60 % humidity.

Temperature

Depending on the original growth habitat and genetic makeup, the temperature response of photosynthesis varies with variety; however, 25–30 °C growth temperature is found to be ideal for most varieties (Chandra et al. 2008, 2011a).

Carbon Dioxide

Doubling the ambient Carbon Dioxide (CO_2) concentration stimulated the rate of photosynthesis and water use efficiency in cannabis by 50 and 111 %, respectively, resulting in increased overall growth (Chandra et al. 2008, 2011b).

Irrigation

The amount and frequency of watering for cannabis is depended on numerous factors such as growth stage, size of the plants and pots, temperature, humidity, etc. During the early seedling or vegetative stage, it is recommended to keep the soil moist; however, the top layer of soil should be dry before established plants are watered.

Air Circulation

Regulation of gas and water vapor exchange rates between leaves and the microenvironment is achieved by having air flow around the leaf surface. This affects the leaf boundary layer thermal conductance, energy budget, and ultimately, the physiology and growth of the plant (Chandra et al. 2008). Electric fans can be employed to manage air circulation for indoor grown plants.

5.7.2.2 Propagation Through Seeds

Propagation through seed is the most popular conventional method of cultivating cannabis. Seeds can be planted in small biodegradable pots (2 in height × 2 in diameter) containing moist, aerated soil. Seedling heat mats can be placed underneath the pots to increase soil temperature and to enhance germination in the winter. Seed germination starts after 4 days, with most of the viable seeds germinating within 15 days, depending on seed variety, age and storage condition, and soil and water temperature. Seedlings should be exposed to cool fluorescent light (18 h photoperiod), followed by transplantation to bigger pots (12 × 12 in) exposed to full spectrum grow light (18 h photoperiod). Plants can be exposed to 12 h photoperiod when appropriate vegetative growth has been achieved, which induces flowering in about 2 weeks. Male flowers appear before female flowers, allowing for separation or culling of the male plants if sinsemilla buds are required. Cuttings from the lower branches of female plants can be used for additional vegetative propagation. If a specific chemical profile is required, chemical analysis can be performed to aid in selection of the appropriate plants.

5.7.2.3 Propagation Through Vegetative Cuttings

Upon screening and selection of a particular clone, a fresh nodal segment (6–10 cm) containing at least two nodes, from the mother plant can be used for vegetative/

Fig. 5.4 **a** Fully rooted in vitro propagated plant of *Cannabis sativa,* **b** Tissue culture raised plants under controlled climatic conditions at vegetative stage, **c**, **d** Hydroponically grown plants at budding stage, **e** Field cultivation of identical clones from a single mother plant, **f** A field grown plant at budding (ready to harvest) stage, **g** Processed *Cannabis sativa* biomass (leaves and buds) for the extraction of phytocannabinoids

conventional propagation in solid (soil) or liquid (hydroponics) medium. In vitro vegetative micropropagation can also be used for mass-propagation of cannabis.

Hydroponics Propagation

A small branch consisting of a growing tip with two or three leaves is cut and immediately dipped in distilled water. Prior to dipping the cutting in a rooting compound, a fresh cut is made just above the first cut. The cuttings are inserted one inch deep into a rockwool cube or a hydrotone clay ball supporting medium. Plants are supplied with vegetative fertilizer formula (Advance Nutrient, Canada) and exposed to a diffused light: dark cycle (18:6) for vegetative growth. Rooting initiates in 2–3 weeks, followed by transplantation to a bigger hydroponic system (Fig. 5.4c and d).

Propagation in Soil

A soft apical branch is cut at a 45° angle immediately below a node and immediately dipped in distilled water. The base of the cutting (2 cm) is subsequently dipped in rooting hormone (Green Light, USA) and planted in pots (2 × 2 in) containing coco natural growth medium and a mixture (1:1) of sterile potting mix and fertilome (Canna Continental, USA). At least one node is covered by soil for efficient rooting. Plants are regularly irrigated and kept under controlled environmental conditions. Rooting initiates in 2–3 weeks, followed by transplantation to bigger pots (12 × 12 in) after 6 weeks. The cuttings can be maintained in a constant vegetative state (18 h photoperiod).

5.7.3 In vitro Propagation

Due to the allogamous nature of Cannabis, seeds raised plants are highly heterozygous and vegetative propagation of a selected mother plant can only produce a certain number of cuttings at a time, thus presenting difficulties when large-scale production of cannabis is needed. Plant tissue culture is recognized as one of the key areas of biotechnology because of its potential use to regenerate elite clones and conserve valuable plant genetic resources. Plant tissue culture offer advantages for multiplication and large-scale production of Cannabis plants.

5.7.3.1 Propagation Through Nodal Segments

In order to maintain true to type high yielding elite clones of *C. sativa*, plant regeneration via direct organogenesis is the preferred means. Since our goal is to develop a reliable source of desirable pharmacologically active chemotypes of *C. sativa* through in vitro propagation, it necessitates the establishment of a tissue culture system that should involve minimal or no callus development in order to reduce the likelihood of induction and recovery of variants. Although, earlier attempts have been made for the propagation of *C. sativa* through tissue culture (Fisse et al. 1981; Mandolino and Ranalli 1999; Feeney and Punja 2003; Slusarkiewicz-Jarzina et al. 2005; Bing et al. 2007) but a considerable effort is still required for the refinement and production of an efficient protocol. In all the previous studies involving shoot regeneration from *C. sativa*, BA was the choice as the plant growth regulator with lower shoot regeneration.

In our laboratory, an attempt has been made to refine a *C. sativa* in vitro propagation protocol for the micropropagation of high THC yielding elite clones using TDZ for multiple shoot induction and plant regeneration. The explants (nodal segments containing axillary buds) were collected from screened and selected high yielding *C. sativa* clone grown in an indoor cultivation facility housed at the Coy-Waller laboratory, University of Mississippi. The explants were cut into small pieces as ~1 cm in length and then were treated with 0.5 % NaOCl (15 % v/v bleach) and 0.1 % Tween 20 for 20 min and later washed in sterile distilled water three times for 5 min prior to inoculation on the culture medium. Further, micropropagation and acclimatization of the micropropagated plants was done following the protocol described by Lata et al. (2009a).

Thidiazuron, is a substituted phenylurea (N-phenyl-1,2,3-thidiazol-5-ylurea) with intrinsic cytokinin-like activity (Mok et al. 1982; Huetteman and Preece 1993). Compared to most other compounds, with cytokinins activity, TDZ can stimulate better shoot proliferation and regeneration (Lata et al. 2009a; Shaik et al. 2009; Parveen and Shahzad 2010). In *C. sativa*, the best response for shoot induction was observed on Murashige and Skoog (MS) medium containing 0.5 μM TDZ (Lata et al. 2009a). Well-developed shoots were then transferred to half strength MS medium activated charcoal supplemented by different concentration of Indole-3-acetic acid (IAA), Indole-3-butyric acid (IBA), and 1-Napthalene-acetic acid (NAA) for rooting. The highest percentage of rooting in micropropagated plants was achieved in half strength Murashige and Skoog (1/2 MS) salts with 500 mg/L activated charcoal supplemented with 2.5 μM IBA. Well rooted micropropagated plants of *C. sativa* are shown in Fig. 5.4a and b.

The propagation of *C. sativa* through direct organogenesis using nodal segments would provide a rapid and reliable system for the production of large number of genetically uniform disease free plantlets and would be a promising approach for the commercial level propagation of high yielding elite varieties for the pharmaceutical industry.

5.7.3.2 Propagation Through Leaf Tissues

Somaclonal variation, defined as the genetic variation induced by in vitro techniques (Larkin and Scowcroft 1981) has been demonstrated among tissue culture regenerants of many species. The phenomenon offers an opportunity to uncover the natural variability in plants and to use this genetic variability for new product development and crop improvement (Collonier et al. 2001; Kashyap et al. 2003). Due to the intervening callus phase, regeneration through indirect shoot organogenesis has been shown to produce more somaclonal variants (Lata et al. 2002; Faisal and Anis 2003).

In our laboratory, we have developed a high frequency plant regeneration system from leaf tissue derived callus of a high yielding variety of *C. sativa* (Lata et al. 2010). The concentration and combination of plant growth regulators are the key factors influencing indirect shoot organogenesis in *C. sativa* cultures. The calli were introduced from leaf explants on MS medium supplemented with different concentrations (0.5, 1.0, 1.5, and 2.0 μM) of IAA, IBA, NAA, and 2,4-dichlorophenoxyacetic acid (2,4-D) in combination with 1.0 μM TDZ for the production of callus. The optimum callus growth and maintenance was in 0.5 μM NAA plus 1.0 μM TDZ. The 2-month-old calli were subcultured in MS media containing different concentrations of cytokinins (BAP, KN, TDZ). The rate of shoot induction and proliferation was highest in 0.5 μM TDZ. Regenerated shoots rooted best on half strength MS medium supplemented with 2.5 μM IBA of the various auxins (IAA, IBA, and NAA) tested.

5.7.3.3 Quality Assurance of Micropropagated Plants

Propagation of Cannabis through in vitro techniques has been reported by several researchers using different explant sources as well as regeneration pathways. (Mandolino and Ranalli 1999; Slusarkiewicz-Jarzina et al. 2005; Bing et al. 2007; Lata et al. 2009a, b, 2010). It is generally believed that the rate of somaclonal variation increases with the time a culture has been maintained in vitro. The longer a culture is maintained in vitro, the greater is the somaclonal variation (Kuznetsova et al. 2006), thus, questioning the very fidelity of their clonal nature. Therefore, it is necessary to assess the genetic uniformity of the micropropagated plants in terms of the production of qualitatively and quantitatively identical plants. Clonal fidelity of micropropagated *C. sativa* plants in our laboratory was tested by comparing them with mother plants and conventionally propagated plants from the same mother plant, using inter sample sequence repeat (ISSR) marker for genetic stability, GC-FID for their cannabinoids content, and using gas and water vapour exchange characteristics for the growth and physiological traits.

Genetic Fidelity

Genetic stability is of a major importance in micropropagation of any crop species. Micropropagated plants raised from explants like shoot tips, axillary buds, and nodal cuttings have been reported to maintain clonal fidelity but the use of plant growth regulators and prolonged duration in culture give rise to somaclonal variation. Due to their uncontrollable and unpredictable nature, occurrence of somaclonal variation is a serious drawback in propagation of the elite species (Rani et al. 1995; Damasco et al. 1996; Wang et al. 2000; Salvi et al. 2001; Kumar et al. 2011). Therefore, the assessment of the genetic stability of in vitro derived clones is an essential step in the application of biotechnology for micropropagation of true to type clones (Eshraghi et al. 2005).

For the production of a secure and stable in vitro clonal repository of elite *C. sativa* germplasm, the maintenance of clonal fidelity is of utmost importance. Molecular techniques are at present a powerful and valuable tools being used in the analysis of genetic fidelity of in vitro propagated plants. In comparison to various morphological, cytological, and protein markers used for detection of variations in tissue cultured plants, molecular markers are stable, highly reproducible, detectible in almost all tissues, reliable, easy to access and fast to assay, independent of development environment, and thus have been successfully employed to assess the genomic stability of regenerated plants (Pathak and Dhawan 2011; Aggarwal et al. 2008).

In our laboratory, a DNA-based ISSR markers have been successfully used to monitor the genetic stability of the micropropagated plants of *C. sativa* up to thirty passages in culture and hardened in soil for 8 months (Lata et al. 2009c). Fifteen ISSR primers generated a total of 115 distinct and reproducible bands. The banding pattern for each primer was uniform and comparable to mother plant from which the cultures had been established. Based on our results, the true to type nature of the in vitro raised clones was confirmed using ISSR markers. The utility of ISSR markers for the analysis of genetic fidelity of micropropagated plants has been well-exemplified in other micropropagated plants such as cauliflower (Leroy et al. 2000), *Swertia chirayita* (Joshi and Dhawan 2007), *Nothapodytes foetida* (Chandrika et al. 2010), apple (Pathak and Dhawan 2011). Based on our results, the micropropagation protocol standardized for *C. sativa* can be used commercially with a minimum possibility of any in vitro induced variability.

Stability in Phytocannabinoids-Chemical Profile and Content

Gas chromatography-flame ionization detection (GC-FID) was used to assess the chemical profile and quantification of cannabinoids to identify the differences, if exists, in the chemical constituents of in vitro propagated plants, vegetatively grown plants, and indoor grown mother plant of a high THC yielding variety of *C. sativa* L. during different developmental stages of growth. In general, THC content in all groups increased with the plant age up to a highest level during budding stage where the THC content reached a plateau before the onset of senescence. The pattern of changes observed in the concentration of other cannabinoids content with plants age has followed a similar trend in all groups of plants. Qualitatively, cannabinoids profile obtained using GC-FID, in an indoor grown mother plant, vegetatively propagated plants, and in vitro propagated plants were found similar to each other and to that of the field grown mother plant of *C. sativa*. Minor differences observed in cannabinoids concentration within and among the groups were not found statistically significant. Our results confirm the clonal fidelity of in vitro propagated plants of *C. sativa* and suggest that the biochemical mechanism used in this study to produce the micropropagated plants does not affect the metabolic content and can be used to for the mass propagation of true to type plants of this species for the commercial pharmaceutical use.

Stability in Ecophysiology and Gross Morphology

Following acclimatization, growth performance of in vitro propagated plants was compared to those of ex vitro vegetatively grown plants from the same mother plant on the basis of selected physiological and morphological parameters. For comparison, all the plants were grown under similar environmental conditions (25 ± 3 °C temperature and 55 ± 5 % RH). Indoor light (18 h photoperiod for vegetative and 12 h photoperiod for the reproductive growth, ~700 ± 24 μmol/m^{2}s at plant canopy level, measured by LI-COR quantum meter, model LI-189, Lincoln, Nebraska, USA) was provided with full spectrum 1000 W HID (high intensity discharge) lamps in combination with 1,000 W HPS bulbs (Sun Systems, CA). All plants were grown in the same size pots containing the same mixture of top soil sand and manure and, were watered equally and regularly to maintain identical growth conditions.

To compare the photosynthetic response, plants grown from both the methods were exposed to a range of photon flux densities (0 to 2,000 μmol/m^{2}s, Lata et al. 2009a), temperature (20–40 °C), and CO_2 concentrations (250–750 μmol/mol) under controlled humidity (55 ± 5 %) conditions. The gas exchange measurements were made on the three upper undamaged, fully expanded, and healthy leaves of the five selected plants of each type with the help of a closed portable photosynthesis system (Model LI-6400; LI-COR, Lincoln, Nebraska, USA) equipped with light, temperature, humidity, and CO_2 controls. Different PPFD levels were provided with the help of an artificial light source (Model LI-6400-02; light emitting silicon diode; LI-COR), fixed on the top of the leaf chamber and were recorded with the help of quantum sensor kept in range of 660–675 nm, mounted at the leaf level. The rate of dark respiration was measured by maintaining the leaf cuvette at zero irradiance. Temperature of the cuvette was controlled by the integrated Peltier coolers, which is controlled by the microprocessor. Different concentrations of CO_2 were supplied to the cuvette of climatic unit (LI-6400-01, LI-COR Inc., USA) by mixing pure CO_2 with CO_2 free air and were measured by infrared gas analyzer. All the measurements for gas and water vapour exchange were first recorded at the lowest PPFD, temperature, and CO_2 condition and subsequently, to the increasing levels of these parameters. Air flow rate (500 mmol/s) was kept constant throughout the experiment. Four gas exchange parameters viz., photosynthetic rate (P_N), transpirational water loss (E), stomatal conductance for CO_2 (g_s) and intercellular CO_2 concentration (Ci) were measured simultaneously at a steady state condition under various permutations and combinations of light and temperature. Our results show that in vitro propagated and hardened plants were functionally comparable to ex vitro plants of the same age in terms of gas and water vapour exchange characteristics. Furthermore, no gross morphological variation (average leaf length, width, and thickness) was observed among plants grown from the two different methods. High multiplication efficiency, good rooting, easy establishment in the soil, and normal growth performance of micropropagated plants, as reported in this study, are features necessary for the adoption of in vitro propagation technology for the large-scale multiplication of a species.

5.7.4 Harvesting and Processing

Identifying the optimum harvesting stage is a critical step in cannabis cultivation to ensure the required chemical profile, e.g., optimum Δ^9-THC content. This can be achieved through daily analysis of the cannabinoid content in different parts of the plant. Δ^9-THC content increase with plant age up to a highest level during budding stage where it reaches a plateau for approximately 1–2 weeks, followed by a decrease with the onset of senescence. Also, Δ^9-THC content is higher before noon. The upper mature buds may be harvested first, allowing time for the lower immature buds to mature. A visual indication of maturity is the color of the stigmas, which shrivels and turns brown as the flowers ripen (Fig. 5.4f).

A commercial tobacco drying barn, such as BulkTobac (Gas-Fired Products, Inc., USA), can be used for drying biomass from large-scale cultivation (12–15 h, 40 °C); a laboratory oven may be used for small samples (overnight, 40 °C). Dead leaves and stems should be removed from mature buds before drying. The whole buds can be dried, or the buds can be cut into smaller pieces. The dried plant material can be hand manicured. Any remaining leaves should be separated from the buds. The buds are carefully rubbed through different sized screens to separate small stems and seeds. Automated plant processing machines can also be used to separate big stems and seeds from the useable biomass.

Adequately dried and processed cannabis can be stored in FDA approved polyethylene bags placed in sealable fiber drums (short term: 18–20 °C; long term: −10 °C, Fig. 5.4g). A major concern when storing cannabis is the stability of the cannabinoids content. Degradation of the Δ^9-THC is negligible during processing especially if once the material is well-dried and sealed; however, it is still extremely sensitive to oxygen and UV light, and slow degradation occurs at room temperature storage through oxidation to CBN. Also, Δ^9-THC readily converts to Δ^8-THC under thermodynamic control. Therefore, the preferred conditions for long-term storage are low temperature in the absence of light.

5.8 Germplasm Conservation

5.8.1 Slow Growth

In vitro techniques provide valuable means for conservation and regeneration of valuable plant genetic resources. Although subculture at certain intervals can be used for the conservation of in vitro cultures, frequent subcultures of many clones is costly and labor-intensive. Conservation of plant germplasm by in vitro technology can be accomplished by using slow growth procedures or cryopreservation (Withers 1986) and has been practiced in many medicinal plant species such as *Ocimum* species (Mandal et al. 2000), *Adhatoda vasica* (Anand and Bansal 2002), *Phyllanthus amarus* (Singh et al. 2006a), *Withania somnifera* (Singh et al. 2006b), *Dioscorea bulbifera* (Narula et al. 2007), *Tylophora indica* (Faisal and Anis 2007), *Rauvolfia serpentina* (Ray and Bhattacharyaa 2008), and *Podophyllum peltatum* (Lata et al. 2009d). Slow growth is usually achieved by reducing the culture temperature, by modifying culture media with supplements of osmotic agents, by the use of growth inhibitors or by removing growth promoters (Dodds and Roberts 1995). This technique has the potential of prolonging the subculture interval, thereby maintaining the elite germplasm and reducing the overall cost of labor. However, for an efficient germplasm storage, the protocols should reflect maximum survival rate as well as chemical and genetic stability with minimum subculture frequency. In our laboratory, we have developed an efficient conservation protocol for *C. sativa* at 15 °C with prolongation of the subculture interval up to 16 weeks (Lata et al. 2012).

5.8.2 Encapsulation Technology

Encapsulation (synthetic seed) technology is considered one of the most promising biotechnological tools for the in vitro conservation of germplasm, possessing the ability of efficient regrowth, rooting, and developing into plantlet for in vitro and ex vitro usage. An efficient synthetic seeds production and plantlet regeneration have been reported for various types of plants, including cereals, vegetables, fruits, ornamentals, aromatic grass, and conifers (Ganapathi et al. 2001; Brischia et al. 2002; Hao and Deng 2003). However, in most cases somatic embryos were used in the encapsulation process. Few reporters (Mathur et al. 1989; Pattnaik and Chand 2000) described the encapsulation of vegetative propagules such as axillary buds or shoot tips which could be used for mass clonal propagation as well as in long-term conservation of germplasm. Beside assuring a high degree of genetic uniformity and stability, the use of vegetative propagules minimizes the occurrence of somaclonal variations.

In *Cannabis sativa*, synthetic seed technology can be successfully used for economical large-scale clonal propagation and germplasm conservation of the screened and selected elite germplasm. To further improve and refine the established protocol (Lata et al. 2009a), we have previously reported a simple and efficient method for in vitro propagation of screened and selected high yielding variety of *C. sativa* using synthetic seed technology (Lata et al. 2009b). In order to achieve that, axillary buds of *C. sativa* isolated from aseptic multiple shoot cultures were successfully encapsulated in calcium alginate beads. The best gel complexation was achieved using 5 % sodium alginate with 50 mM $CaCl_2.2H_2O$. Regrowth and conversion after encapsulation was evaluated both under in vitro and in vivo conditions on different planting substrates. The addition of antimicrobial substance—Plant Preservative Mixture (PPM) had a positive effect on overall plantlet development. Encapsulated explants exhibited the best regrowth and conversion frequency on MS medium supplemented with TDZ (0.5 μM) and PPM (0.075 %) under in vitro conditions. Plantlets regenerated from the encapsulated explants were hardened off and successfully transferred to the soil.

Germplasm conservation of a high Δ^9-Tetrahydrocannabinol yielding variety of *C. sativa* L. was successfully attempted in our laboratory by using synthetic seed technology and media supplemented with osmotic agents (Lata et al. 2012). Explants of nodal segments containing single axillary bud were excised from in vitro proliferated shoot cultures and encapsulated in high density sodium alginate (230 mM) hardened by 50 mM $CaCl_2$. The 'encapsulated' (synthetic seeds) nodal segments were stored at 5, 15, and 25 °C for 8, 16, and 24 weeks and monitored for the regrowth and survival frequency under the tissue culture conditions (16 h photoperiod, 25 °C) on MS medium supplemented with thidiazuron (TDZ 0.5 μM). 'Encapsulated' nodal segments could be stored at low temperature of 15 °C up to 24 weeks with maximum regrowth ability and survival frequency of 60 %. Well developed plantlets regenerated from 'encapsulated' nodal segments were successfully acclimatized inside the growing room with 90 % survival frequency. GC-FID was used to compare the chemical profile and the concentration of the different cannabinoids (CBD, CBC, CBG, CBN, Δ^9-Tetrahydrocannabinol, and Tetrahydrocannabivarin) of the plants grown from 'encapsulated' nodal segments to those of the donor plant. The data showed similar cannabinoid profile and insignificant differences in the cannabinoids content between the two types of plants.

Furthermore, inter simple sequence repeat (ISSR) DNA markers were used to assess the genetic stability of synthetic seeds grown randomly selected *C. sativa* L.

plants following in vitro storage for 6 months at different growing conditions (5, 15, and 25 °C). Of the 14 primers tested, 9 produced 40 distinct and reproducible bands. All the ISSR profiles of in vitro stored plants were monomorphic and comparable to mother plant which confirms the genetic stability of plants derived from synthetic seeds following 6 months of storage (Lata et al. 2011).

5.9 Conclusion

In this chapter, an attempt has been made to provide an overview of the role of biotechnology in cannabis production and our efforts to screen, propagate, and conserve elite *Cannabis sativa* clones for the production of phytocannabinoids. Plants raised through in vitro propagation were compared with mother plants and vegetatively grown plants in terms of their genetic stability, cannabinoid profile, and ecophysiology and gross morphology. Our results confirm that plants raised through micropropagation were highly comparable to the mother plant and the vegetatively propagated plants, therefore the micropropagation and the conservation protocol followed is appropriate and applicable for the clonal mass propagation of elite *C. sativa*.

Acknowledgments This work was supported in part by the National Institute on Drug Abuse (NIDA), National Institute of Health (NIH), Department of Health and Human Services, USA, Contract No. N01DA-10-7773.

References

Abrams DI, Jay CA, Shade SB, Vizoso H, Reda H, Press S, Kelly ME, Rowbotham MC, Petersen KL (2007) Cannabis in painful HIV-associated sensory neuropathy: a randomized placebo-controlled trial. Neurology 68:515–521

Adams R, Baker BR, Wearn RB (1940a) Structure of cannabinol. III. Synthesis of cannabinol, 1-hydroxy-3-amyl-6,6,9-trimethyl-6-dibenzopyran. J Am Chem Soc 62:2204–2207

Adams R, Hunt M, Clark JH (1940b) Structure of cannabidiol, a product isolated from the marihuana extract of Minnesota wild hemp. J Am Chem Soc 62:196–200

Agarwal M, Shrivastava N, Padh H (2008) Advances in molecular marker techniques and their applications in plant sciences. Plant Cell Rep 27:617–631

Ahmed SA, Ross SA, Slade D, Radwan MM, Khan IA, ElSohly MA (2008) Structure determination and absolute configuration of cannabichromanone derivatives from high potency *C. sativa*. Tetrahedron Lett 49:6050–6053

Anand Y, Bansal YK (2002) Synthetic seeds: a novel approach of in vitro plantlet formation in vasaka (*Adhatoda vasica* Nees). Plant Biotech 19:159–162

Bing X, Ning L, Jinfeng T, Nan G (2007) Rapid tissue culture method of *Cannabis sativa* for industrial uses. CN 1887043 A 20070103 Patent (p 9)

Bócsa I, Máthé P, Hangyel L (1997) Effect of nitrogen on tetrahydrocannabinol (THC) content in hemp (*C. sativa* L.) leaves at different positions. J Int Hemp Assoc 4:80–81

Brenneisen R, Egli A, ElSohly MA, Henn V, Spiess Y (1996) The effect of orally and rectally administered Δ^9-tetrahydrocannabinol on spasticity: a pilot study with 2 patients. Int J Clin Pharmacol Ther 34:446–452

Brischia R, Piccioni E, Standardi A (2002) Micropropagation and synthetic seed in M.26 apple rootstock (II): a new protocol for production of encapsulated differentiating propagules. Plant Cell Tissue Organ Cult 68:137–141

Cahn RS (1932) Cannabis *indica* resin. III. Constitution of cannabinol. J Chem Soc 3:1342–1353

Chandra S, Lata H, Khan IA, ElSohly MA (2008) Photosynthetic response of *C. sativa* L. to variations in photosynthetic photon flux densities, temperature and CO_2 conditions. Physiol Mol Biol Plants 14:299–306

Chandra S, Lata H, Khan IA, ElSohly MA (2010a) Propagation of elite *Cannabis sativa* for the production of Δ^9-Tetrahydrocannabinol (THC) using biotechnological tools. In: Arora R (ed) Medicinal plant biotechnology (Chapter 7). CABI, Wallingford, pp 98–114

Chandra S, Lata H, Mehmedic Z, Khan IA, ElSohly MA (2010b) Assessment of cannabinoids content in micropropagated plants of *Cannabis sativa* L. and their comparison with conventionally propagated plants and mother plant during developmental stages of growth. Planta Med 76:743–750

Chandra S, Lata H, Khan IA, ElSohly MA (2011a) Temperature response of photosynthesis in different drug and fiber varieties of *Cannabis sativa* L. Physiol Mol Biol Plants 17(3):297–303

Chandra S, Lata H, Khan IA, ElSohly MA (2011b) Photosynthetic response of *Cannabis sativa* L., an important medicinal plant, to elevated levels of CO_2. Physiol Mol Biol Plants 17(3):291–295

Chandrika M, Ravishankar Rai V, Thoyajoksha (2010) ISSR marker based analysis of micropropagated plantlets of *Nothapodytes foetida*. Biol Plant 54(3):561–565

Collonier C, Mulya K, Fock I, Mariska I, Servaes A, Vedel F, Siljak-Yakovlev S, Souvannavong V, Ducreux G, Sihachakr D (2001) Source of resistance against *Ralstonia solanaceraum* in fertile somatic hybrids of eggplant (*Solanum melongen*a L.) with *Solanum aethiopicum* L. Plant Sci 160:301–313

Damasco OP, Graham GC, Henry RJ, Adkins SW, Smith MK (1996) Random amplified polymorphic DNA (RAPD) detection of dwarf off types in micropropagated Cavendish bananas. Acta Hort 461:157–164

De Backer B, Debrus B, Lebrun P, Theunis L, Dubois N, Decock L, Verstraete A, Hubert P, Charlier C (2009) Innovative development and validation of an HPLC/DAD method for the qualitative and quantitative determination of major cannabinoids in cannabis plant material. J Chromatogr B Anal Technol Biomed Life Sci 877:4115–4124

De Meijer EPM, Van Der Kamp HJ, Van Eeuwijk FA (1992) Characterization of cannabis accessions with regard to cannabinoid content in relation to other plant characters. Euphytica 62:187–200

De Meijer EPM, Bagatta M, Carboni A, Crucitti P, Moliterni VMC, Ranalli P, Mandolino G (2003) The inheritance of chemical phenotype in *C. sativa* L. Genetics 163:335–346

Devane WA, Dysarz FA, Johnson MR, Melvin LS, Howlett AC (1988) Determination and characterization of a cannabinoid receptor in rat brain. Mol Pharmacol 34:605–613

Dodds JH, Roberts LW (1995) Experiments in plant tissue culture. Cambridge University Press, New York

Doyle E, Spence AA (1995) Cannabis as a medicine? British J Anesthesia 74:359–361

ElSohly MA, Slade D (2005) Chemical constituents of marijuana: the complex mixture of natural cannabinoids. Life Sci 78:539–548

ElSohly MA, Ross SA, Mehmedic Z, Arafat R, Yi B, Banahan BF (2000) Potency trends of Δ^9-THC and other cannabinoids in confiscated marijuana from 1980–1997. J Forensic Sci 45:24–30

Emboden WA (1974) Cannabis, a polytypic genus. Econ Bot 28:304–310

Eshraghi P, Zarghami R, Ofoghi H (2005) Genetic stability of micropropagated plantlets in Date Palm. J Sci 16(4):311–315

Faisal M, Anis M (2003) Rapid mass propagation of *Tylophora indica* Merrill via leaf callus culture. Plant Cell Tissue Organ Cult 75:125–129

Faisal M, Anis M (2007) Regeneration of plants from alginate encapsulated shoots of *Tylophora indica* (Burm.f.) Merrill, an endangered medicinal plant. J Hort Sci Biotechnol 82:351–354

Feeney M, Punja ZK (2003) Tissue culture and Agrobacterium-mediated transformation of hemp (*Cannabis sativa* L.). In Vitro Cellular Dev Biology-Plant 39:578–585

Fellermeier M, Zenk MH (1998) Prenylation of olivetolate by a hemp transferase yields cannabigerolic acic, the precursor of tetrahydrocannabinol. FEBS Lett 427:283–285

Fisse J, Braut F, Cosson L, Paris M (1981) In vitro study of the organogenetic capacity of *C. sativa* L. tissues: effect of different growth substances. Plantes Medicinales et Phytotherapie 15:217–223

Flachowsky H, Schuhmann E, Weber WE, Peil A (2001) Application of AFLP for the detection of sex-specific markers in hemp. Plant Breeding 120:305–309

Flores-Sanchez IJ, Verpoorte R (2008a) Secondary metabolism in cannabis. Phytochem Rev 7:615–639

Flores-Sanchez IJ, Verpoorte R (2008b) PKS activities and biosynthesis of cannabinoids and flavonoids in *C. sativa* L. plants. Plant Cell Physiol 49:1767–1782

Formukong EA, Evans AT, Evans F (1989) The medicinal uses of *Cannabis* and its constitutents. J Phytother Res 3:219–231

Galal AM, Slade D, Gul W, El-Alfy AT, Ferreira D, ElSohly MA (2009) Naturally occurring and related synthetic cannabinoids and their potential therapeutic applications. Recent Pat CNS Drug Discov 4:112–136

Ganapathi TR, Srinivas I, Suprasanna P, Bapat VA (2001) Regeneration of plants from alginated-encapsulated somatic embryos of banana cv. Rasthali (Musa spp. AAB group). Biol Plant 37:178–181

Gaoni Y, Mechoulam R (1964) Hashish. III. Isolation, structure, and partial synthesis of an active constituent of hashish. J Am Chem Soc 86:1646–1647

Ghosh R, Todd AR, Wilkinson S (1940) Cannabis *indica*. V. Synthesis of cannabinol. J Chem Soc 4:1393–1396

Gilmore S, Peakall R, Robertson J (2003) Short tandem repeat (STR) DNA markers are hypervariable and informative in *C. sativa*: implications for forensic investigations. Forensic Sci Int 131:65–74

Grinspoon L, Bakalar JB (1993) Marihuana, the forbidden medicine. Yale University Press, New Haven

Guindon J, Hohmann AG (2009) The endocannabinoid system and pain. CNS Neurol Disord Drug Targets 8:403–421

Hammond CT, Mahlberg PG (1977) Morphogenesis of capitate glandular hairs of *Cannabis sativa* (Cannabaceae). Amer J Bot 64:1023–1031

Hao YJ, Deng XX (2003) Genetically stable regeneration of apple plants from slow growth. Plant Cell Tiss Org Cult 72:253–260

Hemphill JK, Turner JC, Mahlberg PG (1980) Cannabinoid content of individual plant organs from different geographical strains of *C. sativa* L. J Nat Prod 43:112–122

Hillig KW (2004) A chemotaxonomic analysis of terpenoid variation in cannabis. Biochem Syst Ecol 32:875–891

Hillig KW (2005) Genetic evidence for speciation in *Cannabis* (Cannabaceae). Genet Resour Crop Evol 52:161–180

Hillig KW, Mahlberg PG (2004) A chemotaxonomic analysis of cannabinoids variation in *Cannabis* (Cannabaceae). Am J Bot 91:966–975

Huetteman A, Preece EJ (1993) Thidiazuron: a potent cytokinin for woody plant tissue culture. Plant Cell Tiss Org Cult 33:105–119

Jarvinen T, Pate DW, Laine K (2002) Cannabinoids in the treatment of glaucoma. Pharmacol Ther 95:203–220

Joshi P, Dhawan V (2007) Assessment of genetic fidelity of micropropagated *Swertia chirayita* plantlets by ISSR marker assay. Biol Plant 51(1):22–26

Kajima M, Piraux M (1982) The biogenesis of cannabinoids in *Cannabis sativa*. Phytochemistry 21:67–69

Kashyap V, Kumar S, Collonier C, Fusari F, Haircour R, Rotino GL, Sihachakr D, Rajam MV (2003) Biotechnology in eggplant. Sci Horticult 97(1):1–25

Kriese U, Schumann E, Weber WE, Beyer M, Brühl L, Matthus B (2004) Oil content, tocopherol composition and fatty acid patterns of the seeds of 51 *C. sativa* L. genotypes. Euphytica 137:339–351

Kumar S, Mangal M, Dhawan AK, Singh N (2011) Assessment of genetic fidelity of micropropagated plants of *Simmondsia chinensis* (Link) Schneider using RAPD and ISSR markers. Acta Pysiol Plant 33:2541–2545

Kuznetsova OI, Ash OA, Gostimsky SA (2006) The effect of the duration of callus culture on the accumulation of genetic alternation in pea *Pisum sativum* L. Russ J Genet 42:555–562

Larkin PJ, Scowcroft WR (1981) Somaclonal variation—a novel source of variability from cell cultures for plant improvement. Theor Appl Genet 60:197–214

Lata H, Bedir E, Hosick A, Ganzera M, Khan I, Moraes RM (2002) In vitro plant regeneration from leaf-derived callus of *Cimicifuga racemosa*. Planta Med 68:912–915

Lata H, Chandra S, Khan IA, ElSohly MA (2009a) Thidiazuron induced high frequency direct shoot organogenesis of *C. sativa* L. In vitro Cellular Dev Biology-Plant 45:12–19

Lata H, Chandra S, Khan I, ElSohly MA, ElSohly MA (2009b) Propagation through Alginate Encapsulation of Axillary Buds of *Cannabis sativa* L.–an important medicinal plant. Physiol Mol Biol Plants 15(1):79–86

Lata H, Chandra S, Techen N, Khan IA, ElSohly MA (2009c) Assessment of genetic stability of micropropagated plants of *C. sativa* L. by ISSR markers. Planta Med 76:97–100

Lata H, Moraes RM, Bertoni B, Pereira AMS (2009d) In vitro germplasm conservation of *Podophyllum peltatum* L. under slow growth conditions. In vitro Cell Dev Biol Plant 46:22–27

Lata H, Chandra S, Khan IA, ElSohly MA (2010) High frequency plant regeneration from leaf derived callus of high Δ^9-tetrahydrocannabinol yielding *Cannabis sativa* L. Planta Med 76:1629–1633

Lata H, Chandra S, Mehmedic Z, Khan IA, ElSohly MA (2011) Molecular analysis of genetic fidelity in *Cannabis sativa* L. plants grown from synthetic seeds following in vitro storage conditions. Biotechnol Lett 33:2503–2508

Lata H, Chandra S, Techen N, Khan IA, ElSohly MA (2012) In vitro germplasm conservation of high Δ^9-tetrahydrocannabinol yielding elite clones of *Cannabis sativa* L. under slow growth conditions. Acta Physiologiae Plantarum 34(2):743–750

Leroy X, Leon K, Charles G, Branchard M (2000) Cauliflower somatic embryogenesis and analysis of regenerants stability by ISSRs. Plant Cell Rep 19:1102–1107

Liang YC, Huang CC, Hsu KS (2004) Therapeutic potential of cannabinoids in trigeminal neuralgia. Curr Drug Targets CNS Neurol Disord 3:507–514

Makriyannis A, Tian X, Guo J (2005) How lipophilic cannabinergic ligands reach their receptor sites. Prostaglandins Other Lipid Mediat 77:210–218

Mandal J, Patnaik S, Chand PK (2000) Alginate encapsulation of axillary buds of *Ocimum americanum* L. (hoary basil), *O. basilicum* L. (sweet basil), *O. gratissimum* L. (shrubby basil), and *O. sanctum* L. (sacred basil). In vitro Cell Dev Biol Plant 36:287–292

Mandolino G, Ranalli P (1999) Advances in biotechnological approaches for hemp breeding and industry. In: Ranalli P (ed) Advances in Hemp research. Haworth Press, New York, pp 185–208

Mandolino G, Carboni A, Forapani S, Faeti V, Ranalli P (1999) Identification of DNA markers linked to the male sex in dioecious hemp (*C. sativa* L.). Theor Appl Genet 98:86–92

Mathur J, Ahuja PS, Lal N, Mathur AK (1989) Propagation of *Valeriana wallichii* DC using encapsulated apical and axial shoot buds. Plant Sci 60:111–116

Matsuda LA, Lolait SJ, Brownstein MJ, Young AC, Bonner TI (1990) Structures of a cannabinoid receptor and functional expression of the cloned cDNA. Nature 346:561–564

Mattes RD, Shaw LM, Eding-Owens J, Egelman K, ElSohly MA (1993) Bypassing the first pass effect for therapeutic use of cannabnoids. Pharmacol Biochem Behav 44:745–747

Mattes RD, Egelman K, Shaw LM, ElSohly MA (1994) Cannabinoids appetite stimulation. Pharmacol Biochem Behav 44:745–747

Mechoulam R (1986) The pharmacohistory of *Cannabis sativa*. In: Mechoulam R (ed) Cannabinoids as therapeutic agents. CRC Press, Boca Raton, pp 1–19

Mechoulam R, Ben-Shabat A (1999) From gan-zi-gun-nu to anandamide and 2-arachidonoylglycerol: the ongoing story of *Cannabis*. Nat Prod Rep 16:131–143
Mechoulam R, Gaoni Y (1967) Absolute configuration of Δ^1-tetrahydrocannabinol, the major active constituent of hashish. Tetrahedron Lett 11:1109–1111
Mechoulam R, Shvo Y (1963) Hashish. I. Structure of cannabidiol. Tetrahedron 19:2073–2078
Mehmedic Z, Chandra S, Slade D, Denham H, Foster S, Patel AS, Ross SA, Khan IA, ElSohly MA (2010) Potency trends of Δ^9-THC and other cannabinoids in confiscated cannabis preparations from 1993–2008. J Forensic Sci 55:1209–1217
Mendoza M, Mills DE, Lata H, Chandra S, ElSohly MA, Almirall J (2009) Genetic individualization of *C. sativa* by an STR multiplex. Anal Bioanal Chem 393:719–726
Mok MC, Mok DWS, Armstrong DJ, Shudo K, Isogai Y, Okamoto T (1982) Cytokinin activity of N-phenyl-N'-1,2,3-thiadiazol-5-ylurea (Thidiazuron). Phytochemistry 21:1509–1511
Mouslech Z, Valla V (2009) Endocannabinoid system: an overview of its potential in current medical practice. Neuroendocr Lett 30:153–179
Narula A, Kumar S, Srivastava PS (2007) Genetic fidelity of in vitro regenerants, encapsulation of shoot tips and high diosgenin content in *Dioscorea bulbifera* L., a potential alternative source of diosgenin. Biotechnol Lett 29:623–629
Parveen S, Shahzad A (2010) TDZ-induced high frequency shoot regeneration in *Cassia sophera* Linn. via cotyledonary node explants. Physiol Mol Biol Plants 16(2):201–206
Pate D (1999) Anandamide structure-activity relationships and mechanisms of action on intraocular pressure in the normotensive rabbit model. PhD thesis, University of Kuopio, Kuopio, Finland
Pathak H, Dhawan V (2011) ISSR assay for ascertaining genetic fidelity of micropropagated plants of apple rootstock Merton 793. In vitro Cell Dev Biol-Plant. doi:10.1007/s11627-011-9385-0
Pattnaik S, Chand PK (2000) Morphogenic response of the alginate encapsulated axillary buds from in vitro shoot cultures of six mulberries. Plant Cell Tissue Organ Cult 64:177–185
Pryce G, Baker D (2005) Emerging properties of cannabinoid medicines in management of multiple sclerosis. Trends Neurosci 28:272–276
Radwan MM, ElSohly MA, Slade D, Ahmed SA, Wilson L, El-Alfy AT, Khan IA, Ross SA (2008) Non-cannabinoid constituents from a high potency *C. sativa* variety. Phytochemistry 69:2627–2633
Raharjo TJ, Chang WT, Choi YH, Peltenburg-Looman AMG, Verpoorte R (2004) Olivetol as product of a polyketide synthase in *Cannabis sativa* L. Plant Sci 166:381–385
Rani V, Parida A, Raina SN (1995) Random amplified polymorphic DNA (RAPD) markers for genetic analysis in micro-propagated plants of *Populus deltoids* Marsh. Plant Cell Rep 14:459–462
Ray A, Bhattacharya S (2008) Storage and plant regeneration from encapsulated shoot tips of *Rauvolfia serpentine*-an effective way of conservation and mass propagation. S Afr J Bot 74:776–779
Ross SA, ElSohly MA (1995) Constituents of *C. sativa* L. XXVIII. A review of the natural constituents: 1980–1994. J Pharm Sci 4:1–10
Ross SA, ElSohly MA (1996) The volatile oil composition of fresh and air-dried buds of *C. sativa* L. J Nat Prod 59:49–51
Sakamoto K, Shimomura K, Komeda Y, Kamada H, Satoh S (1995) A male-associated DNA sequence in a dioecious plant, *C. sativa* L. Plant Cell Physiol 36:1549–1554
Sakamoto K, Abe T, Matsuyama T, Yoshida S, Ohmido N, Fukui K, Satoh S (2005) RAPD markers encoding retrotransposable elements are linked to the male sex in *C. sativa* L. Genome 48:931–936
Salvi ND, George L, Eapen S (2001) Plant regeneration from leaf base callus of turmeric and random amplified polymorphic DNA analysis of regenerated plants. Plant Cell Tiss Org Cult 66:113–119
Shaik NM, Arha M, Nookaraju A, Gupta SK, Srivastava S, Yadav AK, Kulkarni PS, Abhilash OU, Vishwakarma RK, Singh S, Tatkare R, Chinnathambi K, Rawal SK, Khan BM (2009)

Improved method of in vitro regeneration in *Leucaena leucocephala*—a leguminous pulpwood tree species. Physiol Mol Biol Plants 16(2):201–206

Shoyama Y, Hirano H, Oda M, Somehara T, Nishioka I (1975) Cannabis. IX. Cannabichromevarin and cannabigerovarin, two new propyl homologs of cannabichromene and cannabigerol. Chem Pharm Bull 23:1894–1895

Singh AK, Sharma M, Varshney R, Agarwal SS, Bansal KC (2006a) Plant regeneration from alginate to encapsulated shoot tips of *Phyllanthus amarus* Schum and Thonn, a medicinally important plant species. In vitro Cell Dev Biol Plant 42:109–113

Singh AK, Varshney R, Sharma M, Agarwal SS, Bansal KC (2006b) Regeneration of plants from alginate to encapsulated shoot tips of *Withania somnifera* (L.) Dunal, a medicinally important plant species. J Plant Physiol 163:220–223

Sirikantaramas S, Taura F, Tanaka Y, Ishikawa Y, Morimoto S, Shoyama Y (2005) Tetrahydrocannabinolic acid synthase, the enzyme controlling marijuana psychoactivity is secreted into the storage cavity of the glandular trichomes. Plant Cell Physiol 46:1578–1582

Sirikantaramas S, Taura F, Morimoto S, Shoyama Y (2007) Recent Advances in *Cannabis sativa* Research: Biosynthetic Studies and Its Potential in Biotechnology. Curr Pharm Biotechnol 8:237–243

Slatkin NE (2007) Cannabinoids in the treatment of chemotherapy-induced nausea and vomiting: beyond prevention of acute emesis. J Support Oncol 5:1–9

Ślusarkiewicz-Jarzina A, Ponitka A, Kaczmarek Z (2005) Influence of cultivar, explant source and plant growth regulator on callus induction and plant regeneration of *C. sativa* L. Acta Biologica Cracoviensia Series Botanica 47:145–151

Small E (1975a) American law and the species problem in cannabis: science and semantics. Bull Narc 27:1–20

Small E (1975b) Morphological variation of cannabis. Can J Bot 53:978–987

Small E, Cronquist A (1976) A practical and natural taxonomy for cannabis. Taxon 25:405–435

Small E, Marcus D (2002) Hemp: a new crop with new uses for North America. In: Janick J, Whipkey A (eds) Trends in new crops and new uses. ASHS Press, Alexandria, pp 284–326

Small E, Jui PY, Lefkovitch LP (1976) A numerical taxonomic analysis of cannabis with special reference to species delimitation. Syst Bot 1:67–84

Taura F, Tanaka S, Taguchi C, Fukamizu T, Tanaka H, Shoyama Y, Morimoto S (2009) Characterization of olivetol synthase, a polyketide synthase putatively involved in cannabinoid biosynthetic pathway. FEBS Lett 583:2061–2066

Techen N, Chandra S, Lata H, ElSohly MA, Khan IA (2010) Genetic identification of female *C. sativa* plants at early developmental stage. Planta Med 16:1938–1939

Törjék O, Bucherna N, Kiss E, Homoki H, Finta-Korpelová Z, Bócsa I, Nagy I, Heszky LE (2002) Novel male-specific molecular markers (MADC5, MADC6) in hemp. Euphytica 127:209–218

Turner CE, ElSohly MA, Boeren EG (1980) Constituents of *C. sativa* L. XVII. A review of the natural constituents. J Nat Prod 43:169–234

Valle JR, Vieira JEV, Aucélio JG, Valio IFM (1978) Influence of photoperiodism on cannabinoid content of *C. sativa* L. Bull Narc 30:67–68

Viveros MP, Marco EM (2007) Cannabinoids, anxiety and depression. Recent Prog Medicinal Plants 18:225–249

Wang L, Jiang T, Gusfield D (2000) A more efficient approximation scheme for tree alignment. SIAM J Comput 30(1):283–299

Withers LA (1986) In vitro approaches to the conservation of plant genetic resources. In: Withers LA, Alderson PG (eds) Plant tissue culture and its agriculture applications. Butterworths, London, pp 261–276

Wood TB, Spivey WTN, Easterfield TH (1896) Charas: the resin of Indian hemp. J Chem Soc Trans 69:539–546

Wood TB, Spivey WTN, Easterfield TH (1899) Cannabinol. Part I. J Chem Soc 75:20–36

Zuardi AW (2006) History of cannabis as a medicine: a review. Brazilian J Psychiatry 28:153–157

Chapter 6
Epilobium Sp. (Willow Herb): Micropropagation and Production of Secondary Metabolites

Deliu Constantin, Ana Coste and Tămaş Mircea

6.1 Introduction

6.1.1 The Plant

Epilobium is the largest genus of the plant family Onagraceae, with 165 species (185 taxa) worldwide, except Antarctica. The family Onagraceae includes two subfamilies, Ludwigioideae (Ludwigia) and Onagroideae. The latter comprises six tribes: Hauyeae, Circaeeae, Lopezieae, Gongylocarpeae, Epilobieae, and Onagreae (Wagner et al. 2007). The tribe Epilobiae includes two genera, *Chamerion* and *Epilobium*, with north temperate origin (Raven 1976; Baum et al. 1994; Levin et al. 2003, 2004). There were many morphological, phytochemical, and genetical analyses performed to give convincing arguments for treating *Chamerion* as a separate genus from *Epilobium* (Baum et al. 1994; Levin et al. 2004; Strgulc Krajšek 2006; Obradovič et al. 2007; Wagner et al. 2007; Strgulc Krajšek et al. 2008; Strgulc Krajšek et al. 2009; Tămaş et al. 2009). However, authors still use in their works *Epilobium angustifolium* L. instead of the correct denomination of *Chamerion angustifolium* (L.) Holub. Therefore, it was mandatory for us to cite articles that still maintain the denomination of *Epilobium angustifolium* used by some authors, even though considered incorrect by others.

D. Constantin (✉) · A. Coste
Institute of Biological Research, 48 Republicii St, 400015 Cluj-Napoca, Romania
e-mail: deliu_konstantin@yahoo.com

A. Coste
e-mail: ana.coste@icbcluj.ro

T. Mircea
Faculty of Pharmacy, University of Medicine and Pharmacy Iuliu Haţieganu,
8 Victor Babeş St, 400012 Cluj-Napoca, Romania
e-mail: mtbotanica@yahoo.com

S. Chandra et al. (eds.), *Biotechnology for Medicinal Plants*,
DOI: 10.1007/978-3-642-29974-2_6, © Springer-Verlag Berlin Heidelberg 2013

Epilobium genus is remarkable for its morphological, ecological, and cytological diversity (Baum et al. 1994). Taxonomically, *Epilobium* is a very difficult group because of species similarity and frequent interspecific hybridization (Strgulc Krajšek et al. 2006; Akbari and Azizian 2006; Bleeker et al. 2007). The taxonomy of the genus has therefore varied between different botanists (Raven 1976; Hoch and Raven 1992; Baum et al. 1994; Levin et al. 2004). In this context, Wagner et al. (2007) summarized the available morphological and molecular data to underline the new classification of the genus into eight sections, one of which has two subsections. The section *Epilobium* is the largest section in the genus (and in the Onagraceae), comprising 150 species (168 taxa) distributed in cool montane or boreal habitats on all continents, except Antarctica, and extending at high elevations into the tropics (Wagner et al. 2007). In Europe, 28 species of the section *Epilobium* are native or naturalized (Holub 1978; Raven 1980; Snogerup 1982).

The *Epilobium* species are perennial or annual herbs, or subshrubs, stems erect to ascending or decumbent. Leaves are opposite and decussate below inflorescence or only at base and alternate distally, subsessile to petiolate. Flowers are actinomorphic or rarely zygomorphic. Petals four, rose-purple to white, rarely cream-yellow, or rarely orange-red apically notched. Fruit a loculicidal capsule, usually narrowly cylindrical. Seeds many or rarely 1–8 per locule, in 1 (2) rows per locule (Wagner et al. 2007).

6.1.2 Importance of Epilobium Species

The use of *Epilobium* was widespread in Central and Eastern Europe, as well as in certain areas of North America for centuries, to treat a variety of diseases and enhance wound healing (Tyler 1986). *Epilobium* has been used as an antispasmodic for conditions such as hiccups, asthma, and whooping cough. Ancient healers would often use the dried leaves to soothe mucous membranes. *Epilobium* was regularly used to treat diarrhea, bowel problems, and colitis. It has been used successfully for bladder health maintenance, male health maintenance, hormonal imbalances, and urinary system health. An infusion of the leaves will be found beneficial in leucorrhea, menorrhagia, and uterine hemorrhage; and forms an excellent local application for ophthalmia and ulcerations of the mouth and throat (Treben 1984; Bohinsky 1993; Moerman 1998; Wichtl 2004). Young shoots and leaves are traditionally used as a decoction or infusion. Leaves can also be consumed as vegetable or salad (Laurent 1989; Bunney 1992). The commercially available plant drug *Epilobii herba* contains the dried aerial parts of several *Epilobium* species. The classical drug *Epilobii herba* may be obtained from different species collected as plant source, including small-flowered species such as *E. parviflorum* Schreb., *E. montanum* L., *E. roseum* L. and *E. collinum* L., and large-flowered species such as *Chamerion angustifolium* (L.) Holub (syn *E. angustifolium* L.) and *E. hirustum* L. (Ciulei et al. 1993; Wichtl 1994; Tămaş 1997;

Wichtl 2004). It is mostly collected in Central Europe, former Yugoslavia and Romania (Bisset and Wichtl 1994).

There are also a variety of herbal remedies containing *Epilobium* extracts which are claimed to be efficient in treating prostate diseases. *E. parviflorum* was disclosed by Maria Treben (1984) as useful in the treatment of all kinds of bladder, kidney, and particularly prostate disorders. Further studies revealed other *Epilobium* species with curative effects in the case of prostatitis, benign tumor, and hypertrophy of the prostate gland (BHP) (Van Hellemont 1986; Wichtl 1989; Hiermann et al. 1991; Voynova et al. 1991; Bohinsky 1993). *Epilobium* is a potent antimicrobial and anti-inflammatory agent that tones and protects the entire urogenital system. *Epilobium* can be a safe, effective way to help reduce BHP and to improve incontinence and urgency. Initially, pharmacological investigations have been restricted mostly to two species, *E. angustifolium* and *E. parviflorum*, being afterwards extended to some other *Epilobium* species. These pharmacological studies, both in vitro and in vivo, of extracts obtained from aerial parts of *Epilobium* taxa have confirmed their therapeutic properties including anti-inflammatory (Hiermann et al. 1986; Juan et al. 1988; Hiermann et al. 1991; Hiermann et al. 1998; Steenkamp et al. 2006; Hevesi Tóth et al. 2008; Kiss et al. 2010), antiandrogenic (Hiermann and Bucar 1997), antimicrobial (Ivancheva et al. 1992; Rauha et al. 2000; Battinelli et al. 2001; Steenkamp et al. 2006; Borchardt et al. 2008; Bartfay et al. 2011), antimotility (Vitali et al. 2006), antifungal (Jones et al. 2000; Webster et al. 2008), antinociceptive (Tita et al. 2001; Pourmorad et al. 2007), antiproliferative (Ducrey et al. 1997; Vitalone et al. 2001; Vitalone et al. 2003a; Vitalone et al. 2003b; Kiss et al. 2006a), antioxidant (Kähkönen et al. 1999; Shikov et al. 2006; Steenkamp et al. 2006; Stajner et al. 2007; Wojdyło et al. 2007; Hevesi Tóth et al. 2008; Kiss et al. 2010; Smirnova et al. 2010), and antitumour (Voynova et al. 1991) effects.

6.1.3 Compounds in Epilobium Species and Their Effects

The active constituents of *Epilobium* species are not well known, but it is considered that they include two pharmacologically important compounds: flavonols (glycosides of myricetin and quercetin) (Averett et al. 1978, 1979; Hiermann 1983; Slacanin et al. 1991; Dukrey et al. 1995; Bazylko et al. 2007a; Hevesi Tóth et al. 2009), and macrocyclic ellagitannins such as oenothein A (OeA) and B (OeB) (Lesuisse et al. 1996; Ducrey et al. 1997; Bazylko et al. 2007b; Shikov et al. 2010). Additionally, *Epilobium* species synthesize sterols (Hiermann and Mayr 1985; Nowak and Krzaczek 1998a; Pelc et al. 2005; Hevesi Tóth 2009), triterpenes (Glen et al. 1967; Pascual Teresa et al. 1979; Hiermann 1983; Nowak and Krzaczek 1998b), gallic, chlorogenic, and ellagic acids (Hiermann 1983; Ducrey et al. 1995; Hevesi Tóth et al. 2009), fatty acids (Guil-Guerrero et al. 2001; Pelc et al. 2005), and a few other compounds such as a mucilage, sugars, vitamins, essential microelements, and volatile oils. Chemical analysis of *E. hirsutum* revealed the presence of some specific ellagittanins in this species, such as epilobamide-A (Nawwar et al. 1997; Barakat et al. 1997). For *E. angustifolium* monomeric ellagitannins (tellimagrandin I and II and others) (Gupta et al. 1982; Liu et al. 2003) were detected.

Several different methods were used for the separation, identification, and quantification of unknown active constituents in plant extracts of *Epilobium* species. Thus, flavonoids were examined by applying TLC, HPLC, LC, CE, and GC methods (Kim and Lee 2002; Rijke et al. 2006; Stalikas 2007; Liu et al. 2008). The analysis of polyphenolic compounds and hydrolyzable tannins is generally achieved through HPLC. Nevertheless, to determine the total polyphenolic content the spectrophotometry method with Folin Ciocâlteau reagent (Kähkönen et al. 1999; Pourmorad et al. 2007; Wojdyło et al. 2007; Tămaş et al. 2009) is still used. For the analysis of gallotannins and ellagitannins, a method based on the reaction between potassium iodate and hydrolyzable tannins (Hartzfeld et al. 2002) is being used. Initial studies proved that, among all polyphenolic compounds, ellagitannins OeB and OeA are the most active constituents of the plant extracts. Therefore, qualitative and quantitative analyses were mostly performed for these two compounds. The trimer OeA and dimer OeB were first isolated as the main tannins in *Oenothera erythrosepala* Borbás leaves (Miyamoto et al. 1987; Hatano et al. 1990). OeB was identified in *E. parviflorum* extract by means of HPLC and NMR techniques (Lesuisse et al. 1996). Ducrey et al. (1997) also investigated the OeB molecule by NMR and FAB-MS for determining the molar mass. Toward the separation and quantification of OeB, HPTLC- densitometry (Bazylko et al. 2007b) and HPTLC (Shikov et al. 2010) methods were optimized. For the investigation of polyphenol composition of various *Epilobium* species, and exact identification of their chemical composition, Hevesi Tóth et al. (2009) applied the LC–MS/MS method. Several ellagitannins, present in *E. hirsutum*, were identified and confirmed by ^{1}H and ^{13}C NMR spectroscopy (Nawwar et al. 1997; Barakat et al. 1997).

Different methods used to quantify a biologically active compound may yield significantly different results even though the same species is analyzed. Thus, in *Chamerion angustifolium* the total polyphenolic content determined through HPLC was 194.4 mg g^{-1} of dry weight (DW) (Shikov et al. 2007), while using Folin-Ciocalteu procedure, the level of total polyphenols reached only 32.2 mg of GAE/g of DW (Kähkönen et al. 1999). Furthermore, for the same species, the HPTLC-densitometry analysis revealed 152.5 mg g^{-1} of DW of OeB (Bazylko et al. 2007b), while the OeB content reached only 20.2 mg g^{-1} of DW when HPTCL method was used (Shikov et al. 2010). Quantitative analysis may reveal different results even if the same method is applied. For instance, quantitative determination of OeB in *E. angustifolium* through HPTLC-densitometry was 152.5 mg g^{-1} of DW when performed by Bazylko et al. (2007b), while OeB analyzed with the same method in the same species by Kiss et al. (2010) reached 225.8 mg g^{-1} of DW. Therefore, we may conclude that different results occur not only due to different methodological approaches, but also due to the plant material used for subsequent analyses. Thus, plant material may be collected from different sites, in different developmental stages. This may contribute substantially to the variability of the data obtained. Although both oenotheins A and B were detected in the majority of *Epilobium* species, up to the present time, only OeB was quantified. Thus, Ducrey et al. (1997) analyzing 9 *Epilobium* species found out that its content varies considerably from plant to plant, between 3 and 14 %. In the last few years only three species of *Epilobium* were investigated for OeB content: *E.angustifolium* (226 mg $g^{-1}$of DW), *E. parviflorum* (327 mg $g^{-1}$of DW), and *E. hirsutum* (334 mg $g^{-1}$of DW) (Kiss et al. 2010).

In the case of flavonoids the same aspects were observed when different analytical methods were used. Nevertheless, the first analysis performed—even if qualitative (TLC)—provided a suggestive image regarding the flavonoid content for the entire *Epilobium* genus. Thus, Averett et al. (1978, 1979) revealed quercetin 3-O-rhamnoside and glucoside and myricetin 3-O-rhamnoside and glucoside to be present in most species of *Epilobium*, and kaempferol 3-O-rhamnoside to also be present in some populations of some entities in small quantities. Further researches then proved the accuracy of these observations. Thus, investigations carried out by means of LCMS regarding *Epilobium* species showed that among 19 flavonol glycosides, 13 compounds were 3-O-glycosides of kaempferol, quercetin, and myricetin (Ducrey et al. 1995). The quantitative analysis performed on flavonol glycosides in eight *Epilobium* species revealed various proportions of quercitrin, myricitrin, and isomyricitrin, while *E. angustifolium* gives a completely different flavonoid pattern from other species. Instead of having myricitrin as the main constituent, isoquercitrin was identified as the most important flavonoid in the ethanolic extract (Slacanin et al. 1991). As a result of flavonoid composition analysis in five *Epilobium* species, myricetin, quercetin, kaempferol, and their various glycosides were found dominant in samples, but their combination and ratio were distinctive in all cases. The most remarkable differences were registered for *E. angustifolium*, which, in contrast with other species, exclusively contained flavonol-glucuronide components (quercetin-glucuronide; kaempferol-glucuronide), and its main flavonoid compound was identified as quercetin-glucuronide, while in other species, myricitrin was found to be the main flavonoid (Hevesi Tóth et al. 2009). The same was noticed by Kiss et al. (2010), who showed that *E. angustifolium* contains quercetin-3-O-glucuronide as a main flavonoid, while *E. hirsutum* and *E. parviflorum* contain myricetin-3-O-rhamnoside. The results of quantitative analysis were dependent on the analytical method used. Thus, the total flavonoids synthesized by *Epilobium* species and determined by the spectrophotometric method ranged between 1.9 and 4.3 % (the results were expressed in rutoside) (Tămaş et al. 2009) and between 0.69 and 0.83/100 g dried herb (the results were expressed in hyperoside) (Hevesi Tóth 2009). The results obtained after analyzing three *Epilobium* species through the Christ-Müller method revealed a total flavonoid content between 7.1 and 19.2 mg g^{-1} of DW (Kiss et al. 2010).

Experiments performed on *Epilobium extracts* proved that all active compounds comprised within these extracts are responsible for their therapeutic effects. Actually, the pharmacological effect of *Epilobium* could be explained by the presence of steroids (in particular *β*-sitosterol and its esters), triterpenes, fatty acids, macrocyclic tannins, and flavonoids (in particular myricitrin, isomyricitrin, quercitrin, and quercetin-3-O-β-D-glucuronide) in the aerial parts (Barakat et al. 1997; Vitalone et al. 2001; Wojdyło et al. 2007; Kiss et al. 2010; Shikov et al. 2010).

Flavonoids have been shown to possess a variety of biological activities; they were reported to act in the gastrointestinal tract as either antiulcer, antispasmodic, antisecretory, or antidiarrheal agents. Flavonoids have demonstrated to possess effects on intestinal motility both in vitro and in vivo (Vitali et al. 2006). Furthermore, it was proven that *Epilobium* extracts possess a high antioxidant capacity, which is attributed to the high concentration of flavonoids (Arredondo et al. 2004; Shikov et al. 2006). For instance, *E. angustifolium* extract contains flavonoids, such as myricetin, kaempferol, and quercetin, which have antioxidant activity (Hiermann et al. 1991; Bazylko et al. 2007a; Hevesi Tóth et al. 2009). Mostly, the antioxidant activity of flavonoids depends on the structure and substitution pattern of hydroxyl groups. Thus, Wojdyło et al. (2007) revealed that among the flavonoids identified in their study, only myricetin had a galloyl structure in ring B and appeared to be a better antioxidant than quercetin, with a catechol moiety. There are very few studies regarding the antimicrobial effect of compounds comprised of *Epilobium* plant extracts. Nevertheless, Rauha et al. (2000) showed that quercetin and naringenin were effective in inhibiting the growth of bacteria. Investigations on the anti-inflammatory effect of *Epilobium* extracts revealed that myricetin 3-O-glucuronide is a strong inhibitor of cyclooxygenase-1 and -2 (COX-1 and COX-2) and 5-LOX activities (Hiermann et al. 1991; Hiermann et al. 1998), while other flavonoids (e.g., kaempferol, quercetin, miricetin, and their glycosides) inhibit proliferation of human prostate cells (Vitalone et al. 2001; Vitalone et al. 2003b). Moreover, flavonoid contents of the aerial parts of various members of the genus *Epilobium* can be used as a chemotaxonomic marker. Chemotaxonomy based on flavonoids permits identification of the different species, even if botanically very similar (Vitalone et al. 2001; Hevesi Tóth et al. 2006).

Among the ellagitannins synthesized by different *Epilobium* species, the most studied were the dimeric macrocyclic ellagitannin OeB and trimer OeA, due to their multiple pharmacological effects. Thus, the OeA and especially OeB inhibit the activity of 5α-reductase and aromatase, two enzymes which are involved in the etiology of benign prostatic hyperplasia (BPH) (Lesuisse et al. 1996; Ducrey et al. 1997). Previous studies have shown that OeB is a potent inhibitor of herpes simplex virus HSV-1 (Fukuchi et al. 1989), strongly inhibits the replication of human immunodeficiency viruses (HIV) (Okuda et al. 1989), and exhibits significant antitumor activity (Miyamoto et al. 1993a; Miyamoto et al. 1993b; Aoki et al. 1995). Further researches revealed the inhibitory effect of OeB on DNA synthesis in different human cell lines (Vitalone et al. 2003b), as well as a significant antioxidant activity (Yoshimura et al. 2008). Miyamoto et al. (1993a) reported that OeB had potent antitumor activity upon intraperitoneal administration to mice before tumor inoculation and suggested that this may be due to enhancement of the host immune system via macrophage activation. Still, only recent studies—Schepetkin et al. (2009)—showed that extracts of *E. angustifolium* activate functional responses in neutrophils and monocyte/macrophages, and OeB induces a number of phagocyte functions in vitro, including intracellular Ca^{2+} flux, production of reactive oxygen species (ROS), chemotaxis, nuclear factor (NF)-κB activation, and proinflammatory cytokine production. The ability of OeB to modulate phagocyte functions in vitro and in vivo suggests that this compound is responsible for at least part of the therapeutic properties of *E. angustifolium* extracts (Schepetkin et al. 2009). Researches performed on OeB's contribution to the anti-inflammatory activity of *Epilobium* species proved that this compound induces a neutral endopeptidase in prostate cancer cells (Kiss et al. 2006b). Furthermore, OeB inhibits the activity of hyaluronidase and MPO (myeloperoxidase) release from stimulated neutrophils, similar to the anti-inflammatory drug indomethacin, and appears as specific COX-1 inhibitor comparable in activity to indomethacin (Kiss et al. 2010). Besides dimeric and trimeric ellagitannins, it seems that monomeric ellagitannins can reveal important effects as well. Thus, Miyamoto et al. (1993c) showed that among the monomeric ellagitannins tested for antitumor activities, tellimagrandin II was the most active. Furthermore, Yi et al. (2006) proved that tellimagrandin I attenuates the tumor phenotype of human cervical carcinoma HeLa cells in vitro. Additionally, several in vitro and a handful of in vivo studies have shown that 1, 2, 3, 4, 6-penta-O-galloyl-β-D-glucose (PGG) (a prototypical gallotannin) exhibits multiple biological activities which implicate a great potential for PGG in the therapy and prevention of several major diseases including cancer and diabetes (Zhang et al. 2009).

6.2 In Vitro Approaches

6.2.1 Review

There are few reports of in vitro micropropagation of plant species from Onagraceae family. Until now these studies were mainly directed toward the multiplication and secondary metabolites production in nine species from the genus *Oenothera* (Skrzypczak et al. 1998; Thiem et al. 1999; Taniguchi et al. 2006; Ghauri et al. 2008; Ghasemnezhad et al. 2011). Besides these, only one more species of the Onagraceae family, *Ludwigia repen,s* has been introduced in vitro for micropropagation (Öztürk et al. 2004).

To the best of our knowledge, only few studies on the in vitro micropropagation have been undertaken in only three *Epilobium* species: *E. parviflorum* (Akbudak and Babaoglu 2005; Deliu et al. 2007; Vălimăreanu and Deliu 2008), *E. angustifolium* (Turker et al. 2008), and *E. hirsutum* (Vălimăreanu and Deliu 2008; Tămaş et al. 2009) (Table 6.1). The first tissue culture study carried out within the genus *Epilobium* was published by Akbudak and Babaoglu (2005). This study consisted of callus culture induction in *E. parviflorum* from hypocotyl, cotyledon, petiole, and leaf explants, cultured on MS basal medium with different plant hormones. The best results, for cotyledon- and petiole-derived callus, were obtained on MS (Murashige and Skoog 1962) medium supplemented with 1.0 mg l^{-1} 2,4-D and 0.1 mg l^{-1} Kin and 2.0 mg l^{-1} 2,4-D and 0.2 mg l^{-1} Kin. However, no plant regeneration was observed in either callus induction phase or during the subculturing stages (Akbudak and Babaoglu 2005). The first serious report, describing a rapid and highly efficient in vitro regeneration system via direct shoot development from explants of *Epilobium* species, was published in 2008 by Turker et al. This was dedicated to the in vitro micropropagation of the species *E. angustifolium* through explants excised from sterile seedlings cultured on MS medium supplemented with various PGRs (Plant Growth Regulators). Moreover, to prevent explant and adjacent medium browning, due to phenolic oxidation of freshly cut surfaces of explant tissue, 100 mg l^{-1} ascorbic acid (ASA) was added in all culture media. Best shoot proliferation was obtained from root explants (37 shoots/explant) cultured on MS media with 0.1 mg l^{-1} BA and 0.5 mg l^{-1} IAA. Since the regenerated plantlets were rootless, shoots were transferred onto rooting media cultures. Most shoots developed roots on MS medium with 0.5 mg/l IAA (Turker et al. 2008). Our first attempt on *Epilobium* species micropropagation was directed toward the in vitro culture of *E. parviflorum* (Deliu et. al. 2007; Vălimăreanu and Deliu 2008). Plant tissue cultures were initiated from different types of explants (leaf, petiole, root, and stem) excised from sterile seedlings cultured on a MS agar-gelified medium enriched with different concentrations and combinations of PGRs (BA, 2iP, TDZ, Kin, and NAA). Among these, nodal explants produced the highest number of shoots, when cultured on MS media with BA (1.0 mg l^{-1}) and NAA (0.15 mg l^{-1}) (cca 24 shoots/explant). Nevertheless, the plants obtained on this culture media were very small (cca 1.8 cm length) and rootless. Furthermore, nodal explants cultured on MS medium supplemented with 2iP (1.0 mg l^{-1}) and NAA (0.15 mg l^{-1}) resulted in a smaller number of shoots (5–7 shoots/explant) but with higher shoot length (cca 5.0 cm length). Though explant browning was noticed in all experimental variants, the process was low in intensity and did not require the use of antioxidants. Prior the hardening process, neoformed shootlets were transferred on rooting media cultures. The optimum rooting efficiency was obtained on MS media lacking PGR's but supplemented with activated charcoal (2.5 g l^{-1}) (Deliu et. al. 2007; Vălimăreanu and Deliu 2008).

6.2.2 Establishment of Tissue Cultures and Plant Regeneration

6.2.2.1 Micropropagation of *Epilobium* Species

In order to establish an efficient in vitro micropropagation system for *E. parviflorum* (small flowered willow herb) and *E. hirsutum* L. (great hairy willow herb, hairy willow herb, codlins-and-cream, apple-pie), seeds collected from Alexandru Borza Botanical Garden, Cluj-Napoca, were surface sterilized with 10 % H_2O_2,

Table 6.1 Summary of in vitro culture studies on *Epilobium* species

Epilobium species	Explant sources	MS medium with PGRs (mg l^{-1})	Organogenic response	Chemical analysis	References
Gelified medium					
E. parviflorum Schreb	Plantlet parts	2,4-D (1.0) + Kin (0.1) 2,4-D (2.0) + Kin (0.2)	Callus		Akbudak and Babaoglu (2005)
E. angustifolium L	Seedling parts	BA (0.1) + IAA (0.5) BA (0.5) + IAA (0.5)	Multiple shoots		Turker et al. (2008)
E. parviflorum Schreb	Seedling parts	BA (1.0) + NAA (0.15) 2iP (1.0) + NAA (0.15)	Multiple shoots		Deliu et al. (2007)
E. parviflorum Schreb *E. hirsutum* L	Nodal explants	BA (1.0) + NAA (0.15) 2iP (1.0) + NAA (0.15)	Multiple shoots	RAPD	Vălimăreanu and Deliu (2008)
Liquid medium (shake flasks)					
E. hirsutum L	Nodal explants	BA (1.0) + NAA (0.15) 2iP (1.0) + NAA (0.15)	Multiple shoots	HPLC Tannins	Cristea et al. (2009)
E. hirsutum L	Nodal explants	2iP + NAA	Multiple shoots	Folin-Ciocâlteu Total polyphenols	Tămaş et al. (2009)

Fig. 6.1 *E. parviflorum* **a′** and *E. hirsutum* **b′** cultured on MS supplemented with BA (1.0 mg l^{-1}) for *E. parviflorum*, and (0.5 mg l^{-1}) for *E. hirsutum* and NAA (0.15 mg l^{-1}) **a** 2iP (1.0 mg l^{-1}) and NAA (0.15 mg l^{-1}) **b** and without PGRs **c**. Adventive roots (*arrow*). Growth period 6 weeks. *Bar* 1 cm

and cultured on ½ MS medium with 0.7 % agar. The regenerated seedlings were used for further micropropagation experiments as previously described (Deliu et. al. 2007; Vălimăreanu and Deliu 2008; Cristea et al. 2009). The results showed that MS culture medium supplemented with BA (0.5; 1.0 mg l^{-1}) and NAA (0.15 mg l^{-1}) (Fig. 6.1a) was optimal for both species multiplication. The number of regenerated shoots per explant varied in association with BA concentration, resulting in 24–25 shoots per explant at 1.0 mg l^{-1} BA and 34–35 shoots per explant at 0.5 mg l^{-1} BA. Moreover, under the influence of 1.0 mg l^{-1} BA most regenerated shoots were shorter (1–2 cm height) and showed hyperhydric malformations (Fig. 6.1aa′) compared to those grown on MS culture medium with 0.5 mg l^{-1} BA (2–3 cm height) (Fig. 6.1ab). Similar aspects were noticed in other species as well, under the influence of high BA concentrations (Coste et al. 2011). The association of 2iP (1.0 mg l^{-1}) and NAA (0.15 mg l^{-1}) positively affected both the regeneration rate and shoot length: 4–4.5 cm for *E. parviflorum*, and 6–6.5 cm for *E. hirsutum*. Additionally, a poorly developed root system was observed in both species (Fig. 6.1b). The most obvious difference between the two species regarded especially plantlets size, *E. hirsutum* shoots always being taller. Another process that is rarely encountered in in vitro tissue culture experiments but constantly occurring in both *Epilobium* species was the formation of nodal adventitious roots (about 2–3 roots), especially in plantlets with a rudimentary rooting system (Fig. 6.1c). The adventitious roots might have undertaken the functions of the primary root system, by absorbing the humidity from the hyper-hydrated atmosphere of culture vessels, since plantlets showed a full development. These adventitious roots disappeared once plantlets were transferred onto rooting media cultures and developed a normal rooting system. For hardening, the regenerated plantlets were transferred in pots, and then in experimental plots. These plantlets were then analyzed for their genetic stability by means of RAPED analysis. The results showed that *Epilobium* plants regenerated in vitro on different culture media showed no polymorphism (Vălimăreanu and Deliu 2008).

6.2.2.2 Shake-Flask Cultures

Liquid shoot cultures were established by transferring nodal explants, excised from fully developed *Epilobium* plantlets, to MS liquid medium containing the same growth regulators as the solidified MS medium used for micropropagation. These were subcultured every 3 weeks and grown on a rotary shaker (100 rpm) under fluorescent light. Though both species showed multiplication, biomass production was low especially in *E. parviflorum*. Therefore, after 21 days of culture on MS medium with 2iP (1.0 mg l^{-1}) and NAA (0.15 mg l^{-1}), *E. parviflorum* shoot biomass reached only 7.2 g l^{-1} DW (dry weight), while in *E. hirsutum* reached 11.1 g l^{-1} DW. This was especially due to browning of tissues and liquid media culture. The browning of the tissue and the adjacent medium is assumed to be due to the oxidation of polyphenols, exuded from the cut surfaces of the explants, and formation of guinones, which are highly reactive and toxic to plant tissue (Taji and Williams 1996). The browning of the tissue can be overcome by adding antioxidant substances such as ascorbic acid (AAS), citric acid (CTA), or cysteine (CY), and adsorbent material such as polyvinylpyrrolidone (PVP) or activated charcoal, to prevent oxidation (Pati et al. 2006; Abdelwahd et al. 2008).

Our results registered higher yields in *E. hirsutum* than in *E. parviflorum* for both growth rate and secondary metabolite production. Therefore, further experiments were performed only in *E. hirsutum*. The first experiment consisted in shoot transfer on MS liquid medium supplemented with BA and NAA and 2iP and NAA. The maximum biomass production (26.7 g l^{-1} DW) was achieved after 3 weeks when NAA was used in association with 2iP. In addition, another hormonal combination was used: 2iP (1.0 mg l^{-1}), BA (0.1 mg l^{-1}), and NAA (0.1 mg l^{-1}). The second experiment involved an MS modified culture medium (MSM), similar to that used by Taniguchi et al. (2002) and supplemented with the same hormonal combination (2iP, BA and NAA) mentioned above. Moreover, an antioxidant mixture (AM) consisting of AAS, CTA, and CY (all in 100 mg l^{-1} concentration) was added to this culture medium. To reduce explant browning, nodal explants were presoaked for 1 hour in AM and then cultured on MSM with different PGRs and with or without AM. At the end of the culture period (21 days) the best experimental variant proved to be the one including AM pretreatment of nodal explants followed by in vitro culture on MSM culture variant including AM.

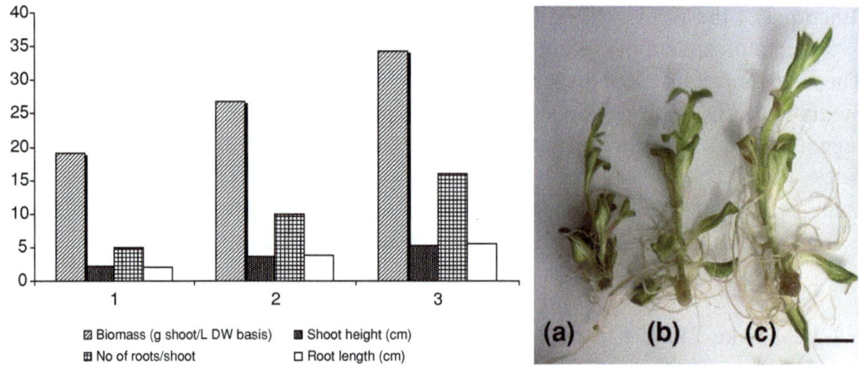

Fig. 6.2 The effect of MSM medium supplemented with 2iP (1.0 mg l^{-1}), BA (0.1 mg l^{-1}), and NAA (0.1 mg l^{-1}) and differential use of an antioxidant mixture (AM) on biomass production, shoot height, and root development in *E. hirsutum* liquid shoot cultures. Nodal explants presoaked in AM and cultured on MSM lacking AM (1; **a**); Nodal explants cultured on MSM culture medium including AM (2; **b**); Nodal explants presoaked in AM and cultured on MSM including AM (3; **c**). Growth period 3 weeks. *Bar* 1 cm

The resulting plantlets proved to be more vigorous, taller, with numerous and longer roots, yielding higher biomass production than other variants used (Fig. 6.2).

Other experimental variants involved the addition of PVP, in different concentrations (100, 250, and 500 mg l^{-1}) as an adsorbent material. This experiment was mainly performed to test the effect of PVP on secondary metabolite production.

6.2.2.3 Secondary Metabolites Content

Chemical analysis of in vitro plantlets revealed that both *E. hirsutum* and *E. parviflorum* have the capacity to produce significant amounts of secondary metabolites. In the beginning TLC was used for the identification of these secondary metabolites. Thus, the presence of phenolic acids (gallic and ellagic acids) and some flavonoids, such as quercitrin and rutin, was detected. Nevertheless, HPLC quantitative analysis showed that flavonoids are produced only in very small amounts. Therefore, further experiments were focused, in both species, on the identification and quantification of tannins.

Ethanolic extracts of *E. hirsutum* and *E. parviflorum*, obtained from both in vitro developed plantlets and plants from the native flora of Transylvania, were analyzed by HPLC Shimadzu system. The method for measuring polyphenolic compounds was developed based on RPLC–MS method. Tannins (ellagitannins and gallotannins) identification was achieved on the generated molecular ion basis. The quantitative determination of whole tannins, in ellagic and gallic acids equivalents, was made after the acidic hydrolysis of plant extracts (Cristea et al. 2009).

HPLC analysis revealed that both *E. hirsutum* and *E. parviflorum*, from either in vivo and/or in vitro experiments, produced galloylglucoses (tetra-, penta-, hexa-, hepta-, and octa-GGs) and monomeric (tellimagrandin I, tellimagrandin II and strictinin), dimeric (oenothein B—OeB), and trimeric (oenothein A—OeA) ellagitannins (Table 6.2). Also, it was determined for the first time the presence of strictinin (monomeric ellagitanin) in native plants and shoots of *E. parviflorum* cultured on MS liquid medium. The same compound was also identified in shoots of *E. hirsutum* cultured on MSM liquid medium. It is important to underline that the presence of strictinin was pointed out in only few species such as *Casuarina stricta* and *Stachyurus praecox* (Okuda et al. 1982), and in *Punica granatum* (Fischer et al. 2011)

The differences regarding hydrolyzable tannins content were multiple and influenced by species, culture medium composition, hormonal variant, and use of the antioxidant mixture or PVP. Thus, the native *E. hirsutum* is richer in both ellagitannins and gallotannins. On the other hand, OeA and OeB, oligomeric ellagitannins with macrocyclic structures and potent antitumor activity, are present in both species. The same aspects were noticed for the in vitro cultured plantlets. Thus, both species revealed richer tannin content when cultured on MS gelified medium suplemented with 2iP and NAA than when cultured on MS gelified medium with BA and NAA. Similarly, shoots cultured on MS liquid medium displayed higher tannin levels than those cultured on MS gelified medium. Nevertheless, none of the two species, cultured on MS medium, synthesize OeB, regardless of hormonal combination, or media culture consistency (liquid or gelified) (Table 6.2). Furthermore, the subcultivation of *E. hirsutum* shoots on MSM liquid medium, with 2iP, BA, NAA, antioxidant mixture, and PVP, stimulated ellagitannins and gallotannins biosynthesis. Thus, those shoots regenerated from nodal explants presoaked in AM and cultured on MSM medium including the same antioxidant mixture revealed the entire set of analyzed tannins, including OeA and OeB, even in a higher ratio than native *E. hirsutum* plants (Table 6.2).

Quantitative analysis of total tannins (Table 6.3) revealed the same aspects as qualitative analysis, irrespective of results being expressed in gallic or ellagic acid. The highest tannin content was registered for both in vivo and in vitro *E. hirsutum* plantlets. Moreover, the subcultivation of *E. hirsutum* shoots on MSM liquid medium with AM and PVP significantly enhanced tannin's production. In the same manner as previously noticed, the shoots regenerated from nodal explants presoaked in AM and cultured on MSM variant including AM exhibited even higher levels of tannins than the native plant. Thus, shoots cultured on the experimental variant including 100 mg l^{-1} PVP (1APVP) accumulated a total tannin content of 19.9 mg g^{-1} DW (ellagic acid equivalent) or 16.0 mg g^{-1} DW (gallic acid equivalent), compared to that found in the native mother plants, respectively, of 16.3 mg g^{-1} DW (ellagic acid equivalent) or 13.6 mg g^{-1} DW (gallic acid equivalent) (Table 6.3).

The results obtained within our study, although lacking the quantitative determination of ellagitannins and gallotannins, are in agreement with those acquired by Taniguchi et al. (2002), at least regarding the composition of

Table 6.2 Presence of tannins (ellagitannins and gallotannins) in *Epilobium* extracts

Epilobium species	MS medium with PGRs (mg l^{-1})	Ellagitannis					Gallotannins				
		OeA	Oe B	Tell I	Tell II	Str	Tetra GG	Penta GG	Hexa GG	Hepta GG	Octa GG
E. hirstum											
Native flora	–	+	+	+	+	–	+	–	+	–	+
In vitro	Gelified MS medium										
	0	+	–	+	–	–	–	–	–	–	–
	BA (1.0) + NAA (0.15)	+	–	+	–	–	+	–	–	–	+
	2iP (1.0) + NAA (0.15)	+	–	+	–	–	+	+	+	–	+
	Liquid MS without antioxidants and not presoaked nodal explants										
	2iP (1.0) + BA (0.1) + NAA (0.1)	+	–	+	+	–	+	+	–	–	+
E. parviflorum											
Native flora	–	+	+	+	–	+	+	–	–	–	–
In vitro	Gelified MS medium										
	0	–	–	+	–	–	+	–	–	–	–
	BA (1.0) + NAA (0.15)	+	–	+	–	–	–	–	–	–	–
	2iP (1.0) + NAA (0.15)	+	–	+	–	–	+	–	–	–	–
	Liquid MS medium without antioxidants and not presoaked nodal explants										
	2iP (1.0) + BA (0.1) + NAA (0.1)	+	–	+	+	+	+	–	–	–	–
E. hirsutum cultured on MSM liquid medium with 2iP (1.0 mg l^{-1}) + BA (0.1 mg l^{-1}) + NAA (0.1 mg l^{-1}) +/_ antioxidant											
1 M		+	+	+	+	+	+	+	+	–	+
2 M		+	+	+	+	+	+ +	+	+	–	+
3 M		++	+	++	+	+	++	++	+	–	+
1APVP		++	++	++	+	+	++	++	+	+	+
2APVP		++	+	+	+	+	++	+	+	–	+
3APVP		+	+	+	+	+	+	+	–	–	–

(++) High ratio; (+) media ratio; (–) minor or absent compound. OeA = oenothein A; OeB = Oenothein B; Tell I = Tellimagrandin I; Tell II = Tellimagrandin II; Str = strictinin. GG = galloylglucose. The tannins identification was achieved on the generated molecular ion basis (Cristea et al. 2009)

(1 M) Nodal explants presoaked in AM and cultured on MSM without AM; (2 M) Nodal explants not presoaked in AM and cultured on MSM culture medium with AM; (3 M) Nodal explants presoaked in AM and cultured on MSM medium with AM; (1APVP) Nodal explants presoaked in AM and cultured on MSM with AM and PVP (100 mg l^{-1}); (2APVP) Nodal explants presoaked in AM and cultured on MSM with AM and PVP (250 mg l^{-1}); (3APVP) Nodal explants presoaked in AM and cultured on MSM with AM and PVP (500 mg l^{-1})

Table 6.3 Content of total taninns of *E. hirsutum* and *E. parviflorum*, in vivo and in vitro

Epilobium species	MS medium with PGRs ($mg\ g^{-1}$)	Ellagic acid equivalent ($mg\ g^{-1}$ DW)	Gallic acid equivalent ($mg\ g^{-1}$ DW)
E hirstum			
Native flora	–	16.3	13.6
In vitro	Gelified MS medium		
	0	2.5	0.9
	BA (1.0) + NAA (0.15)	0.5	0.3
	2iP (1.0) + NAA (0.15)	4.2	1.5
	Liquid MS medium without antioxidants and not presoaked nodal explants		
	2iP (1.0) + BA (0.1) + NAA (0.1)	9.9	93.7
E. parviflorum			
Native flora	–	10.8	9.2
In vitro	Gelified MS medium		
	0	1.8	0.6
	BA (1.0) + NAA (0.15)	0.7	0.1
	2iP (1.0) + NAA (0.15)	3.4	0.8
	Liquid MS medium without antioxidants and not presoaked nodal explants		
	2iP (1.0) + BA (0.1) + NAA (0.1)	5.8	4.2
E. hirsutum cultured on MSM liquid medium with 2iP (1.0 $mg\ l^{-1}$) + BA (0.1 $mg\ l^{-1}$) + NAA (0.1 $mg\ l^{-1}$) +/_ antioxidants			
1 M		12.3	9.8
2 M		14.1	12.4
3 M		16.9	14.0
1APVP		19.9	16.0
2APVP		15.3	13.2
3APVP		11.8	6.8

The quantitative determination of the whole tannins in ellagic and gallic acids equivalents was made after the acidic hydrolysis of those extracts and based on the HPLC method (Cristea et al. 2009)

(1 M) Nodal explants presoaked in AM and cultured on MS medium without AM; (2 M) Nodal explants not presoaked in AM and cultured on MS medium with AM; (3 M) Nodal explants presoaked in AM and cultured on MS medium with AM; (1APVP) Nodal explants presoaked in AM and cultured on MS medium with AM and PVP (100 $mg\ l^{-1}$); (2APVP) Nodal explants presoaked in AM and cultured on MS medium with AM and PVP (250 $mg\ l^{-1}$); (3APVP) Nodal explants presoaked in AM and cultured on MS medium with AM and PVP (500 $mg\ l^{-1}$)

hydrolyzable tannins (galloylglucoses to oligomeric ellagitannins). The authors proved that shoot cultures of *Oenothera tetraptera* Cav. are able to synthesize the same complex set of tannins as the native mother plants. They also determined that the intensity of the process was influenced by the developmental stage of the regenerated plantlets, respectively, by the degree of cellular differentiation. Moreover, the same authors found that OeB production increased with the decrease of shoot vitrification rate (Taniguchi et al. 2002). Our results revealed similar aspects, especially when shoots of both species showed signs of vitrification when BA (1.0 mg l^{-1}) was used, and the production of tannins dropped significantly compared to other hormonal variants (Tables 6.2, 6.3).

6.3 Conclusions

Our micropropagation experiments of *E. hirsutum* and *E. parviflorum* revealed that a single nodal explant can produce up to 24 or 34 shoots. The use of liquid culture system stimulated shoot proliferation and biomass production (26–34 g l^{-1} DW). The regenerated plantlets maintained their genetic stability regardless of the experimental variant used. The use of MSM liquid culture medium, comprising an antioxidant mixture, stimulated shoot development, plantlets being more vigorous with a well-developed rooting system only after 3 weeks of culture. Thus, plantlets were ready for direct ex vitro transfer and acclimatization. Also, *E. hirsutum* shoots cultured on this experimental variant were able to synthesize different types of tannins and produce a larger number of ellagitannins and gallotannins than native mother plants. Similarly, regardless of experimental variants, *E. hirsutum* shoots revealed higher total tannin content when compared with ex vitro donor plants. Furthermore, it was determined for the first time the presence of strictinin in native plants and shoots of *E. parviflorum* cultured on MS liquid medium and in shoots of *E. hirsutum* cultured on MSM liquid medium.

These results show that the micropropagation of both *Epilobium* species, performed either on gelified culture medium or especially on liquid culture medium, can enhance biomass and secondary metabolite production, especially of tannins.

References

Abdelwahd R, Hakam N, Labhilili M, Udupa SM (2008) Use of an adsorbent and antioxidants to reduce the effects of leached phenolics in in vitro plantlet regeneration of faba bean. Afr J Biotechnol 7:997–1002

Akbari RS, Azizian D (2006) Seed morphology and seed coat sculpturing of *Epilobium* L. species (Onagraceae Juss.) from Iran. Turk J Bot 30:435–440

Akbudak MA, Babaoglu M (2005) Callus induction in small flowered willow herb (*Epilobium parviflorum* Schreb). J Plant Biochem Biotechnol 14:189–191

Aoki K, Maruta H, Uchiumi F, Hatano T, Yoshida T, Tanuma S (1995) A macrocircular ellagitannin, oenothein B, suppresses mouse mammary tumor gene expression via inhibition of poly(ADP-ribose) glycohydrolase. Biochem Biophys Res Commun 210:329–337

Arredondo MF, Blasina F, Echeverry C, Morquio A, Ferreiora M, Abin-Carriquiry JA, Lafon L, Dajas F (2004) Cytoprotection by *Achyrocline saturejoides* (Lam) D.C. and some of its main flavonoids against oxidative stress. J Ethnopharmac 91:13–20

Averett JE, Kerr B, Raven PH (1978) Flavonoids of Onagraceae: *Epilobium* sect. Epilobium Amer J Bot 65:567–570

Averett JE, Raven PH, Becker H (1979) The flavonoids of Onagraceae: tribe Epilobieae. Amer J Bot 66:1151–1155

Barakat HH, Hussein SAM, Marzour MS, Merfort I, Linscheid M, Nawwar MAM (1997) Polyphenolic metabolites of *Epilobium hirsutum*. Phytochem 46:935–941

Bartfay WJ, Bartfay E, Green-Johnson J (2011) Gram-negative and gram-positive antibacterial properties of the whole plant extract of willow herb (*Epilobium angustifolium*). Biol Res Nurs. doi:10.1177/1099800410393947

Battinelli L, Tita B, Evandri MG, Mazzanti G (2001) Antimicrobial activity of *Epilobium* spp. extracts. Farmaco 56:345–348

Baum DA, Sytsma KJ, Hoch PC (1994) The phylogeny of *Epilobium* (Onagraceae) based on nuclear ribosomal DNA sequences. Syst Bot 19:363–388

Bazylko A, Kiss AK, Kowalski J (2007a) Densitometric determination of flavonoids in methanolic and aqueous extracts of *Epilobii angustifolii herba* by use of HPTLC. J Plant Chromatogr 20:53–56

Bazylko A, Kiss AK, Kowalski J (2007b) High-performance thin-layer chromatography method for quantitative determination of oenothein B and quercetin glucuronide in aqueous extract of *Epilobii angustifolii herba*. J Chromatogr A 1173:146–150

Bisset NG, Wichtl M (1994) Herbal drugs and phytopharmaceuticals: a handbook for practice on a scientific basis. Medpharm Scientific Publishers, CRC Press, Stuttgart

Bleeker W, Schmitz U, Ristow M (2007) Interspecific hybridisation between alien and native plant species in Germany and its consequences for native biodiversity. Biol Conserv 137: 248–253

Bohinsky P (1993) *Epilobium* spp. as herbal plants. Presented at the conference on cultivation, harvesting and processing of herbs, Nova Lubovna, Slovakia, 15–17 June 1993, p 8

Borchardt JR, Wyse DL, Sheaffer CC, Kauppi KL, Fulcher GR, Ehlke NJ, Biesboer DD, Bey RF (2008) Antimicrobial activity of native and naturalized plants of Minnesota and Wiscons. J Med Plants Res 2:98–110

Bunney S (1992) The illustrated encyclopedia of herbs (their medicinal and culinary uses). Chancellor press, London

Ciulei I, Grigorescu E, Stănescu U (1993) Plante medicinale (Medicinal plants). Fitochimie şi Fitoterapie. Tratat de Farmacognozie, vol. 2. Edit Med, Bucureşti, pp 646–647

Coste A, Vlase L, Halmagyi A, Deliu C, Coldea Gh (2011) Effects of plant growth regulators and elicitors on production of secondary metabolites in shoot cultures of *Hypericum hirsutum* and *Hypericum maculatum*. Plant Cell Tiss Organ Cult. doi:10.1007/s11240-011-9919-5

Cristea V, Deliu C, Oltean B, Butiuc-Keul A, Brummer A, Albu C, Radu GL (2009) Soilless cultures for pharmaceutical use and biodiversity conservation. Acta Hort 843:157–164

Deliu C, Coste A, Tămaş M, Halmagyi A, Butiuc-Keul A (2007) Propagarea in vitro a speciei *Epilobium parviflorum* Schreb.—o importantă plantă medicinală (In vitro propagation of *Epilobium parviflorum* Schreb—an important medicinal plant—in the Romanian languages). In: lucrările celui de al XV-lea Simpozion Naţional de Culturi de Ţesuturi şi Celule Vegetale "Micropropagarea speciilor vegetale" Iaşi, 7 iunie 2006, pp 212–224

Ducrey B, Marston A, Gohring S, Hartmann RW, Hostettmann K (1997) Inhibition of 5 α-reductase and aromatase by the ellagitannins oenothein A and oenothein B from *Epilobium* species. Planta Med 63:111–114

Ducrey B, Wolfender JL, Marston A, Hostettmann K (1995) Analysis of flavonol glycosides of thirteen *Epilobium* species (Onagraceae) by LC-UV and thermospray LC-MS. Phytochemistry 38:129–137

Fischer UA, Carle R, Kammerer DR (2011) Identification and quantification of phenolic compounds from pomegranate (*Punica granatum* L.) peel, mesocarp, aril and differently produced juices by HPLC-DAD–ESI/MS^n. Food Chem 127:807–821

Fukuchi K, Sakagami H, Okuda T et al (1989) Inhibition of herpes simplex virus infection by tannins and related compounds. Antiviral Res 11:285–298

Ghasemnezhad A, Mousavizadeh SJ, Mashayekhi K (2011) A study on evening-primrose (*Oenothera biennis* L.) callus regeneration and somatic embryogenesis. Iranian J Biotechnol 9:31–36

Ghauri EG, Shafi N, Fatima A, Ghani S (2008) High frequency plant regeneration from leaf explants derived callus of evening primrose (*Oenothera biennis*). Pakistan J Biochem Mol Biol 41:158–161

Glen AT, Lawrie W, McLean J, El-Garby Younes M (1967) Triterpenoid constituents of rosebay willow-herb. J Chem Soc C Org chem 26 510–515

Guil-Guerrero JL, García Maroto FF, Giménez Giménez A (2001) Fatty acid profiles from forty-nine plant species that are potential new sources of γ-linolenic acid. JAOCS 78:677–684

Gupta RK, Al-Shafi SM, Layden K, Haslam E (1982) The metabolism of gallic acid and hexahydroxydiphenic acid in plants. Part 2. Esters of (S)-hexahydroxydiphenic acid with Dglucopyranose ($^4C_1$). J Chem Soc Perkin Trans 1:2525–2534

Hartzfeld PW, Forkner R, Hunter MD, Hagerman AE (2002) Determination of hydrolyzable tannins (gallotannins and ellagitannins) after reaction with potassium iodate. J Agric Food Chem 50:1785–1790

Hatano T, Yasuhara T, Matsuda M, Yazaki K, Yoshida T, Okuda T (1990) Oenothein B, a dimereric hydrolysable tannin with macrocyclic structure and accompanying tannins from *Oenothera erythrosepara*. J Chem Soc Perkin Trans 1:2735–2743

Hevesi Tóth B, Balázs A, Vukics V, Szőke É, Kéry Á (2006) Identification of *Epilobium* species and willow-herbs (*Onagraceae*) by HPLC analysis of flavonoids as chemotaxonomic markers. Chromatographia Suppl 63:S119–S123

Hevesi Tóth B, Blazics B, Kéry Á (2009) Polyphenol composition and antioxidant capacity of *Epilobium* species. J Pharm Biomed Anal 49:26–31

Hevesi Tóth B, Houghton PJ, Habtemariam S, Kéry Á (2008) Antioxidant and antiinflammatory effect of *Epilobium parviflorum* Schreb. Phytother Res 23:719–724

Hevesi Tóth B (2009) Phytochemical and in vitro biological evaluation of potentially active compounds in *Epilobium* species. Thesis of doctoral (Ph.D.) dissertation. Semmelweis University, Department of Pharmacognosy, Budapest

Hiermann A (1983) Die Untersuchung potentieller Wirkstoffe in *Epilobium*-Arten.1. Mitteilung: Aufklärung der Flavonoidmuster. Sci Pharm 51:158–167

Hiermann A, Bucar F (1997) Studies of *Epilobium angustifolium* extracts on growth of accessory sexual organs in rats. J Ethnopharmacol 55:179–183

Hiermann A, Juan H, Sametz W (1986) Influence of *Epilobium* extracts on prostaglandin biosynthesis and carrageenin induced oedema of the rat paw. J Ethnopharmacol 17:161–169

Hiermann A, Mayr K (1985) Die Untersuchung potentieller Wirkstoffe in *Epilobium*-Arten. 3. Mitteilung: Uber das Vorkommen von β-Sitosterolderivaten in *Epilobium angustifolium* L. und *Epilobium parviflorum* Schreb. Sci Pharm 53:39–44

Hiermann A, Reidlinger M, Juan H, Samez W (1991) Isolierung des antiphlogistischen Wirkprinzips von *Epilobium angustifolium* (isolation of the antiphlogistic principle from *Epilobium angustifolium*). Planta Med 57:357–360

Hiermann A, Schramm HW, Laufer S (1998) Anti-inflammatory activity of myricetin-3-O-β-D-glucuronide and related compounds. Inflamm Res 47:421–427

Hoch PC, Raven PH (1992) Boisduvalia, a coma-less *Epilobium* (Onagraceae). Phytologia 73:456–459

Holub J (1978) *Epilobium komarovianum* Leveille Československu. Zpr Č Bot Společ 13:77–85

Ivancheva S, Manolova N, Serkedjieva J, Dimov V, Ivanovska N (1992) Polyphenols from Bulgarian medicinal plants with anti-infectious activity. Basic Life Sci 59:717–728

Jones NP, Arnason JT, Abou-Zaid M, Akpagana K, Sanchez-Vindas P, Smith ML (2000) Antifungal activity of extracts from medicinal plants used by First Nations Peoples of Eastern Canada. J Ethnopharmacol 73:191–198

Juan H, Sametz W, Hiermann A (1988) Anti-inflammatory effects of a substance extracted from *Epilobium angustifolium*. Agents Actions 23:106–107

Kähkönen MP, Hopia AI, Vuorela HJ, Rauha JP, Pihlaja K, Kujala TS, Heinonen M (1999) Antioxidant activity of plant extracts containing phenolic compounds. J Agric Food Chem 47:3954–3962

Kim DO, Lee CY (2002) Unit I1.3 HPLC separation of polyphenolics. Published Online: 1 FEB 2003 doi:10.1002/0471142913.fai0103s06

Kiss A, Kowalski J, Melzig MF (2006a) Effect of *Epilobium angustifolium* L. extracts and polyphenols on cell proliferation and neutral endopeptidase activity in selected cell lines. Pharmazie 61:66–69

Kiss A, Kowalski J, Melzig MF (2006b) Induction of neutral endopeptidase activity in PC-3 cells by an aqueous extract of *Epilobium angustifolium* L. and oenothein B. Phytomed 13:284–289

Kiss AK, Bazylko A, Filipek A, Granica S, Jaszewska E, Kiarszus U, Kośmider A, Piwowarski J (2010) Oenothein B's contribution to the anti-inflammatory and antioxidant activity of *Epilobium* sp. Phytomed. doi:10.1016/j.phymed.2010.10.016

Laurent E (1989). Edible & medicinal plants of Britain and Northern Europe. Hamlym publishing group, London

Lesuisse D, Berjonneau J, Ciot C et al (1996) Determination of oenothein B as the active 5-alpha-reductase-inhibiting principle of the folk medicine *Epilobium parviflorum*. J Nat Prod 59: 490–492

Levin RA, Wagne WL, Hoch PC et al (2003) Family-level relationships of Onagraceae based on chloroplast *rbcL* and *ndh*F data. Am J Bot 90:107–115

Levin RA, Wagner WL, Hoch PC et al (2004) Paraphyly in tribe Onagreae: insights into phylogenetic relationships Onagraceae based on nuclear and chloroplast sequence data. Syst Bot 29:147–164

Liu HE, Qi LW, Cao J, Li P, Li CY, Peng YB (2008) Advances of modern chromatographic and electrophoretic methods in separation and analysis of flavonoids. Molecules 13:2521–2544

Liu Y, Wang C, Han Q, Yu B, Ding G (2003) Study on chemical constituents of *Chamaenerion angustifolium* II. Tannins and related polyphenolic compounds. Chin Trad Herb Drugs 34:967–969

Miyamoto K, Kishi N, Koshiura R, Yoshida T, Hatano T, Okuda T (1987) Relationship between the structures and the antitumor activities of tannins. Chem Pharm Bull 35:814–822

Miyamoto K, Murayama T, Nomura M et al (1993a) Antitumor activity and interleukin-1 induction by tannins. Anticancer Res 13:37–42

Miyamoto K, Nomura M, Murayama T et al (1993b) Antitumor activities of ellagitannins against sarcoma-180 in mice. Biol Pharm Bull 16:379–387

Miyamoto K, Nomura M, Sasakura M et al (1993c) Antitumor activity of oenothein B, a unique macrocyclic ellagitannin. Jpn J Cancer Res 84:99–103

Moerman DE (1998) Native American Ethnobotany. Timber Press, Inc., Portland. Oregon

Murashige T, Skoog F (1962) A revised medium for rapid growth and bio-assays with tobacco tissue cultures. Physiol Plant 15:473–497

Nawwar MAM, Marzouk MS, Nigge W, Linscheid M (1997) High performance liquid chromatographic/electrospray ionization mass spectrometric screening for polyphenolic compounds of *Epilobium hirsutum*—The structure of the unique ellagitannin epilobamide-A. J Mass Spectrom 32:645–654

Nowak R, Krzaczek T (1998a) Sterols in the herb of *Epilobium angustifolium* L. Herba Polonica 4:297–299

Nowak R, Krzaczek T (1998b) Pharmacognostical research involving herbs of *Epilobium agustifolium* L. and *Epilobium parviflorum* Schreb. Herba Polonica 1:5–10

Obradovič M, Strgulc Krajšek S, Dermastia M, Kreft S (2007) A new method for the authentication of plant samples by analyzing fingerprint chromatograms. Phytochem Anal 18:123–132

Okuda T, Yoshida T, Hatano T (1989) Ellagitannins as active constituents of medicinal plants. Planta Med 55:117–122

Okuda T, Yoshida T, Yazaki K, Ashida M (1982) Casuariin, stachyurin and strictinin, new ellagitannins from *Casuarina stricta* and *Stachyurus praecox*. Chem Pharm Bull 30:766–769

Öztürk M, Khawar KM, Atar HH, Sancak C, Özcan S (2004) In vitro micropropagation of the aquarium plant *Ludwigia repens*. Asia Pac J Mol Biol Biotechnol 12:21–25

Pascual Teresa J, Corrales B, Grande M (1979) Acidos triterpénicos de *Epilobium hirsutum* L. Acido 23-hidroxitorméntico. An Quím (Madrid) 75:135–139

Pati PK, Rath SP, Sharma M, Sood A, Ahuja PS (2006) In vitro propagation of rose—a review. Biotechnol Adv 24:94–114

Pelc M, Kosakowska O, Weglarz Z, Przybyl J (2005) Sterols and fatty acids in the seeds of evening primrose (Oenothera sp.) and willow herb (*Epilobium* sp.). Herba Polonica 51:20–24

Pourmorad F, Ebrahimzadeh MA, Mahmoudi M, Yasini S (2007) Antinociceptive activity of methanolic extract of *Epilobium hirsutum*. Pak Biol Sci 10:2764–2767

Rauha JP, Remes S, Heinonen M, Hopia A, Kahkonen M, Kujala T, Pihlaja H, Vuorela H, Vuorela P (2000) Antimicrobial effects of Finnish plant extracts containing flavonoids and other phenolic compounds. Int J Food Microbiol 56:3–12

Raven PH (1980) Epilobium L. In: Tutin TG (ed) Flora Europaea 5. Cambridge University Press, Cambridge pp 308–311

Raven PH (1976) Generic and sectional delimitation in Onagraceae, tribe *Epilobieae*. Ann Missouri Bot Gard 63: 326–340. [published 18 Jun 1977]

Rijke E, Out P, Niessen WMA, Ariese F, Gooijer C, Brinkman UATh (2006) Analytical separation and detection methods for flavonoids. J Chromatogr A 1112:31–63

Schepetkin LN, Kirpotina IA, Jakiw L, Khlebnikov AI, Blaskovich CL, Jutila MA, Quinn MT (2009) Immunomodulatory activity of oenothein B isolated from *Epilobium angustifolium*. J Immunol 183:6754–6766

Shikov AN, Poltanov E, Pozharitskaya O, Dorman HJD, Marakov VG, Tikhonov V, Hiltunen R (2007) HPLC evaluation of water-soluble extracts of *Chamaenerion angustifolium* L. and *Pentaphylloides fruticosa* L. Planta Med 73:916–917

Shikov AN, Poltanov EA, Dorman HJ, Makarov VG, Tikhonov VP, Hiltunen R (2006) Chemical composition and in vitro antioxidant evaluation of commercial water-soluble willow herb (*Epilobium angustifolium* L.) extracts. J Agric Food Chem 54:3617–3624

Shikov AN, Pozharitskaya ON, Ivanova SA, Makarov VG, Tikhonov VP, Galambosi B (2010) Improved and validated HPTLC method for quantification of oenothein B and its use for analysis of *Epilobium angustifolium* L. J Planar Chromatogr 23:70–74

Skrzypczak L, Thiem B, Wesołowska M (1998) *Oenothera* species (eving primrose): In vitro regeneration, production of flavonoids, fatty acids, and other secondary metabolites. In: Bajaj YPS (ed) Biotechnol agric for, Med arom plants X, vol 41. Springer, Berlin pp 276–304

Slacanin I, Marston A, Hostettmann K, Delabays N, Darbellay C (1991) Isolation and determination of flavonol glycosides from *Epilobium* species. J Chromatogr 557:391–398

Smirnova GV, Vysochina GI, Muzyka NG, Samoylova ZY, Kukushkina TA, Oktyabrsky ON (2010) Evaluation of antioxidant properties of medical plants using microbial test systems. World J Microbiol Biotechnol doi:10.1007/s11274-010-0417-4

Snogerup S (1982) A new species of *Epilobium* (Onagraceae) from Northern Greece. Wildenowia 14:227–229

Stajner D, Popovic BM, Boza P (2007) Evaluation of willow herb's (*Epilobium angustifolium L.*) antioxidant and radical scavenging capacities. Phytother Res 21:1242–1245

Stalikas CD (2007) Extraction, separation, and detection methods for phenolic acids and flavonoids. J Sep Sci 30:3268–3295

Steenkamp V, Gouws MC, Gulumian M, Elgorashi EE, van Staden J (2006) Studies on antibacterial, anti-inflammatory and antioxidant activity of herbal remedies used in the treatment of benign prostatic hyperplasia and prostatitis. J Ethnopharmacol 103:71–75

Strgulc Krajšek S, Buh P, Žega A, Kreft S (2008) Identification of herbarium whole-leaf samples of *Epilobium* species by ATR-IRspectroscopy. Chem Biodivers 5:310–317

Strgulc Krajšek S, Dermastia M, Jogan N (2006) Determination key for Central European *Epilobium* species based on trichome morphology. Bot Helv 116:169–178

Strgulc Krajšek S, Šegedin T, Jogan N (2009) Revision of *Epilobium* and *Chamerion* in the Croatian herbaria ZA and ZAHO. Nat Croat Zagreb 18:155–167

Taji AM, Williams RR (1996) Overview of plant tissue culture. In: Taji AM, Williams RR (eds) Tissue culture of Australian plants: Past. University of New England Press, Present and Future, Armidale, Australia, pp 1–15

Tămaş M (1997) Plante medicinale şi preparate fitoterapeutice utilizate în afecţiuni ale prostatei. Acta Phytother Rom 4:12–15

Tămaş M, Toiu A, Oniga I, Deliu C, Oltean B, Coldea Gh (2009) Quantitative determination of total polyphenols and flavonoids from indigenous species of *Epilobium* of wild origin and in vitro regenerated plantlets. Contrib Bot Cluj-Napoca 44:119–123

Taniguchi S, Hatano T, Yazaki K (2006) Production of tannin by tissue culture of woody plants and tannin biosynthesis. Mokuzai Gakkaishi 52:67–76

Taniguchi S, Imayoshi Y, Hatano T, Yazaki K, Yoshida T (2002) Hydrolysable tannin production in *Oenothera tetraptera* shoot tissue cultures. Plant Biotechnol 19:357–363

Thiem B, Skrzypczak L, Lamer-Zarawska E (1999) Micropropagation of selected *Oenothera* species and preliminary studies on their secondary metabolites. Acta Soc Bot Pol 68:15–20

Tita B, Abdel-Haq H, Vitalone A, Mazzanti G, Saso L (2001) Analgesic properties of *Epilobium angustifolium*, evaluated by the hot plate test and the writhing test. Farmaco 56:341–343

Treben M (1984) Health through God's pharmacy, advice and experiences with medicinal herbs. Wilhelm Ennsthaler, Steyr, Austria

Turker AU, Mutlu EC, Yıldırım AB (2008) Efficient in vitro regeneration of fireweed, a medicinal plant. Acta Physiol Plant 30:421–426

Tyler VE (1986) Some recent advances in herbal medicine. Pharm Int 7:203–207

Vălimăreanu S, Deliu C (2008) Polymorphism detection of in vitro cultivated *Epilobium* (Onagraceae) species using RAPD method. Contrib Bot Cluj-Napoca 43:113–120

Van Hellemont J (1986) Compendium de phytothérapie. Association pharmaceutique Belge, Brussels, pp 143–144

Vitali F, Fonte G, Saija A, Tita B (2006) Inhibition of intestinal motility and secretion by extract of *Epilobium* in mice. J Etnopharmacol 107:342–348

Vitalone A, Bordi F, Baldazzi C, Mazzanti G, Saso L, Tita B (2001) Anti-proliferative effect on a prostatic epithelial cell line (PZ-HPV-7) by *Epilobium angustifolium* L. Farmaco 56:483–489

Vitalone A, Guizzetti M, Costa LG, Tita B (2003a) Extracts of various species of *Epilobium* inhibit proliferation of human prostate cells. J Pharm Pharmacol 55:683–690

Vitalone A, McColl J, Thome D, Costa LG, Tita B (2003b) Characterization of the effect of *Epilobium* extracts on human cell proliferation. Pharmacology 69:79–87

Voynova E, Dimitrova S, Naydenova E, Karadjov P (1991) Inhibitory action of extracts of *Maclura aurantiaca* and *Epilobium hirsutum* on tumour models in mice. Acta Physiol Pharmacol Bulg 17:50–52

Wagner LW, Hoch CP, Raven HP (2007) Revised classification of the Onagraceae. Syst Bot Monogr 83:1–240

Webster D, Taschereau P, Belland RJ, Sand C, Rennie RP (2008) Antifungal activity of medicinal plant extracts; preliminary screening studies. J Ethnopharmacol 115:140–146

Wichtl M (1989) Teedrogen, Ein Handbuch fur die Praxis auf Wissenschaftlicher Grundlage. Wissenschafliche Verlag, Stuttgart pp 519–521

Wichtl M (ed) (1994) Herbal Drugs and Phytopharmaceuticals, Medpharm, Stuttgart, Germany

Wichtl M (2004) Herbal Drug and Phytopharmaceuticals. Medpharm GmbH, Stuttgart, Germany

Wojdyło A, Oszmiański J, Czemerys R (2007) Antioxidant activity and phenolic compounds in 32 selected herbs. Food Chem 105:940–949

Yi ZC, Liu YZ, Li HX, Yan Y, Zhuang FY, Fan YB, Zhao W (2006) Tellimagrandin I enhances gap junctional communication and attenuates the tumor phenotype of human cervical carcinoma HeLa cells in vitro. Cancer Lett 242:77–87

Yoshimura M, Amakura Y, Tokuhara M, Yoshida T (2008) Polyphenolic compounds isolated from the leaves of *Myrtus communis*. Nat Med (Tokyo) 62:366–368

Zhang J, Li L, Kim SH, Hagerman AE, Junxuan Lü J (2009) Anti-cancer, anti-diabetic and other pharmacologic and biological activities of penta-galloyl-glucose. Pharm Res 26:2066

Chapter 7
Photoelicitation of Bioactive Secondary Metabolites by Ultraviolet Radiation: Mechanisms, Strategies, and Applications

Hélio Nitta Matsuura, Fernanda de Costa, Anna Carolina Alves Yendo and Arthur Germano Fett-Neto

7.1 Introduction

UV radiation is divided into three classes: UV-C, UV-B, and UV-A. Although the highly energetic UV-C (200–280 nm) is completely absorbed by atmospheric gases and UV-A (315–400 nm) is hardly absorbed by ozone, the potentially harmful UV-B (280–320 nm) is only partially absorbed by atmospheric ozone, comprising approximately 4% of terrestrial radiation. In the last 20 years, the depletion of the stratospheric ozone layer, catalyzed by chlorofluorocarbons and other pollutants, resulted in rising levels of the sun's UV-B radiation reaching the Earth's surface. Due to the high energy of UV-B radiation, even modest increases could lead to significant biological damage (Jansen et al. 1998; Frohnmeyer and Staiger 2003).

Elevations of UV-B radiation levels have effects on plant development, morphology, and physiology. Such responses include inhibition of plant growth rates, biomass reduction, increased accumulation of UV absorbing secondary metabolites, and influence on numerous ecological processes. In addition to indirect changes, caused by affecting host plant quality, predators, and pathogens,

H. N. Matsuura · F. de Costa · A. C. A. Yendo · A. G. Fett-Neto (✉)
Department of Botany, Plant Physiology Laboratory, Center for Biotechnology,
Federal University of Rio Grande do Sul (UFRGS), Av. Bento Gonçalves 9500, CP 15005,
Porto Alegre, RS 91501-970, Brazil
e-mail: fettneto@cbiot.ufrgs.br

H. N. Matsuura
e-mail: helio_nitta@yahoo.com.br

F. de Costa
e-mail: fernandadecosta@yahoo.com.br

A. C. A. Yendo
e-mail: anna.yendo@yahoo.com.br

S. Chandra et al. (eds.), *Biotechnology for Medicinal Plants*,
DOI: 10.1007/978-3-642-29974-2_7,

UV-B radiation may directly cause modifications in herbivore behavior and physiological processes (Julkunen-Tiitto et al. 2005; Tuteja et al. 2009).

As sessile organisms, plants are exposed to various environmental factors that lead to changes in physiology and morphology. One of the most relevant of these factors is ultraviolet radiation, on which significant advances have been achieved, both in understanding effects on plants and describing response mechanisms. UV-B radiation can inhibit growth and decrease leaf area, but also increase parameters such as leaf number, secondary branch number, and leaf weight per plant. Other aspects of leaf morphology can also change upon UV-B exposure, including presence of particulate matter in substomatic chambers (Zu et al. 2010), and accumulation of epicuticular wax crystalloids on the leaf surface, reflecting solar radiation (Barnes et al. 1996). UV-A and UV-C radiation affected the structure and ultrastructure of *Capsicum longum*, leading to a decrease in shoot growth and leaf area, whereas stem and leaf thickness, as well as stomata number and size significantly increased. UV treatment also caused thylakoid expansion, starch reduction, and formation of crystals in peroxisomes of mesophyll cells (Sarghein et al. 2011).

Plants have protective mechanisms, both constitutive and induced, or can activate repair responses to cope with UV-B stress. One of the most common protective responses is the accumulation of secondary metabolites capable of absorbing radiation in the ultraviolet wavelength range, such as anthocyanins, flavonols, and flavones, which can also scavenge free radicals, mainly ROS (Hahlbrock and Scheel 1989). Plants can also produce antioxidant enzymes, such as catalases, peroxidases, and superoxide dismutases, to scavenge these free radicals and protect cellular integrity (Heinonen et al. 1998). Differences in defense responses, as well as in their kinetics and intensity, can be the reason for some plants being highly resistant to UV radiation compared to others.

7.2 Non-Specific UV-B Signaling

7.2.1 DNA Damage Signaling

DNA damage can be caused by various chemical agents, for example, the nitrogen mustard, bromouracil, nitrous acid, acridine orange, and psoralens, which have different mechanisms of action (Tuteja et al. 2009). The second cause of DNA damage is radiation, and, among all types of radiation that can act as mutagenic agents, UV-B is the main form that can affect plants. UV-B photons absorbed by DNA can result in the formation of cyclobutane-pyrimidine dimers (CPDs) and, less frequently, pyrimidine (6-4)-pyrimidinone dimers [(6-4) photoproducts] (Jansen et al. 1998), besides the oxidatively damaged and crosslinked lesions. In addition to mutagenicity, DNA modifications can affect cellular metabolism, since DNA- and RNA-polymerase are unable to progress beyond the unrepaired dimers, resulting in impaired DNA replication and gene transcription (Britt 2004).

The ability to repair UV-B damaged DNA is key in the adaptation of plants to UV-B radiation. This repair can be done via light-dependent photoreactivation by DNA photolyases, enzymes that use energy from light at wavelengths from 300 to 500 nm (blue, red, and far-red wavelengths), and by light-independent processes (Jenkins 2009). All photolyases known carry two chromophores—a flavin cofactor and a folate, that function as antenna pigments (Jansen et al. 1998). This class of enzyme has minimal activity in the dark, but, when a photon is absorbed in the presence of UV-A or blue light, the lesion is effectively reversed. If reversion does not take place, transcriptional activity will be blocked by impeding RNA-polymerases (Britt 2004). In yeast and mammalian cells, UV radiation (UV-B and UV-C) initiates DNA damage signaling pathways, leading to DNA repair promotion (Sancar et al. 2004); these signaling processes are, at least partially, also conserved in plants (Jenkins 2009).

The induction of an isoflavonoid pigment, coumestrol, by UV radiation in leaves of *Phaseolus vulgaris*, seems to be mediated by DNA damage in the form of pyrimidine dimers production. Coumestrol was induced by short wavelengths, lower than 280 nm, and could be reversed by treatment with white light, which promotes photoreactivation (Beggs et al. 1985). Another example is suppression of hypocotyl elongation in cucumber (*Cucumis sativus*), elicited by short-term exposure to UV-B, that could be reversed by blue light-enriched white light, indicating that the response is initially mediated by the formation of damaged DNA (Shinkle et al. 2005).

An additional mechanism to repair UV-B damaged DNA is via a light-independent pathway, "the nucleotide excision repair" (NER), in which lesion-specific endonucleases make an incision on each side of a lesion, releasing the damaged oligonucleotide from the double helix as a small oligomer, and finally reconnect the resulting 5' and 3' ends, enabling the excision of dimers, like 6-4 photoproduct. There are two sub-pathways that comprise NER: the global genome repair and the transcription-coupled repair, and both have multistep pathways involving several enzymes (Tuteja et al. 2009).

7.2.2 Reactive Oxygen Species Signaling

In contrast to O_2, the ROS, such as singlet oxygen ($^1O_2$), superoxide anion ($O_2^{\bullet-}$), hydrogen peroxide (H_2O_2), and hydroxyl radical ($HO^{\bullet}$), are highly toxic and reactive, and can cause oxidative damage to cells. On the other hand, ROS play an important role in plant pathogen defense and can accumulate in response to biotic and abiotic stresses. Besides, plants generate ROS as signaling molecules to control cell death, hormone responses, and other cell processes (Apel and Hirt 2004). ROS seem to play multiple roles in plant biology (Mittler et al. 2004).

In plants, ROS are continuously produced mainly in chloroplast and mitochondria, places with intense redox metabolic activity. Under steady-state conditions, there is a balance between ROS production and scavenging carried out by antioxidative defense components (Apel and Hirt 2004). Under excessive amounts of UV-B radiation, the photosynthetic machinery is impaired and ROS may be produced to dissipate the excess of energy (Jansen et al. 1998; Jenkins 2009). Plants also generate ROS by peroxidases and oxidases in response to environmental changes.

It was shown that UV-B led to increases in NADPH-oxidase-like activity in *Arabidopsis thaliana*, a potential source of H_2O_2 in plants (Casati and Walbot 2003). Studies by Kalbina and Strid (2006) showed an involvement of NADPH oxidase and MAP kinase phosphatase (MKP1) in UV-B-dependent stress. Results indicated involvement of MKP1 in gene repression of more general stress response pathways, whereas NADPH oxidase would be more specifically involved in regulation of UV-B related genes.

Scavenging mechanisms through enzymatic or non-enzymatic systems can detoxify increased amount of ROS and alleviate stress (Jansen et al. 1998). UV-B boosts the activities of antioxidant enzymes, such as catalase (CAT), superoxide dismutase (SODs), and ascorbate peroxidases (APX), which play important roles as defenses against these radicals (Wang et al. 2010). Examples of common non-enzymatic systems include: alpha-tocopherol or vitamin E, a key scavenger of membrane free radicals, ascorbic acid, proline, flavonoids, carotenoids, and the efficient quencher of $^1O_2$, glutathione (GSH), which may become oxidized (GSSG) as a result of different environmental stresses (Stratmann 2003; Mishra et al. 2011).

Wang et al. (2010) showed that UV-B radiation decreased pollen germination and tube length in maize (*Zea mays* L.) and increased both $O_2^{\bullet -}$ and H_2O_2 accumulation. Increased ROS levels resulted in lipid peroxidation, membrane damage, and cell death. Gerhardt et al. (2005) showed that UV-B induced cotyledon curling in *Brassica napus* seedlings, and this induction may be mediated by ROS since it occurs associated with an increase in H_2O_2 and a decrease in ascorbate.

Mishra et al. (2008) reported that the exposure to UV-B radiation and the pesticide dimethoate (alone or in combination) causes decline of growth and photosynthesis in *Vigna unguiculata*, characteristics known to be related to ROS production. Further studies (Mishra et al. 2011) demonstrated that this exposure resulted in an increase of $O_2^{\bullet -}$ and H_2O_2, leading to lipid peroxidation and membrane leakage. The content of flavonoids and proline also increased, reinforcing their role as ROS scavengers.

In soybean seedlings (*Glycine max*), UV-B exposure reduced the activities of photosystems I and II (PSI and PSII), enhanced $O_2^{\bullet -}$ and H_2O_2, and proline content. Increases in SOD, CAT, and peroxidase activities were also observed (Prasad et al. 2005). An experiment using *Gunnera magellanica* was performed in a greenhouse, under natural growth conditions, to analyze the impact of UV-radiation in a more realistic situation. The results showed that UV-B reduced leaf expansion and induced damage at molecular level, with the production of CPDs and lipid radicals; however, the degree of lipid peroxidation was not affected (Giordano et al. 2004).

7.2.3 Wound/Defense Signaling

Exposure to UV-B radiation stimulates expression of various genes involved in defense responses (Jenkins 2009). The first kind of defense is the limitation of UV-B radiation tissue penetration. The exposure leads to an accumulation of pigments and UV-B absorbing compounds, including anthocyanins, sinapic acid esters, and flavonoids, which can attenuate UV-B radiation that reaches internal portions of tissues. The accumulation of these phenolics is a result of an increase in enzyme activities of the phenylpropanoid pathway (Mackerness 2000). A number of pathogenesis related (PR) proteins, PR-1, -2, and -5, and a defensin protein, have also been shown to increase in response to UV-B exposure.

Numerous studies have shown that salicylic acid (SA) is a key signaling compound involved in the activation of plant defense responses, including induction of PR genes. The importance of SA was demonstrated through analysis of transgenic *Arabidopsis* plants expressing the bacterial *nahG* gene, which encodes the enzyme salicylate hydroxylase, turning plants unable to accumulate SA. The effect of UV-B on photosynthetic genes was similar in both wild-type and mutant plants, although UV-B-induced up-regulation of *PR1*, *PR2*, and *PR5* transcripts was reduced in mutant plants (Surplus et al. 1998).

Another signaling molecule implicated in plant responses to UV-B stress is jasmonic acid (JA). Induction of genes associated with wounding and pathogen responses were observed with UV-B exposure in both tomato and *Arabidopsis* plants (Conconi et al. 1996; Mackerness 2000). In Solanaceae species, UV-B-induced genes were blocked by inhibitors of the octadecanoid pathway, and several UV-induced responses were not present in JA-deficient mutants (Conconi et al. 1996; Demkura et al. 2010).

Ethylene, a gaseous plant hormone, has also been identified as an important signaling component in wounding and defense. An increase of ethylene with UV-B exposure has been reported in *Arabidopsis thaliana*, as well as activation of the PR genes *PDF1·2* and *PR-1*. Induction of these genes was compromised in the ethylene insensitive mutant *etr1* (Mackerness 2000). Wang et al. (2006) showed that there was a synergistic effect between nitric oxide and ROS in promoting the accumulation of UV-B-induced ethylene in the leaves of maize (*Zea mays* L.) seedlings.

7.3 Specific UV-B Signaling

Responses to low levels of UV-B may be mediated by UV-B-specific pathways that lead to photomorphogenic responses different from general stress responses (e.g. ROS formation, DNA damage signaling, production of signaling molecules), generated under high levels of this radiation.

Photomorphogenic signaling stimulates the expression of genes involved in UV-protection thereby promoting plant survival under UV-B. This signaling process initiates with UV-B perception by a UV-B-specific component, UV RESISTANCE LOCUS8 (UVR8), presenting high functional specificity and acting exclusively in UV-B signaling, as described by Brown et al. (2005). The sequence of UVR8 shows similarity to the eukaryotic guanine nucleotide exchange factor RCC1 (Kliebenstein et al. 2002) and is mainly located in the nucleus, where it associates with chromatin via histones (Brown et al. 2005). However, unlike

RCC, UVR8 has little exchange activity (Jenkins 2009) and 14 tryptophan residues were identified in its structure (in contrast with human RCC that contains four), all located at the top of the predicted UVR8 β-propeller structure (Rizzini et al. 2011); these residues are highly conserved among species, including the moss *Physcomitrella*, suggesting functional importance (Jenkins 2009). UVR8 regulates genes involved in flavonoid biosynthesis, genes encoding type II photolyase PHR1 (required for photoreactivating DNA repair), genes concerned with protection against oxidative stress and photo-oxidative damage, and a number of genes encoding transcription factors such as ELONGATED HYPOCOTYL5 (HY5) (Brown et al. 2005). UVR8 also regulates leaf growth through the control of epidermal cell development, is required for normal progression of endocycles in response to UV-B, and has a regulatory role in stomatal differentiation (Wargent et al. 2009), emphasizing the importance of UVR-8 in responses to UV-B. It has been proposed that monomerization of UVR8 dimers through a tryptophan-based mechanism is the key to UV-B perception (Rizzini et al. 2011).

Absorption of UV-B induces prompt monomerization of UVR8 and interaction with the CONSTITUTIVELY PHOTOMORPHOGENIC1 (COP1) protein (Rizzini et al. 2011), the central regulator of light signaling. In presence of visible light, cryptochrome photoreceptor activity leads to the inactivation and nuclear exclusion of COP1, which acts as an E3 ubiquitin ligase, maintaining low levels of HY5 in the dark. This change allows HY5 stabilization and downstream activation of light-responsive genes (Osterlund et al. 2000). Early inactivation of COP1 by visible light occurs most likely through direct interaction with cryptochromes (Wang et al. 2001; Seo et al. 2004). Mutational analysis indicates that nuclear exclusion of COP1 is a rate-limiting step for the establishment of photomorphogenic development (Subramanian et al. 2004). In spite of the fact that the interaction between UVR8 and COP1 is part of UV-B perception, the role of the latter protein in UV-B mediated responses is quite distinct from that played in visible light responses, possibly not involving the E3 ubiquitin ligase of COP1 (Jenkins 2009).

COP1 regulates a range of low-fluence UV-B-mediated gene expression responses in addition to flavonoid accumulation and inhibition of hypocotyl elongation, playing a crucial role as a positive regulator of the photomorphogenic UV-B response in *Arabidopsis*, in contrast with its repressor function in visible light-induced photomorphogenesis (Oravecz et al. 2006). This role of COP1 in UV-B response is independent of SUPPRESSOR OF PHYTOCHROME A-105 family (SPA) proteins, which are required for COP1 function in dark and visible light (Oravecz et al. 2006).

COP1 and UVR8 interaction is required for the activation of HY5-mediated gene expression under UV-B, and all three proteins accumulate in the nucleus (Oravecz et al. 2006; Kaiserli and Jenkins 2007). The basic leucine-zipper transcription factor HY5 plays a central role in the regulation of a substantial number of genes (Ulm et al. 2004; Brown et al. 2005; Oravecz et al. 2006). The exact mechanism through which the interaction between UVR8 and COP1 activates HY5 transcription remains unknown. Apparently, the UV-B-dependent steps of monomerization of UVR8 and UVR8 monomer interaction with COP1 allow *HY5* transcription. HY5 protein then regulates target genes involved in UV photomorphogenic responses, including UV protection genes, such as those encoding enzymes of the flavonoid biosynthesis pathway (Wade et al. 2001; Brosché and Strid 2003; Ulm et al. 2004; Favory et al. 2009; Jenkins 2009; Rizzini et al. 2011). A summary of the UV-mediated responses in plants, with particular emphasis on consequences for secondary metabolism, is shown in Fig. 7.1.

7.4 Enhancement of Secondary Metabolite Accumulation by UV-B Irradiation

Environmental stress can enhance the formation of ROS such as $O_2^{\bullet-}$ and H_2O_2. In addition to morphological changes, like decreasing leaf number, leaf area (Teramura and Sullivan 1994), chlorophyll concentration, and photosynthetic rate (Kakani et al. 2003), plants can produce UV absorbing compounds that protect from UV damage. Increased input of UV light has been shown to increase ROS production, activity of antioxidative enzymes, and secondary metabolite pathways in different plant species. A list of examples of secondary metabolites that have been shown to accumulate in response to UV-B treatment is shown in Table 7.1.

7.4.1 Phenolic Compounds

Phenolic compounds play a variety of functions in plants, providing mechanical support, as in the case of lignins, or protecting plants against pathogens, herbivores, and ultraviolet radiation. Of all classes of secondary metabolites, phenolics, specifically flavonoids, are regarded as the most relevant for UV protection. Flavonoid biosynthesis and its regulation have been thoroughly investigated. Many of the biosynthetic enzymes of phenolics are activated by UV-B, such as phenylalanine ammonia lyase (PAL) and chalcone synthase (CHS), as well as enzymes involved in initial steps of flavonoid synthesis (e.g. glucose 6-phosphate dehydrogenase, 3-deoxy-D-arabino heptulosonate 7-phosphate synthase, and S-adenosylhomocysteine hydrolase) (Logemann et al. 2000). Coumarins with a furan ring structure are activated by UV-A radiation, leading to phototoxic reactions.

7.4.1.1 Lignins

Lignins are polymers composed of cinnamic acid derivatives, which form a complex three-dimensional structure. Although mostly structural in function, recent studies indicated that this polymer has significant importance in the containment of ROS in the intercellular space, which can be produced by UV action on cells. UV radiation exposure may also result in stem thickening and increased lignin content in plants, presumably attenuating radiation effects on the

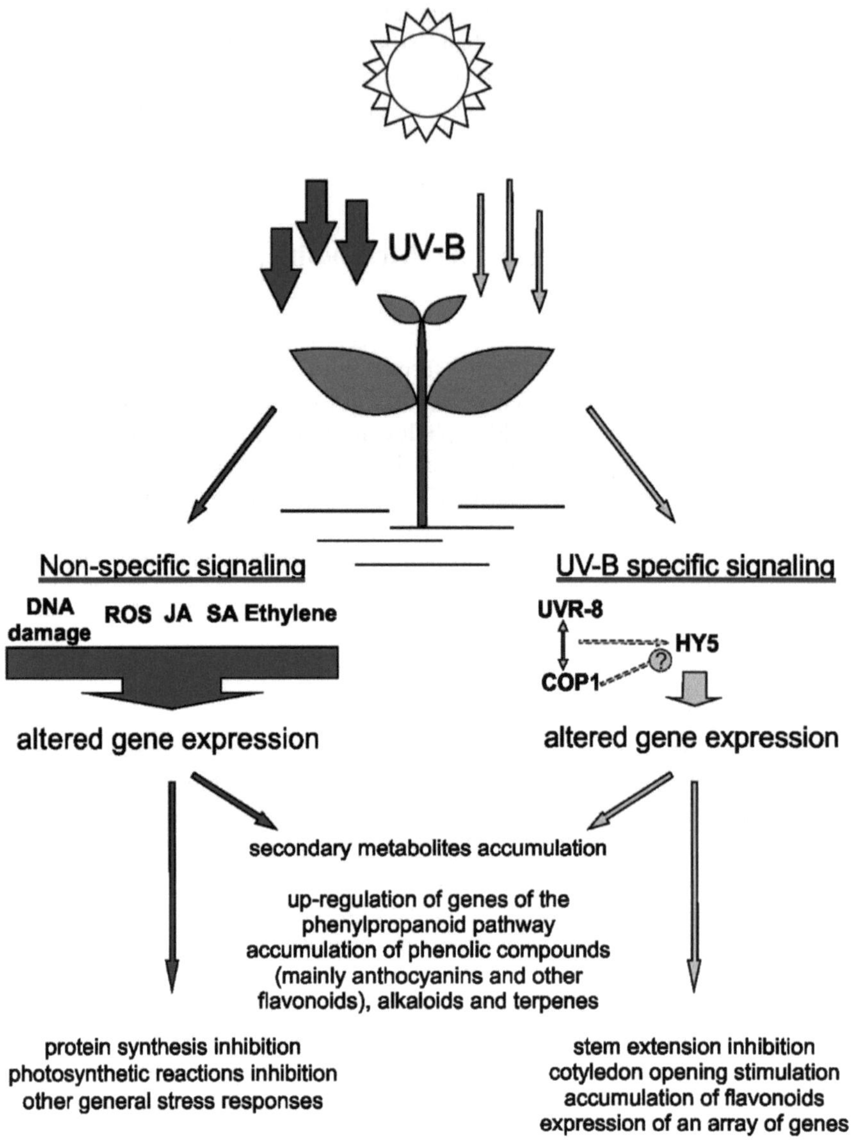

Fig. 7.1 Schematic representation of main UV-B signal transduction pathways in plants. Low levels of UV-B (*light-colored arrows*) may mediate UV-B specific pathways, differently from general stress responses, generated under high levels of radiation (*dark arrows*). These pathways change the expression of an array of genes, with both specific and/or overlapping responses. Dashed lines indicate incompletely understood correlations. Data to date indicates that UVR8 monomerization and UVR8-COP1 interaction are UV-B dependent processes that lead to HY5 gene transcription. HY5 protein then regulates downstream target genes involved in UV photomorphogenic responses, including UV protection genes, such as those encoding enzymes of the flavonoid pathway. ROS: reactive oxygen species; JA: jasmonic acid; SA: salicylic acid; UVR8: UV RESISTANCE LOCUS8; COP1: CONSTITUTIVELY PHOTOMORPHOGENIC1; HY5: ELONGATED HYPOCOTYL5

Table 7.1 Examples of plant secondary metabolites induced by UV-B exposure

Species	Organ/Tissue	Induced metabolite	UV-B intensity	UV-B exposure time	Fold induction	Reference
Acorus calamus L.	whole plant	caryophyllene oxides, aristolene, carvacrol and *p*-cymene	1.8 kJ m^{-2}	3 h/day (80 days)	1.7-fold for caryophyllene oxides, 1.5-fold for aristolene; carvacrol and *p*-cymene detected only on UV-B treatment	Kumari et al. (2009)
Arabidopsis thaliana (L.) Heynh.	leaves	flavonoids	3 pmol sec^{-1}; UV-A complement	8 h	7.5–10-fold on chalcone synthase activity (Landsberg erecta ecotype)	Fuglevand et al. (1996)
Camptotheca acuminata Decne.	callus	camptothecin	5 μmol m^{-2} s^{-1}	3 days	11-fold	Pi et al. (2010)
Catharanthus roseus (L.) G. Don	cultured cells	catharanthine	2.5 cm distance	5 min	3-fold	Ramani and Chelliah (2007)
Cucumis sativus L.	thricomes and cotyledons	polyphenols and lignin	0.57 W m^{-2}	0–15 days	1.5-fold of lignin content on cotyledons and increase of polyphenols on thricomes	Yamasaki et al. (2010)
Fagopyrum tataricum (L.) Gaertn.	leaves	rutin	1.26 mW m^{-2}	30 min	1.2 fold	Suzuki et al. (2005)
Glycyrrhiza uralensis Fisch.	roots	glycyrrhizin	1.13 W m^{-2} or 0.43 W m^{-2}	3 or 15 days respectively	1.5-fold in both treatments	Afreen et al. (2005)
Lactuca sativa L.	leaves	cyanidin and delphinidin	5.5 W m^{-2}	5 min pulse/day (10 days)	induction of chalcone synthase, flavanone 3-hydroxylase and dihydroflavonol 4-reductase genes	Park et al. (2007)
Mentha spicata L.	leaves	essential oil rich in carvone and dihydrocarvone	–	~4 months	1.5-fold	Karousou et al. (1998)
Mentha x *piperita* L.	leaves	essential oil rich in terpenes and phenolic compounds	7.1 kJ m^{-2} day^{-1}	1 h	detailed in respective reference	Dolzhenko et al. (2010)

(continued)

Table 7.1 (continued)

Species	Organ/Tissue	Induced metabolite	UV-B intensity	UV-B exposure time	Fold induction	Reference
Ocimum basilicum L.	leaves	eugenol, methyl eugenol, linalool, 1,8-cineole and trans-β-ocimene	two Philips 20 W; 12 UV-B fluorescent tubes	14 days	4-fold	Johnson et al. (1999)
Pirus malus L.	fruit skin	cyanidin 3-galactoside and cyanidin 3-arabinoside	1.3 W m^{-2}	5 days	8-fold in cyanidin 3-galactoside (cv. 'Iwai') and a slight increase in cyaniding 3-arabinoside (cv. 'Tsugaru')	Ubi et al. (2006)
Pisum sativum L.	leaves	nicotinamide and trigonelline	1.4 W m^{-2}	12 h	3.5-fold and 9.5-fold respectively (cv. Greenfeast)	Kalbin et al. (1997)
Psychotria brachyceras Müll. Arg.	leaves	brachycerine	10.5 kJ cm^{-2}	16 h/day (6 days)	2-fold	Gregianini et al. (2003)
Rosmarinus officinalis L.	leaves	carnosic acid and rosmarinic acid	5.4 and 31 kJ m^{-2} day^{-1}	14 days	5.4 kJ m^{-2} day^{-1}: 1.5-fold and 1.1-fold respectively; 31 kJ m^{-2} day^{-1}: 1.8-fold and 2.3-fold respectively	Luis et al. (2007)
Vitis vinifera L.	grape berries	resveratrol, catechin, and quercetin	UV-B elimination in control plants	15 days	1.8-fold for resveratrol and trend of increase for catechins and quercetin (cv. Malbec)	Berli et al. (2008)
Zea mays L.	leaves	rhamnosylisoorientin and maysin	solar UV-B radiation and supplemental UV-B light (4-fold increase in UV-B at 305 nm)	21 days	28-fold and 38-fold respectively (Mishca genotype)	Casati and Walbot (2005)

cytoplasm of epidermal tissue (Moura et al. 2010). UV-B-induced expansion, endoreduplication, peroxidase activity, and the accumulation of polyphenolic compounds in epidermal cells surrounding the trichomes of *Cucumis sativus* L. cotyledons have been reported (Yamasaki et al. 2010).

7.4.1.2 Flavonones and Flavonols

Flavonoids are the classical UV-B-regulated compounds in plants, and play a major role in protection against this radiation. The structure of flavonoids, which absorb wavelengths between 280 and 350 nm, consists of two condensed rings, and a third ring attached to carbon 2. These phenolics are classified according to their degree of oxidation, comprising anthocyanins, flavones, flavonols, and isoflavones. The chemical diversity, size, three-dimensional conformation, and physical and biochemical properties of flavonoids allow them to interact with targets in different subcellular locations and to influence biological activity in plants, animals, and microbes (Taylor and Grotewold 2005; Peer and Murphy 2007).

A number of early physiological experiments provided some circumstantial evidence that flavonoids were involved in UV-protection. Flavones and flavonols protect cells because they accumulate in epidermal layers of leaves and stems, acting like filters, absorbing radiation in the UV range of the spectrum. UV-B exposure makes plants increase the activity of enzymes of flavonoid synthesis. This occurs not only at epidermal cell level of the adaxial leaf surface, where there is direct exposure to radiation, but also in flavonoids in the leaf wax and in leaf hairs (Harborne and Williams 2000).

7.4.1.3 Anthocyanins

Low fluence UV-B stimulates distinct responses, such as the accumulation of UV-absorbing pigments, including anthocyanins. Accumulation of these pigments was observed in maize, rice, apple fruits, kangaroo paw, apple flowers, *Arabidopsis,* and roses. UV-B increased the accumulation of anthocyanins by stimulating the expression of genes encoding enzymes in the anthocyanin biosynthetic pathway (Fuglevand et al. 1996). Irradiation of UV-B light on lettuce leaves leads to an increase of expression of genes encoding CHS, dihydroflavonol 4-reductase (DFR) and flavanone 3-hydroxylase (F3H) (Park et al. 2007), as well as for stress response proteins (in maize and *Arabidopsis*) and ribosomal protein genes.

UV-A induction of anthocyanin accumulation was observed in grape (Kataoka et al. 2003), *Arabidopsis* (Christie and Jenkins 1996), eggplant (Toguri et al. 1993), carrot cells (Hirner and Satz 2000), and Kalanchöe (Hoffmann 1999). This accumulation was associated with increased expression of anthocyanin biosynthesis genes with exposure time, including those encoding for P450 enzymes, PAL, F3H, anthocyanidin synthase (ANS), CHS, DFR, and glutathione S-transferase (GST) (Xu and Li 2006; Zhou et al. 2007). The production of anthocyanins by UV-A exposure may contribute to protection of plant tissue from the potential damage by further UV absorption, since UV-A did not have a strong impact on gene expression profiles as observed for low fluence UV-B response (Guo et al. 2008).

7.4.1.4 Stilbenes

Stilbenes derive from the same biosynthetic pathway leading to the other phenolic compounds, formed from 4-coumaroyl-CoA and malonyl-CoA by stilbene synthase (STS). These substances, such as picetannol and resveratrol, also have their accumulation promoted by UV. Resveratrol, for example, exists in two forms: *cis* and *trans*, the latter being the first to be formed, and upon exposure to UV-A, isomerization of the structure results in the *cis* form (Zhang and Björn 2009).

7.4.2 Alkaloids

These nitrogen containing compounds are well known as a defense strategy against herbivory. However, several studies have implicated indole and purine alkaloids in protection against UV-B radiation, showing increased accumulation of these metabolites by exposure to UV-B and UV-C. In *Psychotria brachyceras*, a 10-fold increase in brachycerine content in cuttings was seen after exposure for 6 days to 16 h a day of UV-C, whereas a 4 h daily supplementation doubled the amount of the alkaloid in leaves. Exposure to a 5% UV-B source also doubled the alkaloid yield. Brachycerine was shown to quench reactive oxygen species besides absorbing UV (Gregianini et al. 2003). Just as a highly UV-tolerant phenotype was observed for *P. brachyceras*, the same was described for *Psychotria umbellata*, although the latter was shown to have constitutively high concentrations of the main alkaloid, psychollatine, also an effective quencher of ROS and absorbing in the UV wavelength (Paranhos et al. 2009). Camptothecin and 10-hydroxycamptothecin in *Camptotheca acuminata* culture cells was induced by UV-B light, which showed an 11-fold increase in camptothecin concentration (Pi et al. 2010). Plants, multiple shoot cultures, and cell suspensions of *Catharanthus roseus* showed significant increases in the content of precursors for vinblastine and vincristine upon UV-B exposure. This light treatment applied for up to 20 min caused significant yield improvement of lochnericine, serpentine, and ajmalicine, whereas a decrease was observed in hörhammericine (Binder et al. 2009).

7.4.3 Terpenes

Terpenes constitute the largest class of secondary metabolites, including products such as essential oils, triterpenes, and steroids. Terpenes are mostly involved in plant defense against herbivores. Supplementation with UV-B does not always lead to an increase in the content of terpenes, but there are several studies showing increased production of members of this metabolite class upon exposure to such radiation. UV-B can induce differential expression of peppermint genes involved in essential oil biogenesis (Dolzhenko et al. 2010), and promotes the yield of paclitaxel and related taxanes in leaves of *Taxus cuspidata* (Li et al. 2007). These responses can be mediated by alterations in jasmonic acid levels (Creelman and Mulleti 1997). UV-A and UV-C radiations, mainly the latter, increased accumulation of 10-deacetyl baccatin III (DAB III) and paclitaxel in *Taxus baccata* twigs exposed for 48 h (Hajnos et al. 2001).

7.4.4 Glucosinolates and Cyanogenic Glucosides

Glucosinolates and cyanogenic glucosides are well known for their role in defense against herbivores and share similar biochemical aspects, such as the facts that they derive from amino acids, share some common biosynthetic intermediates, localize mostly to vacuoles, and need to be chemically cleaved to become active. Whereas cyanogenic glucosides can be toxic also to humans, glucosinolates and its degradation products, mostly isothiocyanates, are considered important antioxidants, antimicrobials, and nutraceuticals. Experiments with *Trifolium repens* showed that UV-B acclimation improved the cyanogenic capacity of members of at least one population, presumably by increasing the accumulation or potency of cyanogenic glucosides (Lindroth et al. 2000). Different glucosinolates have been shown to increase in concentration upon exposure of various plant species to near natural UV-B. However, the consequences of plant UV treatment for herbivore insect deterrence and human health promoting effects through use as food or supplements are not always clear (Jansen et al. 2008; Kuhlmann and Muller 2009).

7.5 Methodology Considerations in UV Experiments

7.5.1 Indoor Methods

Indoor experiments are usually carried out in growth chambers or greenhouses. In both conditions UV-B effects are exaggerated, being most severe in growth chambers; therefore, results may not be adequate to predict the influence of natural UV-B in plants. These indoor methodologies are useful tools in the study of natural products production enhancement, since higher UV-B doses often stimulate the production of many secondary compounds of pharmaceutical interest (Zhang and Björn 2009).

7.5.2 *Outdoor Methods*

7.5.2.1 UV-B Attenuation

Outdoor models are conducted either by attenuation or enhancement of UV-B incidence and yield more realistic results to evaluate biological effects of solar UV-B radiation in the field. Attenuation methods help understanding responses occurring in near-ambient or lower UV-B levels. The main approach is to use filter sheets that absorb or transmit most UV-B, making this methodology cheap, simple, and without need for electricity (Xu and Sullivan 2010). More complex methods in UV-B reduction involve cuvettes containing ozone placed over the plants (Tevini et al. 1991), providing simulation of a more realistic changing in solar UV-B radiation due to ozone depletion. A disadvantage of all attenuation approaches is that the results cannot be accurately used to predict ozone depletion consequences, since evidence shows that responses are likely to be nonlinear (Ballaré et al. 1996).

7.5.2.2 UV-B Enhancement

Enhancement methods comprise UV-B supplement with fluorescent lamps and are a useful tool to mimic future UV-B enhancement due to ozone depletion. Most UV-B enhancement studies use square-wave delivery systems (SQW), which provide supplementation of UV-B radiation at fixed dose rates for a period of hours centered on solar noon, producing results in a maximum UV-B exposure scenario (clear sky conditions). If not carefully monitored, this can lead to UV-B supplementation greater than would occur due to ozone depletion, with higher UV-B to UV-A and PAR (photosynthetically active radiation) ratios (Fiscus and Booker 1995), which are particularly important in plant responses sensitive to UV-B changes (Caldwell et al. 1994; Krizek 2004).

The problem can be partially fixed by the use of UV-B output from lamps increasing through the morning and declining after solar noon in combination with switching lamps off during completely overcast days, yielding UV-B enhancements similar to those provided by more costly modulated systems (Musil et al. 2002). In modulated field radiation systems (MOD) UV-B supplements are supplied by increments proportionally to the constantly monitored UV-B background conditions (Caldwell et al. 1983; McLeod 1997; Díaz et al. 2006), compensating not only environmental conditions but problems inherent to the method, such as filter photodegradation, lamp aging, and temperature, by the maintenance of proportional supplements to background UV-B. Unfortunately, MOD are technically more complex and expensive than SQW systems and have not been widely used (Xu and Sullivan 2010).

An important issue in UV-B supplementation experiments is that these fluorescent lamps produce some UV-A radiation, which is little modified by ozone, thereby with ecological importance. UV-A supplementation has significant effects on plant growth; thus supplemental studies need to include proper controls for UV-A (Middleton and Teramura 1993; Newsham et al. 1996; Heijari et al. 2006).

7.5.3 Post Harvest and Acute Exposure Methods

In contrast to the many considerations and limitations outlined above regarding experiments that try to mimic realistic environmental scenarios of UV exposure, pre and post harvest experiments with UV treatments to enhance secondary metabolite accumulation are relatively easy to perform. Acute short-term exposures, as well as UV-C treatments, can be applied, leading to faster responses and restricted demand of energy and controlling devices for light sources. In addition, post harvest treatments, such as irradiation of freshly detached leaves, allow the use of localized facilities for light source installations. In all cases, it is important to bear in mind the fact that plant metabolic responses to UV may vary depending on several factors, including light source, intensity, time of treatment, target compound, plant/organ developmental stage, and physiological status, including previous acclimation events. Not surprisingly, standardization of protocols is a key factor for precision of results.

7.6 Conclusions and Perspectives

UV is a powerful tool to modulate secondary metabolism pathways in plants. Understanding of the molecular basis of UV perception, transduction, and metabolic response has advanced considerably in recent years, but many details are still missing. This is particularly true for the effects of natural or near-natural levels of UV-B exposure on secondary metabolic pathways, as well as for competition between UV-stimulated pathways for common precursors, carbon, and nitrogen pools. At ecological level, the impacts of these changes on plant–insect and tritrophic interactions are also beginning to unveil. On a more technical application, besides field level manipulation of UV-B incidence, the use of acute treatments and UV-C irradiance may provide manageable large-scale strategies for boosting yields or modifying metabolic flux of plant secondary compounds of medicinal interest.

References

Afreen F, Zobayed SM, Kozai T (2005) Spectral quality and UV-B stress stimulate glycyrrhizin concentration of *Glycyrrhiza uralensis* in hydroponic and pot system. Plant Physiol Biochem 43:1074–1081

Apel K, Hirt H (2004) Reactive oxygen species: metabolism, oxidative stress, and signal transduction. Annu Rev Plant Biol 55:373–399

Ballaré CL, Scopel AL, Stapleton AE, Yanovsky MJ (1996) Solar ultraviolet-B radiation affects seedling emergence, DNA integrity, plant morphology, growth rate, and attractiveness to herbivore insects in *Datura ferox*. Plant Physiol 112:161–170

Barnes JD, Percy KE, Paul ND, Jones P, McLaughlin CK, Mullineaux PM, Creissen G, Wellburn AR (1996) The influence of UV-B radiation on the physicochemical nature of tobacco (*Nicotiana tabacum* L) leaf surfaces. J Exp Bot 47:99–109

Beggs CJ, Stolzer-Jehle A, Wellmann E (1985) Isoflavonoid formation as an indicator of UV stress in bean (*Phaseolus vulgaris* L.) leaves. Plant Physiol 79:630–634

Berli F, D'Angelo J, Cavagnaro B, Bottini R, Wuilloud R, Silva MF (2008) Phenolic composition in grape (*Vitis vinifera* L. cv. Malbec) ripened with different solar UV-B radiation levels by capillary zone electrophoresis. J Agr Food Chem 56:2892–2898

Binder BYK, Peebles CAM, Shanks JV, San KY (2009) The effects of UV-B stress on the production of terpenoid indole alkaloids in *Catharanthus roseus* hairy roots. Biotechnol Progr 25:861–865

Britt AB (2004) Repair of DNA damage induced by solar UV. Photosynth Res 81:105–112

Brosché N, Strid A (2003) Molecular events following perception of UV-B radiation by plants. Physiol Plantarum 117:1–10

Brown BA, Cloix C, Jiang GH, Kaiserli E, Herzyk P, Kliebenstein DJ, Jenkins GI (2005) A UV-B-specific signaling component orchestrates plant UV protection. P Natl A Sci USA 102:18225–18230

Caldwell MM, Flint SD, Searles PS (1994) Spectral balance and UV-B sensitivity of soybean: a field experiment. Plant Cell Environ 17:267–276

Caldwell MM, Gold WG, Harris G, Ashurst CW (1983) A modulated lamp system for solar UV-B (280–320 nm). Supplementation studies in the field. Photochem Photobiol 37:479–485

Casati P, Walbot V (2003) Gene expression profiling in response to ultraviolet radiation in maize genotypes with varying flavonoid content. Plant Physiol 132:1739–1754

Casati P, Walbot V (2005) Differential accumulation of maysin and rhamnosylisoorientin in leaves of high-altitude landraces of maize after UV-B exposure. Plant Cell Environ 28:788–799

Christie JM, Jenkins GI (1996) Distinct UV-B and UV-A/blue light signal transduction pathways induce chalcone synthase gene expression in *Arabidopsis* cells. Plant Cell 8:1555–1567

Conconi A, Smerdon MJ, Howe GA, Ryan CA (1996) The octadecanoid signaling pathway in plants mediates a response to UV radiation. Nature 383:826–829

Creelman RA, Mulleti JE (1997) Oligosaccharins, brassinolides, and jasmonates: nontraditional regulators of plant growth, development, and gene expression. Plant Cell 9:1211–1223

Demkura PV, Abdala G, Baldwin IT, Ballaré CL (2010) Jasmonate-dependent and -independent pathways mediate specific effects of solar Ultraviolet B radiation on leaf phenolics and antiherbivore defense. Plant Physiol 152:1084–1095

Díaz S, Camilión C, Escobar J, Deferrari G, Roy S, Lacoste K, Demers S, Belzile C, Ferreyra G, Gianesella S, Gosselin M, Nozais C, Pelletier E, Schloss I, Vernet M (2006) Simulation of ozone depletion using ambient irradiance supplemented with UV lamps. Photochem Photobiol 82:857–864

Dolzhenko Y, Bertea CM, Occhipinti A, Bossi S, Maffei ME (2010) UV-B modulates the interplay between terpenoids and flavonoids in peppermint (*Mentha x piperita* L.). J Photoch Photobio B 100:67–75

Favory J, Stec A, Gruber H, Rizzini L, Oravecz A, Funk M, Albert A, Cloix C, Jenkins GI, Oakeley EJ, Seidlitz HK, Nagy F, Ulm R (2009) Interaction of COP1 and UVR8 regulates UV-B-induced photomorphogenesis and stress acclimation in *Arabidopsis*. EMBO J 28:591–601

Fiscus EL, Booker FL (1995) Is UV-B a hazard to crop photosynthesis and productivity? Photosynth Res 43:81–92

Frohnmeyer H, Staiger D (2003) Ultraviolet-B radiation-mediated responses in plants. Balancing damage and protection. Plant Physiol 133:1420–1428

Fuglevand G, Jackson JA, Jenkins GI (1996) UV-B, UV-A and blue light signal transduction pathways interact synergistically to regulate chalcone synthase gene expression in *Arabidopsis*. Plant Cell 8:2347–2357

Gerhardt KE, Wilson MI, Greenberg BM (2005) Ultraviolet wavelength dependence of photomorphological and photosynthetic responses in *Brassica napus* and *Arabidopsis thaliana*. Photochem Photobiol 81:1061–1068

Giordano CV, Galatro A, Puntarulo S, Ballaré CL (2004) The inhibitory effects of UV-B radiation (280–315 nm) on *Gunnera magellanica* growth correlate with increased DNA damage but not with oxidative damage to lipids. Plant Cell Environ 27:1415–1423

Gregianini TS, Silveira VC, Porto DD, Kerber VA, Henriques AT, Fett-Neto AG (2003) The alkaloid brachycerine is induced by ultraviolet radiation and is a singlet oxygen quencher. Photochem Photobiol 78:470–474

Guo J, Han W, Wang MH (2008) Ultraviolet and environmental stresses involved in the induction and regulation of anthocyanin biosynthesis: a review. Afr J Biotechnol 7:4966–4972

Hahlbrock K, Scheel D (1989) Physiology and molecular biology of phenylpropanoid metabolism. Ann Rev Plant Physiol Plant Mol Biol 40:347–369

Hajnos MK, Zobel AM, Glowniak K (2001) The influence of ultraviolet radiation on the content of pharmacologically active taxoids in yew tissues. Phytomedicine 8:139–143

Harborne JB, Williams CA (2000) Advances in flavonoid research since 1992. Phytochemistry 55:481–504

Heijari J, Kivimäenpää M, Hartikainen H, Julkunen-Tiitto R, Wulff A (2006) Responses of strawberry (*Fragaria* × *ananassa*) to supplemental UV-B radiation and selenium under field conditions. Plant Soil 282:27–39

Heinonen IM, Meyer AS, Frankel EN (1998) Antioxidant activity of berry phenolic on human low-density lipoprotein and liposome oxidation. J Agr Food Chem 46:4107–4112

Hirner AA, Satz HU (2000) Isoforms of chalcone synthase in *Daucus carota* L. and their differential expression in organs from the European wild carrot and in ultraviolet-A-irradiated cell cultures. Planta 210:993–998

Hoffmann S (1999) The effect of UV-radiation on colours of leaves and flowers of ornamental plants. Gartenbauwissenschaft 64:88–93

Jansen MAK, Gaba V, Greenberg BM (1998) Higher plants and UV-B radiation: balancing damage, repair and acclimation. Trends Plant Sci 3:131–135

Jansen MAK, Hectors K, O' Brien NM, Guisez Y, Potters G (2008) Plant stress and human health: do human consumers benefit from UV-B acclimated crops? Plant Sci 175:449–458

Jenkins GI (2009) Signal transduction in responses to UV-B radiation. Annu Rev Plant Biol 60:407–431

Johnson CB, Kirby J, Naxakis G, Pearson S (1999) Substantial UV-B-mediated induction of essential oils in sweet basil (*Ocimum basilicum* L.). Phytochemistry 51:507–510

Julkunen-Tiitto R, Häggman H, Aphalo PJ, Lavola A, Tegelberg R, Veteli T (2005) Growth and defense in deciduous trees and shrubs under UV-B. Environ Pollut 137:404–414

Kaiserli E, Jenkins GI (2007) UV-B promotes rapid nuclear translocation of the UV-B-specific signaling component UVR8 and activates its function in the nucleus. Plant Cell 19:2662–2673

Kakani VG, Reddy KR, Zhao D, Sailaja K (2003) Field crop response to ultraviolet-B radiation: a review. Agr Forest Meterol 120:191–218

Kalbin G, Ohlsson AB, Berglund T, Rydström J, Strid A (1997) Ultraviolet-B-radiation-induced changes in nicotinamide and glutathione metabolism and gene expression in plants. Eur J Biochem 249:465–472

Kalbina I, Strid A (2006) The role of NADPH oxidase and MAP kinase phosphatase in UV-B-dependent gene expression in *Arabidopsis*. Plant Cell Environ 29:1783–1793

Karousou R, Grammatikopoulos G, Lanaras T, Manetas Y, Kokkini S (1998) Effects of enhanced UV-B radiation on *Mentha spicata* essential oils. Phytochemistry 49:2273–2277

Kataoka I, Sugiyama A, Beppu K (2003) Role of ultraviolet radiation in accumulation of anthocyanin in berries of 'Gros Colman' grapes (*Vitis vinifera* L.) J Jpn Soc Horticult Sci 72:1–6

Kliebenstein DJ, Lim JE, Landry LG, Last RL (2002) *Arabidopsis* UVR8 regulates UV-B signal transduction and tolerance and contains sequence similarity to human regulator of chromatin condensation 1. Plant Physiol 130:234–243

Krizek DT (2004) Influence of PAR and UV-A in determining plant sensitivity and photomorphogenic responses to UV-B radiation. Photochem Photobiol 79:307–315

Kuhlmann F, Muller C (2009) Independent responses to ultraviolet radiation and herbivore attack in broccoli. J Expt Bot 60:3467–3475

Kumari R, Agrawal SB, Singh S, Dubey NK (2009) Supplemental ultraviolet-B induced changes in essential oil composition and total phenolics of *Acorus calamus* L. (sweet flag). Ecotoxicol Environ Saf 72:2013–2019

Li SM, Sun R, Luo H, Sun S, Wang Y, Fu YJ (2007) The influence of UV radiation on the contents of paclitaxel and cephalomannine in the fresh leaves of *Taxus cuspidata*. Bull Bot Res 4:500–504

Lindroth RL, Hofmann RW, Campbell BD, McNabb WC, Hunt DY (2000) Population differences in *Trifolium repens* L. to response to ultraviolet-B radiation: foliar chemistry and consequences for two lepidopteran herbivores. Oecologia 122:20–28

Logemann E, Tavernaro A, Schulz W, Somssich IE, Hahlbrock K (2000) UV light selectively coinduces supply pathways from primary metabolism and flavonoid secondary product formation in parsley. Proc Natl Acad Sci 97:1903

Luis JC, Martín Pérez R, Valdés González F (2007) UV-B radiation effects on foliar concentrations of rosmarinic and carnosic acids in rosemary plants. Food Chem 101: 1211–1215

Mackerness SAH (2000) Plant responses to ultraviolet-B (UV-B: 280–320 nm) stress: what are the key regulators? Plant Growth Regul 32:27–39

McLeod AR (1997) Outdoor supplementation systems for studies of the effects of increased UV-B radiation. Plant Ecol 128:78–92

Middleton EH, Teramura AH (1993) The role of flavonol glycosides and carotenoids in protecting soybean from ultraviolet-B damage. Plant Physiol 103:741–752

Mishra V, Mishra P, Srivastava G, Prasad SM (2011) Effect of dimethoate and UV-B irradiation on the response of antioxidant defense systems in cowpea (*Vigna unguiculata* L.) seedlings. Pestic Biochem Phys 100:118–123

Mishra V, Srivastava G, Prasad SM, Abraham G (2008) Growth, photosynthetic pigments and photosynthetic activity during seedling stage of cowpea (*Vigna unguiculata*) in response to UV-B and dimethoate. Pestic Biochem Phys 92:30–37

Mittler R, Vanderauwera S, Gollery M, Van Breusegem F (2004) Reactive oxygen gene network of plants. Trends Plant Sci 9:490–498

Moura JCMS, Bonine CAV, Viana JOF, Dornelas MC, Mazzafera P (2010) Abiotic and biotic stresses and changes in the lignin content and composition in plants. J Integr Plant Biol 52:360–376

Musil CF, Björn LO, Scourfield MWJ, Bodeker GE (2002) How substantial are ultraviolet-B supplementation inaccuracies in experimental square-wave delivery systems? Environ Exp Bot 47:25–38

Newsham KK, McLeod AR, Greenslade PD, Emmett BA (1996) Appropriate controls in outdoor UV-B supplementation experiments. Glob Change Biol 2:319–324

Oravecz A, Baumann A, Mate Z, Brzezinska A, Molinier J, Oakeley EJ, Adam E, Schäfer E, Nagy F, Ulm R (2006) CONSTITUTIVELY PHOTOMORPHOGENIC1 is required for the UV-B response in *Arabidopsis*. Plant Cell 18:1975–1990

Osterlund MT, Wei N, Deng XW (2000) The roles of photoreceptor systems and the COP1-targeted destabilization of HY5 in light control of *Arabidopsis* seedling development. Plant Physiol 124:1520–1524

Paranhos JT, Fragoso V, Silveira VC, Henriques AT, Fett-Neto AG (2009) Organ-specific and environmental control of accumulation of psychollatine, a major indole alkaloid glucoside from *Psychotria umbellata*. Biochem Syst Ecol 37:707–715

Park JS, Choung MG, Kim JB, Hahn BS, Kim JB, Bae SC, Roh KH, Kim YH, Cheon CI, Sung MK, Cho KJ (2007) Genes up-regulated during red coloration in UV-B irradiated lettuce leaves. Plant Cell Rep 26:507–516

Peer WA, Murphy AS (2007) Flavonoids and auxin transport: modulators or regulators? Trends Plant Sci 12:556-563

Pi Y, Jiang KJ, Hou R, Gong YF, Lin JA, Sun XF, Tang KX (2010) Examination of camptothecin and 10-hydroxycamptothecin in *Camptotheca acuminata* plant and cell culture, and the affected yields under several cell culture treatments. Biocell 34:139–143. PMID 21443144

Prasad SM, Dwivedi R, Zeeshan M (2005) Growth, photosynthetic electron transport, and antioxidant responses of young soybean seedlings to simultaneous exposure of nickel and UV-B stress. Photosynthetica 43:177–185

Ramani S, Chelliah J (2007) UV-B-induced signaling events leading to enhanced-production of catharanthine in *Catharanthus roseus* cell suspension cultures. BMC Plant Biol 7:61

Rizzini L, Favory JJ, Cloix C, Faggionato D, O'Hara A, Kaiserli E, Baumeister R, Schäfer E, Nagy F, Jenkins GI, Ulm R (2011) Perception of UV-B by the *Arabidopsis* UVR8 protein. Science 332:103–106

Sancar A, Lindsey-Boltz LA, Unsal-Kaçmaz K, Linn S (2004) Molecular mechanisms of mammalian DNA repair and the DNA damage checkpoints. Annu Rev Biochem 73:39–85

Sarghein SH, Carapetian J, Khara J (2011) The effects of UV radiation on some structural and ultrastructural parameters in pepper (*Capsicum longum* A.DC.) Turk. J Biol 35:69–77

Seo HS, Watanabe E, Tokutomi S, Nagatani A, Chua NH (2004) Photoreceptor ubiquitination by COP1 E3 ligase desensitizes phytochrome A signaling. Gene Dev 18:617–622

Shinkle JR, Derickson DL, Barnes PW (2005) Comparative photobiology of growth responses to two UV-B wavebands and UV-C in dim-red-light- and white-light-grown cucumber (*Cucumis sativus*) seedlings: physiological evidence for photoreactivation. Photochem Photobiol 81:1069–1074

Stratmann J (2003) Ultraviolet-B radiation co-opts defense signaling pathways. Trends Plant Sci 8:526–533

Subramanian C, Kim BH, Lyssenko NN, Xu X, Johnson CH, von Arnim AG (2004) The *Arabidopsis* repressor of light signaling, COP1, is regulated by nuclear exclusion: mutational analysis by bioluminescence resonance energy transfer. P Natl A Sci USA 101:6798–6802

Surplus SL, Jordan BR, Murphy AM, Carr JP, Thomas B, Macherness SAH (1998) Ultraviolet-B-induced responses in *Arabidopsis thaliana*: role of salicylic acid and reactive oxygen species in the regulation of transcripts encoding photosynthetic and acidic pathogenesis-related proteins. Plant Cell Environ 21:685–694

Suzuki T, Honda Y, Mukasa Y (2005) Effects of UV-B radiation, cold and desiccation stress on rutin concentration and rutin glucosidase activity in tartary buckwheat (*Fagopyrum tataricum*) leaves. Plant Sci 168:1303–1307

Taylor LP, Grotewold E (2005) Flavonoids as developmental regulators. Curr Opin Plant Biol 8:317–323

Teramura AH, Sullivan JH (1994) Effects of UV-B radiation on photosynthesis and growth of terrestrial plants. Photosynth Res 39:463–473

Tevini M, Braun J, Fiese G (1991) The protective function of the epidermal later of rye seedlings against ultraviolet-B radiation. Photochem Photobiol 53:329–333

Toguri T, Umemoto N, Kobayashi O, Ohtani T (1993) Activation of anthocyanin synthesis genes by white light in eggplant hypocotyl tissues, and identification of an inducible P-450 cDNA. Plant Mol Biol 23:933–946

Tuteja N, Ahmad P, Panda BB, Tuteja R (2009) Genotoxic stress in plants: shedding light on DNA damage, repair and DNA repair helicases. Mutat Res 681:134–149

Ubi BE, Honda C, Bessho H, Kondo S, Wada M, Kobayashi S, Moriguchi T (2006) Expression analysis of anthocyanin biosynthetic genes in apple skin: effect of UV-B and temperature. Plant Sci 170:571–578

Ulm R, Baumann A, Oravecz A, Mate Z, Adam E, Oakeley EJ, Schäfer E, Nagy F (2004) Genome-wide analysis of gene expression reveals function of the bZIP transcription factor HY5 in the UV-B response of *Arabidopsis*. P Natl A Sci USA 101:1397–1402

Wade HK, Bibikova TN, Valentine WJ, Jenkins GI (2001) Interactions within a network of phytochrome, cryptochrome and UV-B phototransduction pathways regulate chalcone synthase gene expression in *Arabidopsis* leaf tissue. Plant J 25:675–685

Wang HY, Ma LG, Li JM, Zhao HY, Deng XW (2001) Direct interaction of *Arabidopsis* cryptochromes with COP1 in light control development. Science 294:154–158

Wang S, Xie B, Yin L, Duan L, Li Z, Eneji AE, Tsuji W, Tsunekawa A (2010) Increased UV-B radiation affects the viability, reactive oxygen species accumulation and antioxidant enzyme activities in maize (*Zea mays* L.) pollen. Photochem Photobiol 86:110–116

Wang Y, Feng H, Qu Y, Cheng J, Zhao Z, Zhang M, Wang X, An L (2006) The relationship between reactive oxygen species and nitric oxide in ultraviolet-B-induced ethylene production in leaves of maize seedlings. Environ Exp Bot 57:51–56

Wargent JJ, Gegas VC, Jenkins GI, Doonan JH, Paul ND (2009) UVR8 in *Arabidopsis thaliana* regulates multiple aspects of cellular differentiation during leaf development in response to ultraviolet B radiation. New Phytol 183:315–326

Xu C, Sullivan JH (2010) Reviewing the technical designs for experiments with UV-B radiation and UV-B impact on photosynthesis, DNA and secondary metabolism. J Integr Plant Biol 52:377–387

Xu ZR, Li YH (2006) Screening the genes associated with anthocyanin biosynthesis in roots of 'Tsuda' turnip using cDNA microarray. Yi Chuan (Hereditas) 28:1101–1106

Yamasaki S, Shimada E, Kuwano T, Kawano T, Noguchi N (2010) Continuous UV-B irradiation induces endoreduplication and peroxidase activity in epidermal cells surrounding trichomes on cucumber cotyledons. J Radiat Res 51:187–196

Zhang WJ, Björn LO (2009) The effect of ultraviolet radiation on the accumulation of medicinal compounds in plants. Fitoterapia 80:207–218

Zhou B, Li Y, Xu Z, Yan H, Homma S, Kawabta S (2007) Ultraviolet A-specific induction of anthocyanin biosynthesis in the swollen hypocotyls of turnip (*Brassica rapa*). J Exp Bot 58:1771–1781

Zu YG, Pang HH, Yu JH, Li DW, Wei XX, Gao YX, Tong L (2010) Responses in the morphology, physiology, and biochemistry of *Taxus chinensis* var. *mairei* grown under supplementary UV-B radiation. J Photochem Photobiol 98:152–158

Chapter 8
Advances in Microspore Culture Technology: A Biotechnological Tool for the Improvement of Medicinal Plants

Alison M. R. Ferrie

8.1 Introduction

Utilization of medicinal plants to treat medical conditions is not new; however, over the last decade there is a greater awareness of healthy foods and using natural products for preventative medicine rather than conventional drugs. This has led to increased harvest of plants, commercialization of the plant or active ingredient, and a demand for consistent uniform product. Many of these medicinal plants are harvested from natural sources or "wild crafted" which in some cases has resulted in over-harvesting of a specific species. As a consequence of this, many medicinal species are listed as endangered or threatened (CITES—Convention on International Trade in Endangered Species of Wild Flora and Fauna www.cites.org, ICUN). Tissue culture methodologies have been developed to address some of the concerns regarding overharvesting of native plants and production of a uniform product for the consumer.

8.2 Doubled Haploidy

Haploid plants are defined as having half the chromosome number of the normal plant or only one complete set of chromosomes. When the chromosome number of this haploid plant is doubled, the resulting plant is termed a doubled haploid (DH) which is a homozygous fertile plant that can be utilized in practical application and basic research. Doubled haploidy techniques are the fastest way in which to fix traits and generate true-breeding plants.

A. M. R. Ferrie (✉)
Plant Biotechnology Institute—National Research Council of Canada,
110 Gymnasium Place, Saskatoon, SK S7N 0W9, Canada
e-mail: Alison.Ferrie@nrc-cnrc.gc.ca

S. Chandra et al. (eds.), *Biotechnology for Medicinal Plants*,
DOI: 10.1007/978-3-642-29974-2_8, © Springer-Verlag Berlin Heidelberg 2013

Haploid plants or doubled haploid plants can be produced by a variety of methods. A naturally occurring haploid from *Datura stramonium* L. (jimson weed) was first described by Blakeslee in 1922 and since then, there have been many reports of other naturally occurring haploids (see review by Dunwell 2010). The disadvantage of spontaneous haploids is the low frequency in which they occur making this method ineffective for breeding purposes or basic research. Haploids can also be produced by wide crosses (interspecific or intergeneric) followed by chromosome elimination (Rines et al. 1996) or parthenogenesis (Sestili and Ficcadenti 1996). The most common method to produce haploid and doubled haploid plants is the in vitro culture of the male (i.e., microspores or anthers) or female (i.e., ovules or ovaries) gametophyte termed androgenesis or gynogenesis, respectively. When embryos or calli are produced from the gametophytes, these can be regenerated to haploid and doubled haploid plants.

Androgenesis, via anther culture, was first described by Guha and Maheshwari (1964) in *Datura*. Most of the early work focused on anther culture; however, with improvements and refinements of the technique, isolated microspore culture is now the preferred method. Isolating the microspores from the anthers prior to culture eliminates the chance that anther tissue (i.e., somatic tissue) can regenerate plants, which is undesirable.

Gynogenic methods for haploid production have been demonstrated for at least 24 species (Bohanec 2006). These methods are usually less efficient than androgenesis and therefore fewer studies have been carried out in gynogenesis. Although gynogenic methods are beneficial when plants do not respond to androgenetic methods, there is a problem with regenerating albino plants from anther culture, or the donor plants are male sterile.

There are many journal articles, books, and review articles on doubled haploidy methodology via androgenesis or gynogenesis (Chen et al. 2011; Ferrie and Caswell 2011; Germanà 2011a, b; Maluszynski et al. 2003).

8.3 Doubled Haploidy in Medicinal Plants

There has been very little research done on genetically improving medicinal plants although herbs, spices, and traditional plants have been used for millennia to treat ailments. Tissue culture methods, i.e., micropropagation and regeneration, have been used for conservation purposes and propagation. However, there has been very little doubled haploidy work conducted on medicinal species compared to agronomically important crops. There is certain interest in improving medicinal plants and there are reviews on doubled haploidy methods in nutraceutical species (Ferrie 2007, 2009; Ferrie et al. 2005). These review papers provide lists of species in which doubled haploidy methods have been attempted and have resulted in calli, embryos, or plants. Doubled haploid plants have been produced from medicinal plants by culturing the male gametophyte [i.e., anther culture (Table 8.1), isolated microspore culture (Table 8.2)], or by culture of the female gametophyte (Table 8.3). This review focuses on androgenesis.

8.4 Factors Influencing Microspore Embryogenic Response

The switch that triggers a microspore to develop into an embryo or callus rather than a mature pollen grain is influenced by many factors. These factors vary depending on the plant species, which means doubled haploidy protocols have to be optimized for each species. Protocols can also differ among genotypes within a species. The factors influencing embryogenic response of a species includes plant-related factors, i.e., genotype of the donor plant, donor plant growing conditions, and the developmental stage of the pollen grain. Culture methods, i.e., pretreatments of the floral organs, media composition, and post isolation conditions also play a major role in influencing embryogenic response.

8.4.1 Genotype

As in most tissue culture systems, genotype plays a role in culture response. A study comparing different *Hypericum* species and genotypes within these species for anther culture response found that callus production was only observed in *H. tetrapterum* Fr. (Square-stalked St. John's wort) and *H. perforatum* L. (common St. John's wort) but not *H. hirsutum* L. (Hairy St. John's wort) (Schulte et al. 1996). Shoots were only produced from the *H. perforatum* genotypes. Individual plants within a species were also compared and there was considerable variation between the plants for callus production (Schulte et al. 1996). In another study, three gentian (*Gentiana triflora* Pall.) genotypes were screened for anther culture response. One genotype, 'Ashiro-no-Aki', produced 21.3 embryos per 100 anthers, 'Ashiro-no-Natsu' yielded 3.7 embryos per 100 anthers and the third genotype, 'Lovely-Ashiro' did not respond (Doi et al. 2010). In a study screening buckwheat (*Fagopyrum esculentum* Moench) genotypes for androgenic ability, eight out of 32 genotypes produced callus and were recorded as responsive (Bohanec et al. 1993). Varietal differences in embryogenic response were also observed in garlic (*Allium sativum* L.), poppy (*Papaver somniferum* L.), and *Bupleurum falcatum* L. (Suh and Park 1986; Dieu and Dunwell 1988; Shon et al. 2004). These results are in contrast to the results from Smýkalová et al. (2009). The authors evaluated ten caraway (*Carum carvi* L.) genotypes for anther culture response but there was little difference among them as all of them responded equally (Smýkalová et al. 2009).

Genotypic differences have also been observed using isolated microspore culture. Five *Saponaria vaccaria* L. (cow cockle) cultivars and accessions were evaluated for isolated microspore culture response (Kernan and Ferrie 2006). All genotypes produced microspore-derived embryos, with the most responsive being 'White Beauty' and accession PI578121 from Finland with an average response of over 300 embryos per 100 buds. Genotypic response was also observed in fennel (*Foeniculum vulgare* Mill.), dill (*Anethum graveolens* L.), and caraway isolated microspore culture (Ferrie et al. 2011).

8.4.2 Donor Plant Conditions

The growth conditions to which the donor plants are subjected can influence the embryogenic response of the microspores. Donor plants can be grown in the field, greenhouse, or in environmentally controlled growth chambers. For most woody species or large plants, it would be impossible to grow plants under greenhouse or artificial conditions and therefore floral buds are harvested from field grown plants. Growth chambers do have an advantage as temperature, photoperiod, light intensity, watering, and fertilizing regime can be controlled. As shown in Table 8.1, the majority of the plants for anther culture studies are grown in the field, although greenhouse and growth chambers are also used. Anther culture response of caraway was improved when donor plants were grown in the greenhouse but exposed to different temperature regimes. The plants grown at 22°C for nine days followed by a cold treatment (4°C) for 19 days were more responsive than the control plants (Smýkalová et al. 2009).

The majority of donor plants used for isolated microspore culture in the author's laboratory are grown in environmentally controlled chambers. For dill and anise (*Pimpinella anisum*) the cooler temperature regime of 10/5°C was beneficial. A comparison between 10/5°C and 20/15°C in *Saponaria vaccaria* showed that the cooler temperature was not beneficial as more embryos were produced from plants grown at 20/15°C for both genotypes evaluated (Kernan and Ferrie 2006). Donor plants of fennel were grown at three temperature regimes (10/5°C, 15/10°C, and 17/13°C) and floral buds were collected for isolated microspore culture. For five of seven genotypes, the temperature of 15/10°C produced the most microspore-derived embryos and for the other two genotypes, 10/5°C yielded the most embryos. The temperature treatment of 17/13°C was not beneficial (Ferrie et al. 2011).

8.4.3 Pretreatment of the Floral Organs

Physical or chemical pretreatments can be given to the plant, floral organs, or excised anthers prior to culture to induce embryogenic response from the anthers or microspores. Low temperature, elevated temperature, EMS (ethyl methane sulfonate), ethanol stress, irradiation, or reduced atmospheric pressure are examples of stresses used as pretreatments to enhance androgenic response (see review Ferrie et al. 1995). Temperature shock is the most widely used pretreatment and for many medicinal plant species, a cold pretreatment of 0–10°C for 12 h–7 days is beneficial (Table 8.1). A heat shock of 35°C for 4 days was used for *Hepatica nobilis* anthers resulting in an embryogenic response of 45 embryos per 30 anthers, which exceeded the response from cold pretreatment of 5°C or a reduced heat shock of 25°C (Nomizu et al. 2004). All these pretreatments were single temperature regimes, but a combination heat/cold shock pretreatment

Table 8.1 Factors influencing anther culture of medicinal plants

Species	Location of donor plants	Pretreatment	Microspore stage	Media for callus/embryo induction	Culture conditions	Response	Reference
Aconitum carmichaeli	Field	–		Solid MS, 3% sucrose 5 mg/l 2,4-D, 1 mg/l K	25°C dark	C, E	Hatano et al. (1987)
Albizzia lebbeck	Field	–	Late uninucleate–early binucleate	B5 2 mg/l K, 0.5 mg/l 2,4-D	27/25°C 16 h	C, P	Gharyal et al. (1983a)
Allium sativum	–	5°C 12 h	–	MS 1 mg/l NAA, 2 mg/l K	25°C 16 h	C, P	Suh and Park (1986)
Atropa belladonna	Field	–	Uninucleate	LS 4 mg/l K, 2 mg/l IAA	–	E, P	Zenkteler (1971)
Azadirachta indica	Field	–	Early–late uninucleate	Solid MS, 9% sucrose 1 μM 2,4-D, 1 μM NAA, 5 μM BA	25°C 8 wk dark	C, P	Chaturvedi et al. (2003)
Boswellia serrata	Field	4°C 24 h	Uni-binucleate	Solid BN, 3% sucrose 0.5 mg/2,4-D, 1 mg/l BA, 200 mg/l CH or Solid BN,3% sucrose 1 mg/l 2,4-D, 0.5 mg/l NAA, 0.5 mg/l K, 200 mg/l CH	25°C 8 h or 24 h dark	C	Prakash and Chand (1999)
Bupleurum falcatum	Field	–	Uninucleate	MS, 3% sucrose 1 mg/l 2,4-D	25°C dark	C, P	Shon and Yoshida (1997)
	Field	10°C 5 d	Uninucleate	Solid MS, 3% sucrose 0.075 mg/l 2,4-D + 0.075 mg/l picloram or Solid MS, 3% sucrose, 0.75 mg/l 2,4-D	30°C dark	C, P	Shon et al. (2004)
Carum carvi	Greenhouse 6°C, 19 days	–	Mid uninucleate to early binucleate	Solid B5, 2% sucrose 0.1 mg/l NAA, 0.1 mg/l 2,4-D, 15%CM	6°C	E, P	Smýkalová et al. (2009)

(continued)

Table 8.1 (continued)

Species	Location of donor plants	Pretreatment	Microspore stage	Media for callus/embryo induction	Culture conditions	Response	Reference
Cassia siamea	Field	–	Late uninucleate to early binucleate	Solid B5, 2 mg/l 2,4-D, 0.5 mg/l K	27/25°C 16 h	C	Gharyal et al. (1983b)
Catharanthus roseus	–	10°C 3–7 d	Uninucleate	Liquid MS, 8% sucrose 0.5 mg/l NAA, 0.05 mg/l BA	30°C 3–7 d	E	George (1985)
Curculigo orchioides	Field	–	Uninucleate	Liquid MS, 2% sucrose 0.5 mg/l 2,4-D	–	E, P	Augustine et al. (2008)
Digitalis obscura	Field	4°C 2 – 7 d	Late tetrad-uninucleate	Callus: Solid MS, 2% sucrose, 2 mg/l 2,4-D, 2 mg/l NAA Embryos: Solid MS, 2% sucrose, 0.5 mg/l IAA or NAA	26°C dark	C, E, P	Pérez-Bermúdez et al. (1985)
Digitalis purpurea	Field	4°C 2–12 d	Late tetrad	Whites 10 mg/l 2,4-D	25 °C 24 h light	C, P	Corduan and Spix (1975)
Digitalis lanata	Field	5°C 5–6 d	–	Solid MS 1 mg/l BA, 0.01 mg/l NAA	25–30 °C 16 h light	C, E, P	Badea et al. (1985)
Datura metel	Greenhouse 26–30°C	45°C/10°C 30 s pulse	Mid-late uninucleate	Solid NN, 3% Sucrose 4.65 μM K	25°C 16 h light	E, P	Iqbal and Wijesekara (2007)
Echinacea purpurea	Net house	8°C 3 d	Uninucleate	Solid N6, 3% Sucrose 2.22 uM BA, 0.054 uM 2,4-D	25°C dark	C, P	Zhao et al. (2006)
Fagopyrum esculentum	Field or greenhouse	–	Uninucleate to early binucleate	Solid MS, 9% maltose, 2.5 mg/l BA, 0.5 mg/l IAA	25°C dark	C, P	Bohanec et al. (1993)
Foeniculum vulgare	Field	–	–	Solid MS, 3% Sucrose 1 mg/l 2,4-D	25°C 16 h light	C	Matsubara et al. (1995)
Gentiana triflora	Field	–	Uninucleate–binucleate	Solid ½ strength NLN 13% sucrose	32.5°C 24 h then 25°C	E, P	Doi et al. (2010)

(continued)

Table 8.1 (continued)

Species	Location of donor plants	Pretreatment	Microspore stage	Media for callus/embryo induction	Culture conditions	Response	Reference
Gentiana triflora	Field	4°C 48 h	Mid–late uninucleate	Solid NN, 3% Sucrose 0.7 mg/l NOA, 1.2 mg/l BA	24°C dark		Pathirana et al. (2011)
Hepatica nobilis	Greenhouse	35 °C 4 d	Uninucleate	Solid NN, 3% sucrose 1% AC, double layer (solid:liquid)	5°C 2d then 25°C dark	E, P	Nomizu et al. (2004)
Hieracium pilosella	Greenhouse 16 h	–	Early uninucleate	Liquid MS, 3% sucrose, 0.5 mg/l 2,4 D, 200 mg/l glutamine	27°C dark	C, P	Bicknell and Borst (1996)
Hyoscyamus sp.	Growth cabinet	–	Uninucleate	Solid BN 0.1–2 mg/l ZR	26°C 12 h light	E, C, P	Raghavan and Nagmani (1989)
Hypericum perforatum	Field or greenhouse	4°C 7 d	–	MS 0.5 mg/l 2,4-D, 0.5 mg/l K	25°C Dark	C, P	Schulte et al. (1996)
Oenothera hookeri	26°C 24 h light	5°C 2 d	Uninucleate	Solid MS, 3% Sucrose 2 mg/1 2,4-D, 2 mg/l BA	25/20°C 16 h light	C, P	Martinez and de Halac (1995)
Panax ginseng	Field	6–9°C 4 d	Mid uninucleate	Solid MS, 6% Sucrose 3 mg/l 2,4-D, 1 mg/l IAA, 1 mg/l IBA	25–28°C	C, P	Du et al. (1987)
Panax quinquifolius	–	–	Mid uninucleate	N6 or Miller, 3% sucrose 1.5 mg/l 2,4-D, 1 mg/l K	–	C, P	Shao and Li (1986)
Papaver somnifernm	Growth cabinet 15°C then 25°C 16 h growth cabinet or field	7°C 7 d	Uninucleate	Solid MS 2 mg/l 2,4-D, 0.5 mg/l IAA 0.5 mg/l BA 1 mg/l K	–	C, P	Dieu and Dunwell (1988)
Physalis ixocarpa	Greenhouse 22°C	3°C 2 d	Late uninucleate	Solid NN, 5 mg/l IAA, 0.2 mg/l K	26°C dark	E, P	Bapat and Wenzel (1982)

(continued)

Table 8.1 (continued)

Species	Location of donor plants	Pretreatment	Microspore stage	Media for callus/embryo induction	Culture conditions	Response	Reference
Salvia sclarea	–	Cold	–	MS	–	C	Bugara (1986)
Stevia rabaudiana	–	–	Uninucleate				Flachsland et al. (1996)
Withania somnifera	Field	–	Uninucleate	MS 1 μM BA	25°C 24 h light	C, E	Vishnoi et al. (1979)
Zingiber officinale	Field	0°C 7 d	Uninucleate	MS, 2% sucrose 2–3 mg/L 2,4-D	22°C 2–3 d dark then 14 h light	C, P	Samsudden et al. (2000)

Abbreviations: Media: B5, Gamborg et al. (1968); BN, Bourgin and Nitsch (1967); LS, Linsmaier and Skoog (1965); MS, Murashige and Skoog (1962); N6, Chu (1978); NN, Nitsch and Nitsch (1969); NLN, Lichter (1982); White, White (1943)

Growth regulators and additives: 2,4-D, 2,4,-dichlorophenoxyacetic acid; AC, Activated charcoal; BA, 6-Benzylaminopurine; CH, Casein Hydrolysate; CM, Coconut milk; IAA, indole acetic acid; IBA, indole-3-butyric acid; K, Kinetin; NAA, Naphtaleneacetic acid; NOA, naphthoxyacetic acid; ZR, Zeatin riboside

Response: C, callus; E, embryos; P, plants

Table 8.2 Factors influencing isolated microspore culture of medicinal plants

Species	Donor plant conditions	Media (liquid)	Culture conditions	Other factors	Response	Reference
Ammi majus	Growth cabinet 20/15°C	NLN, 20–25% sucrose	32°C 3 d then 24 °C, dark		E, P	Ferrie et al. (2011)
Anethum graveolens	Growth cabinet 10/5°C	NLN, 10–25% sucrose	32°C or 24°C, dark		E, P	Ferrie et al. (2011)
Angelica archangelica	Growth cabinet 20/15°C	NLN, 13% sucrose	32°C 3 d then 24°C, dark		E, P	Ferrie et al. (2011)
Calendula officinalis	22/12°C	R92.01, sucrose	24°C, dark		C, P	Ferrie and Caswell (2011)
Carum carvi	Growth cabinet 20/15°C	AT-3, 9% maltose	32°C 3 d then 24°C, dark		E, P	Ferrie et al. (2011)
Codonopsis pilosula	Greenhouse	AT-3	24°C, dark		E	Irmen and Ferrie (unpublished)
Foeniculum vulgare	Growth cabinet 15/10°C	NLN 25% sucrose Brassinosteroids	32°C 3 d then 24°C, dark		E, P	Ferrie et al. (2011)
Gingko biloba	Field	BN ±11.40 μM IAA, 0.93 μM K	–	Electrostimulation of uninucleate microspores	E	Laurain et al. (1993)
Levisticum officinale	Growth cabinet 20/15°C	NLN, 20–25% sucrose	32°C 3 d then 24°C, dark		E, P	Ferrie et al. (2011)
Matricaria recutita		R92.01, maltose			C, P	Ferrie and Caswell (2011)
Pimpinella anisum	Growth cabinet 10/5°C	NLN, 20–25% sucrose	32°C 3 d then 24°C, dark		E, P	Ferrie et al. (2011)
Saponaria vaccaria	Growth cabinet 20/15°C	NLN, 15% sucrose	32°C for 3 d then 24°C, dark		E, P	Kernan and Ferrie (2006)

Abbreviations: Media: AT-3, Touraev and Heberle-Bors (1999); BN, Bourgin and Nitsch (1967); NLN, Lichter (1982); R92.01, Theiler-Hedtrich and Hunter (1996)
Growth regulators: IAA, indole acetic acid; K, Kinetin
Response: E, embryos; P, plants

Table 8.3 List of medicinal species in which gynogenic methods have been reported

Species	Response	Reference
Cynara scolymus	C	Bastar and Bohanec (2006)
Ephedra foliata	C	Singh and Konar (1981)
Fagopyrum esculentum	C, P	Bohanec (1997)
Gentian triflora	E, P	Doi et al. (2011)
Hyoscyamus sp.	E, P	Chand and Basu (1998)
Psoralea corylifolia	C, P	Chand and Sahrawat (2007)
Solidago virgaurea	C	Bastar and Bohanec (2006)
Valeriana officinalis	C, P	Bastar and Bohanec (2006)

Abbreviations
C, callus; E, embryos; P, plants

was described by Iqbal and Wijesekara (2007) in which the *Datura metel* anthers were subjected to 45°C for 30 s followed by 10°C for 30 s. This temperature regime resulted in an embryo yield of 253 embryos per anther whereas the control (no temperature pulse) had a yield of 103 embryos per anther.

8.4.4 Developmental Stage of Pollen Grain

Determining the correct developmental stage of the pollen grain is critical for achieving an embryogenic response. This most responsive stage for embryogenesis is generally the early uninucleate to early binucleate stage. This is also true for the medicinal species (Table 8.1), although late tetrad to uninucleate microspores were used in *Digitalis obscura* and *D. purpurea* (Pérez-Bermúdez et al. 1985, Corduan and Spix 1975) anther culture experiments. Early binucleate microspores have shown to be responsive in *Gentiana triflora* (Pathirana et al. 2011) and *Fagopyrum esculentum* (Bohanec et al. 1993). Du et al. (1987) cultured ginseng (*Panax ginseng*) anthers with microspores at five different developmental stages (tetrad, early uninucleate, mid-uninucleate, late uninucleate, and binucleate). Callus was produced from all developmental stages although the mid-uninucleate stage of development was the most responsive.

8.4.5 Media Constituents

The composition of the medium in which the anthers or microspores are cultured is a major factor influencing the switch from gametophytic development to the sporophytic pathway. The culture of isolated microspores requires a more complex media formulation than what is required for anther culture. Basal medium, carbohydrate source, concentration of carbohydrate, growth regulators, and other additives such as activated charcoal, PEG (polyethylene glycol), and amino acids are all very important components in media formulation. Other factors to consider are liquid or solid media and pH of the media.

A range of basal media and growth regulators are used in the medicinal plants (Tables 8.1, 8.2). Some of the more common basal media used for anther or microspore culture are B5 (Gamborg et al. 1968), BN (Bourgin and Nitsch 1967), MS (Murashige and Skoog 1962), N6 (Chu 1978), NN (Nitsch and Nitsch 1969), and NLN (Lichter 1982). Sucrose is the most commonly used carbohydrate in plant tissue culture, although maltose has been used in buckwheat (*Fagopyrum esculentum*) anther culture (Bohanec et al. 1993) as well as caraway (Ferrie et al. 2011) and German chamomile (*Matricaria recutita*) isolated microspore culture (Ferrie and Caswell 2011). Six carbohydrates (fructose, glucose, maltose, melibiose, melezitose, and sucrose) at four concentrations were evaluated in *Saponaria vaccaria* isolated microspore culture (Kernan and Ferrie 2006). Microspore-derived embryos developed from all carbohydrate sources evaluated although embryogenic response from media with fructose or glucose was very poor (<10 embryos per 100 buds). The highest embryogenic response was with melezitose (130 embryos per 100 buds) or sucrose (186 embryos per 100 buds). The carbohydrate source has a nutritional benefit as well as an osmotic effect. For anther culture, the carbohydrate concentration is usually between 2 and 9% (Table 8.1). For isolated microspore culture, the carbohydrate concentration is generally higher. For example in the Apiaceae species, i.e., *Ammi majus, Anethum graveolens, Foeniculum vulgare, Levisticum officinale, Pimpinella anisum*, elevated levels of sucrose (20–25%) were beneficial (Table 8.2).

Growth regulators are commonly used in anther culture to promote callus or embryo development. Some of the more commonly used growth regulators include BA (6–Benzylaminopurine, IAA (indole acetic acid), K (Kinetin), and 2,4-D (2,4,-dichlorophenoxyacetic acid). The type and most effective concentration of growth regulator can vary depending on the species (Table 8.1). Additives such as activated charcoal, casein hydrolysate, and coconut milk have also been utilized in anther culture (Table 8.1).

Most anther culture studies are carried out using solid media although liquid media has been used with success. A two-phase system which consists of a liquid layer over a solid lower layer was beneficial for anther culture of *Hepatica nobilis* (Nomizu et al. 2004). Isolated microspore culture generally utilizes liquid media (Table 8.2).

pH of the culture media can be manipulated to enhance embryogenesis. Generally a pH of 5.5–5.8 is used, but a high medium pH of 8.0–8.5 for 4–6 days induced multicellular structures in snapdragon (*Antirrhinum majus* L.) and embryos in tobacco (*Nicotiana tabacum* L) (Barinova et al. 2004).

8.4.6 Culture Conditions

The culture conditions in which the microspores are kept post isolation can influence the embryogenic response. Physical treatments such as temperature, photoperiod, and anther/microspore density are the most frequently manipulated in order to generate embryos or calli. Generally cultures are incubated at 22–27°C; although there are exceptions. For many species, a heat shock (>30°C) is required (Table 8.1); this is especially so with microspore culture (Table 8.2). For *Hepatica nobilis*, a 2 day cold treatment (5°C) prior to culture at 25°C was beneficial (Nomizu et al. 2004). Cultures are kept in continuous light, continuous dark, or with a light/dark cycle (Tables 8.1, 8.2).

The density of the microspores or the number of anthers plated/volume of medium can influence both embryo quality and quantity as well as calli production. This can vary depending on the species. For *Saponaria vaccaria*, four microspore densities were evaluated (6,250–100,000 microspores per ml) with the most optimal range being 12,500–25,000 microspores per ml (Kernan and Ferrie 2006).

8.4.7 Haploid/Doubled Haploid Plant Production

Induced microspores can follow two routes of development to form a plant; direct embryogenesis similar to zygotic embryo development (i.e., globular, heart-shaped, torpedo, and cotyledonary stages) or indirectly through a callus phase which can then undergo organogenesis or embryogenesis. The direct route is preferred and has been observed in several of the medicinal plants (Tables 8.1, 8.2). In a few cases, both developmental pathways are observed and can occasionally be manipulated to favor embryos or callus. For example, two different media formulations differing in type and concentration of growth regulators are given for *Digitalis obscura*, one for the production of calli and one for the production of embryos (Pérez-Bermúdez et al. 1985). When cotyledonary embryos or calli have developed, they can be removed from the anther/microspore culture media to a media for regeneration into plants. Regeneration usually takes place on solid media in the light.

Determining the ploidy level of regenerated plants can be accomplished by counting chromosomes or by using a flow cytometer. For some species, there is a high rate of spontaneous chromosome doubling and this saves the added step of using chromosome doubling agents to produce fertile, homozygous, doubled haploid plants. For those species with a low frequency of spontaneous doubling, chromosome doubling agents like colchicine or trifluralin will have to be used.

8.5 Benefits of Doubled Haploidy

The benefits of haploid/doubled haploid plants and the generation of pure lines have been known for a longtime. Doubled haploidy methodology can be used for breeding, mutagenesis, gene transfer, biochemical, and physiological studies as well as genetic and genomic studies. There are many examples of utilization of microspore-derived embryos and doubled haploid plants in the literature (reviews in Maluszynski et al. 2003; Takahata et al. 2005; Ferrie and Möllers 2011).

Doubled haploid plants can be incorporated into breeding programs directly as advanced pure lines or as parental lines for hybrids. The main advantage of incorporating doubled haploids in a breeding program is to accelerate cultivar development. Homozygous true breeding lines are produced in one generation rather than several generations of selfing and backcrossing which can save up to 3–4 years in a breeding program, depending on the species (Ulrich et al. 1984). Other benefits include easier identification of recessive traits, more efficient mutant selection, and easier genetic manipulation at the haploid level. Furthermore, marker assisted breeding is more efficient in haploid embryos and/or plants as only a single allele is present at any given locus.

As mentioned, many of the medicinal plants are harvested from the wild or are landraces that have not been improved genetically. This means, there is potential for variation within the population for agronomic characteristics as well as levels of the active ingredient or compound of interest. This variation can cause problems to growers who are cultivating the plant on a commercial scale. Variation in plant growth habit, maturity dates, or height could make it more difficult to harvest the crop. Variation in the biochemical profile can pose a problem to those conducting clinical trials or to the processors and manufacturers who are extracting the active ingredient and are expecting a consistent product. Uniformity of the product is also important in order to meet regulatory standards.

Field evaluation of doubled haploids of dill, fennel, and caraway identified lines which were shorter, earlier in maturity, or had higher oil contents (Ferrie et al. 2011). The one obvious feature of the field trials was the uniformity of the doubled haploid lines compared to the parental cultivars. Differences in biochemical profiles were observed in seeds of *Saponaria vaccaria* doubled haploids. The type and concentration of saponin varied among lines and therefore it should be possible to select lines with specific saponin profiles (Ferrie 2007).

8.6 Conclusion

The goal of most germplasm improvement programs is to develop cultivars that will perform consistently under various environmental conditions and yield a stable quality and quantity of product whether it is seed, biomass, or beneficial active ingredient. Doubled haploidy methodology can be used as a tool to support such medicinal plant improvement programs just as it has been used in other agronomically important species.

The term “medicinal plants” covers a wide range of plant families grown in diverse locations with diverse modes of reproduction, and therefore it is very difficult to generalize about conditions required for microspore embryogenesis in medicinal plants. Microspore-derived embryos, calli, and haploid/doubled haploid plants have been successfully produced in some medicinal plant species; however, further research and refinements are required to be able to make such protocols useful for applied and basic research.

References

Augustine AC, Nivas S, D'Souza L (2008) Induction of embryos and plantlets from anthers of *Curculigo orchioides* Gaertn. An endangered medicinal herb. Indian J Biotechnol 7:536–540

Badea E, Iordan M, Mihalea A (1985) Induction of androgenesis in anther culture of *Digitalis lanata*. Revue Roumaine de biologie. Ser Biol Veg 30:63–71

Bapat VA, Wenzel G (1982) In vitro haploid plantlet induction in *Physalis ixocarpa* Brot. through microspore embryogenesis. Plant Cell Rep 1:154–156

Barinova J, Clement C, Marting L, Baillieul F, Soukupova H, Heberle-Bors E, Touraev A (2004) Regulation of developmental pathways in cultured microspores of tobacco and snapdragon by medium pH. Planta 219:141–146

Bastar MT, Bohanec B (2006) Screening for gynogenic ability in selected medicinal plant species: progress report. In: Touraev A, Forster B (eds) Haploids in higher plants III, Vienna, Feb 12–15 2006, p 50

Bicknell RA, Borst NK (1996) Isolation of reduced genotypes of *Hieracium pilosella* using anther culture. Plant Cell, Tissue Organ Cult 45:37–41

Blakeslee AF, Belling J, Farnham ME, Bergner AD (1922) A haploid mutant in the Jimson weed, *Datura stramonium*. Science 55:646–647

Bohanec B (1997) Haploid induction in buckwheat (*Fagopyrum esculentum* Moench). In: Jain SM, Sopory SK, Veilleux RE (eds) In Vitro Haploid Production in Higher Plants. Kluwer Academic Publishers, Dordrecht, pp 163–170

Bohanec B (2006) Mechanisms of gynogenesis. In: Touraev A, Forster B (eds) Haploids in higher plants III, Vienna, Feb 12–15 2006, p 18

Bohanec B, Neskovic M, Vujicic R (1993) Anther culture and androgenetic plant regeneration in buckwheat (*Fagopyrum esculentum* Moench). Plant Cell, Tissue Organ Cult 35:259–266

Bourgin JP, Nitsch JP (1967) Obtention de *Nicotiana* haploides a partir d'etamines cultivees in vitro. Ann Physiol Veg 9:377–382

Bugara AM (1986) Embryoidogenesis in anther culture of *Salvia sclarea*. Fiziologiia I Biokhimiia Kulturnykh Rastenii. Physiol Biochem Cultiv Plants 18:381–386

Chand S, Basu P (1998) Embryogenesis and plant regeneration from callus cultures derived from unpollinated ovaries of *Hyoscyamus muticus* L. Plant Cell Rep 17:302–305

Chand S, Sahrawat AK (2007) Embryogenesis and plant regeneration from unpollinated ovary culture of *Psoralea corylifolia*. Biol Plant 51:223–228

Chaturvedi R, Razdan MK, Bhojwani SS (2003) Production of haploids of neem (*Azadirachta indica* A. Juss.) by anther culture. Plant Cell Rep 21:531–537

Chen JF, Cui L, Malik AA, Mbira KG (2011) In vitro haploid and dihaploid production via unfertilized ovule culture. Plant Cell, Tissue Organ Cult 104:311–319

Chu CC (1978) The N6 medium and its applications to anther culture of cereal crops. In Proceedings of symposium on plant tissue culture, Science Press, Beijing, 43–50

Corduan G, Spix S (1975) Haploid callus and regeneration of plants from anthers of *Digitalis purpurea* L. Planta 124:1–11

Dieu P, Dunwell JM (1988) Anther culture with different genotypes of opium poppy (*Papaver somniferum* L.): effect of cold treatment. Plant Cell, Tissue Organ Cult 12:263–271

Doi H, Takahashi R, Hikage T, Takahata Y (2010) Embryogenesis and doubled haploid production from anther culture in gentian (*Gentiana triflora*). Plant Cell, Tissue Organ Cult 102:27–33

Doi H, Yokoi S, Hikage T, Nishihara M, Tsutsumi K, Takahata Y (2011) Gynogenesis in gentians (*Gentiana triflora, G. scabra*): production of haploids and doubled haploids. Plant Cell Rep 30:1099–1106

Du LG, Hou YH, Chang WC, Li AS, Fu ZM, Shao QQ, Chen ZG, Yang ZT (1987) In vitro pollen plant induction and embryoid clone establishment of *Panax ginseng*. Sci Sin B 30:941–945

Dunwell JM (2010) Haploids in flowering plants: origins and exploitation. Plant Biotechnol J 8:377–424

Ferrie AMR (2007) Doubled haploid production in nutraceutical species: a review. Euphytica 158:347–357

Ferrie AMR (2009) Current status of doubled haploids in medicinal plants. In: Touraev A, Forster BP, Jain SM (eds) Advances in haploid production in higher plants. Springer Science + Business Media B.V, Dordrecht, pp 209–217

Ferrie AMR, Bethune T, Kernan Z (2005) An overview of preliminary studies on the development of doubled haploid protocols for nutraceutical species. Acta Physiol Plant 27:735–741

Ferrie AMR, Bethune T, Mykytyshyn M (2011) Microspore embryogenesis in the Apiaceae. Plant Cell, Tissue Organ Cult 104:399–406

Ferrie AMR, Caswell KL (2011) Isolated microspore culture techniques and recent progress for haploid and doubled haploid plant production. Plant Cell, Tissue Organ Cult 104:301–309

Ferrie AMR, Mőllers C (2011) Haploids and doubled haploids in *Brassica* spp. for genetic and genomic research. Plant Cell, Tissue Organ Cult 104:375–386

Ferrie AMR, Palmer CE, Keller WA (1995) Haploid embryogenesis. In: Thorpe TA (ed) In vitro embryogenesis in plants. Kluwer Academic Publishers, Dordrecht, pp 309–344

Flachsland E, Mroginski L, Daviña J (1996) Regeneration of plants from anthers of *Stevia rebaudiana* Bertoni (Compositae) cultivated in vitro. Biocell 20:87–90

Gamborg OL, Miller RA, Ojima K (1968) Nutrient requirements of suspension cultures of soybean root cells. Exp Cell Res 50:151–158

Gharyal PK, Rashid A, Maheshwari SC (1983a) Production of haploid plantlets in anther cultures of *Albizzia lebbeck* L. Plant Cell Rep 2:308–309

Gharyal PK, Rashid A, Maheshwari SC (1983b) Androgenic response from cultured anthers of a leguminous tree, *Cassia siamea* Lam. Protoplasma 118:91–93

George L (1985) Anther culture of *Catharanthus roseus* L.—development of pollen embryoids. Curr Sci 54:641–642

Germanà MA (2011a) Anther culture for haploid and doubled haploid production. Plant Cell, Tissue Organ Cult 104:283–300

Germanà MA (2011b) Gametic embryogenesis and haploid technology as valuable support to plant breeding. Plant Cell Rep 30:839–857

Guha S, Maheshwari SC (1964) In vitro production of embryos from anthers of *Datura*. Nature 204:497

Hatano K, Shoyama Y, Nishioka I (1987) Somatic embryogenesis and plant regeneration from the anther of *Aconitum carmichaeli* Debx. Plant Cell Rep 6:446–448

Iqbal MCM, Wijesekara KB (2007) A brief temperature pulse enhances the competency of microspores for androgenesis in *Datura metel*. Plant Cell, Tissue Organ Cult 89:141–149

Kernan Z, Ferrie AMR (2006) Microspore embryogenesis and the development of a double haploidy protocol for cow cockle (*Saponaria vaccaria*). Plant Cell Rep 25:274–280

Laurain D, Trémouillaux-Guiller J, Chénieux J (1993) Embryogenesis from microspores of *Ginkgo biloba* L., A medicinal woody species. Plant Cell Rep 12:501–505

Lichter R (1982) Induction of haploid plants from isolated pollen of *Brassica napus*. Z Pflanzenphysiol 105:427–434

Linsmaier EM, Skoog F (1965) Organic growth factor requirements of tobacco tissue cultures. Physiol Plant 18:100–127

Maluszynski M, Kasha KJ, Forster BP, Szarejko I (2003) Doubled haploid production in crop plants: a manual. Kluwer Academic Publishers, Dordrecht

Martinez LD, de Halac IN (1995) Organogenesis of anther-derived calluses in long-term cultures of *Oenothera hookeri* de Vries. Plant Cell, Tissue Organ Cult 42:91–96

Matsubara S, Dohya N, Murakami K (1995) Callus formation and regeneration of adventitious embryos from carrot, fennel and mitsuba microspores by anther and isolated microspore cultures. Acta Hortic 392:129–137

Murashige T, Skoog F (1962) A revised medium for rapid growth and bioassays with tobacco tissue cultures. Physiol Plant 15:473–497

Nitsch JP, Nitsch C (1969) Haploid plants from pollen grains. Science 163:85–87

Nomizu T, Niimi Y, Han D-S (2004) Haploid plant regeneration via embryogenesis from anther cultures of *Hepatica nobilis*. Plant Cell, Tissue Organ Cult 79:307–313

Pathirana R, Frew T, Hedderley D, Timmerman-Vaughan G, Morgan E (2011) Haploid and doubled haploid plants from developing male and female gametes of *Gentiana triflora*. Plant Cell Rep 30:1055–1065

Pérez-Bermúdez P, Cornejo MJ, Segura J (1985) Pollen plant formation from anther cultures of *Digitalis obscura* L. Plant Cell, Tissue Organ Cult 5:63–68

Prakash DVSSR, Chand S (1999) In vitro response from cultured anthers of *Boswellia serrata* Roxb. (Burseraceae). In: Plant tissue culture and biotechnology: emerging trends. Universities Press, Hyderbad, pp 226–231

Raghavan V, Nagmani R (1989) Cytokinin effects on pollen embryogenesis in cultured anthers of *Hyoscyamus niger*. Can J Bot 67:247–257

Rines HW, Riera-Lizarazu O, Maquieira SB, Phillips RL (1996) Wide crosses for haploids. In: Scoles G, Rossnagel B (eds) In: Proceedings of V international oat conference and VII international barley genetics symposium. University of Saskatchewan, Saskatoon, pp 213–222

Samsudden K, Babu KN, Divakaran M, Ravindran PN (2000) Plant regeneration from anther derived callus cultures of ginger (*Zingiber officinale* Rosc.). J Hortic Sci Biotechnol 75:447–450

Schulte J, Büter B, Schaffner W, Berger K (1996) Gametic embryogenesis in *Hypericum* spp. In: Pank F (ed) International symposium on breeding research on medicinal and aromatic plants, Quedlinburg, pp 307–310

Sestili S, Ficcadenti N (1996) Irradiated pollen for haploid production. In: Jain SM, Sopory SK, Veilleux RE (eds) In vitro haploid production in higher plants, vol 1 Kluwer Academic Publishers, Dordrecht, pp 263–274

Shao Q, Li A (1986) Somatic embryogenesis and plant regeneration from anther culture of *Panax quinquifolius* (Ginseng). Int Plant Biotechnol Netw 6:2

Shon T-K, Yoshida T (1997) Induction of haploid plantlets by anther culture of *Bupleurum falcatum* L. Jpn J Crop Sci 66:137–138

Shon T-K, Kim S-K, Acquah D, Lee S-C (2004) Haploid plantlet production through somatic embryogenesis in anther-derived callus of *Bupleurum falcatum*. Plant Prod Sci 7:204–211

Singh MN, Konar RN (1981) In vitro induction of haploid roots and shoots from female gametophyte of *Ephedra foliate* Boiss. Beltr Biol Pflanzen 55:169–177

Smýkalová I, Šmirous P Jr, Kubošiová M, Gasmanová N, Griga M (2009) Doubled haploid production via anther culture in annual, winter type of caraway (*Carum carvi* L.). Acta Physiol Plant 31:21–31

Suh SK, Park HG (1986) Studies on the anther culture of garlic (*Allium sativum* L.). J Korean Soc Hortic Sci 27:89–95

Takahata Y, Fukuoka H, Wakui K (2005) Utilization of microspore-derived embryos. In: Palmer CE, Keller WA, Kasha KJ (eds) Biotechnology in agriculture and forestry: haploids in crop improvement II. 56, pp 153–169

Theiler-Hedtrich R, Hunter CS (1996) Microspore culture in chicory (*Cichorium intybus* L.). In: Jain SM, Sopory SK, Veilleux RE (eds) In vitro haploid production in higher plants, vol 3, Kluwer Academic Publishers, Dordrecht, pp 99–113

Touraev A, Heberle-Bors E (1999) Microspore embryogenesis and in vitro pollen maturation in tobacco. In: Hall RD (eds) Methods in molecular biology: Plant cell culture protocols, vol 111. Humana Press, Totowa, pp 281–291

Ulrich A, Furton WH, Downey RK (1984) Biotechnology and rapeseed breeding: some economic considerations. Sci. Counc. Can. Rep, Ottawa

Vishnoi A, Babbar SB, Gupta SC (1979) Induction of androgenesis in anther cultures of *Withania somnifera*. Z Pflanzen 94:169–171

White PR (1943) Handbook of plant tissue culture. Jacques Cattell Press, Lancaster, 277 pp

Zenkteler M (1971) In vitro production of haploid plants from pollen grains of *Atropa belladonna* L. Experientia 27:1087

Zhao F-C, Nilanthi D, Yang Y-H, Wu H (2006) Anther culture and haploid plant regeneration in purple coneflower (*Echinacea purpurea* L.). Plant Cell Rep 86:55–62

Chapter 9
Implications of Cellular Heterogeneity on Plant Cell Culture Performance

Rohan A. Patil and Susan C. Roberts

9.1 Introduction

Traditionally, valuable plant-derived phytochemicals, also referred to as secondary metabolites, have been obtained either by natural harvestation or chemical synthesis. Natural harvestation necessitates destroying the whole plant or selectively harvesting specialized organs; this makes the process expensive, time-consuming, and environmentally unfriendly (McCoy and O'Connor 2008). These issues are compounded further with slow growing, rare or endangered plants, as well as medicinal plants growing in remote areas (Hawkins 2008). Total or partial chemical synthesis is economically viable only for production of relatively simple structures such as aspirin and ephedrine (Li et al. 2010), but impractical for secondary metabolites with complex structures, such as multiple rings and chiral centers (Chemler and Koffas 2008). Over the past decade, genetic engineering approaches to transfer plant pathways into microbial hosts have provided a competitive alternative for production of certain plant natural products [e.g., artemisinin synthesis in *E. coli* (Martin et al. 2003) and *Saccharomyces cerevisiae* (Ro et al. 2006)]. Microbial fermentation processes are well established and offer the advantages of rapid doubling times, shorter production times, easier extraction of the final product, and inexpensive feed stocks for growth. While a number of plant proteins have been heterologously expressed in microbial hosts (Yesilirmak and Sayers 2009), only a handful of plant natural products have been completely produced in microbes (Alper et al. 2005; Ro et al. 2006). Production of a nonnative compound in microbes requires identification

R. A. Patil · S. C. Roberts (✉)
Department of Chemical Engineering, University of Massachusetts,
686 North Pleasant Street, Amherst, MA 01003, USA
e-mail: sroberts@ecs.umass.edu

R. A. Patil
e-mail: patil@ecs.umass.edu

S. Chandra et al. (eds.), *Biotechnology for Medicinal Plants*,
DOI: 10.1007/978-3-642-29974-2_9, © Springer-Verlag Berlin Heidelberg 2013

and successful transfer of all the relevant plant biosynthetic pathway genes. For a number of plant-derived natural products (e.g., paclitaxel), the metabolic pathways leading to product formation *in planta* are very complex, and are either partially or completely unknown (Croteau et al. 2006). Even after a complete pathway for a particular natural product has been identified, production in prokaryotic hosts is still complicated, as they lack cellular compartmentalization that may be necessary for spatial and temporal partitioning of intermediates en route to final product (Vongpaseuth and Roberts 2007; Wu et al. 2006). Technical drawbacks associated with functional expression of native plant enzymes such as cytochrome P450s (Chemler and Koffas 2008) further impede efficient transfer of a complete biosynthetic pathway for production of a desired compound. Thus, despite the attractiveness of synthesis in microbial hosts, significant engineering challenges remain for more complex secondary metabolites.

The use of plant-based in vitro systems for production of specific secondary metabolites provides an attractive alternative to natural harvestation, chemistry-based routes, and microbial engineering (Kolewe et al. 2008; Rao and Ravishankar 2002; Wilson and Roberts 2012). In vitro culture of plant cells is a mature technology with several decades of success, and can be applied to almost any plant species (Wink et al. 2005). As well as offering a realistic option for large-scale production of secondary metabolites, in vitro plant cultures offer a controlled and regulated environment for studies of growth, metabolism, cell-environment interactions, and for establishment of superior plant species through genetic manipulation.

9.2 In vitro Plant Culture

In vitro culture of plants or plant cells can involve various degrees of differentiation. Whole plants or seedlings, organ cultures, or dedifferentiated suspension cultures propagated from callus can be grown aseptically in a defined culture media. These different types of in vitro cultures (see Fig. 9.1 for *Taxus* species) can be interconverted using established techniques, most of which rely on specific phytohormone concentrations. Differentiated plant cell cultures, specifically organ cultures such as roots and shoots, have been shown to accumulate significant levels of secondary metabolites, often comparable with levels quantified in the whole plant (Matkowski 2008; Roberts and Wink 1998). In particular, hairy root cultures, obtained by transforming root cultures with *Agrobacterium rhizogenes*, have been promising for increased capacity of secondary metabolite production (Srivastava and Srivastava 2007). Several medicinal compounds belonging to the alkaloid, terpenoid, and phenolic families have been produced successfully in hairy root cultures. Examples include resveratrol in *Arachis hypogaea*, artemisinin in *Artemisia annua*, indole alkaloids in *Catharanthus roseus*, and camptothecin in *Camptotheca acuminate* and *Ophiorrhiza pumila* (Ono and Tian 2011). In recent years, a number of valuable reviews on differentiated cultures for secondary

Fig. 9.1 Three different types of in vitro culture for *Taxus*. **a** callus culture, **b** hairy root culture (Syklowska-Baranek et al. 2009), and **c** suspension culture

metabolite production have been published (Guillon et al. 2006; Karuppusamy 2009; Kim et al. 2002; Loyola-Vargas and Miranda-Ham 1995; Pistelli et al. 2010; Verpoorte and Memelink 2002), and therefore, we will not discuss them at length in this chapter. Although many promising reactor designs exist at the small scale, it has been difficult to scale up differentiated cultures to large-sized bioreactors primarily due to issues with nutrient delivery, limiting the widespread commercial application of this technology (Georgiev et al. 2007; Mishra and Ranjan 2008).

On the other hand, dedifferentiated suspension cultures can be maintained in batch, semi-continuous or continuous environments, and are more amenable to scale up than hairy root cultures or other differentiated organ cultures (Kieran et al. 1997). Plant cell suspension cultures under strictly controlled conditions provide a rapid and flexible means for production of desired compounds and a number of processes have been commercialized for the production of secondary metabolites [reviewed in Eibl and Eibl (2002) and Kolewe et al. (2008)] including ginseng, shikonin, and berberine. The bioprocessing principles applied to the culture of microbial and mammalian cells also apply to dedifferentiated plant suspension cells (Hellwig et al. 2004); although plant cell culture technology has lagged behind equivalent fermentation/culture systems for microbes, yeast, and animal cells (Evans et al. 2003; Hellwig et al. 2004), as described below.

9.3 Characteristics of Plant Cell Suspension Cultures

Table 9.1 shows comparison of some of the characteristics of microbial, mammalian, and plant cells of relevance to bioprocessing. Unlike other cell culture systems, dedifferentiated plant cells grow slower and are more easily damaged by traditional mechanisms for aeration and agitation that are required for culture maintenance and processing. Typically in microbial and mammalian systems, the growth phase and production phase are uncoupled, and hence optimal conditions for growth and production can be established and applied separately to maximize synthesis of the desired compound. For plant cell cultures, the product accumulation is sometimes associated with the growth phase, and hence two-phase cultures are not always

Table 9.1 A comparison of key characteristics of microbial, mammalian, and plant cells relevant to bioprocessing

Characteristic	Bacterial cells	Mammalian cells	Plant cells
Size (μm)	~1	~10	~20–50
Doubling time	<1 h	~1 day	Several days
Shear sensitivity	Insensitive	Sensitive	Sensitive
Oxygen demand	High	Low	Medium
Product accumulation	Typically extracellular	Typically extracellular	Often cell-associated
Production phase	Uncoupled with growth	Uncoupled with growth	Often growth associated
Variability in accumulation	Low	Low	High
Contamination risk	Low	High	Low
Cell line stability	High	Low	Medium
Product yields	High	High	Low
Post translational processing	Simple	Advanced	Advanced
Compartmentalization	None	Compartmentalized	Compartmentalized
Cryopreservation techniques	Well-developed	Well-developed	Immature

feasible (Roberts 2007). Plant cell cultures also tend to have poor genetic stability, which has been associated with aneuploidy and polyploidy, intrachromosomal rearrangements, and single gene mutations (Cassells and Curry 2001; Phillips et al. 1994), hence affecting culture performance. In addition, many secondary metabolic pathways are only active in differentiated organs, leading to no or very low accumulation in dedifferentiated culture. Even in cultures that successfully produce the compound of interest, yields are often low and variable (Ketchum and Gibson 1996; Roberts 2007). A number of strategies including strain improvement, selection of high-producing lines, medium optimization, elicitation with biotic or abiotic compounds, precursor addition, permeabilization, immobilization, in situ extraction, and genetic engineering have been used with mixed success to increase metabolite yields to suitable levels for commercial production [reviewed in Bourgaud et al. (2001), Dornenburg and Knorr (1995), Kolewe et al. (2008), Shuler (1999), Smetanska (2008), Verpoorte et al. (1999)]. Long-term variability in product yield over successive subcultures has often been observed (Deusneumann and Zenk 1984; Kim et al. 2004; Ogino et al. 1978; Qu et al. 2005). Relatively, little research has been done to understand and ultimately control this variability in secondary metabolite production, which can have a considerable impact on the success of a commercial plant cell culture process. Plant cell culture bioprocessing and metabolism are complicated by the natural tendency of plant cells to form aggregates (Kieran et al. 1997; Roberts 2007). This chapter reviews the causes and effects of cellular aggregation in plant cell suspension cultures, along with approaches used to characterize cellular heterogeneity at both the whole culture and single cell levels. Understanding heterogeneity can lead to successful strategies to optimize and stabilize production, making plant cell culture a more attractive commercial technology for the supply of valuable plant-derived medicinal compounds.

9.4 Aggregation and Heterogeneity of Plant Cell Suspension Cultures

Plant cell cultures are initiated by transferring pieces of dedifferentiated callus tissue from solid media to a liquid suspension operating under suitable conditions for aeration and agitation. Depending on the friability of the callus tissue, either single cells or small aggregates dissociate and begin growing in a medium containing specific types and concentrations of nutrients and growth hormones (Muir et al. 1954). During cell division, the dividing cells often remain connected to each other via their cell walls, and as a result, aggregates ranging from two to a few hundred cells exist in culture; see Fig. 9.2 for typical aggregate morphology in plant cell culture.

The presence of aggregates during large-scale culture affects mixing, as they tend to sediment and/or stick to the reactor surface, sometimes resulting in extensive wall adhesion/growth. Moreover, large aggregates cause rheological problems by creating dead zones in the culture vessel, blocking reactor ports and gas lines, and affecting the operation of probes to monitor culture conditions (e.g., pH and dissolved oxygen) during growth and product formation. Cellular aggregation has also been found to be the primary reason for the high viscosities observed in a number of plant cell suspension systems (Doran 1999; Kato et al. 1978). These problems are further compounded during the later stages of culture where high cell densities prevail and plant cells become more adherent due to secretion of cell wall extracellular polysaccharides.

Cells within aggregates are subject to different microenvironments with respect to light, oxygen and nutrient availability, cell-to-cell signaling, and applied surface shear forces. These microenvironments often lead to biochemical and morphological heterogeneity amongst cells, with some aggregate populations exhibiting vastly different characteristics. For example, in cell cultures of *Arachis hypogoea*, differences in peroxidase and catalase activity were found amongst cell aggregates of different sizes (Verma and Van Huystee 1970b). Differences in activities of several enzymes including dehydrogenases, oxidases, and hydrolases have been correlated to cellular position in aggregates of tobacco suspension cultures during early stages of growth (De Jong et al. 1967). Similarly, rates of protein synthesis and concentrations of free amino acids have also been shown to vary with aggregate size (Verma and Van Huystee 1970a). Cellular aggregation can additionally promote differentiation, where cells in the core of large aggregates can differentiate and form specialized structures such as tracheary elements (Kuboi and Yamada 1978a).

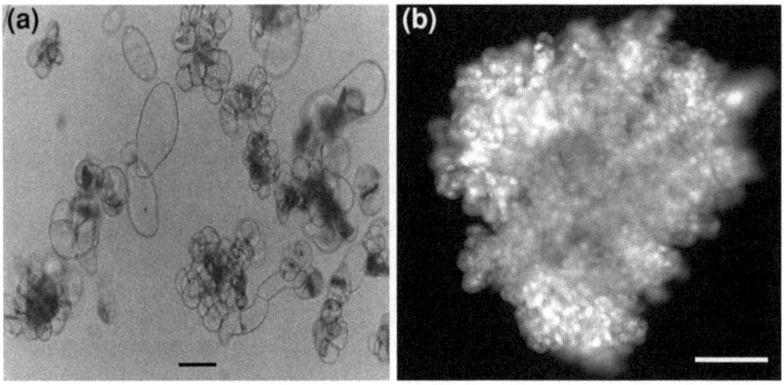

Fig. 9.2 Typical plant cell culture aggregates in suspension culture. **a** *Brightfield image* showing cellular aggregation in *Papaver somniferum* suspension cultures (bar represents 100 μm) (Siah and Doran 1991). **b** fluorescent image stained with fluorescein diacetate to indicate viability, showing a large aggregate in *T. cuspidata* cell cultures (*bar* represents 500 μm)

9.4.1 Factors Affecting the Degree of Cellular Aggregation

The degree of cellular aggregation in suspension cultures varies with the type of plant species. For example, in liquid suspension cultures of *Antirrhinum* (Melchers and Bergmann 1959) and *Saussurea medusa* (Zhao et al. 2003), large aggregates up to 4 mm in diameter have been observed, whereas cultures of *Rosa sp.* Paul's Scarlet (Tulecke 1966), *Anchusa officinalis* and *Nicotiana tabacum* (Su 2006) form highly dispersed suspension cultures with very few large aggregates. Moreover, variation in aggregate size has been observed across different species of the same plant family. For example, in *Tagetes* (marigolds), suspension cultures of *Tagetes patula* exhibited large cell aggregates (diameters up to 20 mm) in contrast to *Tagetes erecta* and *Tagetes minuta* with significantly smaller aggregates (diameters up to 2 mm) (Hulst et al. 1989).

The composition of the basal medium has shown to have a significant influence on the degree of cell separation in suspension cultures. A better separation of cells can be obtained in a purely synthetic medium, where all constituents are precisely known, as opposed to a medium containing undefined components, such as yeast extract, or coconut milk (King and Street 1977). Cellular aggregation is particularly influenced by the concentration of growth regulators in the medium (Lai Keng et al. 2008; Wallner and Nevins 1973). Higher auxin levels generally lead to more dispersed cultures with better cell separation, whereas lower auxin levels increase cell aggregation (Liau and Boll 1971; Machackova et al. 2007; Torrey et al. 1962). Conversely, higher cytokinin levels often induce more aggregation (Kinnersley and Dougall 1980; Kuboi and Yamada 1978b; Zhao et al. 2001; Halperin and Minocha 1973). In some cases, the pH of the medium (Steiner and Dougall 1995) and the frequency of subculture (Meyer et al. 2002) can influence the extent of aggregation. Frequent transfer to fresh medium to maintain cells in an

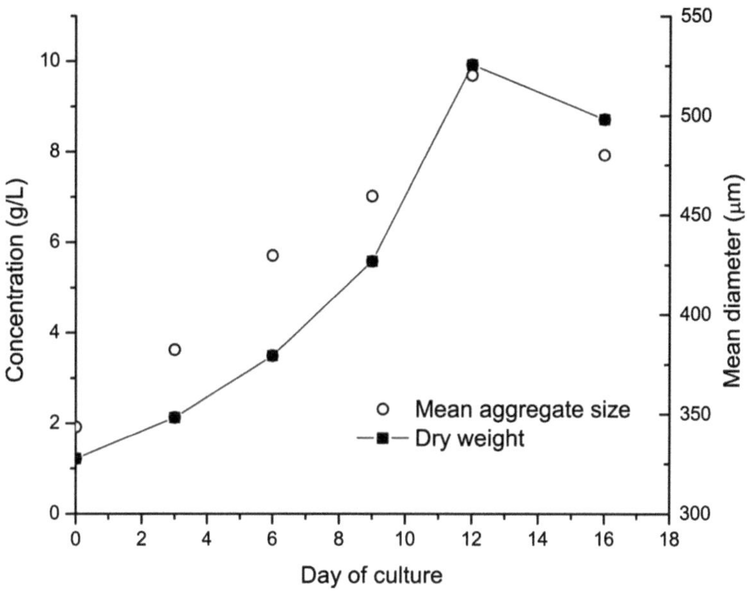

Fig. 9.3 Increase in cell biomass and mean aggregate size during exponential growth in *Taxus* cell cultures (Kolewe et al. 2010)

actively dividing state has shown to increase cell aggregation (Henshaw et al. 1966). In addition, suspension cultures need constant agitation for adequate aeration, which can significantly affect aggregate size distributions (Kieran et al. 2000; Rajasekhar et al. 1971).

9.4.2 *Effect of Aggregation on Typical Culture Parameters*

9.4.2.1 Aggregation and Growth

Plant cells grown in batch culture increase in biomass by cell division until depletion of an essential nutrient shifts cells into stationary phase. Multiple analyses of changes in aggregate size distributions in a batch culture indicate that aggregate size increases during the culture exponential phase, and decreases during the culture stationary phase (Capataz-Tafur et al. 2011; Kolewe et al. 2010; Mavituna and Park 1987; Ranch and Giles 1980; Scragg et al. 1987). Figure 9.3 illustrates changes in biomass and mean aggregate diameter during the culture period of *Taxus cuspidata* suspension cells. Formation of aggregates occurs as a result of cell division, and hence aggregation increases during the period of maximal cell division (i.e., exponential phase). Conversely, during stationary phase, cells are released from aggregates and there is reduced cell division, which leads to a decrease in mean aggregate size.

There is no clear trend in the relationship between culture mean aggregate size and growth rate, and it is very much species dependent. In wheat suspension cultures, finer (less aggregated) suspensions had a higher growth rate than their large aggregate counterparts (Yang et al. 1994). A similar observation was seen with cultures of *Coffea arabica,* where the growth of cultures inoculated with smaller aggregates was significantly superior to those inoculated with larger aggregates (Dubuis et al. 1995). In contrast, studies in celery (Watts et al. 1984) and safflower (Hanagata et al. 1993) suspension cultures show that some degree of aggregation was necessary for rapid growth and cell division, and finer suspensions had slower growth rates compared to aggregated suspensions. Highly dispersed sycamore cultures obtained by incubation with low concentrations of cell wall degrading enzymes show a growth pattern similar to aggregated suspensions (King et al. 1973), indicating that cell aggregation is not essential for high rates of growth and division in this system.

9.4.2.2 Aggregation and Oxygen Consumption

Oxygen requirements in plant cells (typically 1–4 mmol L^{-1} h^{-1}) are comparatively lower than microorganisms (typically 5–90 mmol L^{-1} h^{-1}), due to their lower growth rates (Hellwig et al. 2004; Kobayashi et al. 1989; Taticek et al. 1990). Oxygen supply is known to affect both growth and production of metabolites in plant cell cultures (Huang and Chou 2000; Linden et al. 2001; Thanh et al. 2005). A number of bioreactors have been designed to study the effects of aeration in plant cell cultures, with a primary focus on the influence of mass transfer at the gas–liquid boundary (k_La) (Kieran et al. 1997). However, it has been suggested that the solid–liquid boundary between an aggregate and the culture medium is far more constraining for the delivery of oxygen to the cell than the gas–liquid boundary (Curtis and Tuerk 2006). Oxygen enrichment of the gas phase increases the driving force through bulk liquid phase and can be used to minimize mass transfer limitations of oxygen at the solid–liquid interface. In addition, suspension cultured plant cells must have significant oxygen transport within aggregates to maintain aerobic respiration and desired growth rates. Due to these various constraints, there often exists a critical aggregate size, above which oxygen limitations to the centermost cells of an aggregate may result. Relatively, few studies have been performed to study oxygen transport within aggregates, primarily because experimental measurements within the shear-sensitive aggregates are difficult. The majority of research has been based on development of mathematical models to calculate the critical size of an aggregate, using simple oxygen mass balances (Hulst et al. 1989; Pepin et al. 1999). These models typically assume a zero order reaction with known oxygen uptake rate, diffusion coefficient, and dissolved oxygen concentration to calculate the oxygen concentration profile within an aggregate. These studies suggest that diffusion of oxygen is moderately restricted in the interior of aggregates of 1 mm in diameter, and severe oxygen deficiencies are observed when the aggregates reach approximately

3 mm in size (Hulst et al. 1989; Pepin et al. 1999). A diffusion–reaction model was used to analyze experimentally measured oxygen uptake rates in immobilized cultures of *Solanum aviculare* (Ananta et al. 1995). Results from direct experimental measurement of the aggregate properties revealed that the critical aggregate size at which oxygen limitations occur was significantly larger than that predicted by theory. It was hypothesized that the plasmodesmata, which interconnect cells within aggregates, and the negative pressures created by the gas-filled cavities within the porous aggregates, promote oxygen transfer, factors which were not accounted for in the theoretical analysis (Ananta et al. 1995). The effects of oxygen limitations on culture performance are species dependent, with both increased (Schlatmann et al. 1995) and decreased (Hulst et al. 1989) secondary metabolite synthesis reported in large aggregates.

9.4.2.3 Aggregation and Secondary Metabolite Accumulation

For recombinant protein production, cellular aggregation in plant cell cultures is usually considered undesirable, as it complicates bioreactor operation (Hellwig et al. 2004). However, certain degrees of cellular aggregation and differentiation promote secondary metabolite production in plant cell cultures (Becker 1970; Zhao et al. 2001). Aggregation causes changes in environmental conditions experienced by each cell in an aggregate, altering cellular metabolism, and inducing differential biochemical responses (Verma and Van Huystee 1970b). In some suspension cultures, cells do not just aggregate, but form sophisticated differentiated structures (Ellis et al. 1996; Hoekstra et al. 1990; Kuboi and Yamada 1978a; Xu et al. 1998; Zhao et al. 2001), which can lead to increased secondary metabolite accumulation. Studies in tobacco suspension cultures indicate that all cells in culture have a uniform ability to form aggregates (Kuboi and Yamada 1978a); however, the potential for tracheid differentiation in these studies was higher when cells were located in the center of the aggregate, suggesting that differentiation is a result of environmental circumstances, and not an inherited trait by particular cells.

To date, a number of studies have been performed to understand the effect of cellular aggregation on secondary metabolite accumulation in cell cultures. However, there is no consensus across plant species, and results vary depending on the particular species and secondary metabolite (Table 9.2). Larger aggregates have been shown to have a positive effect on secondary metabolite production, a positive effect up to a critical size, or a negative effect. This lack of trend in metabolite accumulation with aggregate size is not unexpected, as fundamental differences exist amongst distinct plant species, and their metabolic pathways may be differentially regulated.

The majority of these studies aimed at investigating the relationship between aggregate size and secondary metabolite accumulation rely on separation of aggregates based on size using a series of sieves, followed by measurement of biomass associated secondary metabolites. In most cases, cultures were allowed to accumulate secondary metabolites (either by elicitation or through alternate stress-

Table 9.2 Relationship between aggregate size and secondary metabolite accumulation in various plant cell culture systems

System	2° Metabolite	Effect of increasing aggregate size on metabolite accumulation	Reference
Fragaria ananassa (strawberry)	Anthocyanin	Increased	Edahiro and Seki (2006)
Apium graveolens L. (celery)	Phthalides and Terpenoids	Increased	Watts et al. (1984)
Catharanthus roseus (periwinkle)	Ajmalicine	No clear trend	Kessler et al. (1999)
Salvia officinalis (sage)	Ursolic acid	Decreased	Bolta et al. (2003)
Daucus carota (carrot)	Anthocyanin	Decreased	Kinnersley and Dougall (1980)
Vaccinium pahalae (ohelo)	Anthocyanin	Decreased	Pepin et al. (1999)
Carthamus tinctorius L. (safflower)	Anthocyanin	Decreased	Hanagata et al. (1993)
Taxus cuspidata (yew)	Paclitaxel	Decreased	Kolewe et al. (2011)
Daucus carota (carrot)	Anthocyanin	Increased up to critical diameter, then decreased	Madhusudhan and Ravishankar (1996)
Saussurea medusa (snow lotus)	Jaceosidin	Increased up to critical diameter, then decreased	Zhao et al. (2003)
Tagetes patula (marigold)	Anthocyanin	Increased up to critical diameter, then decreased	Hulst et al. (1989)

mediated responses) and then fractionated to measure cell-associated metabolite levels. Though straightforward, this method of fractionation neglects metabolites secreted to the extracellular medium, which can be quite significant for some culture systems. In contrast, some studies were performed by altering the aggregate size distribution at the time of culture initiation (Hanagata et al. 1993; Kinnersley and Dougall 1980; Kolewe et al. 2011). In these studies, cultures are fractionated on the day of inoculation, transferred to fresh media, and maintained independently as small and large aggregate cultures. Over time measurements of total metabolite accumulation (e.g., both cell-associated and extracellular) as a function of aggregate size were made. Results from these studies suggest a possible process optimization strategy, in which rational manipulation of the aggregate size at culture initiation can lead to high yielding cell cultures for use in bioprocesses.

9.4.3 Methods to Study Aggregation and Determine the Aggregated State of a Culture

Quantification of cell growth is essential for monitoring success of both batch and continuous cell culture processes. A number of methods have been established to evaluate growth kinetics in plant cell suspension cultures (Mustafa et al. 2011; Ryu et al. 1990) that include: dry cell weight, fresh cell weight, packed cell volume, cell number (Kolewe et al. 2010; Kubek and Shuler 1978b), culture optical density or turbidity (Eriksson 1965; James and Lee 2000; Sung 1976), electrical conductivity (Hahlbrock and Kuhlen Hahlbrock and Kuhlen 1972), osmolarity (Madhusudhan et al. 1995; Tanaka et al. 1993), nutrient and/or metabolite concentrations (Albiol et al. 1993; Thom et al. 1981), protein (David et al. 1989; Dougall 1964), and nucleic acid content (Lamboursain and Jolicoeur 2005; Nash and Davies 1972), and pH measurements (Nesius and Fletcher 1973). Aggregates of various sizes co-exist in suspension culture and because cells within an aggregate are subject to different local environments, aggregate size can significantly affect cell physiology, resulting in different aggregate growth rates (Mavituna and Park 1987). Information regarding aggregate dynamics can be incorporated into kinetic models to more accurately predict culture growth. Recently, a population balance equation framework to describe plant cell aggregates as particulate system was developed and utilized to quantitatively predict changes in total biomass, mean aggregate size, and aggregate size distributions in a *Taxus* cell suspension batch culture (Kolewe et al. 2012). Such models can be further improved by including information regarding cellular metabolism and product formation. These structured and segregated models can be used to predict operating conditions that optimize culture performance. In order to formulate these models, it is important to quantify aggregation dynamics during batch culture. This section describes the techniques currently used for measuring aggregate size profiles during the cell culture period, along with their advantages and limitations.

9.4.3.1 Wet Sieving and Dry Weight Measurements

The most common technique to measure aggregate size distributions in plant cell cultures is wet sieving, using a series of standard filters with different pore sizes. Fractions are collected and dry weight can be determined for each aggregate size range (Ge et al. 2006; Hulst et al. 1989; Madhusudhan and Ravishankar 1996; Mavituna and Park 1987; Tanaka et al. 1992; Wongsamuth and Doran 1997; Zhao et al. 2003). The dry weight measurement is considered to be a simple, reliable, and inexpensive means of estimating plant cell growth (McDonald et al. 2001). However, measuring dry weight is a time-consuming process, as sample drying times range from 1 to 3 days, minimizing applicability for online measurements and control (Nunez-Palenius et al. 2005). The distribution resolution is limited by the number of filters used and the amount of total biomass needed for analysis.

Additionally, cell cakes can be formed during mechanical sieving, which can interfere with cell separation, leading to potentially erroneous results (Trejo et al. 2003). With highly nonspherical and elongated aggregates, sieving analysis has shown to overestimate particle size (McDonald et al. 2001). For these reasons, filtering and measurement of resulting aggregate fraction dry weight is not feasible for routine monitoring of the aggregate profiles throughout a culture period.

9.4.3.2 Image Analysis

Ex situ image analysis using microscopy to quantify plant cell aggregate size distributions has been employed. Image analysis techniques consist of sample plating, collecting a statistically relevant number of microscopy images, and performing size analysis using established software programs (Ibaraki and Kenji 2001). As plant cell aggregates are not always spherical, the imaging programs need to calculate elliptical form factor (Trejo et al. 2003) or equivalent spherical diameter (Kolewe et al. 2010) in order to calculate the size. The use of image analysis was found to be more precise than mechanical sieving to describe aggregate size distribution changes during growth in *Solanum chrysotrichum* suspension cultures, primarily because mechanical sieving was limited by inefficient separation of aggregates (Trejo et al. 2003). As with sieving, the processing steps increase the likelihood of altering the aggregate size distribution due to potential breakage. Additionally, due to the low-throughput nature of this technique, long and labor intensive procedures are necessary to obtain statistically significant results (Edahiro and Seki 2006), which become impractical for multiple measurements. To minimize culture sampling, imaging systems connected to a bioreactor (i.e., in situ image analysis) have been designed to measure product accumulation (Grand d'Esnon et al. 1989; Smith et al. 1995), growth rate, and cell aggregate distribution (Harrell et al. 1992; McDonald et al. 2001). From a product accumulation point of view, in situ image analysis becomes particularly beneficial when the product formed is a pigment or associated with a pigment, as it allows visualization of changes in cell status over the culture period. Such information can be used for online bioprocess feedback control (Smith and Reid 1996). However, a distinct disadvantage of using an online imaging system for aggregate distribution measurements in bioreactors is that it requires pumping cell aggregates to the imaging cell, which can disturb culture conditions, and alter aggregate size distributions. Image analysis, thus, provides an effective means for studying morphological characteristics and product accumulation in suspension cultures (Ceoldo et al. 2005; Kieran et al. 2000; Takeda et al. 1994; Trejo et al. 2003), but its utility in determining aggregate size distributions is limited (Kolewe et al. 2010).

9.4.3.3 Focused Beam Reflectance Method

Another in situ technique to monitor biomass concentration and aggregate size distribution is the focused beam reflectance method (FBRM). FBRM is an optical technique developed by Lasentec Inc. (Redmond, WA) (schematic shown in Fig. 9.4)

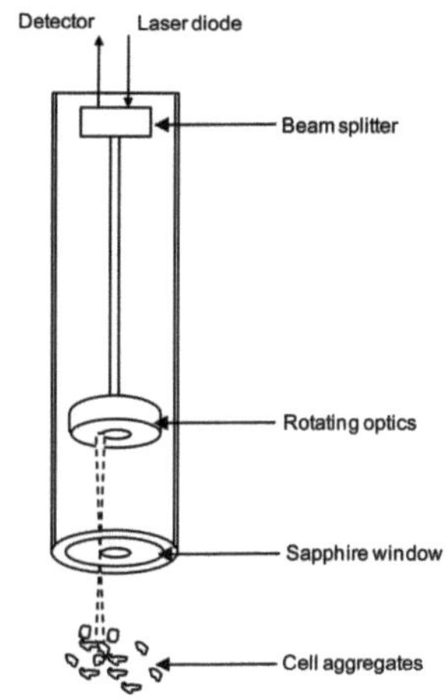

Fig. 9.4 Schematic diagram of the focused beam reflectance method (FBRM) probe

which operates by projecting a highly focused laser beam, through a set of rotating optics and a sapphire window, into a suspension sample. When the beam intersects a particle, light is backscattered and measured. The duration of particle's backscatter correlates directly to the particle size and is expressed in terms of chord length. The number of such chords measured for a given amount of sample and in a specific time period, is related to particle concentration. A detailed description on the working of the FBRM system is presented elsewhere (Barrett and Glennon 1999). This technique has been used for particle size characterization in several chemical processing applications in hydrometallurgy (Richmond et al. 1998), crystallization (Kougoulos et al. 2005), liquid–liquid dispersions, and emulsions (O'Rourke and MacLoughlin 2005; Simmons and Azzopardi 2001). The feasibility of this technique was first demonstrated for the plant cell cultures of *Oryza sativa, Nicotiana benthamiana* and *Trichosanthes kirilowii* (McDonald et al. 2001). The chord length distribution data obtained from FBRM depends on aggregate morphology, density, and reflective properties. The average aggregate diameter of the suspension culture, determined either by image analysis or by sieving measurements, can be related to the average chord length obtained from FBRM measurements. A challenge in applying the FBRM technique to plant cell cultures is to determine the best way to utilize this chord length distribution data to obtain biomass and aggregate size. FBRM has been shown to provide a good representation of the biomass concentration and aggregate size in cell cultures with varying morphological characteristics including cultures with roughly spherical aggregates, irregular aggregates with bulging cells, and cylindrical unbranched aggregates (Jeffers et al. 2003; McDonald et al. 2001). For its use in bioreactors, a probe can be installed for the in situ characterization of particle size and concentration using the FBRM technique; however, issues regarding sterility and optimal probe location/orientation have to be considered (McDonald et al. 2001). A disadvantage as opposed to other biomass and aggregate size measurement techniques is the relatively large sample size required in FBRM technique. For example, both the reports which used

FBRM to measure aggregation characteristics of plant cell cultures used approximately 100 mL of sample volume (Jeffers et al., 2003, McDonald et al. 2001). Also this technique is culture type and culture condition-specific and has to be validated for the suspension culture of interest. However, once validated the FBRM technique is well suited for rapid and sensitive monitoring of aggregate distribution and biomass of plant cell cultures and has the potential for in situ application.

9.4.3.4 Electrical Sensing Zone or Coulter Counter Technique

An alternate method for determining particle size distributions is the electrical resistance pulse sizing technique, commonly known as the Coulter counter (Fig. 9.5) technique. Particles are suspended in an electrolyte solution and pass between electrodes across which a constant current is applied. The measured change in resistivity across two electrodes as particles pass through can be related to the volume of the particles (Graham 2003). This technique has been routinely used for cell counting and particle size characterization in numerous biological systems such as bacteria, yeast, and mammalian cells (Kubitschek 1969). The first application of this technology to plant cell cultures was shown in enzymatically treated cultures of Paul's Scarlet Rose (*Rosa sp.*) and soybean (*Glycine. max L.*) (Kubek and Shuler 1978a). Enzymatic or chemical digestion of the middle lamella of plant cell walls forms either single cells or finely dispersed cultures, allowing them to be analyzed with the Coulter counter (Davis et al. 1984). As this technique was primarily developed for single particles, its application to highly aggregated plant cell cultures has been limited. Recently, using a larger size aperture, this technique was applied to measure both biomass and aggregate size distribution for cell cultures of *T. cuspidata*, which consisted of relatively spherical aggregates up to 2,000 μm in diameter (Kolewe et al. 2010). In this study, a comparison of the aggregate size measured using Coulter counter, filtration, and microscopic image analysis was made. The measurements using the Coulter counter were shown to underestimate the aggregate size compared to the other two methods; however, the relative sizes were found to be consistent. A linear correlation was obtained amongst all three measurements, which allowed aggregate size measured using the Coulter counter to match either filtration or microscopy measurements. The Coulter counter method was found to provide higher resolution than filtration measurements and image analysis, as well as being faster and more reliable. In this case, a sample size of 2×2 mL analyzed for 1 min was sufficient to accurately represent the aggregation distribution for *Taxus* cell cultures (Kolewe et al. 2010). However, the applicability of this technique to cultures with different morphological characteristics such as chain-like aggregates and irregular-shaped aggregates still needs to be investigated.

Fig. 9.5 Schematic diagram illustrating the working principle of the Coulter counter

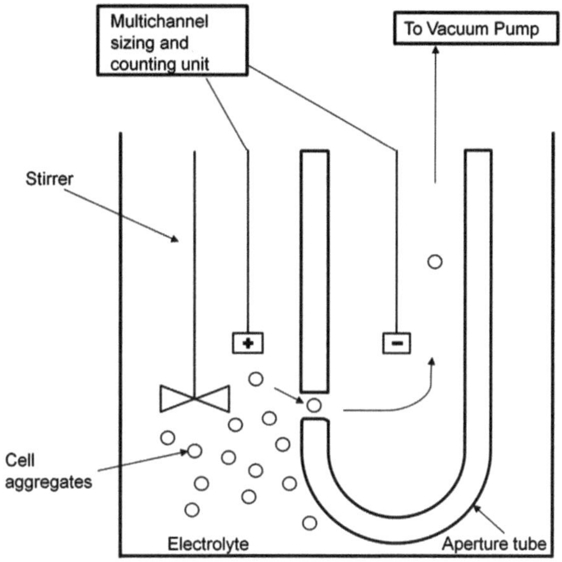

9.4.4 Methods to Control or Manipulate Aggregate Size Distributions in Plant Cell Cultures

9.4.4.1 Chemical and Physical Treatments

Controlling aggregate size in plant cell cultures is important for optimizing both growth and secondary metabolite production. Minimizing the extent of aggregation during plant cell growth has been subject to extensive investigation, primarily due to the negative effects of large aggregates on rheology and mixing in bioreactors. In an effort to maintain cultures as single cell suspensions or in a desired aggregated state, a variety of techniques have been explored, including: the use of chemical treatments such as cell wall degrading enzymes (King et al. 1973; Kubek and Shuler 1978b; Naill and Roberts 2004), addition of hormones (Diwan and Malpathak 2010; Simpkins et al. 1970), application of compounds such as colchicine (Hayashi and Yoshida 1988; Umetsu et al. 1975) or casein hydrolysate (Wallner and Nevins 1973), inclusion of metabolic inhibitors such as L-α-aminooxy-β-phenylpropionic acid (AOPP) (Edahiro and Seki 2006), and lowering of Ca^{2+} ion concentration (Takayama et al. 1977); as well as physical treatments such as filtration using a series of sieves (Dixon 1995; Henshaw et al. 1966), pulses of pressurized air (Kurz 1971), operation in different bioreactor configurations (Tanaka et al. 1988; Yuan et al. 2004), shearing using homogenization (Rodriguez-Monroy et al. 2004; Williams et al. 1988), selective removal of large aggregates (Prenosil and Hegglin 1990), and cell immobilization (Brodelius 1985; Morris and Fowler 1981). Chemical and enzymatic treatments can create cellular metabolic changes by altering gene expression; whereas size distribution cannot be precisely controlled with mechanical methods such as sedimentation and shearing (Morris and Fowler 1981). Some of these treatments can negatively affect cell viability (Doran 1999; Dunlop et al. 1994; Ishii 1988; Kolewe et al. 2011). Use of a particular technique to obtain the desired aggregated state is dependent on both cell type and culture conditions, and generally has to be evaluated individually for each system of interest. In most cases, once fine suspension cultures are established, reversion back to aggregated conditions upon repeated subculture is observed (Morris and Fowler 1981; Naill and Roberts 2005b).

9.4.4.2 Use of Undifferentiated Suspension Cultures

Instead of culturing heterogeneous mixtures of dedifferentiated plant cells, cultures consisting of innately undifferentiated cells derived from the vascular cambium of a plant tissue can be used for production of important metabolites (Lee et al. 2010). Isolation and culture of these undifferentiated cells, referred to as cambial merismatic cells (CMSs) have been shown for a variety of plant species including *T. cuspidata, Panax ginseng* and *Gingko biloba* (Lee et al. 2010). CMCs bypass the dedifferentiation step and grow in suspension primarily as singletons or small aggregates comprised of 2–3 cells. Besides reduced aggregate size, CMCs demonstrated improved performance as compared to dedifferentiated cultures with respect to growth rate, secondary metabolite accumulation, and shear tolerance, at both the laboratory and pilot scales. Furthermore, the ability of CMC cultures to grow in a finely dispersed state has the potential to minimize the variability in growth and secondary metabolite accumulation, which is typically observed in suspension cultures (Roberts and Kolewe 2010). Therefore, CMCs offer distinct advantages over dedifferentiated cultures for production of valuable plant-derived secondary metabolites. However, some secondary metabolites accumulate to much higher extents in differentiated organ cultures such as root or shoot, for example, vinca alkaloids in *C. roseus* hairy root cultures (Leonard et al. 2009), and camptothecin in root cultures (Lorence and Nessler 2004). The benefits of using CMC cultures for the synthesis of such products need to be evaluated.

9.5 Study of Cellular Heterogeneity at the Single Cell Level

Although heterogeneity in plant cell cultures is well accepted, the underlying mechanisms are unclear. Most research relies solely on examination of typical culture parameters, such as biomass accumulation, substrate utilization, and product levels, which are averaged across all cells in a culture. As discussed above, the presence of different local environments can create cell-to-cell differences within an aggregate. Macroscopic analyses under these circumstances give unclear and incomplete results, as they fail to provide information on how local environments affect individual cell metabolism and behavior. Distinct subpopulations of cells with regards to cell cycle participation (Naill and Roberts 2005a;

Yanpaisan et al. 1998), protein content (Naill and Roberts 2005c), and product accumulation (Hall and Yeoman 1987; Naill and Roberts 2005d) have been observed within a single plant cell culture. The occurrence of such subpopulations, along with additional genetic and epigenetic factors relevant in dedifferentiated cell cultures, is considered to be a primary influence on variability in culture metabolite production. Analyzing the properties of individual cells enables identification of distinct subpopulations and differentiation of different cell types existing simultaneously in a culture. Such studies are essential in characterizing the morphological and metabolic heterogeneities observed in plant cell cultures.

9.5.1 Flow Cytometry

The most direct and powerful technique to study cell populations is flow cytometry (FCM). Single particle suspensions are fed through the flow cytometer and interrogated with focused light sources of specific wavelengths. Each individual particle produces light scatter and fluorescence signals, which are detected through a series of filters or photomultiplier tubes, and ultimately displayed in the form of histograms for data analysis. FCM technology and its practice have been reviewed in Shapiro (1994) and Ormerod (1990). The first application of FCM to plants was done using suspensions of field bean nuclei for measurement of DNA content (Heller 1973). Unlike microbial, animal and mammalian cells, plant cells have several unique features that make analyzes using FCM challenging. Cellular aggregates can block the orifice of the FCM, necessitating the preparation of single particle suspensions (Galbraith 1990). Typical plant-derived particles analyzed by FCM include intact single cells, protoplasts—achieved by digesting the plant cell wall under hyperosmotic conditions, and subcellular components such as nuclei, mitochondria, chloroplasts, and chromosomes (Fig. 9.6). Applications include cell counting using cells with compromised membranes (Nicoloso et al. 1994), DNA and RNA analysis using nuclei isolated from protoplasts (Bergounioux et al. 1988b; Dolezel et al. 2007a), secondary metabolite accumulation using protoplasts (Hara et al. 1989) and intact single cells (Naill and Roberts 2005d), measurement of ploidy and genome size (Bennett and Leitch 2005; Sharma et al. 1983), karyotyping (Conia et al. 1987), cell-cycle stage analyses (Naill and Roberts 2005a; Yanpaisan et al. 1998), subcellular organelle (e.g., chloroplasts and mitochondria) investigations (Ashcroft et al. 1986; Petit et al. 1986; Schroder and Petit 1992), cell membrane and cell wall regeneration (Petit 1992), and measurement of cell-specific gene expression (Birnbaum et al. 2003). FCM also allows simultaneous multi-parametric analysis of the physical and/or chemical characteristics of biological particles present in a heterogeneous sample. For example, a sample consisting of viable and dead particles (e.g., plant cells, protoplasts or pollen) can be distinguished by staining with either fluorescein diacetate, which stains only viable cells, or propidium iodide, which stains dead cells (Huang et al. 1986; Schwab and Hulskamp 2008). Simultaneous biparametric analysis of DNA

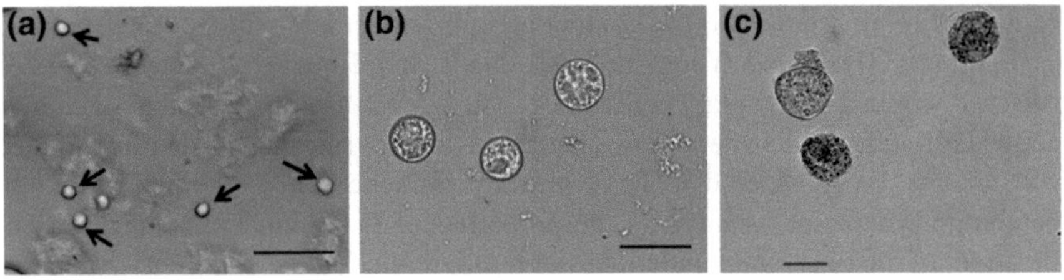

Fig. 9.6 Typical plant suspension culture-derived particles that can be analyzed using flow cytometry. **a** nuclei (indicated by *arrows*), **b** protoplasts, and **c** intact single cells (Naill and Roberts 2004) isolated from *Taxus* cell cultures. *Scale bar* indicates 50 μm

and RNA is possible using Hoescht 33342 dye and pyronin Y, respectively (Darzynkiewicz et al. 2001; Shapiro 1981). When a flow cytometer is equipped with a sorting functionality, it allows isolation of distinct subpopulations of cells with desired characteristics. A detailed review on the use of plant cells in FCM and sorting has been presented recently (Dolezel et al. 2007b). In this chapter, we will limit the discussion to FCM as applied to plant cell cultures in the context of cellular heterogeneity.

9.5.2 Limitations of Flow Cytometry and Cell Sorting in Plant Suspension Cultures

The use of FCM to study cellular heterogeneity requires a liquid suspension of intact single particles. Isolation of intact single cells from plant cell cultures is not trivial, as it requires digestion of the complex extracellular matrix without alteration of cellular physiology or morphology (Kubek and Shuler 1978b; Naill and Roberts 2004). In addition, the presence of the plant cell wall may restrict penetration of various dyes and reagents for labeling, as well as elicit nonspecific binding to fluorescent probes. For these reasons, the use of protoplasts is more common in plant cell FCM. However, hydrophobic metabolites are often stored in the cell wall, for example, paclitaxel in *Taxus* cultures (Aoyagi et al. 2002; Roberts et al. 2003; Russin et al. 1995); hence, removal of the cell wall can delete critical information about the metabolic state of the cell. In addition, the characteristically large size of some plant protoplasts and intact cells creates clogging in conventional flow cytometer chambers, which are typically developed for much smaller microbial and mammalian cells (Dolezel et al. 2007a). To ensure a clog-free run, flow cytometers with special nozzles (ca. 150–200 μm in diameter) are often required for larger plant cells (Galbraith 2004; Gaurav et al. 2010; Harkins and Galbraith 1987). Moreover, the irregular shape of plant cells disturbs the laminar flow pattern and creates flow instabilities that lead to variations in signal detection and incorrect optical measurements (Dolezel et al. 2007a). Flow cytometers with larger nozzle sizes require flow rates and sheath pressures to be maintained at low values to maintain a stable flow stream, which increases the run times for analysis and sorting. As mentioned above, subcellular components (e.g., nuclei, mitochondria, chloroplasts, and chromosomes) can be used for FCM analysis and sorting (Dolezel et al. 2001; Safar et al. 2004). The release of subcellular components of interest is often accompanied by release of cytosol-localized secondary metabolites, which can interfere with analysis, for example, the staining of nuclear DNA (Bennett et al. 2008; Loureiro et al. 2006). Overcoming these limitations can increase utility of FCM for both analysis and sorting of plant cell cultures.

9.5.3 Flow Cytometry to Study Heterogeneity in Plant Suspension Cultures

The ability of FCM to simultaneously measure multiple parameters and analyze a multitude of plant particles has improved our understanding about plants at cellular, subcellular, and molecular levels. Most of this work has been accomplished in the field of plant physiology and molecular biology, with limited applications to plant cell cultures. However, it is important to realize that most of the techniques developed for whole plant FCM can be applied to plant cell cultures. Cell–cell heterogeneity in aggregates of plant cell cultures can lead to unpredictable changes in product accumulation over time (Ketchum and Gibson 1996; Kim et al. 2004). FCM analysis of plant cell cultures can identify populations of cells with common characteristics from potentially different regions of an aggregate. Identification and characterization of such subpopulations can lead both to the design of superior strategies for enhancing secondary metabolite yields in culture and development of a fundamental understanding of aggregation behavior. In the following section, we will discuss the use of FCM and cell sorting to characterize heterogeneity in plant cell cultures.

9.5.3.1 Variable Ploidy Levels

The use of FCM to rapidly determine ploidy levels is by far the most widespread application of plant FCM, with extensive uses in plant breeding (Ochatt et al. 2011), plant ecology, and population biology (Kron et al. 2007). In the context of in vitro cultures, variation in ploidy level is common, and often leads to phenotypic heterogeneity, which affects culture performance. The variable degrees of polyploidy and aneuploidy in plant cell cultures depend on the type and concentration of growth regulators (Mishiba et al. 2001; Weber et al. 2008), the degree of endopolyploidy of the explants (Damato 1985), oxidative stress damage caused to the plant tissues during explant preparation (Cassells and Curry 2001), and the duration of the in vitro cultivation (Evans and Gamborg 1982; Ramulu and Dijkhuis 1986). Using FCM, ploidy levels of cultured plant cells have been correlated to variations in cell growth, cellular differentiation (Schween et al. 2005), and regeneration capacity (Shiba and Mii 2005). Co-existence of subpopulations of different ploidy levels in cultures of *T. x media* has been observed (Baebler et al. 2005). In some instances, changes in ploidy levels have been correlated to secondary metabolite accumulation. For instance, tetraploid cultures of *A. annua* had higher levels of artemisinin than diploids, whereas *Hypericum perforatum* cultures ith higher ploidy levels accumulated less hypericin. Consequently, variation in ploidy levels can have a significant impact on cellular metabolic activity, and careful monitoring of the structure and stability of DNA in large-scale suspension cultures is recommended. Strategic manipulation of ploidy levels using either mitotic inhibitors or growth regulators can be used to increase secondary metabolite levels in suspension cultures (Dhawan and Lavania 1996).

9.5.3.2 Variability in Cell Cycle Analysis and Presence of Noncycling Cells

FCM methods to monitor cell cycle progression in suspension cultures enable a more detailed understanding of culture growth and differentiation. For homogenous populations of cells, where all cells cycle at equal rates, univariate DNA content measurements using FCM are sufficient to understand the cell cycle kinetics (Paau et al. 1977). However, dedifferentiated plant cell cultures are heterogeneous, and often contain quiescent or noncycling cells that rest in the G_0 phase of the cell cycle (Naill and Roberts 2005a; Yanpaisan et al. 1999). In such cases, univariate analysis for DNA content does not resolve every phase of the cell cycle, as cells in G_0 and G_1 both have 2c DNA content, while cells in G_2 and M both have 4c DNA content. Multi-parametric FCM that uses an additional cellular marker such as RNA (Bergounioux et al. 1988a) or protein content (Citterio et al. 1992; Darzynkiewicz et al. 1980), has been used to distinguish between cycling and noncycling cells. Another widely used strategy to distinguish between G_0 and G_1 cells is based on incorporation of the thymidine analog 5-bromo-2-deoxyuridine (BrdU) into DNA during replication and its subsequent detection using a monoclonal antibody. Alternatively, incorporation of another thymidine analog 5-ethynyl-2'-deoxyuridine (EdU), whose detection is based on click chemistry with an azide-conjugated fluorochrome (such as Alexa Fluor 488) and copper (I) as the catalyst, can be used. EdU detection does not rely on bulky antibodies and DNA denaturation steps, which are necessary when using BrdU labeling, providing a more rapid analysis and better preservation of cellular structures (Ayaydin et al. 2010) (Fig. 9.7). Subpopulations of noncycling cells were observed in cell cultures of *Taxus* (Naill and Roberts 2005a) and *S. aviculare* (Yanpaisan et al. 1998) using the BrdU method. The presence of such noncycling cells affects overall culture growth rate and activity of secondary metabolic pathways (Mak and Doran 1993; Yanpaisan et al. 1999). For instance, in *S. aviculare* suspension cultures, the dry weight doubling time was approximately 2 days, whereas the duration of the cell cycle of actively dividing cells was only 1 day, and the proportion of actively

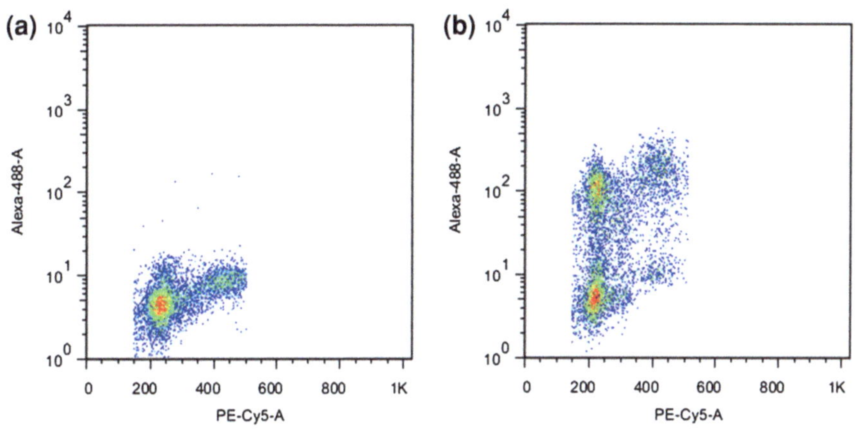

Fig. 9.7 EdU and DNA staining to differentiate between cycling and noncycing cells in *Taxus* suspension cultures. Two parameter *dot plot* for **a** culture not incubated with EdU, and **b** EdU-treated culture. DNA content using 7-aminoactinomycin D (7-AAD) staining is represented on the X-axis and EdU incorporation detected via Alexa Fluor 488 azide is *plotted* on the Y-axis. EdU incorporation in S-phase cells is clearly evident in **b**. EdU incorporation is determined by overlaying and comparing the two parameter *dot plots* of treated and untreated samples

dividing cells never exceeded 52 % of the total population (Yanpaisan et al. 1998). Similar results were found in *Taxus* where approximately 65 % of the suspension cells were found to be noncycling during the exponential growth phase of culture (Naill and Roberts 2005a). Analysis of such noncycling cells can provide information as to why particular groups of cells in aggregates behave differently, and yield important basic information on cell cycle control in suspension cultures.

9.5.3.3 Cell Sorting for Isolation and Analysis of Distinct Populations

While, FCM allows, recognition of distinct cells in a heterogeneous sample based on their optical properties, cell sorting can be employed to allow the selective isolation of subsets of cells for propagation of new cell lines or for further analysis. For instance, FCM analysis of protoplasts derived from embryogenic carrot cell cultures revealed two different subpopulations based on their light scattering properties–one with a higher cytoplasm/vacuole ratio and one with more vacuolated cells. After sorting and reculture, a differential embryogenic potential was observed; the former developed numerous proembryogenic masses and the latter were unable to develop any proembryogenic masses (Guzzo et al. 2002). Given that, plant cell cultures in general exhibit metabolic heterogeneity amongst cells, applying sorting can provide a potential means to enhance yield, and stabilize production of secondary metabolites. FCM sorting has been successfully applied to isolate protoplasts and ultimately generate highly productive cell lines for anthocyanin synthesis in *Aralia cordata* (Sakamoto et al. 1994) and berberine synthesis in *Coptis japonica* suspension cell cultures (Hara et al. 1989). Heterogeneity in paclitaxel accumulation (Naill and Roberts 2005d) and protein content (Naill and Roberts 2005c) has been observed in intact single cells isolated from *Taxus* cultures. The feasibility of reculturing intact single *Taxus* cells has also been demonstrated (Naill and Roberts 2005b). Current work aimed at sorting and reculture of *Taxus* cells based on paclitaxel accumulation to produce high-paclitaxel producing cultures is in progress.

The combination of FCM sorting with the currently available and emerging methods for analysis of global gene expression can provide valuable insights into the regulatory mechanisms leading to a particular cellular state. For example, differential gene expression analysis of isolated subpopulations in cell cultures can reveal the molecular fingerprint for particular cellular phenotypes. FCM and sorting have been used to characterize cell-type specific gene expression in protoplasts isolated from root (Brady et al. 2007; Petersson et al. 2009) and shoot tissues (Yadav et al. 2009) of *Arabidopsis*. Once information on gene expression and operational metabolic pathways within a single cell is known, targeted metabolic engineering strategies using both overexpression and silencing approaches can be defined and applied using suitable gene transfer methods.

9.6 Conclusions and Future Directions

The complexity and cellular heterogeneity associated with plant cell cultures have limited the systematic study of culture behavior. As a result, successful strategies for manipulating culture conditions for increased growth and productivity are still largely empirical. Apart from the engineering and technological issues associated with plant cell aggregation in bioprocesses, structural organization of cells within aggregates results in different local microenvironments, contributing to differences in metabolic activity amongst cells in culture that ultimately influence overall culture success. Morphological heterogeneity due to aggregate formation has been shown to affect growth, oxygen consumption, and secondary metabolite accumulation in a number of plant cell culture systems. However, the diversity of plant species and different conditions under which suspension cultures are maintained makes generalizations difficult; no clear trends have emerged to relate aggregate size directly to important culture parameters. The majority of studies to date have relied on cell-associated biomass metabolite measurements to relate secondary product formation to aggregate size. However, this approach neglects secretion of products to the extracellular medium, which is quite substantial for many systems. Careful consideration regarding experimental design must be taken to ensure accurate interpretation of data and information needs to be collected for each system relevant to the plant species and metabolite of interest. Rational approaches to manipulate and engineer aggregate sizes in cell culture should be developed to provide high-metabolite accumulating cell lines for use in bioprocesses.

Biomass accumulation must be augmented with measurements of the aggregate size distributions for an accurate depiction of culture growth. New techniques for measuring aggregate size distributions rapidly and reliably using a simple Coulter counter allow for studies of aggregation dynamics in batch cultures and analysis of long-term changes in aggregate distributions over multiple subcultures. Population balance equation models to describe heterogeneous cell growth and/or product formation have been developed for some cellular systems which aggregate, like filamentous fungi and hairy root cultures. However, models focusing on plant cell aggregation and culture heterogeneity in suspension cultures are just being developed and need future attention. Information about aggregation dynamics during batch culture along with relevant bioprocess data can be used to develop predictive models to optimize plant cell bioprocess design.

While process conditions can be easily controlled in the bulk culture environment, there are significant differences in the local environment of individual cells dependent on spatial position within an aggregate, about which very little is known. Modern analytical techniques such as flow cytometry allow rapid analysis of an individual cell and/or its components, which can be related to crucial culture parameters. Multi-parametric flow cytometric analysis aimed at correlating distinct subpopulations such as cell cycle participation, size, cellular protein and DNA content, and secondary metabolite accumulation can be used to gain an in-depth understanding of the heterogeneity observed amongst cells in plant cell cultures. Integrating these flow cytometry-based methods for analyzing culture heterogeneity with sophisticated multi-scale modeling can provide a new approach for the optimization of plant cell culture processes. The combination of flow cytometry and cell sorting to monitor and isolate distinct populations present in suspension cultures can be combined with modern methods for metabolic analysis to increase our knowledge of the regulatory networks leading to a particular subpopulation phenotype. Ultimately, if sufficient information is collected from multiple possible cellular states, a complete description of culture metabolic heterogeneity can be obtained. Using information from such system-wide analyses, targeted metabolic engineering strategies can be better designed, enabling high, stable yields of secondary metabolites, and promoting plant cell culture biotechnology for commercial use.

References

Albiol J, Robuste J, Casas C, Poch M (1993) Biomass estimation in plant cell cultures using an extended Kalman filter. Biotechnol Prog 9(2):174–178

Alper H, Miyaoku K, Stephanopoulos G (2005) Construction of lycopene-overproducing E. coli strains by combining systematic and combinatorial gene knockout targets. Nat Biotechnol 23(5):612–616

Ananta I, Subroto MA, Doran PM (1995) Oxygen-transfer and culture characteristics of self-immobilized Solanum Aviculare aggregates. Biotechnol Bioengineering 47(5):541–549

Aoyagi H, DiCosmo F, Tanaka H (2002) Efficient paclitaxel production using protoplasts isolated from cultured cells of *Taxus cuspidata*. Planta Med 68(5):420–424

Ashcroft R, Preston C, Cleland R, Critchley C (1986) Flow-cytometry of isolated-chloroplasts and thylakoids. Photobiochem Photobiophys 13(1–2):1–14

Ayaydin F, Kotogany E, Dudits D, Horvath GV (2010) A rapid and robust assay for detection of S-phase cell cycle progression in plant cells and tissues by using ethynyl deoxyuridine. Plant Methods 6:5

Baebler S, Hren M, Camloh M, Ravnikar M, Bohanec B, Plaper I, Ucman R, Zel J (2005) Establishment of cell suspension cultures of Yew (*Taxus* x Media Rehd.) and assessment of their genomic stability. In Vitro Cell Dev-Pl 41(3):338–343

Barrett P, Glennon B (1999) In-line FBRM monitoring of particle size in dilute agitated suspensions. Part Part Syst Char 16(5):207–211

Becker H (1970) Studies on the formation of volatile substances in plant tissue cultures. Biochem Physiol Pflanzen 161:425–441

Bennett MD, Leitch IJ (2005) Nuclear DNA amounts in angiosperms: progress, problems and prospects. Ann Botany 95(1):45–90

Bennett MD, Price HJ, Johnston JS (2008) Anthocyanin inhibits propidium iodide DNA fluorescence in *Euphorbia pulcherrima*: implications for genome size variation and flow cytometry. Ann Botany 101(6):777–790

Bergounioux C, Perennes C, Brown SC, Gadal P (1988a) Cytometric analysis of growth-regulator-dependent transcription and cell-cycle progression in petunia protoplast cultures. Planta 175(4):500–505

Bergounioux C, Perennes C, Brown SC, Gadal P (1988b) Nuclear-Rna quantification in protoplast cell-cycle phases. Cytometry 9(1):84–87

Birnbaum K, Shasha DE, Wang JY, Jung JW, Lambert GM, Galbraith DW, Benfey PN (2003) A gene expression map of the Arabidopsis root. Science 302(5652):1956–1960

Bolta Z, Baricevic D, Raspor P (2003) Biomass segregation in sage cell suspension culture. Biotechnol Lett 25(1):61–65

Bourgaud F, Gravot A, Milesi S, Gontier E (2001) Production of plant secondary metabolites: a historical perspective. Plant Sci 161(5):839–851

Brady SM, Orlando DA, Lee JY, Wang JY, Koch J, Dinneny JR, Mace D, Ohler U, Benfey PN (2007) A high-resolution root spatiotemporal map reveals dominant expression patterns. Science 318(5851):801–806

Brodelius P (1985) The potential role of immobilization in plant cell biotechnology. Trends Biotechnol 3(11):280–285

Capataz-Tafur J, Trejo-Tapia G, Rodríguez-Monroy M, Sepúlveda-Jiménez G (2011) Arabino-galactan proteins are involved in cell aggregation of cell suspension cultures of *Beta vulgaris* L. Plant Cell Tissue Organ Cult 106(1):169–177

Cassells A, Curry R (2001) Oxidative stress and physiological, epigenetic and genetic variability in plant tissue culture: implications for micropropagators and genetic engineers. Plant Cell Tiss Organ Cult 64(2):145–157

Ceoldo S, Levi M, Marconi AM, Baldan G, Giarola M, Guzzo F (2005) Image analysis and in vivo imaging as tools for investigation of productivity dynamics in anthocyanin-producing cell cultures of *Daucus carota*. New Phytol 166(1):339–352

Chemler JA, Koffas MAG (2008) Metabolic engineering for plant natural product biosynthesis in microbes. Curr Opin Biotechnol 19(6):597–605

Citterio S, Sgorbati S, Levi M, Colombo BM, Sparvoli E (1992) Pcna and total nuclear-protein content as markers of cell-proliferation in Pea tissue. J Cell Sci 102:71–78

Conia J, Bergounioux C, Perennes C, Muller P, Brown S, Gadal P (1987) Flow cytometric analysis and sorting of plant chromosomes from Petunia-Hybrida protoplasts. Cytometry 8(5):500–508

Croteau R, Ketchum R, Long R, Kaspera R, Wildung M (2006) Taxol biosynthesis and molecular genetics. Phytochem Rev 5(1):75–97

Curtis WR, Tuerk AL (2006) Oxygen transport in plant tissue culture systems. In: Gupta SD, Ibaraki Y (eds) Plant tissue culture engineering. Springer, Dordrecht, pp 173–186

Damato F (1985) Cytogenetics of plant-cell and tissue-cultures and their regenerates. CRC Crit Rev Plant Sci 3(1):73–112
Darzynkiewicz Z, Juan G, Srour EF (2001) Differential staining of DNA and RNA. Curr Protoc Cytom 7(7):3
Darzynkiewicz Z, Sharpless T, Staianocoico L, Melamed MR (1980) Subcompartments of the G1 phase of cell-cycle detected by flow cytometry. Proc Nat Acad Sci U S A 77(11):6696–6699
David H, Laigneau C, David A (1989) Growth and soluble proteins of cell cultures derived from explants and protoplasts of Pinus pinaster cotyledons. Tree Physiol 5(4):497–506
Davis DG, Stolzenberg RL, Dusky JA (1984) A comparison of various growth-parameters of cell-suspension cultures to determine phytotoxicity of Xenobiotics. Weed Sci 32(2):235–242
De Jong DW, Jansen EF, Olson AC (1967) Oxidoreductive and hydrolytic enzyme patterns in plant suspension culture cells : local and time relationships. Exp Cell Res 47(1–2):139–156
Deusneumann B, Zenk MH (1984) Instability of indole alkaloid production in *Catharanthus-roseus* cell-suspension cultures. Planta Med 50(5):427–431
Dhawan OP, Lavania UC (1996) Enhancing the productivity of secondary metabolites via induced polyploidy: a review. Euphytica 87(2):81–89
Diwan R, Malpathak N (2010) Histochemical localization in *Ruta graveolens* cell cultures: elucidating the relationship between cellular differentiation and furanocoumarin production. In Vitro Cell Dev Plant 46(1):108–116
Dixon RA (1995) Isolation and maintenance of callus and cell suspension cultures. In: Dixon RA (ed) Plant cell culture. IRL Press, Oxford, pp 1–20
Dolezel J, Greilhuber J, Suda J (2007a) Estimation of nuclear DNA content in plants using flow cytometry. Nat Protoc 2(9):2233–2244
Dolezel J, Greilhuber J, Suda J (2007b) Flow cytometry with plant cells: analysis of genes, chromosomes and genomes. Wiley, Weinheim
Dolezel J, Macas J, Lucretti S (2001) Flow analysis and sorting of plant chromosomes. Curr Protoc Cytom 5(5):3
Doran PM (1999) Design of mixing systems for plant cell suspensions in stirred reactors. Biotechnol Prog 15(3):319–335
Dornenburg H, Knorr D (1995) Strategies for the improvement of secondary metabolite production in plant-cell cultures. Enzyme Microb Technol 17(8):674–684
Dougall DK (1964) A method of plant tissue culture giving high growth rates. Exp Cell Res 33(3):438–444
Dubuis B, Kut OM, Prenosil JE (1995) Pilot-scale culture of *Coffea arabica* in a novel loop fluidised bed reactor. Plant Cell Tissue Organ Cult 43(2):171–183
Dunlop EH, Namdev PK, Rosenberg MZ (1994) Effect of fluid shear forces on plant-cell suspensions. Chem Eng Sci 49(14):2263–2276
Edahiro J, Seki M (2006) Phenylpropanoid metabolite supports cell aggregate formation in strawberry cell suspension culture. J Biosci Bioeng 102(1):8–13
Eibl R, Eibl D (2002) Bioreactors for plant cell and tissue cultures. In: Oksman-Caldentey KM (ed) Plant biotechnology and transgenic plants. Marcel Dekker, New York, pp 163–199
Ellis DD, Zeldin EL, Brodhagen M, Russin WA, McCown BH (1996) Taxol production in nodule cultures of *Taxus*. J Nat Prod 59(3):246–250
Eriksson T (1965) Studies on growth requirements and growth measurements of cell cultures of *Haplopappus gracilis*. Physiol Plantarum 18(4):976–993
Evans DA, Gamborg OL (1982) Chromosome stability of cell suspension cultures of *Nicotiana* spp. Plant Cell Rep 1:104–107
Evans DE, Coleman JOD, Kearns A (2003) Plant cell culture. BIOS Scientific, London
Galbraith DW (1990) Isolation and flow cytometric characterization of plant protoplasts. Methods Cell Biol 33:527–547
Galbraith DW (2004) Cytometry and plant sciences: a personal retrospective. Cytom Part A 58A(1):37–44
Gaurav V, Kolewe ME, Roberts SC (2010) Flow cytometric methods to investigate culture heterogeneities for plant metabolic engineering. Methods Mol Biol 643:243–262

Ge F, Yuan XF, Wang XD, Zhao B, Wang YC (2006) Cell growth and shikonin production of *Arnebia euchroma* in a periodically submerged airlift bioreactor. Biotechnol Lett 28(8):525–529

Georgiev MI, Pavlov AI, Bley T (2007) Hairy root type plant in vitro systems as sources of bioactive substances. Appl Microbiol Biotechnol 74(6):1175–1185

Graham MD (2003) The Coulter principle: foundation of an industry. J Assoc Lab Autom 8:72–81

Grand d'Esnon A, Chee R, Harrell RC, Cantliffe DJ (1989) Qualitative and quantitative evaluation of liquid tissue cultures by artificial vision. Biofutur 76:S3

Guillon S, Tremouillaux-Guiller J, Pati PK, Rideau M, Gantet P (2006) Harnessing the potential of hairy roots: dawn of a new era. Trends Biotechnol 24(9):403–409

Guzzo F, Cantamessa K, Portaluppi P, Levi M (2002) Flow cytometry and sorting of protoplasts from carrot cell cultures reveal two cell subpopulations with different morphogenetic potential. Plant Cell Rep 21(3):214–219

Hahlbrock K, Kuhlen E (1972) Relationship between growth of parsley and soybean cells in suspension cultures and changes in conductivity of culture medium. Planta 108(3):271–278

Hall RD, Yeoman MM (1987) Intercellular and intercultural heterogeneity in secondary metabolite accumulation in cultures of *Catharanthus roseus* following cell line selection. J Exp Bot 38(193):1391–1398

Halperin W, Minocha S (1973) Benzyladenine effects on cell separation and wall metabolism. Can J Botany 51(7):1347–1354

Hanagata N, Ito A, Uehara H, Asari F, Takeuchi T, Karube I (1993) Behavior of cell aggregate of *Carthamus tinctorius* L. cultured cells and correlation with red pigment formation. J Biotechnol 30(3):259–269

Hara Y, Yamagata H, Morimoto T, Hiratsuka J, Yoshioka T, Fujita Y, Yamada Y (1989) Flow cytometric analysis of cellular berberine contents in high-producing and low-producing cell-lines of *Coptis japonica* obtained by repeated selection. Planta Medica 2:151–154

Harkins KR, Galbraith DW (1987) Factors governing the flow cytometric analysis and sorting of large biological particles. Cytometry 8(1):60–70

Harrell RC, Bieniek M, Cantliffe DJ (1992) Noninvasive evaluation of somatic embryogenesis. Biotechnol Bioeng 39(4):378–383

Hawkins B (2008) Plants for life: medicinal plant conservation and botanic gardens. Botanic Gardens Conservation International, Richmond

Hayashi T, Yoshida K (1988) Cell expansion and single-cell separation induced by Colchicine in suspension-cultured soybean cells. Proc Natl Acad Sci U S A 85(8):2618–2622

Heller FO (1973) DNS-Bestimmung an Keimwurzeln von *Vicia faba* L. mit Hilfe der Impulscytophotometrie. Ber Deutsch Bot Ges 86:437–441

Hellwig S, Drossard J, Twyman RM, Fischer R (2004) Plant cell cultures for the production of recombinant proteins. Nat Biotechnol 22(11):1415–1422

Henshaw GG, Jha KK, Mehta AR, Shakesha Dj, Street HE (1966) Studies on growth in culture of plant cells.1. Growth patterns in batch propagated suspension cultures. J Exp Bot 17(51):362–377

Hoekstra SS, Harkes PAA, Verpoorte R, Libbenga KR (1990) Effect of auxin on cytodifferentiation and production of quinoline alkaloids in compact globular structures of *Cinchona ledgeriana*. Plant Cell Rep 8(10):571–574

Huang CN, Cornejo MJ, Bush DS, Jones RL (1986) Estimating viability of plant protoplasts using double and single staining. Protoplasma 135(2):80–87

Huang SY, Chou CJ (2000) Effect of gaseous composition on cell growth and secondary metabolite production in suspension culture of *Stizolobium hassjoo* cells. Bioproc Biosyst Eng 23(6):585–593

Hulst AC, Meyer MMT, Breteler H, Tramper J (1989) Effect of aggregate size in cell cultures of *Tagetes patula* on thiophene production and cell growth. Appl Microbiol Biotechnol 30(1):18–25

Ibaraki Y, Kenji K (2001) Application of image analysis to plant cell suspension cultures. Comput Electron Agric 30(1–3):193–203
Ishii S (1988) Factors influencing protoplast viability of suspension-cultured rice cells during isolation process. Plant Physiol 88(1):26–29
James E, Lee JM (2000) An improved optical technique for monitoring plant cell concentration. Plant Cell Rep 19(3):283–285
Jeffers P, Glennon B, Kieran P (2003) Focussed beam reflectance measurement (FBRM) monitoring of particle size and morphology in suspension cultures of *Morinda citrifolia* and *Centaurea calcitrapa*. Biotechnol Lett 25:2023–2028
Karuppusamy S (2009) A review on trends in production of secondary metabolites from higher plants by in vitro tissue, organ and cell cultures. J Med Plants Res 3(13):1222–1239
Kato A, Kawazoe S, Soh Y (1978) Viscosity of broth of tobacco cells in suspension culture. J Ferment Technol 56(3):224–228
Kessler M, ten Hoopen HJG, Furusaki S (1999) The effect of the aggregate size on the production of ajmalicine and tryptamine in *Catharanthus roseus* suspension culture. Enzyme Microb Technol 24(5–6):308–315
Ketchum REB, Gibson DM (1996) Paclitaxel production in suspension cell cultures of *Taxus*. Plant Cell Tissue Organ Cult 46(1):9–16
Kieran P, Malone D, ML P (2000) Effects of hydrodynamic and interfacial forces on plant cell suspension systems. Adv Biochem Eng Biotechnol 67:139–185
Kieran PM, MacLoughlin PF, Malone DM (1997) Plant cell suspension cultures: some engineering considerations. J Biotechnol 59(1–2):39–52
Kim BJ, Gibson DM, Shuler ML (2004) Effect of subculture and elicitation on instability of Taxol production in *Taxus* sp suspension cultures. Biotechnol Prog 20(6):1666–1673
Kim Y, Wyslouzil BE, Weathers PJ (2002) Invited review: secondary metabolism of hairy root cultures in bioreactors. In Vitro Cell Dev Plants 38(1):1–10
King PJ, Mansfiel Kj, Street HE (1973) Control of growth and cell-division in plant-cell suspension cultures. Can J Botany 51(10):1807–1823
King PJ, Street HE (1977) Growth patterns in cell cultures. In: Street HE (ed) Plant tissue and cell culture, 2nd edn. Blackwell, Oxford, pp 307–387
Kinnersley AM, Dougall DK (1980) Increase in anthocyanin yield from wild-carrot cell-cultures by a selection system based on cell-aggregate size. Planta 149(2):200–204
Kobayashi Y, Fukui H, Tabata M (1989) Effect of oxygen-supply on berberine production in cell-suspension cultures and immobilized cells of *Thalictrum minus*. Plant Cell Rep 8(4):255–258
Kolewe ME, Gaurav V, Roberts SC (2008) Pharmaceutically active natural product synthesis and supply via plant cell culture technology. Mol Pharm 5(2):243–256
Kolewe ME, Henson MA, Roberts SC (2010) Characterization of aggregate size in *Taxus* suspension cell culture. Plant Cell Rep 29(5):485–494
Kolewe ME, Henson MA, Roberts SC (2011) Analysis of aggregate size as a process variable affecting paclitaxel accumulation in *Taxus* suspension cultures. Biotechnol Prog 27(5):1365–1372
Kolewe ME, Roberts SC, Henson MA (2012) A population balance equation model of aggregation dynamics in *Taxus* suspension cell cultures. Biotechnol Bioeng 109(2):472–482
Kougoulos E, Jones AG, Jennings KH, Wood-Kaczmar MW (2005) Use of focused beam reflectance measurement (FBRM) and process video imaging (PVI) in a modified mixed suspension mixed product removal (MSMPR) cooling crystallizer. J Cryst Growth 273(3–4):529–534
Kron P, Suda J, Husband BC (2007) Applications of flow cytometry to evolutionary and population biology. Annu Rev Ecol Evol Syst 38:847–876
Kubek DJ, Shuler ML (1978a) Electronic measurement of plant-cell number and size in suspension culture. J Exp Bot 29(109):511–523
Kubek DJ, Shuler ML (1978b) Generality of methods to obtain single-cell plant suspension cultures. Can J Botany 56(20):2521–2527

Kubitschek HE (1969) Counting and sizing micro-organisms with the Coulter counter. In: Norris R, Ribbons DW (eds) Methods in microbiology. Academic Press, New York, pp 593–610
Kuboi T, Yamada Y (1978a) Changing cell aggregations and lignification in tobacco suspension cultures. Plant Cell Physiol 19(3):437–443
Kuboi T, Yamada Y (1978b) Regulation of enzyme-activities related to lignin synthesis in cell aggregates of tobacco cell-culture. Biochim Biophys Acta 542(2):181–190
Kurz WGW (1971) A chemostat for growing higher plant cells in single cell suspension cultures. Exp Cell Res 64(2):476–479
Lai Keng C, Koay Suan S, Low Poay H, Boey Peng L (2008) Effect of plant growth regulators and subculture frequency on callus culture and the establishment of *Melastoma malabathricum* cell suspension cultures for the production of pigments. Biotechnology 7:678–685
Lamboursain L, Jolicoeur M (2005) Determination of cell concentration in a plant cell suspension using a fluorescence microplate reader. Plant Cell Rep 23(10):665–672
Lee EK, Jin YW, Park JH, Yoo YM et al (2010) Cultured cambial meristematic cells as a source of plant natural products. Nat Biotech 28(11):1213–1217
Leonard E, Runguphan W, O'Connor S, Prather KJ (2009) Opportunities in metabolic engineering to facilitate scalable alkaloid production. Nat Chem Biol 5(5):292–300
Li SY, Yuan W, Yang PY, Antoun MD, Balick MJ, Cragg GM (2010) Pharmaceutical crops: an overview. Pharm Crops 1:1–17
Liau DF, Boll WG (1971) Growth, and patterns of growth and division, in cell suspension cultures of bush bean (*Phaseolus vulgaris* Cv Contender). Can J Botany 49(7):1131–1139
Linden JC, Haigh JR, Mirjalili N, Phisaphalong M (2001) Gas concentration effects on secondary metabolite production by plant cell cultures. Adv Biochem Eng Biotechnol 72:27–62
Lorence A, Nessler CL (2004) Camptothecin, over four decades of surprising findings. Phytochemistry 65(20):2735–2749
Loureiro J, Rodriguez E, Dolezel J, Santos C (2006) Flow cytometric and microscopic analysis of the effect of tannic acid on plant nuclei and estimation of DNA content. Ann Bot 98(3):515–527
Loyola-Vargas VM, Miranda-Ham ML (1995) Root culture as a source of secondary metabolites of economic importance. In: Arnason JT, Mata R, Romeo JT (eds) Phytochemistry of medicinal plants. Plenum Press, New York, pp 217–248
Machackova I, Zazimalova E, George EF (2007) Plant growth regulators I. In: George EF, Hall M, Klerk G (eds) Plant propagation by tissue culture, 3rd edn. Springer Dordrecht, pp 175–204
Madhusudhan R, Rao SR, Ravishankar GA (1995) osmolarity as a measure of growth of plant-cells in suspension-cultures. Enzyme Microb Technol 17(11):989–991
Madhusudhan R, Ravishankar GA (1996) Gradient of anthocyanin in cell aggregates of *Daucus carota* in suspension cultures. Biotechnol Lett 18(11):1253–1256
Mak YX, Doran PM (1993) Effect of cell-cycle inhibition on synthesis of steroidal alkaloids by *Solanum aviculare* Plant Cells. Biotechnol Lett 15(10):1031–1034
Martin VJJ, Pitera DJ, Withers ST, Newman JD, Keasling JD (2003) Engineering a mevalonate pathway in *Escherichia coli* for production of terpenoids. Nat Biotechnol 21(7):796–802
Matkowski A (2008) Plant in vitro culture for the production of antioxidants—a review. Biotechnol Adv 26(6):548–560
Mavituna F, Park JM (1987) Size distribution of plant-cell aggregates in batch culture. Chem Eng J 35(1):B9–B14
McCoy E, O'Connor SE (2008) Natural products from plant cell cultures. Prog Drug Res 65(329):331–370
McDonald KA, Jackman AP, Hurst S (2001) Characterization of plant suspension cultures using the focused beam reflectance technique. Biotechnol Lett 23(4):317–324
Melchers G, Bergmann L (1959) Untersuchungen an Kulturen von haploiden Geweben von Antirrhinum majus. Ber Dtsch Bot Ges 78:21–29
Meyer JE, Pepin MF, Smith MAL (2002) Anthocyanin production from *Vaccinium pahalae*: limitations of the physical micro environment. J Biotechnol 93(1):45–57

Mishiba KI, Okamoto T, Mii M (2001) Increasing ploidy level in cell suspension cultures of Doritaenopsis by exogenous application of 2,4-dichlorophenoxyacetic acid. Physiol Plantarum 112(1):142–148

Mishra BN, Ranjan R (2008) Growth of hairy-root cultures in various bioreactors for the production of secondary metabolites. Biotechnol App Biochem 49:1–10

Morris P, Fowler M (1981) A new method for the production of fine plant cell suspension cultures. Plant Cell Tiss Organ Cult 1(1):15–24

Muir WH, Hildebrandt AC, Riker AJ (1954) Plant tissue cultures produced from single isolated cells. Science 119(3103):877–878

Mustafa NR, de Winter W, van Iren F, Verpoorte R (2011) Initiation, growth and cryopreservation of plant cell suspension cultures. Nat Protoc 6(6):715–742

Naill MC, Roberts SC (2004) Preparation of single cells from aggregated *Taxus* suspension cultures for population analysis. Biotechnol Bioeng 86(7):817–826

Naill MC, Roberts SC (2005a) Cell cycle analysis of *Taxus* suspension cultures at the single cell level as an indicator of culture heterogeneity. Biotechnol Bioeng 90(4):491–500

Naill MC, Roberts SC (2005b) Culture of isolated single cells from *Taxus* suspensions for the propagation of superior cell populations. Biotechnol Lett 27(21):1725–1730

Naill MC, Roberts SC (2005c) Flow cytometric analysis of protein content in *Taxus* protoplasts and single cells as compared to aggregated suspension cultures. Plant Cell Rep 23(8):528–533

Naill MC, Roberts SC (2005d) Flow cytometric identification of paclitaxel-accumulating subpopulations. Biotechnol Prog 21(3):978–983

Nash DT, Davies ME (1972) Some aspects of growth and metabolism of Pauls Scarlet rose cell suspensions. J Exp Bot 23(74):75–91

Nesius KK, Fletcher JS (1973) Carbon-dioxide and ph requirements of non-photosynthetic tissue-culture cells. Physiol Plant 28(2):259–263

Nicoloso FT, Val J, Vanderkeur M, Vaniren F, Kijne JW (1994) Flow-cytometric cell counting and DNA estimation for the study of plant-cell population-dynamics. Plant Cell Tiss Organ Cult 39(3):251–259

Nunez-Palenius H, Cantliffe D, Klee H, Ochoa-Alejo N, Ramirez-Malagon R, Perez-Molphe E (2005) Methods in plant tissue culture. In: Shetty K, Paliyath G, Pometto A, Levin R (eds) Food biotechnology, 2nd edn. CRC Press, Boca Raton, pp 553–603

O'Rourke AM, MacLoughlin PF (2005) A comparison of measurement techniques used in the analysis of evolving liquid–liquid dispersions. Chem Eng Process 44(8):885–894

Ochatt SJ, Patat-Ochatt EM, Moessner A (2011) Ploidy level determination within the context of in vitro breeding. Plant Cell Tiss Organ Cult 104(3):329–341

Ogino T, Hiraoka N, Tabata M (1978) Selection of high nicotine-producing cell lines of tobacco callus by single-cell cloning. Phytochemistry 17(11):1907–1910

Ono NN, Tian L (2011) The multiplicity of hairy root cultures: prolific possibilities. Plant Sci 180(3):439–446

Ormerod MG (1990) Flow cytometry: a practical approach. Oxford University Press, Oxford

Paau AS, Cowles JR, Oro J (1977) Flow-microfluorometric analysis of Escherichia coli, Rhizobium meliloti, and Rhizobium japonicum at different stages of the growth cycle. Can J Microbiol 23(9):1165–1169

Pepin MF, Smith MAL, Reid JF (1999) Application of imaging tools to plant cell culture: relationship between plant cell aggregation and flavonoid production. In Vitro Cell Dev Plants 35(4):290–295

Petersson SV, Johansson AI, Kowalczyk M, Makoveychuk A, Wang JY, Moritz T, Grebe M, Benfey PN, Sandberg G, Ljung K (2009) An auxin gradient and maximum in the arabidopsis root apex shown by high-resolution cell-specific analysis of IAA distribution and synthesis. Plant Cell 21(6):1659–1668

Petit P, Diolez P, Muller P, Brown SC (1986) Binding of concanavalin-A to the outer-membrane of potato-tuber mitochondria detected by flow-cytometry. FEBS Lett 196(1):65–70

Petit PX (1992) Flow cytometric analysis of rhodamine-123 fluorescence during modulation of the membrane-potential in plant-mitochondria. Plant Physiol 98(1):279–286

Phillips RL, Kaeppler SM, O P (1994) Genetic instability of plant tissue cultures: breakdown of normal controls. Proc Nati Acad Sci U S A 91:5222–5226
Pistelli L, Giovannini A, Ruffoni B, Bertoli A, Pistelli L (2010) Hairy root cultures for secondary metabolites production. In: Giardi M, Ria G, Berra B (eds) Bio-farms for nutraceuticals: functional food and safety control by biosensors, 1st edn. Springer, New York, pp 167–184
Prenosil JE, Hegglin M (1990) Self-immobilized plant cell aggregates in a bioreactor system with low shear stress. Ann N Y Acad Sci 613(1):234–247
Qu JG, Zhang W, Yu XJ, Jin MF (2005) Instability of anthocyanin accumulation in *Vitis vinifera* L. var. Gamay Freaux suspension cultures. Biotechnol Bioprocess Eng 10(2):155–161
Rajasekhar EW, Edwards M, Wilson SB, Street HE (1971) Studies on growth in culture of plant cells.11. Influence of shaking rate on growth of suspension cultures. J Exp Bot 22(70):107–117
Ramulu KS, Dijkhuis P (1986) Flow cytometric analysis of polysomaty and in vitro genetic instability in potato. Plant Cell Rep 5(3):234–237
Ranch JP, Giles KL (1980) Factors affecting growth and aggregate dissociation in batch suspension-cultures of *Datura innoxia* (Miller). Ann Bot 46(6):667–683
Rao SR, Ravishankar GA (2002) Plant cell cultures: chemical factories of secondary metabolites. Biotechnol Adv 20(2):101–153
Richmond WR, Jones RL, Fawell PD (1998) The relationship between particle aggregation and rheology in mixed silica-titania suspensions. Chem Eng J 71(1):67–75
Ro DK, Paradise EM, Ouellet M, Fisher KJ et al (2006) Production of the antimalarial drug precursor artemisinic acid in engineered yeast. Nature 440(7086):940–943
Roberts MF, Wink M (1998) Alkaloids: biochemistry, ecology, and medicinal applications. Plenum Press, New York
Roberts SC, Naill M, Gibson DM, Shuler ML (2003) A simple method for enhancing paclitaxel release from *Taxus* canadensis cell suspension cultures utilizing cell wall digesting enzymes. Plant Cell Rep 21(12):1217–1220
Roberts SC (2007) Production and engineering of terpenoids in plant cell culture. Nat Chem Biol 3(7):387–395
Roberts SC, Kolewe M (2010) Plant natural products from cultured multipotent cells. Nat Biotech 28(11):1175–1176
Rodriguez-Monroy M, Trejo-Espino JL, Jimenez-Aparicio A, Morante MDL, Villarreal ML, Trejo-Tapia G (2004) Evaluation of morphological properties of *Solanum chrysotrichum* cell cultures in a shake flask and fermentor and rheological properties of broths. Food Technol Biotech 42(3):153–158
Russin WA, Ellis DD, Gottwald JR, Zeldin EL, Brodhagen M, Evert RF (1995) Immunocytochemical localization of Taxol in *Taxus cuspidata*. Int J Plant Sci 156(5):668–678
Ryu DDY, Lee SO, Romani RJ (1990) Determination of growth-rate for plant-cell cultures—comparative studies. Biotechnol Bioeng 35(3):305–311
Safar J, Noa-Carrazana JC, Vrana J, Bartos J et al (2004) Creation of a BAC resource to study the structure and evolution of the banana (*Musa balbisiana*) genome. Genome 47(6):1182–1191
Sakamoto K, Iida K, Koyano T, Asada Y, Furuya T (1994) Studies on plant-tissue cultures.91. Method for selecting anthocyanin-producing cells by a cell sorter. Planta Med 60(3):253–259
Schlatmann JE, Vinke JL, Tenhoopen HJG, Heijnen JJ (1995) Relation between dissolved-oxygen concentration and Ajmalicine production-rate in high-density cultures of *Catharanthus roseus*. Biotechnol Bioeng 45(5):435–439
Schroder WP, Petit PX (1992) Flow-cytometry of spinach-chloroplasts—determination of intactness and lectin-binding properties of the envelope and the thylakoid membranes. Plant Physiol 100(3):1092–1102
Schwab B, Hulskamp M (2008) Vital stain for plant cytoplasm. CSH Protoc 2008: pdb prot4936
Schween G, Schulte J, Reski R (2005) Effect of ploidy level on growth, differentiation, and morphology in Physcomitrella patens. Bryologist 108(1):27–35
Scragg AH, Bond P, Leckie F, Cresswell R, Fowler MW, Allan EJ (1987) Growth and product formation by plant cell suspensions cultivated in bioreactors. In: Moody JW, Baker PB (eds)

Bioreactors and biotransformations. Elsevier Applied Science Publications, New York, pp 12–25
Shapiro H (1994) Practical flow cytometry. Wiley, New York
Shapiro HM (1981) Flow cytometric estimation of DNA and RNA content in intact cells stained with Hoechst 33342 and pyronin Y. Cytometry 2(3):143–150
Sharma DP, Firoozabady E, Ayres NM, Galbraith DW (1983) Improvement of anther culture in Nicotiana—media, cultural conditions and flow cytometric determination of ploidy levels. Z Pflanzenphysiol 111(5):441–451
Shiba T, Mii M (2005) Visual selection and maintenance of the cell lines with high plant regeneration ability and low ploidy level in *Dianthus acicularis* by monitoring with flow cytometry analysis. Plant Cell Rep 24(10):572–580
Shuler ML (1999) Overview of yield improvement strategies for secondary metabolite production in plant cell culture. In: Fu TJ, Sing G, Curtis WR (eds) Proceedings of the symposium on plant cell and tissue culture for the production of food ingredients. Kluwer Academic, New York, pp 75–83
Siah CL, Doran PM (1991) Enhanced codeine and morphine production in suspended Papaver somniferum cultures after removal of exogenous hormones. Plant Cell Rep 10(6):349–353
Simmons MJH, Azzopardi BJ (2001) Drop size distributions in dispersed liquid–liquid pipe flow. Int J Multiphas Flow 27(5):843–859
Simpkins I, Collin HA, Street HE (1970) The growth of *Acer pseudoplantanus* cells in a synthetic liquid medium. Am J Bot 49:420–425
Smetanska I (2008) Production of secondary metabolites using plant cell cultures. Adv Biochem Eng Biotechnol 111:187–228
Smith MAL, Reid JF (1996) Machine vision and automation in secondary metabolite bioprocess control. In: Misawa M, DiCosmo F (eds) Plant cell culture secondary metabolismtoward industrial application. CRC Press, Boca Raton, pp 53–77
Smith MAL, Reid JF, Hansen AC, Li Z, Madhavi DL (1995) Non-destructive machine vision analysis of pigment-producing cell cultures. J Biotechnol 40:1–11
Srivastava S, Srivastava AK (2007) Hairy root culture for mass-production of high-value secondary metabolites. CRC Critic Rev Biotechnol 27(1):29–43
Steiner HY, Dougall DK (1995) Ammonium uptake in carrot cell structures is influenced by pH-dependent cell aggregation. Physiol Plantarum 95(3):415–422
Su WW (2006) Bioreactor engineering for recombinant protein production using plant cell suspension culture. In: Gupta DS, Ibaraki Y (eds) Plant tissue culture engineering. Springer, Berlin, pp 135–159
Sung ZR (1976) Turbidimetric measurement of plant-cell culture growth. Plant Physiol 57(3):460–462
Syklowska-Baranek K, Pietrosiuk A, Kokoszka A, Furmanowa M (2009) Enhancement of taxane production in hairy root culture of *Taxus* x media var. Hicksii. J Plant Physiol 166(17):1950–1954
Takayama S, Misawa M, Ko K, Misato T (1977) Effect of cultural conditions on growth of *Agrostemma githago* cells in suspension culture and concomitant production of an anti-plant virus substance. Physiol Plantarum 41(4):313–320
Takeda T, Seki M, Furusaki S (1994) Hydrodynamic damage of cultured-cells of *Carthamus tinctorius* in a stirred-tank reactor. J Chem Eng Jpn 27(4):466–471
Tanaka H, Aoyagi H, Jitsufuchi T (1992) Turbidimetric measurement of cell biomass of plant-cell suspensions. J Ferment Bioeng 73(2):130–134
Tanaka H, Semba H, Jitsufuchi T, Harada H (1988) The effect of physical stress on plant cells in suspension cultures. Biotechnol Lett 10(7):485–490
Tanaka H, Uemura M, Kaneko Y, Aoyagi H (1993) Estimation of cell biomass in plant-cell suspensions by the osmotic-pressure measurement of culture broth. J Ferment Bioeng 76(6):501–504

Taticek RA, Mooyoung M, Legge RL (1990) Effect of bioreactor configuration on substrate uptake by cell-suspension cultures of the plant *Eschscholtzia californica*. App Microbiol Biot 33(3):280–286
Thanh NT, Murthy HN, Yu KW, Hahn EJ, Paek KY (2005) Methyl jasmonate elicitation enhanced synthesis of ginsenoside by cell suspension cultures of Panax ginseng in 5-l balloon type bubble bioreactors. Appl Microbiol Biot 67(2):197–201
Thom M, Maretzki A, Komor E, Sakai W (1981) Nutrient uptake and accumulation by sugarcane cell cultures in relation to the growth cycle. Plant Cell Tiss Organ Cult 1(1):3–14
Torrey JG, Merkel N, Reinert J (1962) Mitosis in suspension cultures of higher plant cells in a synthetic medium. Am J Cardiol 10(4):420–425
Trejo TG, Hernandez TR, Trejo EJL, Jimenez AA, Rodriguez MM (2003) Analysis of morphological characteristics of *Solanum chrysotrichum* cell suspension cultures. World J Microbiol Biotechnol 19(9):929–932
Tulecke W (1966) Continuous cultures of higher plant cells in liquid media—advantages and potential use of a phytostat. Ann N Y Acad Sci 139(A1):162–175
Umetsu N, Ojima K, Matsuda K (1975) Enhancement of cell separation by colchicine in cell-suspension cultures of soybean. Planta 125(2):197–200
Verma D, Van Huystee R (1970a) Cellular differentiation and peroxidase isozymes in cell cultures of peanut cotyledons. Can J Botany 48:429–431
Verma D, Van Huystee R (1970b) Relationship between peroxidase, catalase and protein synthesis during cellular development in cell cultures of peanut. Can J Biochem 48:444–449
Verpoorte R, Memelink J (2002) Engineering secondary metabolite production in plants. Curr Opin Biotechnol 13(2):181–187
Verpoorte R, van der Heijden R, ten Hoopen HJG, Memelink J (1999) Metabolic engineering of plant secondary metabolite pathways for the production of fine chemicals. Biotechnol Lett 21(6):467–479
Vongpaseuth K, Roberts SC (2007) Advancements in the understanding of Paclitaxel metabolism in tissue culture. Curr Pharm Biotechnol 8(4):219–236
Wallner SJ, Nevins DJ (1973) Formation and dissociation of cell aggregates in suspension cultures of Pauls Scarlet Rose. Am J Botany 60(3):255–261
Watts MJ, Galpin IJ, Collin HA (1984) The effect of growth-regulators, light and temperature on flavor production in celery tissue-cultures. New Phytol 98(4):583–591
Weber J, Georgiev V, Pavlov A, Bley T (2008) Flow cytometric investigations of diploid and tetraploid plants and in vitro cultures of *Datura stramonium* and *Hyoscyamus niger*. Cytom Part A 73A(10):931–939
Williams PD, Wilkinson AK, Lewis JA, Black GM, Mavituna F (1988) A method for the rapid production of fine plant-cell suspension-cultures. Plant Cell Rep 7(6):459–462
Wilson SA, Roberts SC (2012) Recent advances towards development and commercialization of plant cell culture processes for the synthesis of biomolecules. Plant Biotechnol J 10(3):249–268
Wink M, Alfermann AW, Franke R, Wetterauer B et al (2005) Sustainable bioproduction of phytochemicals by plant in vitro cultures: anticancer agents. Plant Genetic Res 3(02):90–100
Wongsamuth R, Doran PM (1997) The filtration properties of *Atropa belladonna* plant cell suspensions; effects of hydrodynamic shear and elevated carbon dioxide levels on culture and filtration parameters. J Chem Technol Biotechnol 69(1):15–26
Wu SQ, Schalk M, Clark A, Miles RB, Coates R, Chappell J (2006) Redirection of cytosolic or plastidic isoprenoid precursors elevates terpene production in plants. Nat Biotechnol 24(11):1441–1447
Xu JF, Xie J, Han AM, Feng PS, Su ZG (1998) Kinetic and technical studies on large-scale culture of *Rhodiola sachalinensis* compact callus aggregates with air-lift reactors. J Chem Technol Biotechnol 72(3):227–234
Yadav RK, Girke T, Pasala S, Xie MT, Reddy V (2009) Gene expression map of the Arabidopsis shoot apical meristem stem cell niche. Proc Natl Acad Sci U S A 106(12):4941–4946

Yang YM, He DG, Scott KJJ (1994) Cell aggregates in wheat suspension cultures and their effects on isolation and culture of protoplasts. Plant Cell Rep 13(3):176–179

Yanpaisan W, King NJC, Doran PM (1998) Analysis of cell cycle activity and population dynamics in heterogeneous plant cell suspensions using flow cytometry. Biotechnol Bioeng 58(5):515–528

Yanpaisan W, King NJC, Doran PM (1999) Flow cytometry of plant cells with applications in large-scale bioprocessing. Biotechnol Adv 17(1):3–27

Yesilirmak F, Sayers Z (2009) Heterelogous expression of plant genes. Int J Plant Genomics 2009:296482

Yuan X, Zhao B, Wang Y (2004) Cell culture of *Saussurea medusa* in a periodically submerged air-lift bioreactor. Biochem Eng J 21(3):235–239

Zhao D, Huang Y, Jin Z, Qu W, Lu D (2003) Effect of aggregate size in cell cultures of Saussurea medusa on cell growth and jaceosidin production. Plant Cell Rep 21(11):1129–1133

Zhao J, Zhu WH, Hu Q, Guo YQ (2001) Compact callus cluster suspension cultures of *Catharanthus roseus* with enhanced indole alkaloid biosynthesis. In Vitro Cell Dev Plants 37(1):68–72

Chapter 10
Biosynthetic Potential of Hairy Roots for Production of New Natural Products

Yuriy Sheludko and Iryna Gerasymenko

10.1 Introduction

Plants represent a unique and undiminishing source of natural products—complex organic compounds, a considerable part of which possesses multitudinous biological activities. Several schemes have been proposed for classification of natural products based on their pharmacological, biogenetic, taxonomic, or chemical properties (Ikan 2008). One of the most general classification of pharmacologically valuable plant natural compounds allows to define a group of substances with low molecular weight, including first of all plant secondary metabolites (usually up to 3,000 Da) (Williams et al. 1989), and a group of biopolymers with high molecular weight. Among the last, a group of peptides and proteins could be segregated which are the products of heterologously expressed genes.

Analysis of officially approved new drugs exhibited that almost 50 % of low molecular weight chemicals introduced in pharmacy between 1981 and 2002 were natural products or compounds derived from them (Newman et al. 2003). This tendency remained unchanged later on (Newman and Cragg 2007). Moreover, it is considered that nearly 25 % of current drugs in the world are of plant origin (Schmidt et al. 2007; Raskin et al. 2002). Among 225 drugs based on natural compounds, which underwent clinical trials in 2008, almost 50 % (108) originated from plants (Harvey 2008).

Y. Sheludko (✉) · I. Gerasymenko
Lab. of Natural Product Biosynthesis Systems, Dept. of Genetic Engineering,
Institute of Cell Biology and Genetic Engineering, 148 Zabolotnogo Street,
Kiev 03143, Ukraine
e-mail: ysheludko@ukr.net

I. Gerasymenko
e mail: i-gerasimenko@ukr.net

S. Chandra et al. (eds.), *Biotechnology for Medicinal Plants*,
DOI: 10.1007/978-3-642-29974-2_10, © Springer-Verlag Berlin Heidelberg 2013

The amount of low molecular weight compounds synthesized in plants is estimated from over 100,000 (Dixon 2001) to 500,000 (Hadacek 2002) structures. One should take into consideration that a therapeutic effect may be influenced by a synergy action of several components of a complex extract (Ulrich-Merzenich et al. 2010) or metabolic activation of plant compounds in unusual biochemical surrounds (Chen et al. 2011). From this point of view, plants still can be regarded as a valuable and comparatively low exhausted source of novel pharmaceuticals.

An intensive work on the screening of biological activity of plant natural products during the recent decades, including high throughput pharmacological screening, makes actual searching for or development of alternative origins for biodiversity. For example, the plant collection program of The Natural Cancer Institute of USA started in 1960 resulted up to 1995 in screening for anticancer activities of organic and water extracts prepared from over 45,000 plant samples from temperate, tropical, and subtropical regions (Cragg et al. 1996). A similar program for Asia plants resulted in testing for anti-inflammatory activities of 1,000 extracts from 449 species belonging to 68 plant families (Dey et al. 2008).

Such alternative origins, on the one hand, could be species from regions which are difficult of access like marine organisms. Up to 2005, about 16,000 natural products from marine organisms including algae, fungi, bacteria, and marine invertebrates have been isolated (Bhakuni and Rawat 2005). On the other hand, biosynthetic potential of in vitro plant systems grows together with development of biotechnology and genetic engineering protocols for manipulations with metabolic pathways. It is obvious that economic viability of working with in vitro plant cell technologies, which are more expensive compared to intact plants, increases together with exhausting of abundant and ready plant materials. One can enumerate additional benefits of in vitro plant systems as a platform for novel natural products search:

(a) changes in growth/development mechanisms in comparison to intact plant and somaclonal variability can cause induction of metabolic pathways and detection of new intermediates;
(b) independence from climatic factors, and cell cultivation period often shorter than nature plant development cycle;
(c) possibility for controlled manipulations with desired metabolic pathways (elicitation, feeding with intermediates, etc.);
(d) targeted introduction of foreign genes or regulatory elements by genetic engineering protocols without risk of escaping GMO to natural environment.

Problems and perspectives of undifferentiated cell culture applications for new pharmaceutical search and production were discussed in the recent reviews (Littleton et al. 2005; Roberts 2007; Dornenburg 2008; Weathers et al. 2010; Xu et al. 2011). Here, we describe biotechnological applications of another in vitro plant system, hairy roots, for production of novel compounds including secondary metabolites and products of heterologous gene expression.

10.2 *Agrobacterium rhizogenes* Plant Transformation and Characteristics of Hairy Roots

Agrobacterium rhizogenes and *A. tumefaciens* are gram-negative soil bacteria belonging to Rhizobiaceae family. Both are able to infect wounded plant tissues after activation with plant phenolic compounds. After bacterium attachment to plant cell, transfer of a part of their big plasmids (T-DNA of Ri- and Ti-plasmid, respectively, see for review Suzuki et al. 2009) occurs with further stable integration into plant genome. It causes formation of plant tumors called "hairy root" or "crown gall", correspondingly. These tumors are able to produce specific compounds, opines, supporting bacterial growth as nutrients. Hairy root disease and its pathogen *A. rhizogenes* have been described in the first half of the last century (Riker et al. 1930). Molecular mechanisms of T-DNA transfer are similar for both bacteria, although some differences were found, e.g., GALLS protein instead of VirE2 takes part in T-DNA transport into nucleus in some *A. rhizogenes* strains (Hodges et al. 2004). Details of T-DNA transfer are discussed in the recent reviews (Tzfira and Citovsky 2006; Citovsky et al. 2007; Dafny-Yelin et al. 2008; Magori and Citovsky 2011 and references cited therein) and are continuously updating.

Recent taxonomic studies suggested revision of genus *Agrobacterium* and reclassification of bacteria with plasmid-borne tumorigenic and rhizogenic capabilities into species *Rhizobium radiobacter* and *R. rhizogenes* (Young et al. 2001).

10.2.1 Peculiarities of Agrobacterium rhizogenes T-DNA and T-DNA Gene Functions

Ri-plasmids of *A. rhizogenes* have size of approximately 200 kbp. All pathogenic plasmids contain regions with genes responsible for virulence, opine catabolism, DNA replication, conjugation, and T-DNA. Initially *A. rhizogenes* strains were classified as mannopine and agropine depending on the type of opine biosynthesis genes they possessed in Ri-plasmid (Sinkar et al. 1987). Current classification includes additional opine types (see for example Table 10.1).

The most common agropine-type Ri-plasmids (e.g., pRiA4 etc.) contain two stretches of T-DNA (Huffman et al. 1984; White et al. 1985). T-DNAs always have at the ends direct repeats of 25 bp, called right (RB) and left border (LB) (Yadav et al. 1982). RB is necessary and essential for T-DNA transfer, while the function of LB is termination of single-strand DNA formation in *Agrobacterium* cell (Wang et al. 1984).

On TL-DNA, 18 ORFs have been identified (Slightom et al. 1986). Among them, the basic role in the induction of neoplastic root growth is ascribed to *rol* genes (root loci), although other genes (*ORF13* and *14*) may have synergistic function in this process (Capone et al. 1989). There are four *rol* genes in TL-DNA (*rolA*, *B*, *C*, and *D*), three of them, *rolA*, *B*, and *C*, are sufficient for production of hairy roots (Spena et al. 1987).

Table 10.1 Examples of Ri-plasmids found in *A. rhizogenes*

Plasmid	Opine production	T-DNA	Key reference
pRi8196	Mannopine	T-DNA	Hansen et al. (1991)
pRi1724	Mikimopine	T-DNA	Moriguchi et al. (2001)
pRi2659	Cucumopine	T-DNA	Serino et al. (1994)
pRiA4	Agropine	T_L- and T_R-DNA	Bouchez and Tourneur (1991)
pRi1855	Agropine	T_L- and T_R-DNA	Pomponi et al. (1983)

The exact biochemical function was defined only for *rolD*, which encodes ornithine cyclodeaminase (Trovato et al. 2001). RolA protein was supposed to be DNA-binding (Rigden and Carneiro 1999), RolB possesses tyrosine phosphatase activity (Filippini et al. 1996), and RolC was proposed to act as a non-specific glucosidase (Faiss et al. 1996), but the exact biochemical scenario of hairy root induction remains to be elucidated. TR-DNA carries *aux1* and *aux2* genes similar in sequence and function to *tms1* and *tms2* genes of Ti-plasmid *A. tumefaciens* encoding enzymes of auxin biosynthesis (Huffman et al. 1984; White et al. 1985; Offringa et al. 1986), and the genes responsible for opine synthesis (De Paolis et al. 1985). The *aux* genes are supposed to play an ancillary role in hairy root induction (Cardarelli et al. 1987). It was assumed that increased auxin sensitivity conferred on plant cells by transformation with *rolABC* genes is essential for hairy root development, while phytohormone concentrations in the cells remain basically unchanged (Spano‘ et al. 1988). Plant mutations, leading to low auxin sensitivity, diminish hairy root initiation by *Agrobacterium* (Lima et al. 2009). However, *rol* genes alter only some cellular responses to auxin. Tobacco *rolABC*-transformed protoplasts showed increased sensitivity in test for auxin-induced membrane hyperpolarization, but for division they needed approximately the same auxin concentrations as control protoplasts (Maurel et al. 1991). These results are supported by experiments with tobacco BY-2 cells, which are able to divide in auxin-free medium after transformation with *aux* but not *rol* genes (Nemoto et al. 2009). Moreover, in some studies rhizogenicity of *Agrobacterium* was strongly diminished (Offringa et al. 1986) or even eliminated (White et al. 1985) by mutations in *aux* genes. It is possible that expression of *aux* genes is crucial for hairy root initiation, but is dispensable at later stages. For induction of hairy roots in carrot both *rol* and *aux* genes were necessary, but the majority of established root cultures contained only *rol* and no *aux* genes (Capone et al. 1989). Root formation was also observed when TL-DNA genes *ORF13* and *14* were co-transformed with *rol* genes instead of TR-DNA (Capone et al. 1989). It was shown for some plant species that presence of *aux* genes is associated with callus-like morphology of transgenic roots (Moyano et al. 1999; Hong et al. 2006a; Mirjalili et al. 2009), and absence of *aux* loci may result in lower levels of secondary metabolites (Mallol et al. 2001; Mirjalili et al. 2009).

The strains of *A. rhizogenes* that produce other opines, such as mannopine, cucumopine, or mikimopine (Table 10.1), have a single T-DNA fragment, homologous to the agropine TL-DNA but without the *rolD* gene (Meyer et al. 2000). It can be speculated that absence of *aux* genes in these strains is compensated by conserved *ORF13, 14* (Capone et al. 1989) and *ORF8*, which encodes a protein with tryptophan 2-monooxygenase activity (Lemcke et al. 2000).

10.2.2 Characteristics of Hairy Root Cultures

Hairy roots demonstrate fast hormone-independent growth, intensive lateral branching with numerous adventitious roots, and plagiogeotropism (growth at an oblique angle to the stimulus of gravity) (Tepfer 1984; Shanks and Morgan 1999; Ono and Tian 2011). Being cultivated at light, hairy roots are able to form chloroplasts (Flores et al. 1993; Nagatome et al. 2000). Such green roots might be further adapted to photoautotrophy that can influence their physiological and biochemical characteristics (Flores et al. 1993). Transgenic plants can be regenerated from hairy roots. They often have so-called hairy root phenotype with shorter internodes leading to dwarfism, wrinkled leaves, heterostyly, and increased root system. Regeneration may be spontaneous or induced with phytohormones (Tepfer 1984; Casanova et al. 2005; Christensen et al. 2008).

Many authors indicate a high level of genetic stability in hairy roots in contrast to undifferentiated cell cultures (see for review Flores et al. 1987; Shanks and Morgan 1999; Giri and Narasu 2000 and references cited therein). Screening of *Datura stramonium* hairy root lines for growth characteristics and tropane alkaloid production for over 5 years showed remarkable stability of transgenic cultures in comparison to normal roots (Maldonado et al. 1993). It was supported with cytogenetic studies of karyotype and chromosome numbers (Baiza et al. 1999). On the other hand, reduction of chromosome number and loss of regeneration ability of hairy root cultures was reported (Xu and Jia 1996).

10.3 Secondary Metabolism in Hairy Roots

A broad range of substances belonging to all secondary metabolite groups was isolated from hairy root cultures. It is a well-established fact that production of secondary metabolites in hairy roots for the most transformed species is similar to or even surpasses that of intact plants and, especially, plant cell cultures (see for review Flores et al. 1987; Sevon and Oksman-Caldentey 2002; Pistelli et al. 2010; Ono and Tian 2011). A lot of protocols were developed for stimulation of secondary compound biosynthesis and accumulation in hairy root cultures. During the last decade, some progress was achieved for large-scale cultivation of hairy roots in different types of bioreactors (first of all, airlift and mist reactors). The most attention was focused upon scaled up cultivation of *Artemisia annua*, *Beta vulgaris*, *Panax ginseng*, and several Solanaceae species (*Hyoscyamus*, *Atropa*, *Scopolia*, etc.) (Kim et al. 2002; Mishra and Ranjan 2008). The Switzerland company ROOTec (http://www.rootec.com) started large-scale industrial isolation of valuable metabolites from hairy root cultures. Recently, ginsenosides production in hairy roots in bioreactor was achieved even higher than that in roots of intact plants (ROOTek press release 2010).

10.3.1 Influence of Rol Genes on Secondary Metabolism in Hairy Roots

Increased levels of secondary compounds are found after transformation with wild-type Ri-plasmids as well as with individual *rol* genes or their combinations. Influence of single *rol* genes on secondary biosynthesis in plant cell cultures was reviewed recently (Bulgakov 2008). The *rolB* gene is the most effective inducer of secondary metabolism, but it also strongly suppresses cell growth (Shkryl et al. 2008). For *Vitis amuriensis,* transformation with *rolB* led to more than 100-fold increase in resveratrol level (Kiselev et al. 2007). Individual *rolA* and *rolC* can also induce biosynthesis of different classes of secondary compounds (reviewed in Bulgakov 2008), although there exist data about the inhibitory effect of *rolC* on secondary metabolism (Bulgakov et al. 2005). The *rol* genes are able to stimulate not only secondary metabolism but other plant defense systems as well. Transformation of ginseng cells with *rolC* led to transcriptional activation of *β*-1,3-glucanase, a pathogenesis-related protein (Kiselev et al. 2006). The exact mechanism of *rol* genes influence on transcription of plant defense genes is unclear, but it is proposed that Rol proteins may intervene with processes of protein phosphorylation–dephosphorylation, reactive oxygen species production, and Ca^{2+}-dependent signaling and/or mediate uncommon signal transduction pathways in plant cell (Bulgakov 2008 and references cited therein).

10.3.2 New Compounds from Hairy Root Cultures

Since the time of the first publications on isolation of secondary metabolites from hairy root cultures (e.g., Hamill et al. 1986), more than 400 plant species were considered to have been transformed with *Agrobacterium rhizogenes* (Ono and Tian 2011). For most species the aim of transformation was establishment of hairy root cultures with enhanced level of secondary metabolites typical for intact plants. For example, because of low productivity of undifferentiated cell cultures, hairy roots of a number of Solanacea species were induced to achieve enhanced accumulation of the anticholinergic tropane alkaloids, hyoscyamine and scopolamine. Current genetic engineering approaches for biotechnology of scopolamine production in hairy roots and other types of cultures are described in the recent reviews (Oksman-Caldentey 2007; Sheludko 2010). However, in spite of the general similarity of secondary compound profiles in hairy roots and intact plants, numerous publications reported detection of novel substances, which are not characteristic of the intact plant. This suggests that secondary metabolism in hairy roots cannot be regarded as a simple reflection of that in the parental species, but may be the result of the influence of several factors:

(a) transgene integration into plant genome;
(b) changes in biochemical and physiological parameters in hairy roots caused by transgene expression;
(c) metabolic peculiarities of roots detached from whole plant.

Discussing the factors mentioned above, one can say that limited information is currently available about sites of T-DNA integration into plant chromosomes. Very low rate of homologous recombination between T-DNA and plant genomic DNA was detected, and attracting function of chromosomal double-strand breaks as well as chromatin modifications for T-DNA integration has been suggested. Apparently, host systems of DNA reparations take place in T-DNA integration (Tzfira and Citovsky 2006; Magori and Citovsky 2011 and references cited therein). It means that T-DNA integration may potentially influence expression of any functional or regulatory gene and cause variations in secondary metabolite profiles (Abhyankar et al. 2005).

Influence of integrated *rol* genes on secondary metabolism in hairy roots is already discussed in Sect. 10.3.1. The last factor (c) may be important when secondary compounds are transported from the site of their biosynthesis (for example, leaves) to the point of their accumulation/conversion (roots) (e.g., Croteau et al. 1987; Dixit and Srivastava 2000). For example, antimicrobial meroterpene bakuchiol was identified mainly in green tissues of *Psoralea drupacea* in vitro plants and partially in their roots. Hairy roots induced from this species did not accumulate detectable amounts of bakuchiol and their HPLC profiles considerably differed from those of normal roots (Lystvan et al. 2010).

Cultivation parameters may strongly change secondary metabolite profiles in hairy root cultures and stimulate de novo biosynthesis or increase accumulation of minor compounds. For example, it was proven that growing of hairy roots in light often influences secondary metabolite spectrum (Flores et al. 1993; Ikenaga et al. 1995; Bhadra et al. 1998). The most pronounced results were achieved with elicitation by different chemical compounds or stress factors. Since publication by Gundlach et al. in 1992 of their experiments on stimulation of secondary biosynthesis by jasmonates, these compounds transmitting stress signal in plant cell became "classical" inducers, which have been successfully applied to numerous plant cell cultures. For plant cell cultures, about 40-fold increase of single compound contents was detected (Gundlach et al. 1992; Sheludko et al. 1999). For hairy roots with initially high level of secondary biosynthesis, 2–10-fold induction was observed (see for example Stojakowska et al. 2002; Syklowska-Baranek et al. 2009). Application of diverse approaches for stimulation and modulation of secondary metabolism is discussed in the recent review (Shilpa et al. 2010).

Taking into account all the above-mentioned, hairy roots may be regarded as very perspective origin of secondary metabolite biodiversity and as an object for scaled up screening programs. Biotransformation after untypical precursor feeding may additionally increase the spectrum of new secondary compounds from hairy roots. Below we summarize the information about isolation of novel secondary compounds from hairy root cultures (Table 10.2). Compounds of mixed origin are mentioned once in one of the groups.

10.3.3 New Indole Alkaloids of *Rauvolfia serpentina* Hairy Root Cultures as an Example of Alkaloid Metabolism Change

The indian medicinal plant *Rauvolfia serpentina* Benth. ex Kurz (Apocynaceae) is a source of pharmaceutically valuable monoterpenoid indole alkaloids with neuroleptic, antihypertensive, and antiarrhythmic activities. Indole alkaloid pattern of *R. serpentina* hairy roots differed from that in the intact roots as well as in cultured undifferentiated cells of this plant. In addition to isolation of 16 known alkaloids, four new indole alkaloids were characterized (Sheludko et al. 2002a, b). Three of them belonged to 19(S),20(R)-dihydroperaksine type: a novel subgroup of sarpagine alkaloids specific for *R. serpentina* hairy root culture (Sheludko et al. 2002b). Further researches showed that a key enzyme for new alkaloid formation was the esterase, performing deacetylation of indolenine alkaloid perakine (**1**) or raucaffrinoline (**3**) producing the first 19(S),20(R)-dihydroperaksine type alkaloids (Fig. 10.1). Incubation of (**3**) with a crude enzyme preparation from *R. serpentina* hairy roots led to formation of 19(S),20(R)-dihydroperaksine-17-al (**4**) (Sheludko et al. 2002b). Characterization of partially purified enzyme showed that it specifically accepts indolenine type alkaloids as well as several simple O-acetylated compounds with similar ring systems. Reduction of 1–2 double bond in indolenine alkaloids led to formation of ajmalan skeleton and inhibition of the esterase activity possible because of change in the molecule geometry (Sheludko et al., unpublished data). The native precursor of raucaffrinoline (**3**), perakine (**1**), was converted by the esterase giving a new product with two aldehyde groups at C-17 and C-20 (**2**), 19(S),20(R)-dihydroperaksine-17,21-al. The obtained data led to biosynthetic scheme for 19(S),20(R)-dihydroperaksine type alkaloid formation: perakine (**1**) is deacetylated by the mentioned esterase (EST) giving (**2**), which may be further reduced to (**4**) by common *Rauvolfia* secondary metabolism enzymes (Fig. 10.1). This was proved by cloning and characterization of perakine reductase. This enzyme accepted perakine (**1**) and 19(S),20(R)-dihydroperaksine-17,21-al (**2**) converting them into raucaffrinoline (**3**) or 19(S),20(R)-dihydroperaksine-17-al (**4**), respectively (Sun et al. 2008). Additionally, the direct formation of (**4**) from raucaffrinoline (**3**) can also take place. One can suppose that conversion of (**4**) into (**5**) and (**6**) is performed by enzymes participating in formation of typical structurally related *Rauvolfia* alkaloids of sarpagine group (Yu et al. 2002). This example displayed how activation of one specific key enzyme in hairy root cultures may induce formation of a new branch of secondary biosynthesis.

Table 10.2 Summary data of new compounds isolated from hairy root cultures

Plant species	New compounds (number in brackets)	Reference
Terpenes		
Ambrosia trifida	Sesquiterpenoids (3)	Lu et al. (1993)
Artemisia annua	Sesquiterpene (1)	Zhai et al. (2010)
Astragalus membranaceus	Triterpene (1)	Hirotani et al. (1994a)
Astragalus membranaceus	Triterpene (1)	Hirotani et al. (1994b)
Astragalus membranaceus	Triterpenes (2)	Zhou et al. (1995)
Catharanthus roseus	Diterpene glycosides (2)	Chung et al. (2008)
Galphimia glauca	Triterpenoids (3)	Nader et al. (2004)
Hyoscyamus albus	Sesquiterpenoids (4)[a]	Kuroyanagi et al. (1998)
Hyoscyamus albus	Sesquiterpenoids (4)[a]	Kawauchi et al. (2010)
Lactuca virosa	Sesquiterpenes (5)	Kisiel et al. 91995)
Plocama pendula	Triterpenes (2)	Fraga et al. (2006)
Salvia broussonetii	Diterpenes (2)	Fraga et al. (2005)
Tripterygium wilfordii	Diterpenoid (1) Sesquiterpenoid (1)	Nakano et al. (1998)
Valeriana officinalis	Monoterpen (1)	Granicher et al. (1995)
Phenolics		
Beta vulgaris	Betaxanthins (2)	Hempel and Bohm (1997)
Catharanthus roseus	Flavonoids (4)	Chung et al. (2009)
Catharanthus roseus	Anthracene derivatives (2)	Chung et al. (2007a)
Eupatorium cannabinum	Benzofuran (1)	Siebertz et al. (1989)
Glycyrrhiza glabra	Flavonoids (1)	Li et al. (1998)
Glycyrrhiza glabra	Flavonoids (7)	Li et al. (2000)
Glycyrrhiza glabra	Flavonoids (2)	Asada et al. (1998)
Glycyrrhiza glabra	Flavonoids (1)	Asada et al. (1999)
Glycyrrhiza pallidiflora	Flavonoids (2)	Li et al. (2001)
Glycyrrhiza pallidiflora	Flavonoids (3)	Li et al. (2002)
Leontopodium alpinum	Benzopyran (1)	Comey et al. (1997)
Linum austriacum	Lignan (1)	Mohagheghzadeh et al. (2002)
Lithospermum erythrorhizon	Quinones (1)	Fukui et al. (1998)
Lithospermum erythrorhizon	Quinones (1)	Fukui et al. (1999)
Scutellaria baicalensis	Flavonoids (1)	Nishikawa et al. (1999)
Scutellaria baicalensis	Flavonoids (1)	Zhou et al. (1997)
Sesamum indicum	Anthraquinone (1)	Furumoto and Jindai (2008)
Sesamum indicum	Anthraquinones (2)	Ogasawara et al. (1993)
Swertia japonica	Phenyl glucosides (2)	Ishimaru et al. 1990a
Swertia japonica	Xanthone (1)	Ishimaru et al. (1990b)
Tagetes patula	Benzofuran (1)	Menelaou et al. (1991)
Alkaloids		
Atropa baetica	Tropane alkaloid (1)	Zárate (1999)
Catharanthus trychophyllus	Indole alkaloids (5)	Davioud et al. (1989)
Datura stramonium	Tropane alkaloid (1)	Berkov et al. (2003)
Hyoscyamus albus	Piperidone alkaloid (1)	Sauerwein et al. (1991)
Rauvolfia serpentina	Indole alkaloid (1)	Falkenhagen et al. (1993)
Rauvolfia serpentina	Indole alkaloid (1)	Sheludko et al. (2002a)

(continued)

Table 10.2 (continued)

Plant species	New compounds (number in brackets)	Reference
Rauvolfia serpentina	Indole alkaloids (3)	Sheludko et al. (2002b)
Other groups		
Armoracia rusticana	Thiol peptides	Kubota et al. (2000)
Catharanthus roseus	Aliphatic glycoside (1)	Chung et al. (2007b)
Lobelia inflata	Polyacetylenes (2)	Ishimaru et al. (1991)
Lobelia inflata	Polyacetylene (1)	Ishimaru et al. (1992)
Pratia nummularia	Polyacetylenes (2)	Ishimaru et al. (2003)
Biotransformation products		
Lobelia sessilifolia	Phenolics (2)	Yamanaka et al. (1995)
Panax ginseng	Triterpenes (5)	Kawaguchi et al. (1990)
Panax ginseng	Triterpenoids (3)	Asada et al. (1993)

[a] Compounds were isolated from elicited cultures

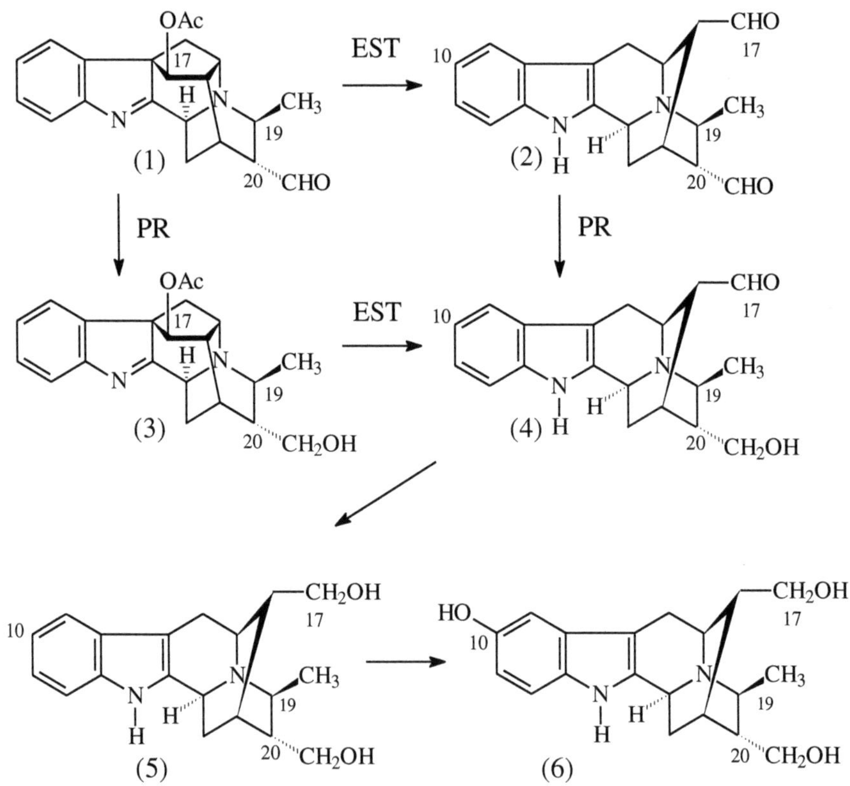

Fig. 10.1 A putative biosynthetic pathway in *R. serpentina* hairy root culture leading from perakine (**1**) or raucaffrinoline (**3**) to 19(S),20(R)-dihydroperaksine-17,21-al (**2**), 19(S),20(R)-dihydroperaksine-17-al (**4**) 19(S),20(R)-dihydroperaksine (**5**), and 10-hydroxy-19(S),20(R)-dihydroperaksine (**6**); *PR* perakine reductase; *EST* esterase

10.4 Hairy Roots Expressing Recombinant Genes

10.4.1 Recombinant Proteins from Hairy Root Cultures

A wide group of natural products includes peptides and proteins which are the products of heterologous gene expression in plant systems. Accumulation of hepatitis B surface antigen (HBsAg) in potato hairy roots reached 97.1 ng/g FW, while in transgenic plants transformed by *A. tumefacience* or regenerated from hairy roots HBsAg level was considerably lower (19.1 and 13.4 ng/g FW, respectively) (Kumar et al. 2006). Remarkably high level of human acetylcholinesterase, up to 3.3 % TSP, has been reported for *N. benthamiana* hairy roots derived by transformation with *A. rhizogenes* from transgenic plants expressing the corresponding gene. Acetylholinesterase accumulation in parental plants was about 3-fold lower than in hairy roots (Woods et al. 2008). Murine interleukin 12 was produced in tobacco hairy roots grown in the mist reactor at 5.3 μg/g FW level (Liu et al. 2009).

A useful property of hairy roots is their ability to secrete heterologous proteins into hydroponic medium. The model protein human secreted alkaline phosphatase was released from tobacco hairy roots at a rate of 28 μg/g root dry weight per day, which is considerably higher than from adventitious roots of transgenic plants (Gaume et al. 2003; Komarnytsky et al. 2004). This approach has been applied to produce single-chain and full-size antibodies in tobacco hairy roots. To improve the rhizosecretion, native signal peptides of immunoglobulins were replaced by signal peptide of plant origin. Target protein degradation was reduced by heterologous protease inhibitor cosecreted with immunoglobulins. Accumulation rates as high as 36.4 μg/g root dry weight per day for single-chain and 21.8 μg/g root dry weight per day for full-size immunoglobulins were achieved (Komarnytsky et al. 2006).

Transient expression of recombinant genes in plants is an efficient technology for production of proteins of interest in high quantities within a short time (Gleba et al. 2007; Sheludko 2008). Levels of transient expression in hairy root cultures are rather low (Collens et al. 2007), but regeneration of plants with hairy root phenotype may help to improve biomass yield of species used for transient expression without affecting target protein accumulation (Sindarovska et al. 2005).

10.4.2 Metabolic Engineering of Hairy Roots for Production of New Metabolites

Metabolic engineering is usually aimed at high levels of known valuable compounds. For this purpose, genes of enzymes located at the beginning of biosynthetic pathways are often introduced into hairy roots in the hope to increase the total metabolic flux in the desirable direction. Effectiveness of this approach varies for different plant species. As an example, putrescine N-methyltransferase (PMT) forms the first specific precursor of both tropane and pyridine-type alkaloids. Heterologous overexpression of this enzyme from *N. tabacum* in hairy roots of *Duboisia* hybrid (Moyano et al. 2002) and *Atropa belladonna* (Rothe et al. 2003) did not alter alkaloid content. However, in hairy roots of *Datura metel* and *Hyoscyamus muticus* overexpression of tobacco *pmt* resulted in increase in tropane alkaloid levels. Both hyoscyamine and scopolamine production was improved in hairy root cultures of *D. metel*, whereas in *H. muticus* only hyoscyamine content grew higher. The total tropane alkaloid and scopolamine levels in *H. muticus* hairy roots with tobacco *pmt* could be further increased by methyl jasmonate treatment (Zhang et al. 2007). Other examples are induction of apigenin and total flavonoid level in hairy roots of *Saussurea involucrata* expressing heterologous chalcone isomerase gene (Li et al. 2006) and extensive work on metabolic engineering of *Catharanthus roseus* hairy roots for production of terpenoid indole alkaloids (TIAs) (Hughes et al. 2004; Hong et al. 2006b; Goklany et al. 2009; Peebles et al. 2011). In *C. roseus* hairy roots genes coding for transcription factor (Peebles et al. 2009; Wang et al. 2010) and other potential regulator of TIA biosynthesis (Jaggi et al. 2011) were overexpressed. In our opinion, there is a chance that in such experiments some new compounds may have appeared but have not been described, because the main attention was paid to the levels of known valuable substances.

On the other hand, if some biosynthetic step is known to be limiting, the corresponding enzyme is overexpressed in hairy roots, e.g., hyoscyamine 6-hydroxylase (H6H) converting hyoscyamine into scopolamine. This method proved to be successful for obtaining of *Duboisia* hybrid and *H. niger* hairy root lines with high scopolamine content (Palazón et al. 2003; Zhang et al. 2004), although it gives little hope for finding new compounds. However, introduction of *h6h* into *N. tabacum* hairy roots allowed them to produce scopolamine (not found in tobacco) from exogenously applied hyoscyamine at a rate higher than *H. niger* transgenic roots overexpressing *h6h* (Hakkinen et al. 2005).

Genetically engineered hairy roots have been used for production of polymers and polymer precursors that are not synthesized in corresponding plants. Polyhydroxyalkanoates (PHAs) are naturally occurring polymers produced in various bacteria. They may be used for production of plastic bags, fibers, and films, and can be decomposed by an enzyme PHA-depolymerase detected in some bacteria and fungi. The best characterized PHA is poly-3-hydroxybutyrate (PHB). It is synthesized in bacteria, e.g., *Ralstonia eutropha*, but the costs of bacterial fermentation are rather high. Plant systems including hairy roots are considered as an alternative source of PHB. Three enzymes are necessary for synthesis of this polymer: β-ketothiolase that forms acetoacetyl-CoA from two acetyl-CoA moieties, acetoacetyl-CoA reductase producing R-3-hydroxybutyryl-CoA, and PHA synthase that polymerizes (R)-(-)-3-hydroxybutyryl-CoA to PHB. The first enzyme is present in plant cells but not in plastids, which proved to be the best cell compartment for PHB production. Three enzymes of PHB biosynthesis were expressed in plants in fusion with transit peptide that ensured their transport to plastids. In leaves of *Arabidopsis thaliana,* up to 14 % (Nawrath et al. 1994) or even up to 40 % of dry weight (DW) of PHB were accumulated, although in the latter case this negatively influenced plant growth (Bohmert et al. 2000). Three recombinant genes of *R. eutropha* coding for enzymes of PHB biosynthesis fused to plastid transit peptide were introduced into hairy roots of sugar beet. Transgenic root cultures produced up to 5.5 % DW PHB in leucoplasts (Menzel et al. 2003).

Another example is production in hairy roots of p-hydroxybenzoic acid (pHBA) that can be used as polymer feedstock. New conjugates of pHBA were synthesized in transgenic roots of *Datura stramonium* (Mitra et al. 2002) and sugarbeet (Rahman et al. 2009) by heterologously expressed bacterial p-hydroxycinnamoyl-CoA hydratase/lyase (HCHL) followed by inherent plant dehydrogenase and glucosyltransferases. In sugar beet hairy roots up to 14 % DW pHBA glucose ester were accumulated, that is the best yield achieved in plant systems so far.

Examples of metabolic engineering of hairy roots for production of totally new substances are few. Hairy roots of Lotus *corniculatus* have been used in experiments on investigation and modification of condensed tannin (CT) biosynthesis. CTs are an important component of forage crops fed to ruminants because they are able to bind and stabilize proteins during the digestion process (Min et al. 2003). The protective effect of CTs increases with increasing polymer molecular weight and number of hydroxyl groups. *L. corniculatus* is a popular object for studies on modifications of CT biosynthesis. This species is characterized by significant biomass that can be harvested from an individual plant, but the average molecular weight and degree of hydroxylation of CTs are lower than in *L. pedunculatus*. Hairy root cultures of *L. corniculatus* were established with heterologous genes encoding enzymes involved in CT biosynthesis, namely dihydroflavonol reductase (DFR) from *Antirrhinum majus* (Bavage et al. 1997) and flavonoid 3′5′ hydroxylase (F3′5′H) from *Eustoma grandiflorum* (Robbins et al. 2005). In one transgenic line carrying *A. majus DFR* gene, elevated CT level was accompanied by significant changes in CT hydroxylation pattern (Bavage et al. 1997). It is interesting to note that in transgenic plants regenerated from *ADFR*-hairy roots no significant changes in CT hydroxylation were observed, although CT levels were higher than in control plants (Robbins et al. 2005). Introduction of *E. grandiflorum F3′5′H* led to increase in CT content in two hairy root lines, but did not change the structure. However, one *EF3′5′H*-line displayed lower level of CTs with changed hydroxylation pattern compared to controls. This line also had unusual growth kinetics (Robbins et al. 2005).

Using RNA interference for silencing of target gene(s) is a perspective method for secondary metabolism modulation by rearrangement of metabolic fluxes in transgenic plants. Suppression of carotenoid cleavage dioxygenase in *Medicago truncatula* hairy roots by RNA interference caused synthesis of the new apocarotenoic acid derivatives (Floss et al. 2008). In hairy root cultures of *Catharanthus roseus* a number of new substances, fluorinated indole alkaloids, were produced when RNA-induced silencing of tryptophan decarboxylase (TDC) was combined with feeding of unnatural precursor, 5-fluorotryptamine. Tryptophan decarboxylation to tryptamine is the reaction that links primary and secondary metabolism in plants producing monoterpenoid indole alkaloids. The content of all major alkaloids was substantially decreased in *C. roseus* hairy roots with silenced TDC. These suppressed lines provided opportunity for mutasynthesis—replacement of natural alkaloids with unnatural analogues. After feeding with 5-fluorotryptamine, four new substances (fluoro-ajmalicine, fluoro-serpentine, fluoro-catharanthine and fluoro-tabersonine), were isolated from *C. roseus* hairy roots (Runguphan et al. 2009).

These examples display a perspective of experiments on modification of secondary metabolism using recombinant DNA technologies in hairy roots.

10.5 Conclusions

In conclusion, hairy roots represent a valuable source of natural products: from secondary compounds to recombinant proteins. Secondary metabolism in hairy root cultures is usually more intensive than in undifferentiated cells. However, metabolite profile of transgenic roots may differ from that of parental plant and include new secondary compounds with valuable activities. The reason for it can be a complex effect of transgene integration into plant genome, peculiarities of hairy root physiology and cultivation parameters. Activation of one enzyme specific to secondary metabolite conversion may lead to appearance of novel metabolic branch. Among more than 400 plant species which have been transformed with *Agrobacterium rhizogenes*, new substances belonging to all major groups of plant secondary metabolites were isolated from about 10 % of them. One should take into account that some new compounds may have appeared but have not been characterized, because the main attention was paid to increasing levels of known valuable substances. Application of metabolic engineering approaches additionally enhances biodiversity of natural compounds in hairy root cultures.

References

Abhyankar G, Reddy VD, Giri CC, Rao KV, Lakshmi VV, Prabhakar S, Vairamani M, Thippeswamy BS, Bhattacharya PS (2005) Amplified fragment length polymorphism and metabolomic profiles of hairy roots of *Psoralea corylifolia* L. Phytochemistry 66:2441–2457

Asada Y, Saito H, Yoshikawa T, Sakamoto K, Furuya T (1993) Biotransformation of 18 beta-glycyrrhetinic acid by ginseng hairy root culture. Phytochemistry 34:1049–1052

Asada Y, Li W, Yoshikawa T (1998) Isoprenylated flavonoids from hairy root cultures of *Glycyrrhiza glabra*. Phytochemistry 47:389–392

Asada Y, Li W, Yoshikawa T (1999) The first prenylated biaurone, licoagrone from hairy root cultures of *Glycyrrhiza glabra*. Phytochemistry 50:1015–1019

Baiza AM, Quiroz MA, Ruiz JA, Loyola VVM (1999) Genetic stability of hairy root cultures of *Datura stramonium*. Plant Cell Tissue Organ Cult 59:9–17

Bavage AD, Davies IG, Robbins MP, Morris P (1997) Expression of an *Antirrhinum* dihydroflavonol reductase gene results in changes in condensed tannin structure and accumulation in root cultures of *Lotus corniculatus* (bird's foot trefoil). Plant Mol Biol 35:443–458

Berkov S, Pavlov A, Kovatcheva P, Stanimirova P, Philipov S (2003) Alkaloid spectrum in diploid and tetraploid hairy root cultures of *Datura stramonium*. Z Naturforsch C 58:42–46

Bhadra R, Morgan JA, Shanks JV (1998) Transient studies of light-adapted cultures of hairy roots of *Catharanthus roseus*: growth and indole alkaloid accumulation. Biotechnol Bioeng 60:670–678

Bhakuni DS, Rawat DS (2005) Preface. In: Bhakuni DS, Rawat DS (eds) Bioactive marine natural products. Springer, New York

Bohmert K, Balbo I, Kopka J, Mittendorf V, Nawrath C, Poirier Y, Tischendorf G, Trethewey RN, Willmitzer L (2000) Transgenic *Arabidopsis* plants can accumulate polyhydroxybutyrate to up to 4% of their fresh weight. Planta 211:841–845

Bouchez D, Tourneur J (1991) Organization of the agropine synthesis region of the T-DNA of the Ri plasmid from *Agrobacterium rhizogenes*. Plasmid 25:27–39

Bulgakov VP (2008) Functions of *rol* genes in plant secondary metabolism. Biotechnol Adv 26:318–324

Bulgakov VP, Veselova MV, Tchernoded GK, Kiselev KV, Fedoreyev SA, Zhuravlev YN (2005) Inhibitory effect of the *Agrobacterium rhizogenes rolC* gene on rabdosiin and rosmarinic acid production in *Eritrichium sericeum* and *Lithospermum erythrorhizon* transformed cell cultures. Planta 221:471–478

Capone I, Spanò L, Cardarelli M, Bellincampi D, Petit A, Costantino P (1989) Induction and growth properties of carrot roots with different complements of *Agrobacterium rhizogenes* T-DNA. Plant Mol Biol 13:43–52

Cardarelli M, Spano‘ L, Mariotti D, Mauro L, van Sluys MA, Costantino P (1987) The role of auxin in hairy root induction. Mol Gen Genet 208:457–463

Casanova E, Trillas MI, Moysset L, Vainstein A (2005) Influence of *rol* genes in floriculture. Biotechnol Adv 23:3–39

Chen XW, Serag ES, Sneed KB, Zhou SF (2011) Herbal bioactivation, molecular targets and the toxicity relevance. Chem Biol Interact 192:161–176

Christensen B, Sriskandarajah S, Serek M, Muller R (2008) Transformation of *Kalanchoe blossfeldiana* with *rol*-genes is useful in molecular breeding towards compact growth. Plant Cell Rep 27:1485–1495

Chung IM, Ali M, Yang YM, Peebles CAM, Chun SC, Lee SJ, San KY, Ahmad A (2007a) Identification of new compounds from *Catharanthus roseus* hairy root cultures. Bull Korean Chem Soc 28:1294–1298

Chung IM, Park HY, Ali M, San KY, Peebles CAM, Hong SB, Ahmad A (2007b) A new chemical constituent from the hairy root cultures of *Catharanthus roseus*. Bull Korean Chem Soc 28:229–234

Chung I, Ali M, Chun S, San K, Peebles C, Ahmad A (2008) New catharanthusopimaranoside A and B from hairy root cultures of *Catharanthus roseus*. Chem Nat Compd 44:458–462

Chung IM, Ahmad A, Ali M, Lee OK, Kim MY, Kim JH, Yoon DY, Peebles CA, San KY (2009) Flavonoid glucosides from the hairy roots of *Catharanthus roseus*. J Nat Prod 72:613–620

Citovsky V, Kozlovsky SV, Lacroix B, Zaltsman A, Dafny-Yelin M, Vyas S, Tovkach A, Tzfira T (2007) Biological systems of the host cell involved in *Agrobacterium* infection. Cell Microbiol 9:9–20

Collens JI, Mason HS, Curtis WR (2007) *Agrobacterium*-mediated viral vector-amplified transient gene expression in *Nicotiana glutinosa* plant tissue culture. Biotechnol Prog 23: 570–576

Comey N, Hook I, Sheridan H, Walsh J, James P (1997) Isolation of (S)-(-)-2,3-Dihydro-2,6-dimethyl-4H-benzopyran-4-one from roots of *Leontopodium alpinum*. J Nat Products 60: 148–149

Cragg GM, Simon JE, Jato JG, Snader KM (1996) Drug discovery and development at the National Cancer Institute: Potential for new pharmaceutical crops. In: Janick J (ed) Progress in new crops. ASHS Press, Arlington

Croteau R, El-Bialy H, Dehal SS (1987) Metabolism of Monoterpenes : Metabolic Fate of (+)-Camphor in Sage (*Salvia officinalis*). Plant Physiol 84:643–648

Dafny-Yelin M, Levy A, Tzfira T (2008) The ongoing saga of *Agrobacterium*-host interactions. Trends Plant Sci 13:102–105

Davioud E, Kan C, Quirion J-C, Das BC, Husson H-P (1989) Epiallo-yohimbine derivatives isolated from in vitro hairy-root cultures of *Catharanthus trichophyllus*. Phytochemistry 28:1383–1387

De Paolis A, Mauro ML, Pomponi M, Cardarelli M, Spano L, Costantino P (1985) Localization of agropine synthesizing functions in the TR region of the root inducing plasmid of *Agrobacterium rhizogenes* 1855. Plasmid 13:1–7

Dey M, Ripoll C, Pouleva R, Dorn R, Aranovich I, Zaurov D, Kurmukov A, Eliseyeva M, Belolipov I, Akimaliev A, Sodombekov I, Akimaliev D, Lila MA, Raskin I (2008) Plant extracts from central Asia showing antiinflammatory activities in gene expression assays. Phytother Res 22:929–934

Dixit D, Srivastava NK (2000) Partitioning of 14C-photosynthate of leaves in roots, rhizome, and in essential oil and curcumin in turmeric (*Curcuma longa* L.). Photosynthetica 38:275–280

Dixon RA (2001) Natural products and plant disease resistance. Nature 411:843–884

Dornenburg H (2008) Plant cell culture technology-harnessing a biological approach for competitive cyclotides production. Biotechnol Lett 30:1311–1321

Faiss M, Strnad M, Redig P, Doležal K, Hanuš J, Van Onckelen H, Schmülling T (1996) Chemically induced expression of the *rolC*-encoded *β*-glucosidase in transgenic tobacco plants and analysis of cytokinin metabolism: *rolC* does not hydrolyze endogenous cytokinin glucosides in planta. Plant J 10:33–46

Falkenhagen H, Stöckigt J, Kuzovkina IN, Alterman IE, Kolshorn H (1993) Indole alkaloids from hairy roots of *Rauwolfia serpentina*. Can J Chem 71:1–3

Filippini F, Rossi V, Marin O, Trovato M, Costantino P, Downey PM, Lo Schiavo F, Terzi M (1996) A plant oncogene as a phosphatase. Nature 379:499–500

Flores HE, Hoy MW, Pickard JJ (1987) Secondary metabolites from root cultures. Trends Biotechnol 5:64–69

Flores HE, Dai Y, Cuello JL, Maldonado-Mendoza IE, Loyola-Vargas VM (1993) Green roots: photosynthesis and photoautotrophy in an underground plant organ. Plant Physiol 101:363–371

Floss DS, Schliemann W, Schmidt J, Strack D, Walter MH (2008) RNA interference-mediated repression of MtCCD1 in mycorrhizal roots of *Medicago truncatula* causes accumulation of C27 apocarotenoids, shedding light on the functional role of CCD1. Plant Physiol 148: 1267–1282

Fraga BM, Diaz CE, Guadano A, Gonzalez-Coloma A (2005) Diterpenes from *Salvia broussonetii* transformed roots and their insecticidal activity. J Agric Food Chem 53: 5200–5206

Fraga BM, Di-az CE, Quintana N (2006) Triterpenes from natural and transformed roots of *Plocama pendula*. J Nat Products 69:1092–1094

Fukui H, Feroj Hasan AFM, Ueoka T, Kyo M (1998) Formation and secretion of a new brown benzoquinone by hairy root cultures of *Lithospermum erythrorhizon*. Phytochemistry 47:1037–1039

Fukui H, Hasan AFMF, Kyo M (1999) Formation and secretion of a unique quinone by hairy root cultures of *Lithospermum erythrorhizon*. Phytochemistry 51:511–515

Furumoto T, Jindai A (2008) Isolation and photoisomerization of a new anthraquinone from hairy root cultures of *Sesamum indicum*. Biosci Biotechnol Biochem 72:2788–2790

Gaume A, Komarnytsky S, Borisjuk N, Raskin I (2003) Rhizosecretion of recombinant proteins from plant hairy roots. Plant Cell Rep 21:1188–1193

Giri A, Narasu ML (2000) Transgenic hairy roots recent trends and applications. Biotechnol Adv 18:1–22

Gleba Y, Klimyuk V, Marillonnet S (2007) Viral vectors for the expression of proteins in plants. Curr Opin Biotechnol 18:134–141

Goklany S, Loring RH, Glick J, Lee-Parsons CW (2009) Assessing the limitations to terpenoid indole alkaloid biosynthesis in *Catharanthus roseus* hairy root cultures through gene expression profiling and precursor feeding. Biotechnol Prog 25:1289–1296

Granicher F, Christen P, Kamalaprija P, Burger U (1995) An iridoid diester from *Valeriana officinalis var. sambucifolia* hairy roots. Phytochemistry 38:103–105

Gundlach H, Muller MJ, Kutchan TM, Zenk MH (1992) Jasmonic acid is a signal transducer in elicitor-induced plant cell cultures. Proc Natl Acad Sci U S A 89:2389–2393

Hadacek F (2002) Secondary metabolites as plant traits: current assessment and future perspectives. CRC Crit Rev Plant Sci 21:273–322

Hakkinen ST, Moyano E, Cusido RM, Palazon J, Pinol MT, Oksman-Caldentey KM (2005) Enhanced secretion of tropane alkaloids in *Nicotiana tabacum* hairy roots expressing heterologous hyoscyamine-6beta-hydroxylase. J Exp Bot 56:2611–2618

Hamill JD, Parr AJ, Robins RJ, Rhodes MJC (1986) Secondary products formation by cultures of *Beta vulgaris* and *Nicotiana rustica* transformed with *Agrobacterium rhizogenes*. Plant Cell Rep 5:111–114

Hansen G, Larribe M, Vaubert D, Tempe J, Biermann BJ, Montoya AL, Chilton MD, Brevet J (1991) *Agrobacterium rhizogenes* pRi8196 T-DNA: mapping and DNA sequence of functions involved in mannopine synthesis and hairy root differentiation. Proc Natl Acad Sci U S A 88:7763–7767

Harvey AL (2008) Natural products in drug discovery. Drug Discov Today 13:894–901

Hempel J, Bohm H (1997) Betaxanthin pattern of hairy roots from *Beta vulgaris var. lutea* and its alteration by feeding of amino acids. Phytochemistry 44:847–852

Hirotani M, Zhou Y, Lui H, Furuya T (1994a) Astragalosides from hairy root cultures of *Astragalus membranaceus*. Phytochemistry 36:665–670

Hirotani M, Zhou Y, Rui H, Furuya T (1994b) Cycloartane triterpene glycosides from the hairy root cultures of *Astragalus membranaceus*. Phytochemistry 37:1403–1407

Hodges LD, Cuperus J, Ream W (2004) *Agrobacterium rhizogenes* GALLS protein substitutes for *Agrobacterium tumefaciens* single-stranded DNA-binding protein VirE2. J Bacteriol 186:3065–3077

Hong SB, Peebles CA, Shanks JV, San KY, Gibson SI (2006a) Terpenoid indole alkaloid production by *Catharanthus roseus* hairy roots induced by *Agrobacterium tumefaciens* harboring *rol ABC* genes. Biotechnol Bioeng 93:386–390

Hong SB, Peebles CA, Shanks JV, San KY, Gibson SI (2006b) Expression of the *Arabidopsis* feedback-insensitive anthranilate synthase holoenzyme and tryptophan decarboxylase genes in *Catharanthus roseus* hairy roots. J Biotechnol 122:28–38

Huffman GA, White FF, Gordon MP, Nester GW (1984) Hairy root inducing plasmid: physical map and homology to tumor inducing plasmid. J Bacteriol 157:269–276

Hughes EH, Hong SB, Gibson SI, Shanks JV, San KY (2004) Metabolic engineering of the indole pathway in *Catharanthus roseus* hairy roots and increased accumulation of tryptamine and serpentine. Metab Eng 6:268–276

Ikan R (2008) The origin and the nature of natural products. In: Ikan R (ed) Selected topics in the chemistry of natural products. World Scientific Publishing, Singapore

Ikenaga T, Oyama T, Muranaka T (1995) Growth and steroidal saponin production in hairy root cultures of *Solanum aculeatissimum*. Plant Cell Rep 14:413–417

Ishimaru K, Sudo H, Satake M, Shimomurat K (1990a) Phenyl glucosides from a hairy root culture of *Swertia japonica*. Phytochemistry 29:3823–3825

Ishimaru K, Sudo H, Satake M, Matsunaga Y, Hasegawa Y, Takemoto S, Shimomura K (1990b) Amarogentin, amaroswerin and four xanthones from hairy root cultures of *Swertia japonica*. Phytochemistry 29:1563–1565

Ishimaru K, Yonemitsu H, Shimomura K (1991) Lobetyolin and lobetyol from hairy root culture of *Lobelia inflata*. Phytochemistry 30:2255–2257

Ishimaru K, Sadoshima S, Neera S, Koyama K, Takahashi K, Shimomura K (1992) A polyacetylene gentiobioside from hairy roots of *Lobelia inflata*. Phytochemistry 31: 1577–1579

Ishimaru K, Osabe M, Yan L, Fujioka T, Mihashi K, Tanaka N (2003) Polyacetylene glycosides from *Pratia nummularia* cultures. Phytochemistry 62:643–646

Jaggi M, Kumar S, Sinha AK (2011) Overexpression of an apoplastic peroxidase gene CrPrx in transgenic hairy root lines of *Catharanthus roseus*. Appl Microbiol Biotechnol 90:1005–1016

Kawaguchi K, Hirotani M, Yoshikawa T, Furuya T (1990) Biotransformation of digitoxigenin by ginseng hairy root cultures. Phytochemistry 29:837–843

Kawauchi M, Arima T, Shirota O, Sekita S, Nakane T, Takase Y, Kuroyanagi M (2010) Production of sesquiterpene-type phytoalexins by hairy roots *Hyoscyamus albus* co-treated with cupper sulfate and methyl jasmonate. Chem Pharm Bull 58:934–938

Kim Y, Wyslouzil B, Weathers P (2002) Secondary metabolism of hairy root cultures in bioreactors in vitro cellular and development biology. Plant 38:1–10

Kiselev KV, Kusaykin MI, Dubrovina AS, Bezverbny DA, Zvyagintseva TN, Bulgakov VP (2006) The *rolC* gene induces expression of a pathogenesis-related β-1,3-glucanase in transformed ginseng cells. Phytochemistry 67:2225–22231

Kiselev KV, Dubrovina AS, Veselova MV, Bulgakov VP, Fedoreyev SA, Zhuravlev YN (2007) The *rolB* gene-induced overproduction of resveratrol in *Vitis amurensis* transformed cells. J Biotechnol 128:681–692

Kisiel W, Stojakowska A, Malarz J, Kohlmuzer S (1995) Sesquiterpene lactones in *Agrobacterium rhizogenes*-transformed hairy root culture of *Lactuca virosa*. Phytochemistry 40: 1139–1140

Komarnytsky S, Gaume A, Garvey A, Borisjuk N, Raskin I (2004) A quick and efficient system for antibiotic-free expression of heterologous genes in tobacco roots. Plant Cell Rep 22:765–773

Komarnytsky S, Borisjuk N, Yakoby N, Garvey A, Raskin I (2006) Cosecretion of protease inhibitor stabilizes antibodies produced by plant roots. Plant Physiol 141:1185–1193

Kubota H, Sato K, Yamada T, Maitani T (2000) Phytochelatin homologs induced in hairy roots of horseradish. Phytochemistry 53:239–245

Kumar GBS, Ganapathi TR, Srinivas L, Revathi CJ, Bapat VA (2006) Expression of hepatitis B surface antigen in potato hairy roots. Plant Sci 170:918–925

Kuroyanagi M, Arakawa T, Mikami Y, Yoshida K, Kawahar N, Hayashi T, Ishimaru H (1998) Phytoalexins from hairy roots of *Hyoscyamus albus* treated with methyl jasmonate. J Nat Prod 61:1516–1519

Lemcke K, Prinsen E, van Onckelen H, Schmulling T (2000) The *ORF8* gene product of *Agrobacterium rhizogenes* TL-DNA has tryptophan 2-monooxygenase activity. Mol Plant Microbe Interact 13:787–790

Li W, Asada Y, Yoshikawa T (1998) Antimicrobial flavonoids from *Glycyrrhiza glabra* hairy root cultures. Planta Med 64:746–747

Li W, Asada Y, Yoshikawa T (2000) Flavonoid constituents from *Glycyrrhiza glabra* hairy root cultures. Phytochemistry 55:447–456

Li W, Asada Y, Koike K, Hirotani M, Rui H, Yoshikawa T, Nikaido T (2001) Flavonoids from *Glycyrrhiza pallidiflora* hairy root cultures. Phytochemistry 58:595–598

Li W, Koike K, Asada Y, Hirotani M, Rui H, Yoshikawa T, Nikaido T (2002) Flavonoids from *Glycyrrhiza pallidiflora* hairy root cultures. Phytochemistry 60:351–355

Li F-X, Jin Z-P, Zhao D-X, Cheng L-Q, Fu C-X, Ma F (2006) Overexpression of the *Saussurea medusa* chalcone isomerase gene in *S. involucrata* hairy root cultures enhances their biosynthesis of apigenin. Phytochemistry 67:553–560

Lima JE, Benedito VA, Figueira A, Peres LE (2009) Callus, shoot and hairy root formation in vitro as affected by the sensitivity to auxin and ethylene in tomato mutants. Plant Cell Rep 28:1169–1177

Littleton J, Rogers T, Falcone D (2005) Novel approaches to plant drug discovery based on high throughput pharmacological screening and genetic manipulation. Life Sci 78:467–475

Liu C, Towler MJ, Medrano G, Cramer CL, Weathers PJ (2009) Production of mouse interleukin-12 is greater in tobacco hairy roots grown in a mist reactor than in an airlift reactor. Biotechnol Bioeng 102:1074–1086

Lu T, Parodi FJ, Vargas D, Quijano L, Mertooetomo ER, Hjortso MA, Fischer NH (1993) Sesquiterpenes and thiarubrines from *Ambrosia trifida* and its transformed roots. Phytochemistry 33:113–116

Lystvan K, Belokurova V, Sheludko Y, Ingham J, Prykhodko V, Kishchenko O, Paton E, Kuchuk M (2010) Production of bakuchiol by in vitro systems of Psoralea drupacea Bge. Plant Cell Tissue Organ Cult 101:99–103

Magori S, Citovsky V (2011) Epigenetic control of *Agrobacterium* T-DNA integration. Biochim Biophys Acta 1809:388–394

Maldonado MIE, Ayora TT, Loyola VVM (1993) Establishment of hairy root cultures of *Datura stramonium*: Characterization and stability of tropane alkaloid production during long periods of subculturing. Plant Cell Tissue Organ Cult 33:321–329

Mallol A, Cusidó RM, Palazón J, Bonfill M, Morales C, Piñol MT (2001) Ginsenoside production in different phenotypes of Panax ginseng transformed roots. Phytochemistry 57:365–371

Maurel C, Barbier-Brygoo H, Spena A, Tempe J, Guern J (1991) Single *rol* genes from the *Agrobacterium rhizogenes* T(L)-DNA alter some of the cellular responses to auxin in *Nicotiana tabacum*. Plant Physiol 97:212–216

Menelaou MA, Fronczek FR, Hjortso MA, Morrison AF, Foroozesh M, Thibodeaux TM, Flores HE, Fisher NH (1991) NMR spectral data of benzofurans and bithiophenes from hairy root cultures of *Tagetes patula* and the molecular structure of isoeuparin. Spectroscopy Lett 24:1405–1413

Menzel G, Harloff HJ, Jung C (2003) Expression of bacterial poly(3-hydroxybutyrate) synthesis genes in hairy roots of sugar beet (*Beta vulgaris* L.). Appl Microbiol Biotechnol 60:571–576

Meyer AD, Tempe' J, Costantino P (2000) Hairy root: A molecular overview. Functional analysis of *Agrobacterium rhizogenes* T-DNA genes. In: Stacey G, Keen NT (eds) Plant-microbe interactions. APS Press, St. Paulo

Min BR, Barry TN, Attwood GT, McNabb WC (2003) The effect of condensed tannins on the nutrition and health of ruminants fed fresh temperate forages: a review. Anim Feed Sci Tech 106:3–19

Mirjalili MH, Fakhr-Tabatabaei SM, Bonfill M, Alizadeh H, Cusido RM, Ghassempour A, Palazon J (2009) Morphology and withanolide production of *Withania coagulans* hairy root cultures. Eng Life Sci 9:197–204

Mishra BN, Ranjan R (2008) Growth of hairy-root cultures in various bioreactors for the production of secondary metabolites. Biotechnol Appl Biochem 49:1–10

Mitra A, Mayer MJ, Mellon FA, Michael AJ, Narbad A, Parr AJ, Waldron KW, Walton NJ (2002) 4-Hydroxycinnamoyl-CoA hydratase/lyase, an enzyme of phenylpropanoid cleavage from *Pseudomonas*, causes formation of C(6)-C(1) acid and alcohol glucose conjugates when expressed in hairy roots of *Datura stramonium* L. Planta 215:79–89

Mohagheghzadeh A, Schmidt TJ, Alfermann AW (2002) Arylnaphthalene lignans from in vitro cultures of *Linum austriacum*. J Nat Prod 65:69–71

Moriguchi K, Maeda Y, Satou M, Hardayani NS, Kataoka M, Tanaka N, Yoshida K (2001) The complete nucleotide sequence of a plant root-inducing (Ri) plasmid indicates its chimeric structure and evolutionary relationship between tumor-inducing (Ti) and symbiotic (Sym) plasmids in Rhizobiaceae. J Mol Biol 307:771–784

Moyano E, Fornale S, Palazon J, Cusido RM, Bonfill M, Morales C, Pinol MT (1999) Effect of *Agrobacterium rhizogenes* T-DNA on alkaloid production in Solanaceae plants. Phytochemistry 52:1287–1292

Moyano E, Fornale S, Palazon J, Cusido RM, Bagni N, Pinol MT (2002) Alkaloid production in *Duboisia* hybrid hairy root cultures overexpressing the *pmt* gene. Phytochemistry 59:697–702

Nader BL, Cardoso Taketa AT, Iturriaga G, Pereda-Miranda R, Villarreal ML (2004) Genetic transformation of *Glaphimia glauca* by *Agrobacterium rhizogenes* and the production of norfriedelanes. Planta Med 70:1174–1179

Nagatome H, Tsutsumi M, Kino-Oka M, Taya M (2000) Development and characterization of a photoautotrophic cell line of pak-bung hairy roots. J Biosci Bioeng 89:151–156

Nakano K, Yoshida C, Furukawa W, Takaishi Y, Shishido K (1998) Terpenoids in transformed root culture of *Tripterygium wilfordii*. Phytochemistry 49:1821–1824

Nawrath C, Poirier Y, Somerville C (1994) Targeting of the polyhydroxybutyrate biosynthetic pathway to the plastids of *Arabidopsis thaliana* results in high levels of polymer accumulation. Proc Natl Acad Sci U S A 91:12760–12764

Nemoto K, Hara M, Suzuki M, Seki H, Oka A, Muranaka T, Mano Y (2009) Function of the *aux* and *rol* genes of the Ri plasmid in plant cell division in vitro. Plant Signal Behav 4:1145–1147

Newman DJ, Cragg GM (2007) Natural products as sources of new drugs over the last 25 years. J Nat Prod 70:461–477

Newman DJ, Cragg GM, Snader KM (2003) Natural products as a source of new drugs over the period 1981–2002. J Nat Prod 66:1002–1037

Nishikawa K, Furukawa H, Fujioka T, Fujii H, Mihashi K, Shimomura K, Ishimaru K (1999) Flavone production in transformed root cultures of *Scutellaria baicalensis* Georgi. Phytochemistry 52:885–890

Offringa IA, Melchers LS, Regensburg-Tuink AJG, Costantino P, Schilperoort RA, Hooykaas PJJ (1986) Complementation of *Agrobacterium tumefaciens* tumor-inducing *aux* mutants by genes from the TR-region of the Ri plasmid of *Agrobacterium rhizogenes*. Proc Natl Acad Sci USA 83:6935–6939

Ogasawara T, Chiba K, Tada M (1993) Production in high-yield of a naphthoquinone by a hairy root culture of *Sesamum indicum*. Phytochemistry 33:1095–1098

Oksman-Caldentey KM (2007) Tropane and nicotine alkaloid biosynthesis-novel approaches towards biotechnological production of plant-derived pharmaceuticals. Curr Pharm Biotechnol 8:203–210

Ono NN, Tian L (2011) The multiplicity of hairy root cultures: prolific possibilities. Plant Sci 180:439–446

Palazón J, Moyano E, Cusidó RM, Bonfill M, Oksman-Caldentey K-M, Piñol MT (2003) Alkaloid production in *Duboisia* hybrid hairy roots and plants overexpressing the *h6 h* gene. Plant Sci 165:1289–1295

Peebles CA, Hughes EH, Shanks JV, San KY (2009) Transcriptional response of the terpenoid indole alkaloid pathway to the overexpression of ORCA3 along with jasmonic acid elicitation of *Catharanthus roseus* hairy roots over time. Metab Eng 11:76–86

Peebles CA, Sander GW, Hughes EH, Peacock R, Shanks JV, San KY (2011) The expression of 1-deoxy-D-xylulose synthase and geraniol-10-hydroxylase or anthranilate synthase increases terpenoid indole alkaloid accumulation in *Catharanthus roseus* hairy roots. Metab Eng 13:234–240

Pistelli L, Giovannini A, Ruffoni B, Bertoli A (2010) Hairy root cultures for secondary metabolites production. Adv Exp Med Biol 698:167–184

Pomponi M, Spano L, Sabbadini MG, Costantino P (1983) Restriction endonuclease mapping of the root-inducing plasmid of *Agrobacterium rhizogenes*1855. Plasmid 10:119–129

Rahman L, Kouno H, Hashiguchi Y, Yamamoto H, Narbad A, Parr A, Walton N, Ikenaga T, Kitamura Y (2009) HCHL expression in hairy roots of *Beta vulgaris* yields a high accumulation of p-hydroxybenzoic acid (pHBA) glucose ester, and linkage of pHBA into cell walls. Bioresour Technol 100:4836–4842

Raskin I, Ribnicky DM, Komarnytsky S, Ilic N, Poulev A, Borisjuk N, Brinker A, Moreno DA, Ripoll C, Yakoby N, O'Neal JM, Cornwell T, Pastor I, Fridlender B (2002) Plants and human health in the twenty-first century. Trends Biotechnol 20:522–531

Rigden D, Carneiro M (1999) A structural model for the *rolA* protein and its interaction with DNA. Proteins 37:697–708

Riker AJ, Banfield WM, Wright WH, Keitt GW, Sagen HE (1930) Studies on infectious hairy root of nursery apple trees. J Agricult Res 41:507–540

Robbins MP, Bavage AD, Allison G, Davies T, Hauck B, Morris P (2005) A comparison of two strategies to modify the hydroxylation of condensed tannin polymers in *Lotus corniculatus* L. Phytochemistry 66:991–999

Roberts SC (2007) Production and engineering of terpenoids in plant cell culture. Nat Chem Biol 3:387–395

ROOTec Press Release (2010) Ginseng produced in Witterswil (Switzerland). http://www.rootec.com/news/20100721~Ginseng_from_Witterswil~Press_Release.pdf. Accessed 7 June 2011

Rothe G, Hachiya A, Yamada Y, Hashimoto T, Drager B (2003) Alkaloids in plants and root cultures of *Atropa belladonna* overexpressing putrescine N-methyltransferase. J Exp Bot 54:2065–2070

Runguphan W, Maresh JJ, O'Connor SE (2009) Silencing of tryptamine biosynthesis for production of nonnatural alkaloids in plant culture. Proc Natl Acad Sci U S A 106:13673–13678

Sauerwein M, Ishimaru K, Shimomura K (1991) A piperidone alkaloid from *Hyoscyamus albus* roots transformed with *Agrobacterium rhizogenes*. Phytochemistry 30:2977–2978

Schmidt BM, Ribnicky DM, Lipsky PE, Raskin I (2007) Revisiting the ancient concept of botanical therapeutics. Nat Chem Biol 3:360–366

Serino G, Clerot D, Brevet J, Costantino P, Cardarelli M (1994) *rol* genes of *Agrobacterium rhizogenes* cucumopine strain: sequence, effects and pattern of expression. Plant Mol Biol 26:415–422

Sevon N, Oksman-Caldentey KM (2002) *Agrobacterium rhizogenes*-mediated transformation: root cultures as a source of alkaloids. Planta Med 68:859–868

Shanks JV, Morgan J (1999) Plant 'hairy root' culture. Curr Opin Biotechnol 10:151–155

Sheludko YV (2008) *Agrobacterium*-mediated transient expression as an approach to production of recombinant proteins in plants. Recent Pat Biotechnol 2:198–208

Sheludko YV (2010) Recent advances in plant biotechnology and genetic engineering for production of secondary metabolites. Tsitol Genet 44:65–75

Sheludko Y, Gerasimenko I, Unger M, Kostenyuk I, Stoeckigt J (1999) Induction of alkaloid diversity in hybrid plant cell cultures. Plant Cell Rep 18:911–918

Sheludko Y, Gerasimenko I, Kolshorn H, Stockigt J (2002a) Isolation and structure elucidation of a new indole alkaloid from *Rauvolfia serpentina* hairy root culture: the first naturally occurring alkaloid of the raumacline group. Planta Med 68:435–439

Sheludko Y, Gerasimenko I, Kolshorn H, Stockigt J (2002b) New alkaloids of the sarpagine group from *Rauvolfia serpentina* hairy root culture. J Nat Prod 65:1006–1010

Shilpa K, Varun K, Lakshmi BS (2010) An alternate method of natural drug production: Elciting secondary metabolite production using plant cell culture. J Plant Sci 5:222–247

Shkryl YN, Veremeichik GN, Bulgakov VP, Tchernoded GK, Mischenko NP, Fedoreyev SA, Zhuravlev YN (2008) Individual and combined effects of the *rolA*, *B*, and *C* genes on anthraquinone production in *Rubia cordifolia* transformed calli. Biotechnol Bioeng 100:118–125

Siebertz R, Proksch P, Wray V, Witte L (1989) Accumulation and biosynthesis of benzofurans in root cultures of *Eupatorium cannabinum*. Phytochemistry 28:789–793

Sindarovska YR, Gerasymenko IM, Sheludko YV, Komarnytskyy IK, Bannikova MA, Kuchuk NV (2005) Transgenic plants regenerated from hairy roots of *Nicotiana benthamiana*: a promising host for transient expression of foreign proteins. Tsitol Genet 39:9–14

Sinkar VP, White FF, Gordon MP (1987) Molecular biology of Ri-plasmid : a review. J Biosci 11:47–57

Slightom JL, Durand-Tardif M, Jouanin L, Tepfer D (1986) Nucleotide sequence analysis of TL-DNA of *Agrobacterium rhizogenes* agropine type plasmid. Identification of open reading frames. J Biol Chem 261:108–121

Spano' L, Mariotti D, Cardarelli M, Branca C, Costantino P (1988) Morphogenesis and auxin sensitivity of transgenic tobacco with different complements of Ri T-DNA. Plant Physiol 87:479–483

Spena A, Schmulling T, Koncz C, Schell JS (1987) Independent and synergistic activity of *rol A*, *B* and *C* loci in stimulating abnormal growth in plants. EMBO J 6:3891–3899

Stojakowska A, Malarz J, Kisiel W (2002) Salicylate and methyl jasmonate differentially influence diacetylene accumulation pattern in transformed roots of feverfew. Plant Sci 163:1147–1152

Sun L, Ruppert M, Sheludko Y, Warzecha H, Zhao Y, Stockigt J (2008) Purification, cloning, functional expression and characterization of perakine reductase: the first example from the AKR enzyme family, extending the alkaloidal network of the *plant Rauvolfia*. Plant Mol Biol 67:455–467

Suzuki K, Tanaka K, Yamamoto S, Kiyokawa K, Moriguchi K, Yoshida K (2009) Ti and Ri Plasmids. In: Schwartz E (ed) Microbial Megaplasmids. Springer, Berlin

Syklowska-Baranek K, Pietrosiuk A, Kokoszka A, Furmanowa M (2009) Enhancement of taxane production in hairy root culture of *Taxus x media var* Hicksii. J Plant Physiol 166:1950–1954

Tepfer D (1984) Transformation of several species of higher plants by *Agrobacterium rhizogenes*: sexual transmission of the transformed genotype and phenotype. Cell 37:959–967

Trovato M, Maras B, Linhares F, Costantino P (2001) The plant oncogene *rolD* encodes a functional ornithine cyclodeaminase. Proc Natl Acad Sci U S A 98:13449–13453

Tzfira T, Citovsky V (2006) *Agrobacterium*-mediated genetic transformation of plants: biology and biotechnology. Curr Opin Biotechnol 17:147–154

Ulrich-Merzenich G, Panek D, Zeitler H, Vetter H, Wagner H (2010) Drug development from natural products: exploiting synergistic effects. Indian J Exp Biol 48:208–219

Wang K, Herrera-Estrella L, Van Montagu M, Zambryski P (1984) Right 25 bp terminus sequence of the nopaline T-DNA is essential for and determines direction of DNA transfer from *Agrobacterium* to the plant genome. Cell 38:455–462

Wang CT, Liu H, Gao XS, Zhang HX (2010) Overexpression of G10H and ORCA3 in the hairy roots of *Catharanthus roseus* improves catharanthine production. Plant Cell Rep 29:887–894

Weathers PJ, Towler MJ, Xu J (2010) Bench to batch: advances in plant cell culture for producing useful products. Appl Microbiol Biotechnol 85:1339–1351

White FF, Taylor BH, Huffman GA, Gordon MP, Nester EW (1985) Molecular and genetic analysis of the transferred DNA regions of the root-inducing plasmid of *Agrobacterium rhizogenes*. J Bacteriol 164:33–44

Williams DH, Stone MJ, Hauck PR, Rahman SK (1989) Why are secondary metabolites (natural products) biosynthesized? J Nat Prod 52:1189–1208

Woods RR, Geyer BC, Mor TS (2008) Hairy-root organ cultures for the production of human acetylcholinesterase. BMC Biotechnol 8:95

Xu ZQ, Jia JF (1996) The reduction of chromosome number and the loss of regeneration ability during subculture of hairy root cultures of *Onobrychis viciaefolia* transformed by *Agrobacterium rhizogenes* A-4. Plant Sci 120:107–112

Xu J, Ge X, Dolan MC (2011) Towards high-yield production of pharmaceutical proteins with plant cell suspension cultures. Biotechnol Adv 29:278–299

Yadav NS, Vanderleyden J, Bennett DR, Barnes WM, Chilton MD (1982) Short direct repeats flank the T-DNA on a nopaline Ti plasmid. Proc Natl Acad Sci U S A 79:6322–6326

Yamanaka M, Shimomura K, Sasaki K, Yoshihira K, Ishimaru K (1995) Glucosylation of phenolics by hairy root cultures of *Lobelia sessilifolia*. Phytochemistry 40:1149–1150

Young JM, Kuykendall LD, Martinez-Romero E, Kerr A, Sawada H (2001) A revision of *Rhizobium* Frank 1889, with an emended description of the genus, and the inclusion of all species of *Agrobacterium* Conn 1942 and *Allorhizobium* undicola de Lajudie et al. 1998 as new combinations: *Rhizobium radiobacter*, *R. rhizogenes*, *R. rubi*, *R. undicola* and *R. vitis*. Int J Syst Evol Microbiol 51:89–103

Yu B, Ruppert M, Stockigt J (2002) Deoxysarpagine hydroxylase—a novel enzyme closing a short side pathway of alkaloid biosynthesis in *Rauvolfia*. Bioorg Med Chem 10:2479–2483

Zárate R (1999) Tropane alkaloid production by *Agrobacterium rhizogenes* transformed hairy root cultures of *Atropa baetica* Willk. (Solanaceae). Plant Cell Rep 18:418–423

Zhai DD, Supaibulwatana K, Zhong JJ (2010) Inhibition of tumor cell proliferation and induction of apoptosis in human lung carcinoma 95-D cells by a new sesquiterpene from hairy root cultures of *Artemisia annua*. Phytomedicine 17:856–861

Zhang L, Ding R, Chai Y, Bonfill M, Moyano E, Oksman-Caldentey KM, Xu T, Pi Y, Wang Z, Zhang H, Kai G, Liao Z, Sun X, Tang K (2004) Engineering tropane biosynthetic pathway in *Hyoscyamus niger* hairy root cultures. Proc Natl Acad Sci U S A 101:6786–6791

Zhang L, Yang B, Lu B, Kai G, Wang Z, Xia Y, Ding R, Zhang H, Sun X, Chen W, Tang K (2007) Tropane alkaloids production in transgenic *Hyoscyamus niger* hairy root cultures over-expressing putrescine N-methyltransferase is methyl jasmonate-dependent. Planta 225:887–896

Zhou Y, Hirotani M, Rui H, Furuya T (1995) Two triglycosidic triterpene astragalosides from hairy root cultures of *Astragalus membranaceus*. Phytochemistry 38:1407–1410

Zhou Y, Hirotani M, Yoshikawa T, Furuya T (1997) Flavonoids and phenylethanoids from hairy root cultures of *Scutellaria baicalensis*. Phytochemistry 44:83–87

Chapter 11
Molecular Biology and Biotechnology of Quinolizidine Alkaloid Biosynthesis in Leguminosae Plants

Somnuk Bunsupa, Kazuki Saito and Mami Yamazaki

11.1 Introduction

Plants produce a vast variety of secondary metabolites which play a major role in the adaptation of plants to their environment (Bourguad et al. 2001). Among these secondary metabolites, alkaloids are found in approximately 20 % of plant species and many of the 12,000 known alkaloids produced by the plants exhibit potential biological activity and several are widely used as pharmaceuticals (Facchini 2001). Over the past decade, much progress has been made on the isolation of gene encoding enzymes involved in the biosynthesis of several alkaloids, including those of the isoquinoline, indole, tropane, and purine classes (Facchini 2001). However, the molecular mechanism underlying quinolizidine alkaloid (QA) biosynthesis is poorly understood. QA, a cadaverine-derived alkaloid, forms a major group of plant alkaloids that contains several structurally related compounds distributed in *Leguminosae* plants (Ohmiya et al. 1995). Some of the QAs exhibit beneficial pharmacological properties and some of QA-containing plants have been used as herbal traditional medicine (Tang and Eisenbrand 1992). From the ecological point of view, QAs are important for the survival of plants as a defense compound against pathogenic organisms or predators and allelopathic metabolites for competing with other plant species (Roberts and Wink 1998). The occurrences,

S. Bunsupa · K. Saito (✉) · M. Yamazaki
Graduate School of Pharmaceutical Sciences,
Department of Molecular Biology and Biotechnology,
Chiba University, Inohana 1-8-1, Chuo-ku, Chiba 260-8675, Japan
e-mail: ksaito@faculty.chiba-u.jp

S. Bunsupa
e-mail: Bunsupa_s@chiba-u.jp

M. Yamazaki
e-mail: mamiy@faculty.chiba-u.jp

S. Chandra et al. (eds.), *Biotechnology for Medicinal Plants*,
DOI: 10.1007/978-3-642-29974-2_11,

chemistry, biochemistry, and chemotaxonomy of QAs have been recently reviewed (Joseph 2008; Saito and Murakoshi 1995; Ohmiya et al. 1995; Wink 1992). This review outlines an overview of QAs and recent studies focusing on molecular biology and biotechnology of QA biosynthesis in *Leguminosae* plants.

11.1.1 Structure and Occurrence

Quinolizidine alkaloids (QA) alkaloids are a group of alkaloids possessing quinolizidine ring or piperidine ring in molecules (Saito and Murakoshi 1995). The main QAs can be divided into the common types of bicyclic alkaloids (lupinine-type), tricyclic alkaloids (cytisine-type), and tetracyclic alkaloids of the matrine and sparteine/lupanine types and into a number of unusual types of alkaloids, such as tsukushinamine-type (Ohmiya et al. 1995) (Fig. 11.1). More than 200 naturally occurring QAs are known that represent ~2 % of the known 12,000 known alkaloids in nature (Facchini 2001; Saito and Murakoshi 1995). QAs occur mainly in the family *Leguminosae*, especially of the genera *Lupinus*, *Baptisia*, *Thermopsis*, *Genista*, *Cytisus*, and *Sophora* (Wink et al. 1995). QAs are commonly referred to as 'lupin(e) alkaloids', since they occur throughout species of the genus *Lupinus*. Out of more than 500 *Lupinus* species that have been described, roughly 200–300 species are distributed in North and South America (Wink et al. 1995).

11.1.2 Biological and Ecological Importance

QAs are of importance to mankind as potential sources of medicine. For instance, some QA-containing plants, such as *Sophora flavescens*, have been used as sources of crude drugs in Chinese–Japanese traditional medicine (KAMPO), and QAs have been shown to be the principal constituents responsible for the pharmacological activities of this medicinal plant (Tang and Eisenbrand 1992). QAs show a broad range of pharmacological properties, such as cytotoxic, oxytocic, antipyretic, antibacterial, antiviral, and hypoglycemic activities, by in vivo pharmacological screening (Saito and Murakoshi 1995). Furthermore, in terms of chemical ecology, QAs are assumed to play important roles in the survival of plants that produce them as defense compounds against pathogenic organisms and predators (Wink 2003). QAs are bitter and toxic, which are undesirable for human or animal consumption, and the bitterness also reduces their palatability as animal feed (Wink 2003). Many of lupin species, such as *Lupinus albus*, *Lupinus Luteus*, *Lupinus mutabilis*, and *Lupinus angustifolius*, have been bred to produce the 'sweet' varities which accumulate less amounts of QA (Muzquiz et al. 1994; Wink 1990). The sweet varieties are useful for animal consumption but have the ecological disadvantage that they can only be cultivated successfully if predators are kept away by fences and pesticides (Muzquiz et al. 1994).

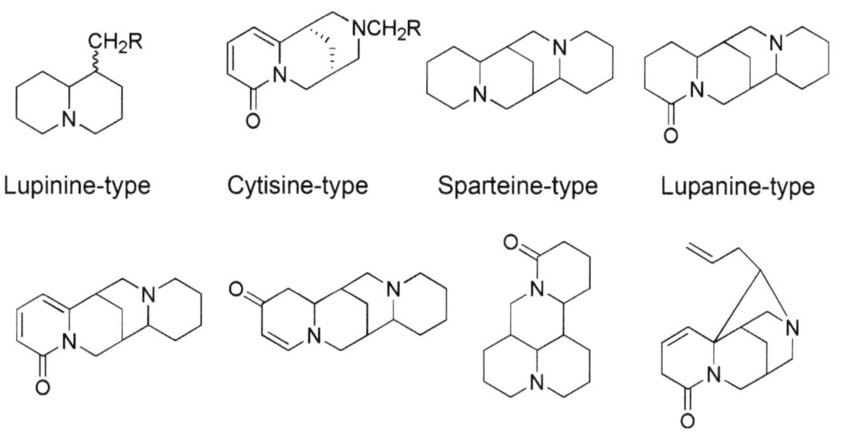

Fig. 11.1 Types of structure of quinolizidine alkaloids in leguminous plants

11.2 Biochemistry and Cell Biology

11.2.1 Biosynthesis

The first step committed in the QA biosynthesis is the decarboxylation of L-lysine to cadaverine by lysine decarboxylase (LDC). Then, three units of cadaverine are subjected to oxidative cyclization via a presumable enzyme-bound intermediate to form cyclic alkaloids, such as lupinine, sparteine, lupanine, and martrine (Fig. 11.2). These cyclic alkaloids are further modified by hydroxylation, esterification, and glycosylation to yield a variety of alkaloids (Ohmiya et al. 1995). The enzymatic studies on LDC reveal that LDC activity is found in lupin plant cell cultures and intact plant, and this enzyme appears to occur within the mesophyll chloroplasts of green leaves (Hartmann et al. 1980; Wink and Hartmann 1982). The activity of this enzyme and the alkaloid content are related to leaf regreening (Saito et al. 1989a, b).

QAs esters are mainly distributed in the genera *Lupinus*, *Cytisus*, *Pearsonia*, *Calpurnia*, and *Rothia* as the esters of tiglic acid, *p*-coumaric acid, acetic acid, and ferulic acid (Suzuki et al. 1996). In *Lupinus* plants, they accumulate two types of QAs including lupinine and lupanine type; however, the precise function in the plants of these ester alkaloids is not yet clarified. These alkaloids are assumed to be end products of biosynthesis and storage forms (Ohmiya et al. 1995). Two acyltransferases (ATs) involved in QA ester biosynthesis have been purified and characterized. First, tigloyl-CoA: (−)-13α-hydroxymultiflorine/(+)-13α-hydroxylupanine *O*-tigloyltransferase (HMT/HLT) catalyzes the transfer of a tigloyl group from tigloyl-CoA to the 13α-hydroxyl of (−)-13α-hydroxymultiforine or (−)-13α-hydroxylupanine (Fig. 11.2). The HMT/HLT activity was detected in *Lupinus hirsutus*, *Lupinus polyphyllus*, and *Lupinus termis* (Saito et al. 1993a, b; Suzuki et al. 1994). HMT/HLT is associated with the presumable mitochondrial fractions and localized in

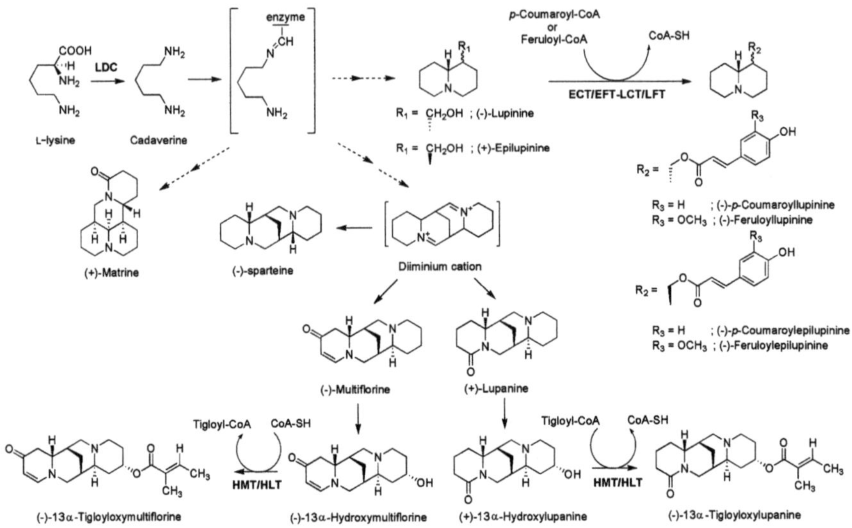

Fig. 11.2 Biosynthetic pathway of ester-type quinolizidine alkaloids in *Lupinus* plants. LDC, lysine decarboxylase; HMT/HLT, tigloyl-CoA: (−)-13α-hydroxymultiflorine/(+)-13α-hydroxylupanine *O*-tigloyltransferase; ECT/EFT-LCT/LFT, *p*-coumaroyl-CoA/feruloyl-CoA: (+)-epilupinine/(−)-lupinine *O*-coumaroy/feruloyltransferase

the matrix by the marker enzymes assay, where tigloyl-CoA is supposed to be synthesized from isoleucine (Suzuki et al. 1996). Second, *p*-coumaroyl-CoA/feruloyl-CoA: (+)-epilupinine/(−)-lupinine *O*-coumaroy/feruloyltransferase (ECT/EFT-LCT/LFT) catalyzes the transfer of *p*-coumaroyl-CoA or feruloyl-CoA to the hydroxyl moiety of (−)-lupinine or (+)-epilupinine (Fig. 11.2). The ECT/EFT activity was detected in *L. hirsutus* (Saito et al. 1993a, b) and *L. luteus* (Suzuki et al. 1994). ECT is predominantly distributed in cytosolic fraction, rather than in chloroplasts, mitochondria, and microsomes (Suzuki et al. 1996). However, these enzymatic activities show no differences in the cell-free extracts of bitter and sweet cultivar plants, which exhibit distinct alkaloid accumulation profiles (Suzuki et al. 1994). The ATs in *Lupinus* plants seem to define the pattern of alkaloids in each species (Ohmiya et al. 1995).

11.2.2 Identification of Quinilizidine Alkaloid Biosynthetic Genes

11.2.2.1 Molecular Cloning and Characterization of HMT/HLT from *L. albus*

The purified HMT/HLT protein was isolated by using several chromatographic techniques, digested with protease, and subjected to sequencing. This allowed design primers for amplification of cDNA fragments. Then the 5′- and 3′-rapid amplification of cDNA ends was used to obtain the full length cDNA (Okada et al.

2005). This HMT/HLT belongs to the BAHD superfamily, a plant acyl-CoA-dependent AT gene family. The BAHD family enzymes have been characterized as transferring a range of different acyl groups from their CoA esters to various plants metabolites, mostly secondary products (D'Auria 2006). The recombinant HMT/HLT enzyme catalyzes the acyl transfer reaction from tigloyl-CoA to both (−)-13α-hydroxymultiflorine and (+)-13α-hydroxylupanine. Benzoyl-CoA also serves efficiently as an acyl donor for these hydroxylated alkaloids. HMT/HLT is mainly expressed in roots and hypocotyls but not in cotyledons and leaves (Okada et al. 2005), which is in agreement with the represent about 84 % of the HMT/HLT enzymatic activity in these plant organs (Suzuki et al. 1994).

11.2.2.2 Identification of an Acyltransferase-Like Gene from *L. angustifolius*

The alkaloid-rich 'bitter' cultivar and the alkaloid-poor 'sweet' cultivar of *Lupinus* plants are useful experimental materials for understanding the regulatory mechanism of QA biosynthesis (Saito et al. 1993a). The bitter cultivar 'Fest' and sweet cultivar 'Uniharvest' of *L. angustifolius* have a similar genetic background, but the sweet cultivar is homozygous for the mutant *iuc* (*iucundus*) allele, which results in a low alkaloid phenotype (Oram 1983). These two cultivars are similar in the case of traits such as flowering time and pod shattering but differ for the color of flower and seed coat (Oram 1983). The investigation of the genetic variations between these two cultivars by using random amplified polymorphic DNA (RAPD) technique indicates that the bitter and sweet cultivars of *L. angustifolius* have only a little polymorphism with total DNA (Hirai et al. 2000). Further experiment by using a cDNA-amplified fragment length polymorphism (cDNA-AFLP) analysis has been conducted to isolate the genes specifically expressed in the bitter plant. However, no bitter-specific gene was isolated (Hirai et al. 2000).

Recently, we have performed differential gene expression profiling of the bitter and sweet cultivars of *L. angustifolius* by using PCR-select subtraction technique (Bunsupa et al. 2011). This method has been previously applied to the characterization of chemo varieties of *Perilla frutescens* to identify anthocyanin biosynthetic genes (Yamazaki et al. 2008). We have obtained 71 and 43 clones specific to the bitter and sweet cultivars, respectively. Among the bitter cultivar-specific clones, three genes exhibit homologies to ornithine decarboxylase, copper amine oxidase, and AT, representing potential candidates for structural genes encoding QA biosynthetic enzymes. In contrast, of the 43 sweet cultivar-specific clones, no clone represents a gene apparently related to QA biosynthesis. We further investigated on the gene that shows homology to AT, which is designed as *LaAT* for *Lupinus angustifolius acyltransferase*. The deduced amino acid sequences of LaAT show the strongest homology to the *Arabidopsis thaliana* BAHD acyltransferases involved in the formation of conjugated polyamines, including spermidine disynapoyl transferase (SDT) and spermidine dicoumaroyl transferase (SCT) (Luo et al. 2009). Based on its close homology to SDT and SCT, LaAT might be either involved in the formation of QAs esters or in the formation of *N*-acylated polyamine conjugates. The expression of *LaAT* may be regulated at a steady state of mRNA level, because it occurs in both cultivars but highly expresses in the bitter cultivar, whereas *LaAT* expression is undetectable in the sweet cultivars (Bunsupa et al. 2011). Although the enzymatic function of LaAT has not been confirmed, this work represents a valuable step toward a better understanding of the enzymes involved in QA biosynthetic pathway.

11.2.3 Accumulation, Transport, and Storage

The biosynthesis of QA occurs mainly in the green part of the plants, where the enzymes for lysine, cadaverine, and lupanine formations take place (Wink and Hartmann 1979, 1982; Wink and Mende 1987). The end products of QA biosynthesis are thought to accumulate particularly within the vacuoles of the epidermis of *Lupinus* plants, where the defensive roles can be most effectively exploited by plant (Wink and Roberts 1998). QAs are particularly accumulated in the reproductive organs. The highest alkaloid concentrations are usually reached in mature seeds, which can contain up to 5 % alkaloid per dry weight (Wink 1987a). Recent study also confirms that the majority of QAs are synthesized in shoot tissues and half of the QAs that accumulate in seeds are due to synthesis in situ and half to translocation principally by phloem (Lee et al. 2007). The amount of QAs in seed is decreased during seed germination and development and then will be increased as soon as the new leaves have been formed (Wink and Witte 1985).

11.3 Chemotaxonomy

The investigation on the distribution of the QAs in *Leguminosae* reveals that the genera containing matrine are the most primitive, followed by the genera accumulating sparteine- and cytisine-type alkaloids, but not matrine bases (Ohmiya et al. 1995). The most advanced genera are those containing lupinine-type alkaloids (Ohmiya et al. 1995). Recently, molecular phylogenies of *Leguminosae* have been reconstructed from nucleotide sequences of the *ribulose-biphosphate carboxylase* gene (*rbc*L) of over 450 legumes and employed as a framework to map and to interpret the distribution of QAs (Wink and Mohamed 2003). The distribution of QAs is restricted to the 'genistoid alliance s.1.' of the subfamily *Papilionoideae*, however, remarkable exception can be observed, in that QAs are absent (or present) in a given taxon, although all the neighboring and ancestral taxa express (or do not express, respectively), the particular traits. It is likely that the genetic capacity to synthesize QAs must be present in the very early members of the *Papilionoideae* (Wink 2003).

11.4 Genetic Mapping of *Lupinus* Plants

The first genetic linkage maps of *L. angustifolius* based on the technique microsatellite-anchored fragment length polymorphism markers has been reported (Boersma et al. 2005). Later, the gene-based map of *L. angustifolius* and its comparison to the partially sequenced genome of *Medicago truncatula* has been reported (Nelson et al. 2006). This enabled the mapping of five major genes controlling key domestication traits in *L. angustifolius*, including alkaloid production, water permeability, pod shattering, flowering time, and pigment production in seed (Nelson et al. 2006). Furthermore, a preliminary comparison between *L. albus* and *M. truncatula* (Phan et al. 2006) and the genetic linkage map of *L. albus* by using orthologous gene-based codominant and AFLP markers (Phan et al. 2007) have been recently reported, and the genetic mapping of three important agronomic traits: anthracnose resistance, flowering time, and alkaloid content allowed loci governing these traits have been identified.

The studies on genetic mapping and cross-comparison of genome models in each of agronomical and economically important traits in *Lupinus* plants serve the basis of resource for both lupin and other legume research community (Phan et al. 2007). Those genetic mapping would be valuable information for molecular breeding and evolutionary study of *Leguminous* plants.

11.5 Tissue Culture and Genetic Engineering

QAs accumulate only in the green callus and its concentrations in the callus are correlated to the amount of chlorophyll in the cells (Saito et al. 1989a, b). Cell suspension cultures of *Lupinus* and other legumes accumulate QAs about 100–1000 times lower than that of concentration in leaves of the respective differentiated plants (Wink and Hartmann 1980). Lupanine is the main compound in the cell suspension culture of *L. polyphyllus* and *Thermopsis lupinoides* (Saito et al. 1989a; Wink and Hartmann 1980), and matrine is produced in green callus culture and in multiple shoots of *Sophora flavescens* (Saito et al. 1989b). It is likely that the production of QAs is related to the tissue differentiation, in particular, in the formation of chloroplast, where the QAs are synthesized. In the cell suspension cultures the storage tissue is probably repressed; thus, the produced QAs cannot be stored in large quantities and are degraded rapidly either inside the cells or in the cell culture medium (Wink 1987a, b). Thus, in vitro tissues and cell cultures of lupine plants are not suitable systems for the study of the biosynthesis of QAs.

Since *Lupinus sp.* is the major grain legume in Australia, the production of lupin lines incorporating novel traits, such as herbicide resistant, resistant to certain pathogens, increased pod, and seed set, and improved seed quality by using genetic engineering, needs a routine transform system in the target species (Pigeaire et al. 1997). Many of the transformation systems for *Lupinus sp.* have been developed. The transformed roots of *Lupinus* plants have been reported from *L. albus*, *Lupinus polyohyllus*, and *Lupinus hartwegii* (Berlin et al. 1991; Mugnier 1988). However, the attempts to generate shoots from transformed roots have not succeeded. Nevertheless, limited success in shoot regeneration are reported in *L. angustifolius* (Pigeaire et al. 1997), *L. luteus* (Li et al. 2000), and *L. mutabilis*

(Babaoglu et al. 2000) by using *Agrobacterium*-mediated gene transfer to preorganized meristematic tissue combined with axillary regeneration. However, the transformation efficiency is poor and difficult in some *Lupinus* species. Recently, it has been reported that transformed roots of *L. mutabilis* synthesizes isoflavones, including genistein and 2′-hydroxygenistein, but not QAs (Babaoglu and Davey 2004). Genetic engineering of *L. angustifolius* by overexpressed *sunflower seed albumin* gene has succeeded to enhance methionine levels and increase nutritive value of seeds of transgenic lupins (Molvig et al. 1997).

11.6 Conclusion and Future Prospects

QAs have been discovered and studied over a century; however, little is known about the molecular mechanism underlying its biosynthesis despite the importance of QAs as a large class of secondary plant products. So far, only HMT/HLT has been isolated and characterized. During the last 5 years, the DNA sequencing technology is rapidly progressing and a number of plant genomes are now available (Delseny et al. 2010). The next generation high-throughput DNA sequencing techniques are opening the new genomic era in plant sciences (Ansorge 2009; Delseny et al. 2010). Comparative and integrative analysis of transcriptomic and metabolomic data will facilitate determination of gene function, as well as pinpoint the rate-limiting steps and factors in the biosynthetic pathways of secondary metabolites. Such information will assist in metabolic engineering of target secondary metabolites. We have recently been starting the deep transcriptome sequencing of *L. angustifolius* in both cultivars to obtain more information regarding the genes and transcription factors that are involved in the production of QAs. Further research on the QA biosynthesis should continue not only to target both enzymes and transcription factors involved in the regulation of pathway enzymes but also in the cellular and subcellular localization of enzymes, and the inter- and intracellular transport of pathway intermediates and products. These are all crucial factors to control the production of target metabolites in plant (Facchini et al. 2004).

References

Ansorge WJ (2009) Next-generation DNA sequencing techniques. Nat Biotechnol 25:195–203

Babaoglu M, Davey MR (2004) Transformed roots of *Lupinus mutabilis*: induction, culture and isoflavone biosynthesis. Plant Cell Tiss Org 78:29–36

Babaoglu M, McCabe MS, Power JB, Davey MR (2000) *Agrobacterium*-mediated transformation of *Lupinus mutabilis* L. using shoot apical explants. Acta Physiol Plant 22:111–119

Berlin J, Fecker LF, Rugenhagen C, Sator C, Strack D, Witte L, Wray V (1991) Isoflavone glycoside formation in transformed and non-transformed suspension and hairy cultures of *L. polyphyllus* and *L. hartwegii*. Z Naturforsch 46c:725–734

Boersma JG, Pallotta M, Li C, Buirchell BJ, Sivasithamparam K, Yang H (2005) Construction of a genetic linkage map using MFLP and identification of molecular markers linked to domestication genes in narrow-leafed lupin (*Lupinus angustifolius* L.). Cell Mol Biol Lett 10:331–344

Bourguad F, Gravot A, Milesi S, Gontier E (2001) Production of plant secondary metabolites: a historical perspective. Plant Sci 161:839–851

Bunsupa S, Okada T, Saito K, Yamazaki M (2011) An acyltransferase-like gene obtained by differential gene expression profiles of quinolizidine alkaloid-producing and nonproducing cultivars of *Lupinus angustifolius*. Plant Biotechnol 28:89–94

D'Auria JC (2006) Acyltransferases in plants: a good time to be BAHD. Curr Opin Plant Biol 9:331–340

Delseny M, Han B, Hsing YL (2010) High throughput DNA sequencing: the new sequencing revolution. Plant Sci 179:407–422

Facchini JP (2001) Alkaloid biosynthesis in plants: Biochemistry, cell biology, molecular regulation, and metabolic engineering application. Annu Rev Plant Physiol Plant Mol Biol 52:29–66

Facchini PJ, Bird DA, St-Pierre B (2004) Can *Arabidopsis* make complex alkaloids? Trends Plant Sci 9:116–122

Hartmann T, Schoofs G, Wink M (1980) A chloroplast-localized lysine decarboxylase of *Lupinus polyphyllus*: the first enzyme in the biosynthetic pathway of quinolizidine alkaloids. FEBS Lett 115(1):35–38

Hirai MY, Suzuki H, Yamazaki M, Saito K (2000) Biochemical and partial molecular characterization of bitter and sweet forms of *Lupinus angustifolius*, an experimental model for study of molecular regulation of quinolizidine alkaloid biosynthesis. Chem Pharm Bull (Tokyo) 48:1458–1461

Joseph PM (2008) Indolizidine and quinolizidine alkaloids. Nat Prod Rep 25:139–165

Lee MJ, Pate JS, Harris DJ, Atkins CA (2007) Synthesis, transport and accumulation of quinolizidine alkaloids in *Lupinus albus* L. and *L. angustifolius* L. J Exp Bot 58:935–946

Li H, Wylie SJ, Jones MGK (2000) Transgenic yellow lupin (*Lupinus luteus*). Plant Cell Rep 19:634–637

Luo J, Fuell C, Parr A, Hill L, Bailey P, Elliott K, Fairhurst SA, Martin C, Michael AJ (2009) A novel polyamine acyltransferase responsible for the accumulation of spermidine conjugates in *Arabidopsis* seed. Plant Cell 21:318–333

Molvig L, Tabe LM, Eggum BO, Moore AE, Craig S, Spencer D, Higgins TJ (1997) Enhanced methionine levels and increased nutritive value of seeds of transgenic lupins (*Lupinus angustifolius* L.) expressing a sunflower seed albumin gene. Proc Natl Acad Sci U S A 94:8393–8398

Mugnier J (1988) Establishment of new axenic hairy root lines by inoculation with *Agrobacterium rhizogenes*. Plant Cell Rep 7:9–12

Muzquiz M, Cuadrado C, Ayet G, de la Cuadra C, Burbano C, Osagie A (1994) Variation of alkaloid components of lupin seeds in 49 genotypes of *Lupinis albus* L. from different countries and locations. J Aric Food Chem 42:1447–1450

Nelson MN, Phan HT, Ellwood SR, Moolhuijzen PM, Hane J, Williams A, O'Lone CE, Fosu-Nyarko J, Scobie M, Cakir M, Jones MG, Bellgard M, Ksiazkiewicz M, Wolko B, Barker SJ, Oliver RP, Cowling WA (2006) The first gene-based map of *Lupinus angustifolius* L.-location of domestication genes and conserved synteny with *Medicago truncatula*. Theor Appl Genet 113:225–238

Ohmiya S, Saito K, Murakoshi I (1995) Lupine alkaloids. In: Cordell GA (ed) The Alkaloids. Academic Press, London

Okada T, Hirai MY, Suzuki H, Yamazaki M, Saito K (2005) Molecular characterization of a novel quinolizidine alkaloid O-tigloyltransferase: cDNA cloning, catalytic activity of recombinant protein and expression analysis in *Lupinus* plants. Plant Cell Physiol 46:233–244

Oram RN (1983) Selection for higher seed yield in the presence of the Deleterious low alkaloid allele *Iucundus* in *Lupinus angustifolius* L. Field Crops Res 7:169–180

Phan HTT, Ellwood SR, Ford R, Thomas S, Oliver R (2006) Differences in syntenic complexity between *Medicago truncatula* with *Lens culinaris* and *Lupinus albus*. Funct Plant Biol 33:775–782

Phan HT, Ellwood SR, Adhikari K, Nelson MN, Oliver RP (2007) Differences in syntenic complexity between *Medicago truncatula* with *Lens culinaris* and *Lupinus albus*. DNA Res 14:59–70

Pigeaire A, Abernethy D, Smith PM, Simpson K, Fletcher N, Lu CY, Atkins CA, Cornish E (1997) Transformation of a grain legume (*Lupinus angustifolius* L.) via *Agrobacterium tumefaciens*-mediated gene transfer to shoot apices. Mol Breed 3:341–349

Roberts MF, Wink M (1998) Chemical ecology of alkaloids. In: Roberts MF, Wink M (ed) Alkaloids biochemistry, ecology, and medicinal applications. Plenum, New York

Saito K, Murakoshi I (1995) Chemistry, biochemistry and chemotaxonomy of lupin alkaloids in the Leguminosae. In: Rahman AU (ed) Studies in natural products chemistry: structure and chemistry (Part C), vol 15. Elsevier Science Publishers, Amsterdam

Saito K, Koike Y, Suzuki H, Murakoshi I (1993a) Biogenic implication of lupin alkaloid biosynthesis in bitter and sweet forms of *Lupinus luteus* and *L. albus*. Phytochemistry 34:1041–1044

Saito K, Suzuki H, Takamatsu S, Murakoshi I (1993b) Acyltransferases for lupin alkaloids in *Lupinus hirsutus*. Phytochemistry 32:87–91

Saito K, Yamazaki M, Takamatsu S, Kawaguchi A, Murakoshi I (1989a) Greening induced production of (+)-lupanine in tissue culture of *Thermopsis lupinoides*. Phytochemistry 28:2341–2344

Saito K, Yamazaki M, Yamakawa K, Fujisawa S, Takamatsu S, Kawaguchi A, Murakoshi I (1989b) Lupin alkaloids in tissue culture of *Sophora flavescens* var. *angustifolius*: greening induced production of matrine. Chem Pharm Bull 37:3001–3004

Suzuki H, Koike Y, Murakoshi I, Saito K (1996) Subcellular localization of acyltransferases for quinolizidine alkaloid biosynthesis in *Lupinus*. Phytochemistry 42:1557–1562

Suzuki H, Murakoshi I, Saito K (1994) A novel *O*-tigloyltransferase for alkaloid biosynthesis in plants. Purification, characterization, and distribution in *Lupinus* plants. J Biol Chem 269:15853–15860

Tang W, Eisenbrand G (1992) Chinese drugs of plant origin, chemistry, pharmacology, and use in traditional and modern medicine. Springer, Berlin

Wink M (1987a) Quinolizidine alkaloids: biochemistry, metabolism, and function in plants and cell suspension cultures. Planta Med 53(6):509–514

Wink M (1987b) Why do lupin cell cultures fail to produce alkaloids in large quantities. Plant Cell Tiss Org 8:103–111

Wink M (1990) Physiology of secondary product formation in plants. In: Charlwood BV, Rhodes MJC (eds) Secondary products from plant tissue culture. Clarendon Press, Oxford

Wink M (1992) The role of quinolizidine alkaloids in plant insect interactions. In: Bernays EA (ed) Insect-plant interactions, Vol IV. CRC Press, Boca Raton

Wink M (2003) Evolution of secondary metabolites from an ecological and molecular phylogenetic perspective. Phytochemistry 64:3–19

Wink M, Hartmann T (1979) Cadaverine-pyruvate transamination: the principal step of enzymatic quinolizidine alkaloid biosynthesis in *Lupinus polyphyllus* cell suspension cultures. FEBS Lett 101:343–346

Wink M, Hartmann T (1980) Production of quinolizidine alkaloids by photomixotrophic cell suspension cultures: biochemical and biogenetic aspects. Planta Med 40:149–155

Wink M, Hartmann T (1982) Localization of the enzymes of quinolizidine alkaloid biosynthesis in leaf chloroplasts of *Lupinus polyphyllus*. Plant Physiol 70:74–77

Wink M, Mende P (1987) Uptake of lupanine by Alkaloid-storing epidermal cells of *Lupinus polyphyllus*. Planta Med 53:465–469

Wink M, Mohamed GIA (2003) Evolution of chemical defense traits in the Leguminosae: mapping of distribution patterns of secondary metabolites on a molecular phylogeny inferred from nucleotide sequences of the *rbc*L gene. Biochem Syst Ecol 31:897–917

Wink M, Roberts MF (1998) Compartmentation of alkaloid synthesis, transport, and storage. In: Roberts MF, Wink M (eds) Alkaloids biochemistry, ecology, and medicinal applications. Plenum, New York

Wink M, Witte L (1985) Quinolizidine alkaloids as nitrogen source for lupin seedlings and cell suspension cultures. Z Naturforsch 40c:767–775

Wink M, Meißner C, Witte L (1995) Patterns of quinolizidine alkaloids in 56 species of the genus *Lupinus*. Phytochemistry 38:139–153

Yamazaki M, Shibata M, Nishiyama Y, Springob K, Kitayama M, Shimada N, Aoki T, Ayabe S, Saito K (2008) Differential gene expression profiles of red and green forms of *Perilla frutescens* leading to comprehensive identification of anthocyanin biosynthetic genes. FEBS J 275:3494–3502

Chapter 12
Metabolomics in Medicinal Plant Research

Kandan Aravindaram and Ning-Sun Yang

12.1 Introduction

Plants provide the source material for over half the drugs currently prescribed as antidiabetics, anticancer, antirheumatism, and other therapeutics. Definitions and uses of herbal medicine vary from country to country: in China and Taiwan there is "traditional Chinese medicine" (TCM), in India Ayurvedic Medicine (AM); some Western countries have also been actively engaged in medicinal plant research in recent decades. TCM has a relatively long history as a systematic and holistic approach to health care: aiming to harmonize the body, mind, and spirit. A complex and integrated medical science, TCM reflects strong traditional Chinese cultural and philosophical principles, embodying rich dialectical thinking and approaches, treating the human body as a finely engineered system to be observed, monitored, and adjusted into a healthy status (Zhang et al. 2010). TCM confers advantages in early intervention, combinational therapies, personalized medicines, among other health care measures. By taking advantage of the health-promoting and maintenance components as well as the valuable clinical experiences and trials of TCM, future efficacy mechanism studies are expected to shed new light on TCM research and development, bringing it into the frame of modern medicines to contribute to health care globally.

K. Aravindaram
Division of Plant Quarantine, National Bureau of Plant Genetic Resources (NBPGR), Pusa Campus, New Delhi 110 012, India
e-mail: genekannz@gmail.com

N.-S. Yang (✉)
Agricultural Biotechnology Research Center, Academia Sinica, 128 Academia Road, Section 2, Nankang, Taipei 115, Taiwan
e-mail: nsyang@gate.sinica.edu.tw

S. Chandra et al. (eds.), *Biotechnology for Medicinal Plants*,
DOI: 10.1007/978-3-642-29974-2_12,

A major obstacle for R&D of Chinese and other traditional medicines has been the lack of appropriate experimental approaches for evaluating the efficacy and toxicity at the cellular and molecular levels. The methodology used in past TCM research followed the path of reductionist and partitioning/fractionation analyses, which are often unable to capture the practicality of characteristics of TCM prescriptions: a holistic and dynamic approach to treating diseases and the interactions among multiple biological components. Much of TCM is still practiced in its original form. Although it may be effective in treating certain conditions, especially chronic ones, TCM typically lacks well-defined molecular mechanisms and even, sometimes, fundamental descriptions at the physiological and tissue levels. Typically, the medicines prescribed in TCM are mixtures of several plant products; their composition is often not well-defined. A key factor in developing TCM toward its future application to health care will be to establish systemic analyses of the human body and its maladies given a holistic consideration. Many diseases, including Alzheimer's, cancer, cardiovascular disease, diabetes, and depression result not from a single defective gene or protein/enzyme but from multiple molecular abnormalities. In many cases, to single out an individual molecular target is likely to be inconclusive. Cells often compensate for a single factor/protein whose activity is affected by a drug: a phenomenon known as redundancy. Hence, using a so-called "magic shotgun" to attack multiple targets in complex systems could reap greater therapeutic rewards than fully blocking one target. The idea of the "magic bullet" continues to be a good one, but in practice it may not offer the best approach for treating complex diseases (Lee et al. 2005). TCM can be inspirational when creating a drug discovery strategy using numerous components to systematically control a complex network system (Kitano 2007). Moreover, in order to reveal working mechanisms and further explore metabolic profile features responsible for bioactivity exhibited, we believe that techniques such as metabolomics will be essential to the modernization of TCM or any herbal medicinal practice.

The field of metabolomics is emerging from metabolite profiling, as a targeted subset of related compounds and/or as a mapping of all extractable metabolites. The approach is expected to have diverse applications in toxicology, disease diagnosis, drug discovery and development, and herbal medicine research (Eliss et al. 2007; Claudino et al. 2007; Ulrich-Merzenich et al. 2007; Griffin and Kauppinen 2007; Shyur and Yang 2008). The word "metabolomics" was coined in analogy with transcriptomics and proteomics; as is the case for transcriptome and the proteome. The definition of metabolome may also be dynamic, changing significantly, depending upon factors of time and tissue-type. Although the metabolome can be defined conceptually, it is currently not technically possible to reveal, observe, and analyze it by a single method, e.g., via a 2D exhibition mode showing the entire spectrum of a metabolome from the crude extract of a test tissue or organ (Shyur and Yang 2008). Scientists have recently completed the first draft of a human metabolome. As recently reported, approximately 2,500 metabolites, 1,200 drugs, and 3,500 food components are detectable in the human body (Kell 2004; Eliss and Goodacre 2006; Wishart et al. 2007). This information is now available from the Human Metabolome Database (www.hmdb.ca). Some metabolomics studies have adopted a "top-down" strategy to reflect the function and metabolic changes of organisms caused by interventions in a holistic context. This approach coincides with the holistic thinking of TCM and may have an impact on our evaluation of TCM theory. Hopefully, metabolomics will lead to a global, comprehensive understanding of the metabolite complements of integrated living systems and/or dynamic responses to the changes of both endogenous and exogenous factors. Metabolomics is positioned to open up the possibility of studying the effect of various herbal medicines, in complex biological systems, bridging it with Western medicines (Wang et al. 2005). We are hopeful, particularly, that advancement in high throughput, improved (e.g., hyphenated) technologies and clustered or targeted analyses of metabolomics will provide new strategies for analysis of bio-active components in formulations of various phytomedicines in vivo. In this chapter, we will discuss applications of specific metabolomics techniques for research into herbal medicine.

12.2 Metabolomics Techniques Used or Applicable for Herbal Medicine Research

In metabolomics technology we perform quantitative and qualitative measuring and/or profiling of metabolite components of a specified, integrated living system: its dynamic responses to the changes in both endogenous (i.e., physiological and developmental) and exogenous (i.e., xenobiotic and environmental) factors (Tang and Wang 2006; Shyur and Yang 2008). Currently, two complementary approaches are popularly used for assessment and employment in metabolomic investigations: metabolic profiling and metabolic fingerprinting. For the latter, the intent is not to identify each metabolite, but to compare patterns or "fingerprints" of metabolites that change in response to test disease, toxin exposure, environmental or genetic alterations. In metabolic profiling, on the other hand, a group of metabolites undergoes analysis, as related to either a specific metabolic pathway or to a class of compounds. In metabolomics, unlike genomics and proteomics, there is no single analytical technique for profiling all low molecular weight metabolites of a test cell (Dunn and Ellis 2005; Dunn 2008; Shyur and Yang 2008). Among the various techniques for metabolome analysis, nuclear magnetic resonance (NMR) and mass spectrometry (MS) coupled with liquid- or gas chromatographic separation are the most popular techniques for simultaneous analysis of a large number of metabolites. Although the technology is remarkably sophisticated, there are disadvantages to using metabolomics. Due to the great diversity among chemical structures and the degree of variability in abundance of most biological systems in vitro or in vivo, there remains extensive impracticability in analyzing the entire metabolome at once. Therefore, a number of complementary approaches must be established in assay sequence for extraction, detection, quantification, and identification of as many metabolites as possible (Roessner and Beckles 2009; Villas-Bôas et al. 2007).

Among these approaches, analytical techniques commonly used in TCM studies include: liquid chromatography (LC) for quality control, high throughput fingerprint, qualitative and quantitative determination of active components, NMR for biological fingerprint and elaboration of complicated theories, gas chromatography (GC) for establishment of experimental animal models, capillary electrophoresis (CE) for pharmacokinetic study, and MS (often combined with other methods, such as GC, LC).

12.2.1 High-Performance Liquid Chromatography

Metabolite analysis of phytoextract mixtures requires complex and multifaceted strategies for example, adequate tissue sampling, homogenization, optimal sample extraction, metabolite separation, detection and identification, single phytochemicals or groups of compounds, automation of data gathering, handling, analysis and quantification (Hall et al. 2002). Of the various techniques, High-Performance Liquid Chromatography (HPLC) is relatively inexpensive, easy to use, and highly sensitive. For the most part, it is not limited by sample volatility or/ and stability. It also shows favorable separating power. It has thus been widely used in phytochemistry, herbal extract analyses, and TCM research.

Recent advances and improvements in HPLC technology significantly facilitate fingerprint analysis of TCM and other spheres of phytomedicine-related research, including ultra high-performance liquid chromatography (UPLC) for identification of active or index compounds (Liu et al. 2007), monolithic columns for biological fingerprinting and analysis of TCM (Wang et al. 2007), capillary HPLC (CHPLC) for qualitative and quantitative determination of active components in TCM Zhao et al. (2008a), and hydrophilic interaction chromatography for comprehensive quality control of TCM (Jin et al. 2009). In addition, there are numerous effective multidimensional liquid separation systems, such as multidimensional countercurrent chromatography system used in studies of *Ginkgo biloba* L. and *Hippophae rhamnoides* L (Yang et al. 1998), comprehensive two-dimensional (2D) capillary LC and CE for resolution of neutral components in Cheng-Qi-Tang (Zhang et al. 2001), and the two-dimensional LC for analysis of in *Rheum palmatum* L. (Hu et al. 2006).

12.2.2 Gas Chromatography

Gas Chromatography (GC) systems have been effectively employed for investigating volatile compounds and the related biological samples in test humans or animals: e.g., urine, blood serum, blood plasma, etc. GC analysis sample pretreatment is a crucial step. Studies have reported various novel sample pretreatment techniques which confer significant advantage over conventional methods (Dong et al. 2007; Tam et al. 2007). Presently, structural information and selectivity available from MS makes the combination of MS and GC, coupled with chemometric analytical method, a highly effective technique for analysis of TCM (Liang et al. 2005).

These advantages apparently cannot be obtained by LC- (Boernsen et al. 2005) and NMR-based methods (Pasikanti et al. 2008), due to the poor overall reproducibility and low sensitivity, respectively. In addition, the two-dimensional GC (GC $\times$ GC) system showed some advantages over other GC-based analytical techniques, likely due to its superior resolution and very sensitive detection capacity (Adahchour et al. 2006).

12.2.3 Nuclear Magnetic Resonance

NMR is a spectroscopic technique that exploits the magnetic properties of the atomic nucleus. In NMR, a sample is immersed in a strong external magnetic field where transitions between the nuclear magnetic energy levels are induced by a suitably oriented radiofrequency field. In theory, any molecule containing one atom with a nonzero nuclear spin may be "visible" via NMR. It is well established that the NMR technique, coupled with pattern recognition (PR) methods, provides valuable information about both the endogenous and exogenous factors induced by biochemical perturbations via analysis of the biofluids and tissues (Wu et al. 2005) of test samples. High resolution ^{1}H NMR spectroscopy has repeatedly been shown to be uniquely suited to detect a large range of endogenous low-molecular weight metabolites in an organism. Moreover, this rapid, noninvasive technique yields a rich array of structural and quantitative information. It also allows metabolites to be analyzed simultaneously (Nicholson and Wilson 1989). Multivariate statistical analyses, such as principle components analysis (PCA) and partial least squares (PLS) can simplify the NMR data sets and aid in toxicity site identification while assessing processes of toxic lesions and characterizing novel biomarker combinations (Xu 2004). In comparison with MS and HPLC, NMR-based method has evident advantages (Lenz and Wilson 2007), as follows: it (a) is noninvasive and non destructive to samples, facilitating in vivo and in situ studies; (b) is a quantitative and simultaneous detection method, thus unbiased for virtually all molecular species in test samples; (c) has high throughput and thus can measure up to 400 samples per 24 h with flow-injection technology; (d) generates rich dynamic molecular information including metabolite structure, concentration, molecular dynamics, interactions, pH and compartmentation when diffusion editing techniques are employed; (e) usually requires little or no sample preparation; and (f) yields data with good resolution and reproducibility.

^{1}H NMR-based metabolomics has been used to study the biochemical effects of epicatechin in rats' urine samples and to determine the metabolic differences between acute phytochemical intake, chamomile, green and black tea consumption in human subjects (Walsh et al. 2007). ^{1}H NMR analysis has the advantages of being non-destructive, applicable to intact biomaterial samples while it produces a high yield of valuable structural information. It is through this quantitative technique that assessments and reports can be made on numerous compounds in a single measurement. A combination of NMR and LC–DAD–MS analyses has revealed comprehensive metabolomic variations and distinctions among three phenotypic cultivars of plant extracts from *Salvia miltiorrhiza* Bunge (Dai et al. 2010). NMR-based metabolomics is an attractive method for nonselective and comprehensive analysis of *Ginkgo* extracts, which are complex phytochemical mixtures prepared from raw leaf extracts by a series of extraction and pre-purification steps (Agnolet et al. 2010). Liu et al. (2010) reported that urinary metabolic perturbations associated with liver toxicity induced by Huang-yao-zi (root of *Dioscorea bulifera* L.) were revealed using NMR, as determined by the correlations between metabolomic profiling and the histopathologic/biochemical properties detected. The same study revealed biomarkers for liver toxicity, indicating metabolic changes observed in urine samples in response to the Huang-yao-zi treatment in test subjects. In addition, the mechanism associated with oxidative injury of hepatic mitochondria was investigated (Liu et al. 2010).

Siwutang, a TCM prescription consisting of four plant species, namely *Paeonia lactiflora*, *Angelica sinensis*, *Rehmannia glutinosa*, and *Rhizoma Chuanx-iong*, was used to treat cyclophosphamide-induced "blood deficiency" in mice. The damaging effect on mitochondria, disorder of energy metabolism and osmoregulation was evaluated in this model by an NMR-based metabolomics approach. Treatment with Siwutang was shown to improve these deficiencies/disorders in test mice (Wang et al. 2010). Coronary heart disease (CHD) with Xin-blood stasis syndrome (XBSS) is related to activities of the glyco-metabolism and lipid metabolism. It is also associated with stress induced by hypoxia and agonia. Luo et al. (2007) determined that a change of plasma metabolic phenotype in rats corresponds with chronic restraint stress (rats with *qi* stagnation of the liver and spleen deficiency syndrome). NMR spectra of metabolites in the rats' blood were found to be differentially changed after chronically stressful conditions. Specific and characteristic metabolic products can be identified by metabolomics analyses, leading to an interpretation of biological character of Chinese medicine syndromes. Discovery of such biomarkers from the metabolome of specific Chinese medicine syndromes will facilitate modernizing TCM study and systems, promote the quantitative and mechanistic elucidation of TCM activities, help predict disease onset, and provide an enhanced understanding of the pharmacological activities of various TCMs. The NMR-based metabolomic analysis is an approach with sound prospects for the advancement of studies of Chinese medicine syndromes (Luo et al. 2007).

12.2.4 Mass Spectrometry

Mass Spectrometry (MS) is playing an increasingly important role in advancing metabolomics research into the mechanisms of TCM. MS is frequently coupled with other separative or analytical techniques; therefore, in most cases, it is an analysis technique rather than a separative one. Since no existing technique including MS is capable of complete or 2D exhibition detection and analysis, MS methods are briefly critiqued as follows: Tang and Wang (2006) have cited the following disadvantages: (i) extensive sample preparations; (ii) invasive assays destructive to samples; (iii) pre-knowledge about samples required; (iv) recurring exorbitant expenditure; and (v) decidedly expensive equipment. On the other hand, merits to MS are listed as follows: (i) although requiring laborious sample pre-treatment (hydrolyzation, derivatization), when hyphenated with GC, MS-based methods provide reliable identification and quantification of volatile and thermally stable components (Ceglarek et al. 2009); (ii) problems can be resolved by online link with LC (Dunn et al. 2005); (iii) system is available for ionization of polar to non-polar components (Baumann et al. 2005); and (iv) offers sensitive and selective detection, multi-analyte analysis, and ability to generate structural information (Ackermann et al. 2006).

As a detection mode, MS is sensitive, relatively non-selective, and capable of identifying compounds with reference to databases of mass spectra obtained from authentic and internal standards. These properties make MS the method of choice for metabolite detection in complex. This technique is particularly powerful when allied to prior chromatographic resolution of such compound mixtures. Under optimized conditions, GC coupled with MS (GC–MS) and LC–MS can detect several hundred component chemicals including sugars, sugar alcohols, organic acids, amino acids, amino acids, fatty acids, and a wide range of diverse secondary metabolites. In the process of identifying volatile components of several essential oils derived from the *Clematis* species (Zeng et al. 2007), microwave-assisted extraction coupled with gas chromatography–mass spectrometry (MAE-GC–MS) was used to successfully reveal the chromatographic fingerprint as well as the temperature-programmed retention indices (PTRIs) of *Fructus xanthii* (Ruan and Li 2007).

It was also useful in pressurized hot water extraction followed by solid-phase microextraction and gas chromatography–mass spectrometry (PHWE-SPME–GC–MS) in three cases, namely: quality control and quantitative determination of volatile active compounds in TCM (Dong et al. 2007), HPLC-GS-MS fingerprint analysis (Li et al. 2007b), and electrospray ionization time-of-flight mass spectrometry (ESITOF-MS) with high-performance liquid chromatography-diode array evaporative light scattering detector (HPLC–DAD-ELSD) for screening of TCM interactive components (Qi et al. 2006).

12.2.4.1 Gas Chromatography–Mass Spectrometry

Gas chromatography–mass spectrometry (GC–MS), a relatively low-cost alternative among technologies, was one of the major analytical tools employed in early developments of metabolomic studies. It provides remarkably fine chromatographic resolution and high separation efficiencies. Samples must be both volatile and thermostable. Intrinsically non-volatile compounds can be volatized by chemical derivatization, but at a cost of additional time and variability.

Volatile compounds can be subjected directly to GC–MS analysis, while metabolites of molecular mass less than 300–400 Da can be analyzed only after appropriate chemical derivatization. These processes allow detection of various chemical classes including amino and organic acids, fatty acids, certain lipids, sugars, sugar alcohols and phosphates, amines, amides, and thiol containing metabolites. GC–MS can be performed with affordable single quadrupole mass analyzers applicable to complex mixtures. Using an automated mass spectral deconvolution and identification system improves effective identification of overlapping chromatographic peaks (Halket et al. 1999).

One key advantage of GC–MS is that it combines the high chromatographic separation power with a universal detection system, resulting in excellent sensitivity and selectivity. Gas chromatography time-of-flight mass spectrometry (GC-TOF–MS) offers an attractive alternative to quadrupolar systems, providing greater m/z accuracy. Specifically, this technique increases the number of metabolites detectable by GC–MS in TCM crude extracts, in some cases from 500 to 1,000 (Fiehn 2002). GC-EI–MS (gas chromatography-electron ionization–mass spectrometry) can be employed to identify or characterize a metabolite using commercially available mass spectral/retention index libraries. This represents a major advantage since searchable libraries of spectra are severely limited in LC–MS applications. In GC–MS as an alternative, chemical ionization (CI) is applicable, producing a 'soft' ionization and yielding mainly molecular ion species and minimal fragmentation.

Two typical, solid examples of GC–MS in TCM as utilized in research are described as follows: Artemisinin has been proven to be an effective antimalarial compound, especially for chloroquine-resistant and cerebral malaria. In order to get new clues about artemisinin biosynthesis, metabolic profiling by GC–MS was applied to compare the secondary metabolites of two *Artemisia annua* L. genotypes. Obvious differences were shown in terpenoids and artemisinin metabolism between different growth periods or plant development stages and the specific genotypes (Wang et al. 2009). Using the GC–MS , essential oils present in three Curcuma species (*C. phaeocaulis, C. kwangsiensis,* and *C. wenyujin*) were identified and compared with respect to their relatively quantitative characteristics (PCA and PLS-DA). It was concluded that a combination of multivariate statistical technique and the GC–MS data mining provided a powerful tool for metabolomics study, specifically in the discrimination and quality control of TCM (Xiang et al. 2011).

12.2.4.2 Liquid Chromatography–MS

Liquid Chromatography (LC) LC–MS is a powerful option offering high selectivity and good sensitivity. Unlike GC–MS a derivatization step is not required. Fragmentation conditions are more gentle, (ionization by electrospray rather than electron impact), allowing clear detection of molecular ions. Main targeting can be achieved by the extraction method or the type of column/solvent used. For metabolomics, the technique can be applied to a significant range of metabolites extending to larger MW compounds up to 1,000 Da, thus presenting various physicochemical properties. A predominant challenge in LC–MS involves high liquid flow rates from HPLC, which are not compatible with the high vacuum required for the mass spectrometer. The overwhelming popularity of LC–MS, however, is largely due to atmospheric pressure ionization (API) interfaces, including electrospray ionization (ESI) and atmospheric pressure chemical ionization (APCI). A classic example of LC–MS applied to TCM research is as follows: An LC–MS-based metabolomics approach was employed to characterize the anti-aging effect of total flavones of *Epimedium*, a TCM. Results indicated that the anti-aging effects of total flavones of *Epimedium* were possibly due to the intervention on lipid metabolism and its property of anti-oxidation. Notably, *Epimedium* flavanoids could reverse this age-dependent change at different levels synchronously (Yan et al. 2009; Wu et al. 2008). This work suggests that the metabolomic approach is potentially a powerful tool for exploring the therapeutic basis and clarifying the possible action mechanism of TCM. As an example: a combined GC/MS and LC/MS metabolic profiling strategy indicated that a high dosage of *Tripterygium wilfordii* caused a time-dependent toxic effect as revealed by the perturbed metabolic regulatory network involving disorders in energy metabolism, elevated amino acid, and choline metabolism pathways, as well as altered structure of gut flora (Chen et al. 2008).

Another technique, namely ultra-performance liquid chromatography–mass spectrometry UPLC–MS, is frequently employed in an effort to obtain the maximum biochemical profile information subset. It is a sensitive tool that can be used to characterize, identify, and quantify numerous compounds in a biological sample where metabolite concentrations cover a broad range of information about disease pathophysiology (Field et al. 2009; Fiehn 2008). UPLC–MS was used to evaluate the efficacy and to study the action mechanism of Xindi soft capsules, consisting of sea buckthorn flavanoids, (e.g., quercetin, kaempferol, and isorhamnetin): a TCM preparation against blood stasis (Zhao et al. 2008b). Compound Danshen tablets, a herbal compound preparation (*Salvia miltiorrhiza* Bge.), presented protective effects on myocardial ischemia by reversing potential biomarkers to sham levels, especially for the four metabolites in the pathway of purine metabolism (hypoxanthine, xanthine, inosine, and allantoin) (Lv et al. 2010). Metabolite profiling of five medicinal Panax herbs, (*Panax ginseng* (Chinese ginseng), *Panax notoginseng* (Sanchi), *Panax japonicus* (Rhizoma Panacis Majoris), *Panax quinquefolium* L. (American ginseng), and *P. ginseng* (Korean ginseng), was performed using UPLC—MS and a multivariate statistical analysis technique (Xie et al. 2008). Aristolochic acids, naturally present in *Aristolochia* plant species that have been used in Chinese herbal and mineral medicine containing a mixture of varying herb species, were identified by UPLC–MS-based methodology (Jacob et al. 2007).

12.2.4.3 Capillary Electrophoresis–MS

Capillary electrophoresis (CE) is a powerful technique for separating either polar or charged metabolites, offering high-analyte resolution. The separation mechanism fundamentally differs from LC, therefore, CE provides additional or complementary information about the composition of a biological sample. CE separations can be achieved remarkably efficiently without extensive sample pretreatment. When in-capillary pre-concentration techniques are combined with MS detection, rapid, efficient, and remarkably sensitive analysis is possible. Therefore, CE–MS has been recognized as an attractive complementary technique for metabolomic studies (Ramautar et al. 2009).

12.2.4.4 Direct Injection MS

Direct injection analysis (DIMS) describes the injection or infusion of a sample into the ionization source of the mass spectrometer without prior separation. This technique is commonly used with API techniques, predominantly ESI. Direct injection is a high-throughput approach, allowing a sample to be processed within a few minutes. The brief period of time increases intersample reproducibility and improves the accuracy of subsequent cluster analysis (Dettmer et al. 2007).

A syringe pump or autosamplers may be used to introduce a sample as a plug into a flowing stream supplied by flow injection or LC pumping systems. A range of powerful mass spectrometers, which generally operate with high mass resolution and mass accuracy, provide information by the detection and preliminary identification of a numerous metabolites. High mass accuracy generates molecular formula information thus increasing the probability of identifying metabolites (Dunn 2008).

12.3 Application of Metabolomics in TCM Against Cancer and Its Toxicity Research

To release natural product research from its stalemate, we need a new strategy for detecting active compounds. Every year the decreasing number of improved drugs inspires renewed interest in developing novel approaches. Metabolomics has been applied in many fields, including the quality control of medicinal plants (Roos et al. 2004; Kim et al. 2005; Yang et al. 2006), the monitoring of the biochemical response of plants to diverse stress conditions (Choi et al. 2004b; Liang et al. 2006; Leiss et al. 2009), and the search for differences between genetically modified plants and controls (Choi et al. 2004a; Le Gall et al. 2005), among others. Measuring metabolite changes resulting from the response of cells upon treatment (with natural products or pure compounds) may help us not only to monitor the response of the cells but also eventually to detect activities of the treated materials themselves. To date, however, the metabolomic approach has been focused more on finding system phenomena while the detection of lead compounds is still in a premature state. Metabolomic analysis has been applied in many different areas of cancer research including the detection of biomarkers for diagnosis (Brindle et al. 2002), the monitoring of drug response (Lindskog et al. 2004), and treatments or prediction of their toxicity (Robertson 2005). NMR-based metabolomics is considered to be a particularly fast, noninvasive method for identifying new biomarkers for clinical diagnosis (Lindon et al. 2004).

Metabolic activity of several pathways was increased in breast cancer epithelial cell lines, leading to the upregulation of fatty acid synthesis among other changes. Similar observations in liver and prostate cancer cells reveal a distinct metabolic profile characterized by high tCho and phosphocholine levels, along with an increase in the glycolytic products lactate and alanine (Spratlin et al. 2009; Swanson et al. 2006; Yang et al. 2004). Lindskog and associates reported applying ^{1}H-NMR to the prediction of resistance/response to chemotherapy by a neuroblastoma. Several cytotoxic drugs, namely cisplatin, etoposide, and irinotecan, were used to treat cell lines with sensitivity to these. The cells were subsequently examined using ^{1}H-NMR (Morse et al. 2007). Several anticancer agents were tested on the breast cancer cell lines showing different hormonal response and metastatic potential. The ^{31}P NMR spectra of perfuse cells revealed a correlation between the mode of action of anticancer drugs and the observed changes in cell metabolic profiles (Sterin et al. 2001).

Following, a brief review of two metabolomics techniques/approaches which have been applied to lead-finding in other fields of natural products: Roos and his co-workers (2004) attempted to correlate metabolomes of St. John's wort extract with opioid receptor binding activity. In this study, they showed that NMR combined with multivariate data analysis could effectively predict the activity of preparations, based on a database they constructed. This was suitable for the quality control of plant preparations since it was correlated with its efficacy, but they did not attempt to identify the corresponding active compound/s. In these applications, the use of an appropriate chemometric tool is critical since it determines the main output from the huge data sets. Similarly, Rantalainen and his co-workers (2006) adopted direct correlation and PLS based approaches in an integrated analysis of NMR metabolic profiles and 2D differential in gel electrophoresis (DIGE) proteomic data from a murine cancer xenograft model. Metabolites strongly correlating positively or negatively to proteins can be determined thus. The general procedure of traditional TCM safety assessment is to medicate a certain number of experimental animals or a certain amount of analyte, and thereafter observe the characteristics of toxic effects induced by the dosed TCM.

In particular circumstances, the essential task of metabolomics in TCM research is to assess the TCM toxicity level (Fig. 12.1). This process can be outlined in the following steps: (i) TCM preparation, (ii) Animal treatment, (iii) Pre-clinical sample collection, (iv) Data analysis and detection, (v) Data mining, (vi) Toxicity reassessment, and (vii) Systems biology milieu.

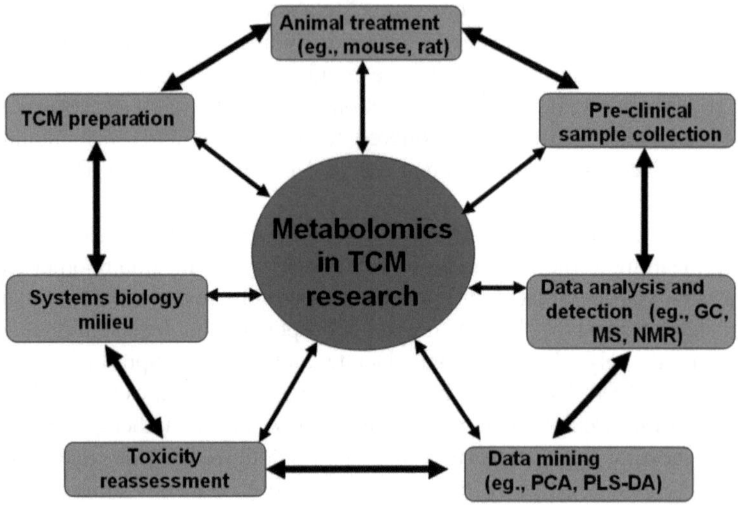

Fig. 12.1 Schematic flowchart of interactions and role of metabolomics investigation in TCM research. *TCM* traditional Chinese medicine, *GC* gas chromatography, *MS* mass spectrometry, *NMR* nuclear magnetic resonance, *PCA* principal components analysis, *PLS-DA* partial least squares discriminant analysis

12.3.1 TCM Preparation

The pharmacological activity of TCM fractions, either as single or crude extracts, depends on variable factors concerning collection, preparation, and prescription. Collecting in various habitats and at different times of the year, prescriptions with different compatibility programs or active fractions may yield widely differing or even opposite effects (Li et al. 2008).

12.3.1.1 Animal Treatment

Establishment of experimental animal model is an intensive and decisive step for TCM research because the so-called "symptom" in TCM science e.g., *Qi* deficiency syndrome, kidney deficiency, blood stasis syndrome, is a complicated multi-factor, interactional, dialectic result induced at the holistic level (Li et al. 2007a). Therefore, employing a single/few compounds or methods will never exactly reproduce an authentic toxic-pathological status in animal model. Accordingly, metabolomics in TCM toxicity research can provide valuable feedback information for reassessment of animal model.

12.3.1.2 Pre-Clinical Sample Collection

After TCM treatment to the animal model, collection of various organs depends upon the diseases or syndromes of the experimental model. Most frequently collected samples are biofluids which include urine, blood plasma and serum, etc. However, there is a wide range of available biofluids that can and have been investigated, including seminal-, lung aspirate and dialysis-, blister- and cyst-, digestive-, synovial-, amniotic- and cerebrospinal fluids (Lenz and Wilson 2007).

12.3.1.3 Data Analysis and Detection

For data acquisition and detection, three different methods are typically adopted: chromatography-, MS-, and NMR-based methods. As we have mentioned above, each of these methods has its own advantages and disadvantages in TCM research. However, the combination of a few metabolomics approaches in detection and analysis may improve the quality of these aspects. Up to now, this problem fails to be overcome commendably; further studies are needed to improve these methods.

12.3.1.4 Data Mining

Combination of different analytical techniques may offer a better strategy for identifying a broadened spectrum of important metabolites associated with TCM perturbation. Among the chemometric tools used, principal components analysis (PCA) enjoys the most extensive use. (Rousseau et al. 2008). In a classic example, Chan and Cai (2008) used LC–MS metabonomic technique coupled with PCA to find variations in urine profiles between the control and the dosed rats. PCA data showed clear differences in metabolic profiles between control- and dosed rats analyzed by LC–MS. Two potential biomarkers, kynurenic and hippuric acid, were identified by high-resolution mass measurement and MS–MS analyses on a Qq-TOF. It was believed that changes in dosed urine profile could have occurred before symptoms were observed. But, at present, lack of standardization is the most disillusionary obstructor for real data mining. Further, standardization of either single analytical technique or combination may be available to the public domain to inviting additionally fine-tuning of data mining.

12.3.1.5 Toxicity Reassessment

In order to develop a program for assessing the toxicity of TCM from test animal model or clinical patients with optimal outcome, and to facilitate our studies on toxicity and therapy with TCM formulations, it becomes advisable to work with combinations of different detection and analytical techniques. One such attempted example was reported by Ran et al. (2006), who applied LC coupled with ion trap

MS combined with PCA to studies of TCM scutellarin in rat urine. Using this methodology, they detected two potential drug metabolites of scutellarin and found nine ions responsible for the gender variation and one ion for the dosage variation.

12.4 Metabolomics in a Systems Biology Milieu

Principally, metabolomics aims to evaluate and analyze all small molecular mass in a cell or tissue, in an integrated and related manner. It is mainly and sometimes solely a portion of the cellular metabolomics products of the whole organism. For a systems biology approach, metabolomics provides the measurement of the intermediate metabolites in a biological system. Yet, systems biology aims not merely to measure all components of a life system, including DNA, mRNA, proteins, lipids, metabolites, and structural elements such as cell walls/membranes, but also to determine the interactions between those elements as part of the system's adaptations to environmental or genetic perturbation.

Although metabolomic measurements are key contributors to our understanding of both overall and specific metabolic networking in a given biological system, a sound systems biology approach will require the integration of metabolomic activities in a time-dependent, tissue, or organ-defined concentrations of other types of cellular and molecular components. The availability of protein microarrays existing for the global population or a special targeted group of proteins of an organism can be screened systematically for drug or metabolite binding; this may make a major contribution to integrative biology. A particularly exciting possibility is that proteins may be purified directly from target organisms using a mass spectrometer approach without cloning them, thereby producing microarrays directly. Essential to the success of the systems biology approach is the ability to create an environment which is accessible to virtually all of the high-throughput platforms needed to obtain and measure the properties and elements of the system of interest. Also, hopefully, an effective systems biology approach will offer the opportunity and scale for rapid development and employment of new global technologies and powerful computational tools allowing the gathering, classifying, analyzing, integrating, and ultimately, good modeling of biological information addressed by specific biological questions.

12.5 Advantages of Metabolomics Approach

While metabolomics technology is a functional level tool useful for investigating complex networking and interactions among metabolites, it can also address the regulatory roles of metabolites where they interact with genes, transcripts, and proteins, (e.g., allosteric regulation). One major benefit of metabolomics in studying disease and drug therapy is that informative metabolic profiling is often achieved using urine or plasma samples. The easy accessibility of these body fluid samples makes for ideal large-scale and systematic research. Integration and work flow of metabolomics studies include sample collection, pretreatment, data analysis, metabolic variation interpretation, and bioinformatics analysis. Metabolomics can therefore be seen as a bridge between the genotypes and phenotypes, providing a comprehensive view of cell-function, as well as identifying novel or striking changes in specific metabolisms and the constituent metabolites. Analysis and mining of metabolomic data sets and their metadata are hence also expected to lead to new hypotheses and new targets for biotechnology.

Three analytical platforms of GC, HPLC, MS, and NMR spectroscopy have driven forward the disciplines of metabolomics. These platforms each have advantages and disadvantages. Currently, there are new opportunities for combining these technologies for specific requirements in metabolomic studies. Metabolomics is designed to pursue a comprehensive, noninvasive analysis of metabolic biomarkers, intermediates, and signaling/functional systems that could help detect early stage disease, identify residual disease post-surgery or post-treatment and monitor treatment response. The novel or striking changes in specific metabolites detected from a study system may also yield valuable information and knowledge of basic biology. Metabolomics will, hopefully, become useful in medical research areas as an aid to disease diagnosis or staging and as a tool to predict and/or monitor treatment response or toxicity. The development of new or improved analytical techniques, either alone or in combination, along with the emergence of technologies in other "omics" studies is expected to have a heavy impact on research into TCM.

12.6 Conclusion and Future Perspective

Metabolomics is a relatively new conceptual approach. Promising tools for the detection of bio-activities from natural products, especially medicinal plants, are under development. Major areas for improvement will involve sensitivity, universality and discrimination of employed instruments, new approaches, and better deconvolution. Structural (i.e., chemical) identification of the many uncharacterized secondary plant metabolites is a serious problem which remains unsolved (Shyur and Yang 2008). MS-based metabolomics enables the sensitive fingerprinting and profiling of a spectrum of important metabolites. The recent progress in hyphenated analytical methods, instrument resolution, reproducibility, and simultaneous development of databases has already been wildness as responsible for some of the successes obtained. The improved use of mass spectrometry in plant metabolomics may continue to grow massively in the research field, and it needs to be accompanied by parallel improvement in raw data filtering and deconvolution, and in "automated" metabolite identification. Similarly, NMR data can now be stored permanently and systematically, and whenever new data is generated, it can be included for use in a new database for comparable and parallel analyses.

The goal of metabolomics is the overall non-biased quantification and identification or an integrated view of the component metabolites present in a given or test biological system. This cannot be achieved with one single analytical method. The only way to circumvent this issue is to combine different technological platforms and employ multiple sample preparations. Only recently, researchers have started to combine the advantages of MS and NMR spectroscopy. Another important aspect to be considered is that the assay system required for metabolomics approach needs to be fast, simple, and suitable for high-throughput screening. By integrating the metabolic models in genomically characterized organisms with their experimentally determined metabolomes, we aim to improve our global and integrated views of the systems biology of organisms under our study. Hopefully, the effects of complex mixtures such as multi-herb, traditional medicinal preparations in specific and complex biological systems of human body in relation to the molecular pharmacology can be understood with the perspective of metabolomics, thus contributing to the development of evidence-based herbal medicines. We expect that consecutive study along these lines will deepen our knowledge and reduce its 'grey scale', gradually increasing transparency, thereby improving the quality assessment of TCM and other phytomedicines. Interdisciplinary study among chemometrics, bioinformatics, and systems biology, will be essential for sustained research into various traditional medicines, ultimately benefiting public health and societal needs.

References

Ackermann BL, Hale JE, Duffin KL (2006) The role of mass spectrometry in biomarker discovery and measurement. Curr Drug Metab 7:525–539

Adahchour M, Beens J, Vreuls RJJ, Brinkman UATh (2006) Recent developments in comprehensive two-dimensional gas chromatography (GC × GC): I. Introduction and instrumental set-up. TrAC Trend Anal Chem 25:438–454

Agnolet S, Jaroszewski JW, Verpoorte R, Staerk D (2010) H NMR-based metabolomics combined with HPLC-PDA-S-SPE-NMR for investigation of standardized *Ginkgo biloba* preparations. Metabolomics 6:292–302

Baumann S, Ceglarek U, Fiedler GM, Lembcke J, Leichtle A, Thiery J (2005) Standardized approach to proteome profiling of human serum based on magnetic bead separation and matrix-assisted laser desorption/ionization time-of-flight mass spectrometry. Clin Chem 51:973–980

Boernsen KO, Gatzek S, Imbert G (2005) Controlled protein precipitation in combination with chip-based nanospray infusion mass spectrometry. An approach for metabolomics profiling of plasma. Anal Chem 77:7255–7264

Brindle JT, Antti H, Holmes E, Tranter G, Nicholson JK, Bethell HWL, Clarkec S, Schofield PM, McKilligin E, Mosedale DE, Grainger DJ (2002) Rapid and non-invasive diagnosis of the presence and severity of coronary heart disease using 1H-NMR-based metabonomics. Nat Med 8:1439–1444

Ceglarek U, Leichtle A, Brügel M, Kortz L, Brauer R, Bresler K et al (2009) Challenges, developments in tandem mass spectrometry based clinical metabolomics. Mol Cell Endocrinol 301:266–271

Chan W, Cai ZW (2008) Aristolochic acid induced changes in the metabolic profile of rat urine. J Pharm Biomed Anal 46:757–762
Chen M, Ni Y, Duan H, Qiu Y, Guo C, Jiao Y, Shi H, Su M, Jia W (2008) Mass spectrometry-based metabolic profiling of rat urine associated with general toxicity induced by the multiglycoside of *Tripterygium wilfordii* Hook F. Chem Res Toxicol 21:288–294
Choi HK, Choi YH, Verberne M, Lefeber AWM, Erkelens C, Verpoorte R (2004a) Metabolic fingerprinting of wild type and transgenic tobacco plants by 1H NMR and multivariate analysis technique. Phytochem 65:857–864
Choi YH, Tapias EC, Kim HK, Lefeber AWM, Erkelens C, Verhoeven JTJ, Brzin J, Zel J, Verpoorte R (2004b) Metabolic discrimination of *Catharanthus roseus* leaves infected by phytoplasma using 1H-NMR spectroscopy and multivariate data analysis. Plant Physiol 135:2398–2410
Claudino WM, Quattrone A, Biganzoli L, Pestrin M, Bertini I, Di Leo A (2007) Metabolomics: available results, current research projects in breast cancer, and future applications. J Clin Oncol 25:2840–2846
Dai H, Xiao C, Liu H, Hao F, Tang H (2010) Combined NMR and LC–DAD–MS analysis reveals comprehensive metabonomic variations for three phenotypic cultivars of *Salvia miltiorrhiza* Bunge. J Proteome Res 9:1565–1578
Dettmer K, Aronov PA, Hammock BD (2007) Mass spectrometry-based metabolomics. Mass Spectrom Rev 26:51–78
Dong L, Deng CH, Wang JY, Shen XZ (2007) Fast determination of paeonol in plasma by headspace solid-phase microextraction followed by gas chromatography–mass spectrometry. Anal Chim Acta 585:76–80
Dunn WB (2008) Current trends and future requirements for the mass spectrometric investigation of microbial, mammalian and plant metabolomes. Physical Biol 5:1–24
Dunn WB, Ellis DI (2005) Metabolomics: current analytical platforms and methodologies. Trends Anal Chem 24:285–294
Dunn WB, Bailey NJC, Johnson HE (2005) Measuring the metabolome: current analytical technologies. Analyst 130:606–625
Eliss DI, Goodacre R (2006) Metabolic fingerprinting in disease diagnosis: biomedical applications of infrared and Raman spectroscopy. Analyst 131:875–885
Eliss DI, Dunn WB, Griffin JL, Allwood JW, Goodacre R (2007) Metabolic fingerprinting as a diagnostic tool. Pharmacogenomics 8:1243–1266
Fiehn O (2002) Metabolomics: the link between genotypes and phenotypes. Plant Mol Biol 48:155–171
Fiehn O (2008) Extending the breadth of metabolite profiling by gas chromatography coupled to mass spectrometry. Trends Anal Chem 27:261–269
Field D, Sansone SA, Collis A, Booth T, Dukes P, Gregurick SK, Kennedy K, Kolar P, Kolker E, Maxon M, Millard S, Mugabushaka AM, Perrin N, Remacle JE, Remington K, Rocca-Serra P, Taylor CF, Thorley M, Tiwari B, Wilbanks J (2009) Megascience. Omics data sharing. Science 326:234–236
Griffin JL, Kauppinen RA (2007) Tumour metabolomics in animal models of human cancer. J Proteome Res 6:498–505
Halket JM, Przyborowska A, Stein SE, Mallard WG, Down S, Chalmers RA (1999) Deconvolution gas chromatography-mass spectrometry of urinary organic acids-potential for pattern recognition and automated identification of metabolic disorders. Rapid Commun Mass Spectrom 13:279–284
Hall R, Beale M, Fiehn O, Hardy N, Sumner L, Bino R (2002) Plant metabolomics: the missing link in functional genomics strategies. Plant Cell 14:1437–1440
Hu LH, Li X, Feng S, Kong L, Su XY, Chen XG et al (2006) Comprehensive two-dimensional HPLC to study the interaction of multiple components in *Rheum palmatum* L. with HSA by coupling a silicabonded HSA column to a silica monolithic ODS column. J Sep Sci 29:881–888

Jacob SS, Smith NW, Legido-Quigley C (2007) Assessment of Chinese medicinal herb metabolite profiles by UPLC-MS- based methodology for the detection of aristolochic acids. J Sep Sci 30:1200–1206

Jin Y, Liang T, Fu Q, Xiao YS, Feng JT, Ke YX et al (2009) Fingerprint analysis of *Ligusticum chuanxiong* using hydrophilic interaction chromatography and reversed-phase liquid chromatography. J Chromatogr A 1216:2136–2141

Kell DB (2004) Metabolomics and systems biology: making sense of the soup. Curr Opin Microbiol 7:296–307

Kim HK, Choi YH, Erkelens C, Lefeber AWM, Verpoorte R (2005) Metabolic fingerprinting of *Ephedra* species using 1H-NMR spectroscopy and principal component analysis. Chem Pharm Bull 53:105–109

Kitano H (2007) A robustness-based approach to systems-oriented drug design. Nat Rev Drug Discov 6:202–210

Le Gall G, Metzdorff SB, Pedersen J, Bennett RN, Colquhoun IJ (2005) Metabolite profiling of *Arabidopsis thaliana* (L.) plants transformed with an antisense chalcone synthase gene. Metabolomics 1:181–198

Lee MG, Wynder C, Cooch N, Shiekhattar R (2005) An essential role for CoREST in nucleosomal histone 3 lysine 4 demethylation. Nature 437:432–435

Leiss KA, Choi YH, Abdel-Farid IB, Verpoorte R (2009) NMR metabolomicsmetabolomics of thrips (*Frankliniella occidentalis*) resistance in Senecio hybrids. J Chem Ecol 35:219–229

Lenz EM, Wilson ID (2007) Analytical strategies in metabonomics. J Proteome Res 6:443–458

Li L, Wang JN, Ren JX, Xiang JF, Tang YL, Liu JX et al (2007a) Metabonomics analysis of the urine of rats with Qi deficiency and blood stasis syndrome based on NMR techniques. Chinese Sci Bull 52:3068–3073

Li YP, Hu Z, He LC (2007b) An approach to develop binary chromatographic fingerprints of the total alkaloids from *Caulophyllum robustum* by high performance liquid chromatography/ diode array detector and gas chromatography/mass spectrometry. J Pharm Biomed Anal 43:1667–1672

Li WF, Jiang JG, Chen J (2008) Chinese medicine and its modernization demands. Arch Med Res 39:246–251

Liang MJ, He LC, Yang GD (2005) Screening, analysis and in vitro vasodilatation of effective components from *Ligusticum Chuanxiong*. Life Sci 78:128–133

Liang YS, Choi YH, Kim HK, Linthorst HJM, Verpoorte R (2006) Metabolomic analysis of methyl jasmonate treated *Brassica rapa* leaves by two dimensional NMR spectroscopy and multivariate analysis. Phytochem 67:2503–2511

Lindon JC, Holmes E, Nicholson JK (2004) Metabonomics and its role in drug development and disease diagnosis. Expert Rev Mol Diagn 4:189–199

Lindskog M, Spenger C, Jarvet J, Gräslund A, Kogner P (2004) Predicting resistance or response to chemotherapy by proton magnetic resonance spectroscopy in neuroblastoma. J Natl Cancer Inst 96:1457–1466

Liu M, Li YG, Chou GX, Cheng XM, Zhang M, Wang ZT (2007) Extraction and ultra-performance liquid chromatography of hydrophilic and lipophilic bioactive components in a Chinese herb Radix Salviae Miltiorrhizae. J Chromatogr A 1157:51–55

Liu Y, Huang R, Liu L, Peng J, Xiao B, Yang J, Miao Z, Huang H (2010) Metabonomics study of urine from Sprague-Dawley rats exposed to Huangyao-zi using 1H NMR spectroscopy. J Pharm Biomed Anal 52:136–141

Luo HG, Ding J, Yue GX, Chen JX (2007) Metabonomic study of syndrome of liver qi stagnation and spleen deficiency in rats. Zhong Xi Yi Jie He Xue Bao 5:307–313

Lv Y, Liu X, Yan S, Liang X, Yang Y, Dai W, Zhang W (2010) Metabolomic study of myocardial ischemia and intervention effects of Compound Danshen Tablets in rats using ultra-performance liquid chromatography/quadrupole time-of-flight mass spectrometry. J Pharm Biomed Anal 52:129–135

Morse DL, Raghunand N, Sadarangani P, Murthi S, Job C, Day S, Howison C, Gillies RJ (2007) Response of choline metabolites to docetaxel therapy is quantified in vivo by localized 31P

MRS of human breast cancer xenografts and in vitro by high-resolution ^{31}P NMR spectroscopy of cell extracts. Magn Reson Med 58:270–280
Nicholson JK, Wilson ID (1989) High-resolution proton magnetic resonance spectroscopy of biological fluids. Prog NMR Spectrosc 21:449–501
Pasikanti KK, Ho PC, Chan ECY (2008) Gas chromatography/mass spectrometry in metabolic profiling of biological fluids. J Chromatogr B 871:202–211
Qi LW, Li P, Li SL, Sheng LH, Li RY, Song Y et al (2006) Screening and identification of permeable components in a combined prescription of Danggui Buxue decoction using a liposome equilibrium dialysis system followed by HPLC and LC-MS. J Sep Sci 29:2211–2220
Ramautar R, Somsen GW, de Jong GJ (2009) CE-MS in metabolomics. Electrophoresis 30: 276–291
Ran XR, Si SZ, Liang QL, Luo GA (2006) Application of the reversed phase liquid chromatography coupled to ion trap mass spectrometry with principal components analysis for metabonomics studies of scutellarin in rat urine. Chinese Chem Lett 17:1225–1228
Rantalainen M, Cloarec O, Beckonert O, Wilson ID, Jackson D, Tonge R, Rowlinson R, Rayner S, Nickson J, Wilkinson RW, Mills JD, Trygg J, Nicholson JK, Holmes E (2006) Statistically integrated metabonomic-proteomic studies on a human prostate cancer xenograft model in mice. J Proteom Res 5:2642–2655
Robertson DG (2005) Metabonomics in toxicology: a review. Toxicol Sci 85:809–822
Roessner U, Beckles DM (2009) Metabolite measurements. In: Schwender J (ed) Plant metabolic networks. Springer, New York
Roos G, Röseler C, Berger-Bűter K, Simmen U (2004) Classification and correction of St. John's wort extracts by nuclear magnetic resonance spectroscopy, multivariate data analysis and pharmacological activity. Planta Med 70:771–777
Rousseau R, Govaerts B, Verleysen M, Boulanger B (2008) Comparison of some chemometric tools for metabonomics biomarker identification. Chemometr Intell Lab 91:54–66
Ruan GH, Li GK (2007) The study on the chromatographic fingerprint of *Fructus xanthii* by microwave assisted extraction coupled with GC–MS. J Chromatogr B 850:241–248
Shyur LF, Yang NS (2008) Metabolomics for phytomedicine research and drug development. Curr Opin Chem Biol 12:66–71
Spratlin JL, Serkova NJ, Eckhardt SG (2009) Clinical applications of metabolomics in oncology: a review. Clin Cancer Res 15:431–440
Sterin M, Cohen JS, Mardor Y, Berman E, Ringel I (2001) Levels of phospholipids metabolites in breast cancer cells treated with antimitotic drugs: a ^{31}P magetic resonance spectroscopy study. Cancer Res 61:7536–7543
Swanson MG, Zektzer AS, Tabatabai ZL, Simko J, Jarso S, Keshari KR, Schmitt L, Carroll PR, Shinohara K, Vigneron DB, Kurhanewicz J (2006) Quantitative analysis of prostate metabolites using 1H HR-MAS spectroscopy. Magn Reson Med 55:1257–1264
Tam CU, Yang FQ, Zhang QW, Guan J, Li SP (2007) Optimization and comparison of three methods for extraction of volatile compounds from *Cyperus rotundus* evaluated by gas chromatography–mass spectrometry. J Pharm Biomed Anal 44:444–449
Tang HR, Wang YL (2006) Metabonomics: a revolution in progress. Prog Biochem Biophys 33:401–417
Ulrich-Merzenich G, Zeitler H, Jobst D, Panek D, Vetter H, Wagner H (2007) Application of the ''-omic-'' technologies in phytomedicine. Phytomedicine 14:70–82
Villas-Bôas SG, Roessner U, Hansen M, Smedsgaard J, Nielsen J (2007) Metabolome analysis: an introduction. Wiley, Hoboken
Walsh MC, Brennan L, Pujos-Guillot E, Sebedio JL, Scalbert A, Fagan A, Higgins DG, Gibney MJ (2007) Influence of acute phytochemical intake on human urinary metabolomic profiles. Am J Clin Nutr 86:1687
Wang M, Lamers RJ, Korthout HA, van Nesselrooij JH, Witkamp RF, van der Heijden R, Voshol PJ, Havekes LM, Verpoorte R, van der Greef J (2005) Metabolomics in the context of systems biology: bridging traditional Chinese medicine and molecular pharmacology. Phytother Res 19:173–182

Wang Y, Kong L, Hu LH, Lei XY, Yang L, Chou GX et al (2007) Biological fingerprinting analysis of the traditional Chinese prescription Longdan Xiegan Decoction by on/off-line comprehensive two-dimensional biochromatography. J Chromatogr B 860:185–194

Wang H, Ma C, Ma L, Du Z, Wang H, Ye H, Li G, Liu B, Xu G (2009) Secondary metabolic profiling and artemisinin biosynthesis of two genotypes of *Artemisia annua*. Planta Med 75:1625–1633

Wang M, Rang W, Zhang Q, Huo C, Ma Z, Wang Y, Yan X, Gao Y (2010) NMR- spectroscopy-based metabonomic approach to analysis of Siwutang, a novel prescription, treated blood deficiency in mice. Zhongguo Zhong Yao Za Zhi 35:630–634

Wishart DS, Tzur D, Knox C et al (2007) HMDB: the human metabolome database. Nucleic Acids Res 35:D521–D526

Wu HF, Zhang XY, Li XJ, Wu YJ, Pei FK (2005) Acute biochemical effects of $La(NO_3)_3$ on liver and kidney tissues by magic-angle spinning ^1H nuclear magnetic resonance spectroscopy and pattern recognition. Anal Biochem 339:242–248

Wu B, Yan S, Lin Z, Wang Q, Yang Y, Yang G, Shen Z, Zhang W (2008) Metabonomic study on ageing: NMR-based investigation into rat urinary metabolites and the effect of the total flavone of epimedium. Mol Biosyst 4:855–861

Xiang Z, Wang WQ, Cai XJ, Zeng S (2011) Metabolomics study on quality control and discrimination of three *Curcuma* species based on gas chromatograph–mass spectrometry. Phytochem Anal 22:411–418

Xie G, Plumb R, Su M, Xu Z, Zhao A, Qiu M, Long X, Liu Z, Jia W (2008) Ultra-performance LC/TOF MS analysis of medicinal Panax herbs for metabolomic research. J Sep Sci 31: 1015–1026

Xu L (2004) Methods of chemometrics. Science Press, Beijing, pp 221–227

Yan S, Wu B, Lin Z, Jin H, Huang J, Yang Y, Zhang X, Shen Z, Zhang W (2009) Metabonomic characterization of aging and investigation on the anti-aging effects of total flavones of epimedium. Mol Biosyst 5:1204–1213

Yang FQ, Quan J, Zhang TY, Ito Y (1998) Multidimensional countercurrent chromatographic system and its application. J Chromatogr A 803:298–301

Yang J, Xu G, Zheng Y, Kong H, Pang T, Lva S, Yang Q (2004) Diagnosis of liver cancer using HPLC-based metabonomics avoiding false-positive result from hepatitis and hepatocirrhosis diseases. J Chromatogr B 813:59–65

Yang SY, Kim HK, Lefeber AWM, Erkelens C, Angelova N, Choi YH, Verpoorte R (2006) Application of two-dimensional nuclear magnetic resonance spectroscopy to quality control of ginseng commercial products. Planta Med 72:364–369

Zeng YX, Zhao CX, Liang YZ, Yang H, Fang HZ, Yi LZ et al (2007) Comparative analysis of volatile components from *Clematis* species growing in China. Anal Chim Acta 595:328–339

Zhang XM, Hu HL, Xu SY, Yang XH, Zhang J (2001) Comprehensive two-dimensional capillary LC and CE for resolution of neutral components in traditional Chinese medicines. J Sep Sci 24:385–391

Zhang A, Sun H, Wang Z, Sun W, Wang P, Wang X (2010) Metabolomics: towards understanding traditional Chinese medicine. Planta Med 76:2026–2035

Zhao J, Yu QT, Li P, Zhou P, Zhang YJ, Wang W (2008a) Determination of nine active components in Radix Hedysari and Radix Astragali using capillary HPLC with diode array detection and MS detection. J Sep Sci 31:255–261

Zhao X, Zhang Y, Meng X, Yin P, Deng C, Chen J, Wang Z, Xu G (2008b) Effect of a traditional Chinese medicine preparation Xindi soft capsule on rat model of acute blood stasis: a urinary metabonomics study based on liquid chromatography–mass spectrometry. J Chromatogr B Analyt Technol Biomed Life Sci 873:151–158

Chapter 13
Antioxidants in Medicinal Plants

Indra D. Bhatt, Sandeep Rawat and Ranbeer S. Rawal

13.1 Introduction

Considering the potential role in food industry and human health, antioxidants are gaining popularity all across the globe. Antioxidants are defined as a substance that even in small amounts, is capable of preventing or delaying the oxidation of easily oxidizable materials. Antioxidant are also defined as a substance which are capable of inhibiting a specific oxidising enzymes or a substance that reacts with oxidizing agents prior to causing damage to other molecules or a substance that sequesters metal ions or even a substance capable of repairing system such as iron transporting protein (Brewer 2011). As such, production of free radicals and other reactive oxygen species in the human body by numerous physiological and biochemical processes is reported (Halliwell 1994); however, overproduction of these could lead toward development of diseases. In this context, an antioxidant can acts at different levels by (i) decreasing localized oxygen concentration, (ii) preventing chain initiation by scavenging initiating radicals, (iii) decomposing lipid peroxides to peroxyl and alkoxyl radicals, (iv) decomposing peroxides by converting them to non radicals products, and (v) chain breaking to prevent continued hydrogen abstraction (Miguel 2010).

The natural antioxidants, more recently, have attracted considerable attention of users and researchers largely on account of adverse toxicological reports on some

I. D. Bhatt (✉) · S. Rawat · R. S. Rawal
Biodiversity Conservation and Management Theme, G.B. Pant Institute of Himalayan Environment and Development, Kosi-Katarmal, Almora, 263 643, Uttarakhand, India
e-mail: id_bhatt@yahoo.comidbhatt@gbpihed.nic.in

S. Rawat
e-mail: sandeep_rawat@rediffmail.com

R. S. Rawal
e-mail: ranbeerrawal4@gmail.com

S. Chandra et al. (eds.), *Biotechnology for Medicinal Plants*,
DOI: 10.1007/978-3-642-29974-2_13, © Springer-Verlag Berlin Heidelberg 2013

synthetic antioxidants and growing awareness among consumers (Ramalakshmi et al. 2007). In this context, among others, medicinal plants are being viewed as easily available and potent source of antioxidants as they contain a mixture of different chemical compounds that may act individually or in synergy to cure disease and improve health. In fact, a single plant may have diversity of phytochemicals ranging from bitter compounds that stimulate digestion system, phenolic compounds for antioxidant and many other pharmacological properties, antibacterial, and antifungal, tannins that work as natural antibiotics, diuretic substances, alkaloids, etc. (Miguel 2010).

Over the years, researches on antioxidants and free radicals for understanding their role has grown exponentially. A search of Pubmed and 'Science Direct' database using the term 'antioxidants', 'medicinal plants', and 'antioxidant and medicinal plants' revealed availability of plethora of literature. While prior to 1940 there was not even a single reference on the term antioxidants, about 1,178 hits were obtained for the duration of 1941–1950 when searched on Pubmed database. The number increased considerably to 166,802 hits during 2001–2011. The term 'medicinal plants' when used alone records 84 hits for 1941–1950, thereafter a sharp increase was revealing. In combination of term antioxidants and medicinal plants, the numbers of hits were only 2701 in Pubmed search for 2001–2011; however, 10,374 hits were obtained from Science Direct (Fig. 13.1). These results are indicative of the fact that in a recent years research on antioxidants and medicinal plants has gained enormous popularity. While considering the antioxidant activity in Indian medicinal plants, giving the term 'medicinal plants, antioxidants and India', only 320–356 hits were obtained in Pubmed and Science Direct, respectively. This indicates that, however, a large number of plants across the world have been analyzed for their antioxidant activity, more focus toward Indian medicinal plants is needed.

Keeping the above in view the present chapter focuses on the (i) diversity of antioxidant phytochemicals in medicinal plants; (ii) role of antioxidants in disease prevention, and (iii) review of work on antioxidants in selected medicinal plants.

13.2 Diversity of Antioxidant Phytochemicals

A wide range of diversity of naturally occurring antioxidants are found in medicinal plants which are different in their composition, physical and chemical properties and site of action. Among these, phenolics and flavonoids are reported powerful antioxidants and have been consistently protective through scavenging numerous diverse reactive oxygen species, including hydroxyl radical, peroxyl radical, hypochlorous acids, superoxide anion, and peroxynitrite in various in vitro cellular model (Halliwell 2007). Similarly, antioxidant activity of polyphenols in cardiovascular diseases, hepatoprotective, antiocarcinogenic, antimicrobial, antiviral, and antiinflammatory effects are well investigated (Nijveldt et al. 2001; Serrano et al. 2009). Anthocyanins are the antioxidants known to inhibit chemically-induced cancer and turn off genes involved with proliferation, inflammation and angiogenesis (Karlsen

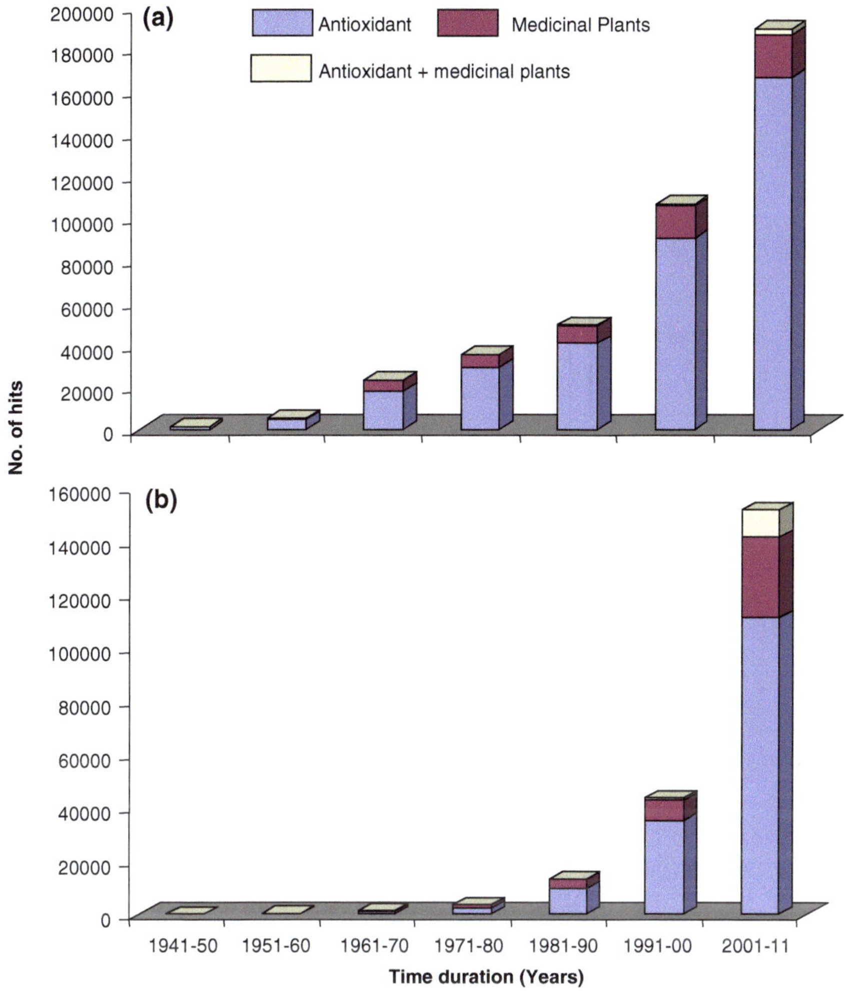

Fig. 13.1 Results of Pubmed (**a**) and ScienceDirect (**b**) search on antioxidant, medicinal plants and 'Antioxidant and medicinal plants' over the years

et al. 2007). Also, these are known to prevent a key step in atherogenesis and effectively protect against LDL oxidation in test-tube studies and considered more potent antioxidant than vitamin C or BHT (Wang and Jiao 2000). Among the carotenoids, β-Carotene is reported to prevent cognitive decline in a study of over 4,000 physicians at mean treatment duration of 18 years (Artemis and Gopalan 2003). Ascorbic acids play an important role in ameliorating the oxidative stress of photosynthesis. In addition, it has a number of other roles in cell division and protein modification. Ascorbic acid behaves not only as an antioxidant but also as a pro-oxidant (McGregor and Biesalski 2006). Ascorbic acid has been shown to reduce transition metals, such as cupric ions (Cu_2+) to cuprous (Cu_1+), and ferric ions (Fe_3+)

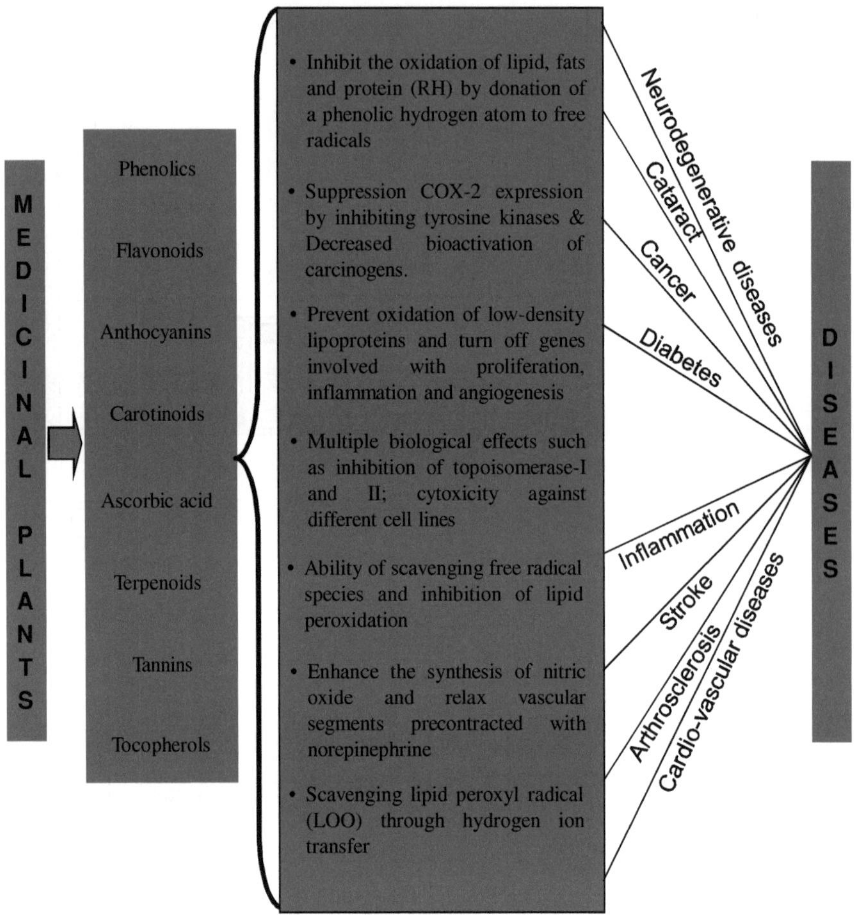

Fig. 13.2 Antioxidant phytochemicals and their mechanism of action to prevent different diseases

to ferrous (Fe_2+) during conversion from ascorbate to dehydroascorbate in in vitro (Satoh and Sakagami 1997). Vitamin C functions as an antioxidant and is necessary for the treatment and prevention of scurvy. Besides, a number of phytochemicals are present in plants which have their own mechanism of action for prevention and curing of diseases. The details of some antioxidant phytochemicals and their mechanism of action are presented (Fig. 13.2).

13.3 Role of Antioxidants in Disease Prevention

Generally, reactive oxygen species (ROS), reactive nitrogen species (RNS) and free radicals in the body are generated through exogenous (radiation, cigarette smoke, atmospheric pollutants, toxic chemicals, over nutrition, changing food habits, etc.)

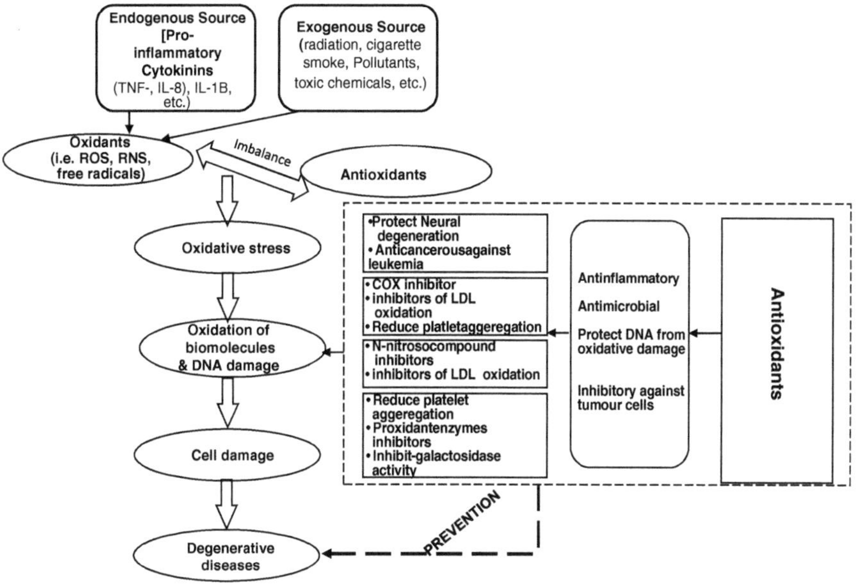

Fig. 13.3 Hypothetical diagram explaining potential of antioxidants for preventing oxidation of biomolecules, DNA damage and degenerative diseases

and/or endogenous sources [Proinflammatory cytokines (tumor necrosis factor-α (TNF- α), interleukin-8 (IL-8), interleukin-1B (IL-1B)], etc. (Devasangayam et al. 2004). Although, free radicals are known to maintain homeostasis at the cellular level and work as signaling molecules but the excess of these are reported for oxidative stress (Sies 1997) and cause of various degenerative diseases. In this context, antioxidant play an important role in prevention, interception and repairing of the body through stopping the formation of ROS, radical scavenging, and repairing the enzymes involved in the process of cellular development (Fig. 13.3).

A number of natural and synthetic antioxidants have been reported to prevent various diseases in different experimental trial, and epidemiological studies suggest that natural antioxidants of plant origin may be a deterrent to degenerative diseases caused due to oxidative stress (Miguel 2010). The use of phytoconstituents as drug therapy to scavenge free radicals and to treat disorders leads to oxidative stress has proved to be clinically effective and relatively less toxic than the existing drugs. Some of the diseases where antioxidants with plant origin found effective are described in detail.

13.3.1 Cardiovascular Diseases

Cardiovascular disease is reported to have the world's largest killers, claiming 17.1 million lives a year (http://www.who.int). Risk factors for heart disease and stroke include raised blood pressure, cholesterol and glucose levels, smoking, inadequate intake of fruit and vegetables, overweight, obesity, and physical inactivity. Cardiovascular diseases are mainly referred to as congestive heart failure, systolic hypertension, angina pectoris, atherosclerosis, cerebral insufficiency, venous insufficiency, arrhythmia, etc. numerous medicinal plants, such as *Digitalis lanata*, *D. purpurea*, *Apocynum cannabinum*, *Calotropis procera*, *Carissa spectabilis*, *Nerium oleander*, *Urginea rubra*, etc. contain potent cardioactive glycosides and have positive inotropic actions on the heart (Tyler 1994). The drug digitoxin, digoxin reported from *Digitalis lanata* and *D. purpurea* has been used in the treatment of congestive heart failure for many decades (Mashour et al. 1998). Besides, increased intake of antioxidant, extracellular and long-lived proteins, such as elastin, laminin, collagen, etc., are reported to prevent cardiovascular diseases. Exogenous antioxidant from natural compounds , i.e., curcumin, baicalen, and resveratol prevent arthrosclerosis formation by exhibiting radical scavenging effects. A number of flavonoids, including quercetin, morin, gossypetin, chrystin, myrecetin, rutin, catechin, and its derivatives and some oligomeric proanthocyanidins are reported to inhibit the oxidation of LDL in in vitro studies (Prochazkova et al. 2011). Some of the flavonoids obtained from leaves of *Morus alba*, such as quercetin 3-(6-malonyglucoside) are reported to attenuate the arthrosclerosis lesion development in LDL receptor in deficient mice through enhancement of LDL resistance to oxidation modification (Enkhmaa et al. 2005).

13.3.2 Neurological Disorders

Neurological diseases are associated with the loss of nerve cells from brain and spinal cord leading to either functional loss (ataxia) or sensory dysfunction (dementia). Mitochondrial dysfunction and excitotoxicity and finally apoptosis have been reported as pathological cause for aging and neurodegenerative diseases, such as Parkinson disease, Alzheimer's disease, Multiple sclerosis, and neurodegeneration (Mark 2004). Oxidative stress catalyzed by redox metal has been shown to play pivotal role in regulating redox reactions in in vivo contributing RNS and ROS as main culprits in neurodegeneration (Emerit and Edeas 2004). Dietary antioxidant may have a potential therapeutic agent for delaying the onset as well as preventing the aging of population with neurodegeneration, i.e., Parkinson disease, Alzheimer's disease, multiple sclerosis, and neurodegeneration. Dietary intake contains variety of antioxidants vitamins supplements which pay a vital role in neuroprotection in variety of neurological disorder (Peter et al. 2004). For example, carotenoids and tocopherol level is positively correlated with maintenance of memory status (Stahelin 1999). Similarly, an extract of *Ginkgo biloba* has been found to improve the symptoms and slow the progression of Alzheimer's disease (Kanowski and Hoerr 1997). *Vinca minor*, major source of Vinpocetine used as a treatment for memory loss and mental impairment (Szatmari and Whitehouse 2003). A significant improvement in moderate Alzheimer's disease particularly in cognition after 16 weeks of treatment is reported from *Salvia officinalis* extract (Akhondzadeh et al. 2003). *Hypericum perforatum* is one of the best known botanical to treat depression (Kim et al. 1999; Gaster and Holroyd 2000). Besides, a number of plants are reported to be used in various neurological disorders. In spite of the usefulness of medicinal plants, many other antioxidative approaches may be used for neurological disorders (Esposito et al. 2002).

13.3.3 Diabetes

Diabetes is a serious human ailment and proving to be a major health problem. It is a chronic disorder of carbohydrate, fat and protein metabolism characterized by increased fasting post prandial blood sugar level. Reports indicated that the diabetes cases are estimated to increase from 4 % in 1995 to 5.4 % in 2025 (Modak et al. 2007). In India, about 33 million adults are suffering from diabetes and this number is likely to increase 57.2 million by the year 2025 (Modak et al. 2007). As such, diabetes is considered a complex metabolic disorder resulting from insulin insufficiency or insulin dysfunction; however, involvement of free radicals in the pathogenesis and development of diabeteic complications is reported (Matteucci and Giampietro 2000; Oberlay 1988; Baynes and Thorpe 1997). In this context, antioxidants are found capable to neutralize free radicals and are effective in preventing experimentally-induced diabetes in animal models (Naziroglo and Cay 2001) as well as reducing the severity of diabetic complications (Kubish 1997). Medicinal plants have been viewed as one of the potential source for suitable antidiabetic and antioxidant therapy. Over 400 plants have been used traditionally for diabetes treatment but only a few has been scientifically evaluated. Among others, *Acacia arabica* plant extract was found to release insulin and induce hypoglycemia in control rat but not in alloxanized animals (Wadood et al. 1989). Aqueous extract of *Aegle marmelous* leaves are reported to improve digestion and reduce blood sugar and urea, serum cholesterol (Karunanyake et al. 1984). The antiinflammatory and antidiabetic property of *Aloe vera* and *A. barbedensis* is reported (Davis and Maro 1998). Hydroalcoholic extract of *Azadirachta indica* showed antihyperglycemic activity in rat (Chattopadhyay et al. 1987). Aqueous and alcoholic extract as well as lyophilized powder of *Eugenia jambolana* showed reduction in blood glucose level (Sheela and Augusti 1992). Water extract of *Morus alba* leaves have been reported good potential of inhibiting the alpha-glucosidase (Yogisha and Rabeesha 2009). Besides number of medicinal plants, such as *Momordica charantia*, *Ocimum sanctum*, *Phyllanthus amarus*, *Pterocarpus marsupium*, *Tinospora cordifolia* (Wadkar et al. 2008; Jain et al. 2008), etc. are known for their antidiabetic properties.

13.3.4 Immunological Disorders

Autoimmunity is considered the result of a breakdown in self-tolerance and is a physiological process that becomes pathological when the number of autoreactive cells and particularly the avidity of their receptors for autoantigens increase.

Antigen presenting cells (APCs) are involved in the fighting against antigens. These antigens could be fungi, bacteria, viruses which are processed by APCs presented to T-cells for further processing. Phagocytic cells which are involved in immune system are neutrophils, basophils, eosinophils, and monocytes they engulf and destroy the antigens or foreign substances with their intercellular mechanisms (Ranjith et al. 2008).

The uses of plant products as immunostimulants or immunosuppressants have a traditional history and are ancient as human civilization (Rates 2001). Treatments of many diseases were done by modulating the immune system or function by using medicinal plants and their products was also the fundamental principal of therapeutic approach (Ismail and Asad 2009). The therapeutic potential of immunomodulatory agents from plant products and the Ayurvedic concepts of preventive healthcare have been highlighted by many researchers (Dahanukar et al. 2000). It has been reported that the immunomodulatory properties of *Embelica officinalis* and *Evolvulus alsinoides* were evaluated in adjuvant-induced arthritic rat model (Gunju et al. 2003). Ethanolic extracts of *Acorus calamus* inhibited proliferation of mitogen and antigen-stimulated human proliferated blood mononuclear cells. Also, *A. calamus* extract inhibited growth of several cell lines of mouse and human origin, inhibit production of nitric oxide, interleukin-2, tumour necrosis factor-α (Mehrotra et al. 2003). *Tinospora cordifolia* is reported to protect against lipopolysaccharide-induced mortality (Desai et al. 2007). Ethanolic extract of *Boerhavia diffusa* significantly inhibit the cell proliferation (Mungantiwar et al. 1999). Besides, number of other plants, such as *Centella asiatica*, *Eclipta alba*, *Nyctanthus* sp. has been reported an effective immunomodulator.

13.3.5 Cancer

Cancer is one of the leading cause of death all around the world, an estimated 6.7 million people killed by cancer and another 15 million people will be diagnosed annually by 2020 (Aggarwal et al. 2006). Several chemotherapeutic, cytotoxic and immunomodulatory agents are available to treat cancer but besides enormous expensive, these drugs are associated with serious side effects and morbidity. In this context, number of medicinal plants and their active component were found effective for anticancer activity (Aggarwal et al. 2006). Different phytochemicals, such as tea catechins, soy genistein, fiber bytyrate, etc., are efficient inducer of programmed cell death in tumour cells (Ferrary et al. 2000). Crucifer such as cabbage, cauliflower has reported to decrease breast cancer. Antimutagenic effect of different Ayurvedic medicinal plants, i.e., *Acorus calamus, Hemidesmus indicus, Holorrhena antidysenterica,* and *Plunbaco zeylanica* have also been reported (Aqil et al. 2006). A detailed study on the molecular targets of chemopreventive agents, such Guggulstrone *from Commiphora mukul*, Curcumin from *Curcuma longa*, Withanolide from *Withania somnifera*, Boswellic acid from *Boswellia serrata* have shown their potential as anticancer agents (Aggarwal et al. 2006,

2007). This has further illustrated that the tumour cells use multiple cell survival pathways to prevail and agents that can suppress these multiple pathways have great potential in the treatment of cancer.

13.4 Antioxidants from Medicinal Plants

Medicinal plants are still the mainstay of about 70–80 % of the world population, largely in developing countries, for primary health care needs because of better cultural acceptability, better compatibility with the human body and lesser side effects (Kamboj 2000). The chemical constituents present in the medicinal plants are a part of the physiological functions of living cells and hence they are believed to have better compatibility with human body. Also, the plants are in use for centuries to cure various ailments. The main preoccupation with the use of synthetic drugs is the side effects which can be even more dangerous than the diseases they claim to cure. In contrast, plant-derived medicines are based upon the premise that they contain natural substances that can promote health and alleviate illness and proved to be safe, better patient tolerance, relatively less expensive and globally competitive.

India is recognized to have a rich system of traditional medicines and one of the major raw materials producing nation. Over 8,000 plant species of medicinal value, which accounts for nearly 50 % of all the higher flowering plant species, are reported from India. India has its own system of medicine widely known as 'Ayurveda' which is predominantly plant based. Ayurveda is supposed to be the oldest among the organized traditional system of medicine. In India, around 25,000 effective plant-based formulations are used in folk medicine, and over 1.5 million practitioners of traditional medical system and over 7,800 medical drug manufacturing units consume about 2,0 tons herbs annually (Ramakrishnappa 2002). The market of Ayurvedic medicine is estimated to be expanding at 20 % annually.

The major groups of the phytochemicals obtained from plant showed antioxidant activities and are known to prevent several degenerative diseases. A number of Indian medicinal plants are investigated for their antioxidant activity and found effective in various diseases (Table 13.1); however, selected medicinal plants growing in Indian Himalayan region (IHR) with their potential antioxidant activity are described in details.

13.4.1 Hedychium Spicatum

Hedychium spicatum, family Zingiberace, is generally grows in subtropical and temperate Himalayan region up to 2,800 m asl. The rhizome is stomachic, carminative, bronchodilator, stimulant, and tonic and traditionally used in dyspepsia, nausea, vomiting, liver complaint, etc. Species is used in preparation of PADAM-28, a Tibetan formulation used to cure the cancer. Also, the species is used in

Table 13.1 List of selected Indian medicinal plants investigated for their antioxidant activity

Name of plant/common name/family	Parts used	Medicinal uses/properties	Mechanism of action	Reference
Abrus precatorius Crab's eye Fabaceae	Seeds	Cough, cold, inflammation, wound healing and leprosy	Reduces ROS generation in hepatocarcinogenesis and promote antioxidant defense system; protect the kidney against alcohol-induced parenchymal injury	Pal et al. (2009), Kartik et al. (2010)
Acacia arabica Indian gum, babool Fabaceae	Bark, stem, gum	Premature ejaculation, asthma, bronchitis, diabetes, dysentery, diarrhea, skin diseases	Potent free radical scavenger and protects TBH-induced lipid peroxidation and CCl_4-induced hepatic damage	Sundaram and Mitra (2007)
Acacia catechu Black catechu, khair Fabaceae	Bark, stem, gum	Asthma, bronchitis, diabetes, dysentery, diarrhea, skin diseases	Catechins and epicatechins possess antimicrobial, anti-inflammatory, antiviral and antioxidant properties	Wojdylo et al. (2007), Ismail and Asad (2009)
Acorus calamus Sweet flag Araceae	Rhizomes	Bronchitis, epilepsy, insanity	Phenolic antioxidants contain anticholinesterase activity	Manikanandan et al. (2005), Ahmad et al. (2009)
Aegle marmelos Stone apple Rutaceae	Fruit, seed	Constipation, diarrhea, peptic ulcer, ear diseases, respiratory disorders, diabetes	Antidiabetic, anti-lipid peroxidative and radioprotective activity in hepatic and renal tissues in diabetic rats; protect DNA damage and genomic instability, scavenge radical-induced damage	Kamalakkannan and Prince (2004)
Allium cepa Onion Lilliaceae	Bulb, leaf	Jaundice, asthma, epilepsy, paralysis	Diallyl disulphide inhibits stomach cancer; quercetin may cure many cancers; oil is an effective antioxidant against the oxidative damage caused by nicotine	Campos et al. (2003)
Allium sativum Garlic Liliaceae	Bulb, leaf	Cardiovascular diseases	Allicin inhibits cancers of stomach, liver, colon, breast, endometrium; sulfur compounds inhibit cancer cells proliferation; anti-inflammatory, anti-diabetic	Surveswaran et al. (2006)

(continued)

Table 13.1 (continued)

Name of plant/common name/family	Parts used	Medicinal uses/properties	Mechanism of action	Reference
Aloe vera Aloe, ghirt kumari Aloeaceae	Leaf	Cancer, inflammation, asthma, constipation, diabetes, skin ailments	Strong radical scavenging and reducing properties, Mucopolysaccharides are stimulates to release TNF, interferons and interleukins; aloctin-A act as immunomodulatory	Hu et al. (2003)
Alpinia galanga Galangal Zingiberaceae	Rhizome	Indigestion, bronchial catarrh, rheumatism, anti-helmintic, anti-diuretic, anti-ulcerative, anti-dementia, fever.	Radical scavenging activity, reducing properties and metal ion chelating and beta carotene bleaching property	Mahae and Chaiseri (2009), Wong et al. (2009)
Amaranthus paniculatus Rajgriha Amaranthaceae	Leaves, seeds	Boils, eczema	Rich source of beta carotene, ascorbic acid and folates; free radical scavenger or hydrogen donor	Amin et al. (2006), Ali et al. (2008)
Annona squamosa Custard apple, seetaphal Annonaceae	Fruits	Constipation, burning sensation, anemia, vomiting, cough, malignancy, general tonic	Reduces the lipid peroxidation and increases the activity of antioxidant enzymes and strong superoxide radicals and singlet oxygen quenchers	Surveswaran et al. (2006), Kaleen et al. (2006)
Aristolochia bracteolata Indian birthwort Aristolochiaeae	Seeds, whole plant	Intestinal worms, constipations, inflammation, amenorrhea, dysmenorrhea, foul ulcer, boils, painful joints, skin disease, eczema	Positive effect on wound healing, significant increase in the level of two powerful antioxidant enzymes, superoxide dismutase and catalase, in the granuloma tissue	Shirwaikar et al. (2003),
Thirugnanasampandan et al. (2008)				
Asparagus racemosus Satavari Liliaceae	Roots	Boils, small pox, diarrhea, dyspepsia, spasm	Strong scavenging, metal chelation, reduction power and inhibition of lipid peroxidation	Kamat et al. (2000)
Berberis aristata Kilmora Berberidaceae	Roots, bark, fruit	Skin diseases, jaundice, infections of eyes	Radical scavenging properties	Singh and Kakkar (2009), Andola et al. (2011)

(continued)

Table 13.1 (continued)

Name of plant/common name/family	Parts used	Medicinal uses/properties	Mechanism of action	Reference
Berberis asiatica Daruhaldi Berberidaceae	Roots, bark, fruit	Skin diseases, jaundice, infections of eyes, malaria, healing ulcers	Reducing and radical scavenging properties	Surveswaran et al. (2006), Andola et al. (2011)
Boerhavia diffusa Spreading hogweed, punarnava Nyctaginaceae	Root	Inflammations, leucorrhoea, ophthalmia, lumbago, myalgia, scabies, cardiac disorders, jaundice, anemia, cough, gonorrhea, dyspepsia, constipation, bronchitis, general debility	Decrease level of thiobarbituric acid reactive substances (TBARS) and increases the activity of glutathione peroxidase (GPX) and glutathione-S- transferase	Satheesh and Pari (2004), Surveswaran et al. (2006)
Brassica oleracea Cabbage Brassicaceae	Leaf	Biliousness, urinary diseases	Anticancer against cancers of bladder, lung, stomach, colon, rectum, breast, cardiotonic, antiinflammatory	Kusznierewicza et al. (2008)
Camellia sinensis Green tea, chai Brassicaceae	Leaf	Hemicranias, pain in heart, eye trouble, piles, inflammation, blood purifiers	Radical scavenging and antioxidant activity; Catechins are useful for heart and lung diseases	Surveswaran et al. (2006)
Capsicum annum Red chili Solanaceae	Fruit	Peptic ulcer, cardiac diseases	Stimulant, stomachic given in neuralgia, rheumatism, sore throat, diarrhea	Materska and Perucka (2005), Sun et al. (2007)
Carum copticum Kusum phool Apiaceae	Seed	Gastric trouble, amenorrhea, blood vomiting with bile, debility, dyspepsia eye wash, fever, dysuria	Radical scavenging and antioxidant activity	Wojdylo et al. (2007)
Centella asiatica Brahmi Apiaceae	Leaves	Insomnia, epilepsy, asthma, bronchitis, abdominal disorders, mental retardation, insanity, fever	Phenolics and ascorbic acid posses strong radical scavenging and antioxidant properties	Jayashree et al. (2003), Gupta and Prakash (2009)
Chenopodium album Goose foot, bathua Chenopodiaceae	Whole plant, leaf, seed	Peptic ulcer, cardiac diseases	Radical scavenging and antioxidant activity	Zhang and Hamauzu (2004), Chludil et al. (2008)

(continued)

Table 13.1 (continued)

Name of plant/common name/family	Parts used	Medicinal uses/properties	Mechanism of action	Reference
Coccinia indica Kovai Cucurbitaceae	Fruits, leaf	Diabetes mellitus, skin diseases, gonorrhea	Hypoglycemic activity; decrease the activity of the enzyme glucose-6-phosphatase; protected lipid peroxidation and enhance effect on cellular antioxidant defence against oxidative damage in streptozotocin diabetes	Venkateswaran and Pari (2002a)
Coscinium fenestratum Tree turmeric Menispermaceae	Whole plant	Diabetes, dyspepsia	Antioxidant activity due to the presence of berberine and phenolic compounds	Punitha et al. (2005)
Cucumis sativus Cucumber, kheera Cucurbitaceae	Fruit, seed	Bronchitis, jaundice, tumor, hemorrhage	Hemostatic, diuretic, tonic, astringent, antipyretic	Mongkolsilp et al. (2004)
Curculigo orchioides Black musali Cucurbitaceae	Rhizome	Hepatitis, piles, irritable bowel syndrome, erectile dysfunction, low libido, low sperm count, low sperm, diuretic, skin diseases	Radical scavenging and antioxidant activity	Surveswaran et al. (2006), Wojdylo et al. (2007)
Curcuma longa Turmeric, haldi Zinziberaceae	Rhizome	Common cold, leprosy, liver disease, dropsy, inflammation, wound healing contraception, swelling, whooping cough, skin disease	Active against esophagus, colon, liver, bladder and prostate cancer, leukemia, fibrosarcoma, stomach papilloma; hepatoprotective, hypo-lipidemic cardiotonic, antidiabetic	Ramadan et al. (2011), Aggarwal et al. (2007)
Curcuma zedoaria Zedoary Zinziberaceae	Rhizome	Menstrual disorders, abdominal discomfort, ulcers, dyspepsia, flatulence, indigestion, vomiting, coughs, blood circulation, liver diseases	Radical scavenging and antioxidant activity	Mongkolsilp et al. (2004), Surveswaran et al. (2006)
Cynodon dactylon Doob grass Poaceae	Whole plant	Hysteria, epilepsy, insanity	Antioxidant properties using ABTS assay	Auddy et al. (2003)

(continued)

Table 13.1 (continued)

Name of plant/common name/family	Parts used	Medicinal uses/properties	Mechanism of action	Reference
Cyperus rotundus Coco grass Cyperaceae	Tuber	Inflammation, degenerative disorders, slowing the aging process, spasms, stomach disorder, inflammatory diseases	Antimutagenic and radical scavenging property	Shanmugasundaram et al. (1994), Kilani et al. (2005)
Decalepis hamiltonii – *Asclepiadaceae*	Roots	Blood purifier, appetiser, indigestion and constipation	2- hydroxyl-4-methoxy benzaldehyde showed radical scavenging activity and up-regulate the expression of genes encoding the enzyme	Murthy et al. (2006)
Diospyros malabarica Gab, tinduk Ebenaceae	bark	Intermittent fever and healing of wound ulcer	Strong reactive nitric species inhibitor	Surveswaran et al. (2006), Mondal et al. (2005)
Diospyros peregrine – Ebenaceae	Fruits	Skin diseases, wounds and ulcers, blood disorders and hyperacidity, digestive disorders	Hypoglycemic and hypolipidemic activity through antioxidant defence mechanism	Dewanjee et al. (2009)
Evolvulus alsinoides Shankhpushpi Convolvulaceae	Whole plant	Insanity, epilepsy, nervous debility	Inhibition of lipid peroxidation and ABTS scavenging activity	Auddy et al. (2003)
Ficus racemosa Umbar Moraceae	Stem bark	Dysentery, piles, diarrhea	Radioprotective activity	Veerapur et al. (2009)
Gentiana kurroo Gentian Gentianaceae	Root, leaf	Tonic, stomachic, urinary infections	Radical scavenging and antioxidant activity	Wojdylo et al. (2007)
Glycyrrhiza glabra Liquorice, mulethi Fabaceae	Root	Eyelid disorder, sight loss, remove biliousness, hiccough, vomiting, bronchitis, abdominal pain, snake bite	Inhibit radiation induced lipid peroxidation	Naik et al. (2003), Surveswaran et al. (2006)
Hedychium spicatum Van haldi Zingiberaceae	Rhizome	Stimulant, tonic, carminative, stomachic, dyspepsia, nausea, vomiting, diarrhea, pain, inflammation	Radical scavenging and reducing antioxidant properties	Joshi et al. (2008), Rawat et al. (2011a)

(continued)

Table 13.1 (continued)

Name of plant/common name/family	Parts used	Medicinal uses/properties	Mechanism of action	Reference
Hygrophila auriculata *Talamakhana* Acanthaceae	Leaf, fruits	Diuretic, aphrodisiac, asthma, jaundice, liver ailments	Antitumor, hypoglycaemic antibacterial, free radical scavenging and lipid peroxidation activities	Singh and Handa (1995), Mazumdar et al. (1997)
Juglans regia Akhrot Juglanadaceae	Fruit kernel leaf	Laxative, anthelmintic, herpes, eczema, scrofula, syphilis	Radical scavenging and antioxidant activity	Surveswaran et al. 2006; Jain et al. (2007)
Lycopersicon esculentum Tomato, tamatar Solanaceae	Fruit	Blood purifier, digestive, antiseptic, lung, asthma, bronchitis, dyspepsia, heart diseases	Antioxidant, anticancer, free radical scavenger	Raffo et al. (2006), Sbartai et al. (2011)
Mentha arvensis *Japanese mint* Lamiaceae	Leaf	Kidney, liver, spleen diseases, asthma, pain in joints, jaundice, vomiting fever, indigestion, cephalagia	Essential oil and plant extract showed antioxidant and radical scavenging activity	Saini et al. (2007)
Mentha spicata Spearmint Laminaceae	Leaf	Fever, bronchitis preventing vomiting, neuralgia, indigestion, diarrhea	Antioxidant, antiinflammatory, antiulcerogenic	Arumugam and Ramesh (2009), Mandana et al. (2011)
Momordica charantia Bitter melon Cucurbitaceae	Leaf, fruit, seed	Asthma, bronchitis	Active against colon, breast, bladder and prostate cancers, lymphoma, leukemia; antidiabetic	Naik et al. (2003)
Momordica dioica Sweet melon Cucurbitaceae	Roots, leaves	Asthma, bronchitis, liver disease	Flavonoid have hepatoprotective activity	Jain et al. (2008)
Murraya koenigii Curry leaf Rutaceae	Leaf	Stomachic, carminative, hysteria, diarrhea, dysentery	Phenolics and ascorbic acid posses strong radical scavenging and antioxidant properties	Gupta and Prakash (2009)
Myrica esculanta Box myrtle, kafal Myricaceae	Fruits	Diarrhea, dysentery, chronic bronchitis	Strong antioxidant and radical scavenging properties	Rawat et al. (2011b)

(continued)

Table 13.1 (continued)

Name of plant/common name/family	Parts used	Medicinal uses/properties	Mechanism of action	Reference
Nardostachys jatammansi Jatamansi Valerianaceae	Root	Hysteria, hair loss, dysmenorrhea, insomnia, skin disease, throat trouble, lumbago, ulcers, rheumatism, paralysis, knee pain, heart disease, snake bite and burn	Phenolic antioxidants contain anticholinesterase activity	Ahmad et al. (2009)
Nelumbo nucifera Kamal, lotus Nymphaeaceae	Plant, seed	Neurasthenia, uterine hemorrhage, leprosy, liver diseases	Antioxidant, antipyretic, demulcent, diuretic	Surveswaran et al. (2006), Rai et al. (2006)
Ocimum basilicum Krishna tulsi Lamiaceae	Leaf	Heart diseases, blood biliousness, leucoderma, burning sensation, itch asthma, chronic pain of joints, inflammation, enlarge spleen, dysentery	Radical scavenging and antioxidant activity	Aqil et al. (2006), Surveswaran et al. (2006)
Ocimum sanctum Tulsi Lamiaceae	Whole plant, seed	Heart diseases, blood biliousness, leucoderma, burning sensation, itch asthma, chronic pain of joints, inflammation, enlarge spleen, dysentery	Radical scavenging and antioxidant activity	Gupta et al. (2006), Wojdylo et al. (2007)
Origanum vulgare *Van tulsi* Lamiaceae	Whole plant	Cough influenza, toothache, fever, diarrhea, colic rheumatism, hysteria, epilepsy, headache, inflammation, paralysis, bronchitis	Radical scavenging and antioxidant activity	Surveswaran et al. (2006)
Phaseolus vulgaris French bean Fabaceae	Fruit	Diabetes	Antioxidant, anticancer, urine flow stimulator	Venkateswaran and Pari (2002b)
Phyllanthus emblica Amla Euphorbiaceae	Fruit	liver disorders	Radical scavenging and antioxidant reducing activity	Poltanov et al. (2009), Khopde et al. (2001)

(continued)

Table 13.1 (continued)

Name of plant/common name/family	Parts used	Medicinal uses/properties	Mechanism of action	Reference
Phyllanthus niruri Bhumyamalki Euphorbiaceae	Fruit	Hepatitis, viral disease	Inhibition of the carbon tetrachloride induced formation of lipid peroxides in the liver of rats; preventive and curative activities against chemically induced oxidative stress in mice	Chatterjee et al. (2006)
Phyllanthus urinaria Pitirishi, budhatri Euphorbiaceae	Fruit	Soar throat, boils, infantic cheeks, tongue trash, arthralgia, snake, centipede bite, fever, ophthalmia, hepatobiliary disease	Antioxidant activity, free radical scavenging, superoxide anion radical scavenging, hydrogen peroxide scavenging, nitric oxide scavenging, reducing power and metal ion chelating activities	Kumaran and Karunakaran (2007)
Picrorhiza kurrooa Kutki Scrophulariaceae	Root	Burning sensation, constipation, anorexia, cough, asthma, bronchitis, jaundice, anemia, headache, cholera, dropsy, diabetes, inflammation, bile trouble, leucoderma, leprosy	Therapeutic action on gastric ulcers predominantly by its antioxidant properties	Ray et al. (2002), Chauhan et al. (2008)
Plumbago zeylanica Chitrak Plumbaginaceae	Flowers	Skin diseases, Fertility effect, intestinal trouble, vesicant	Chemopreventive agent against intestinal neoplasia; strong ABTS radical scavenger; effective against superoxide and nitric oxide production	Sivakumar and Niranjali (2006)
Punica granatum Anardana Punicaceae	Seed, pericarp, rind	Bronchiticpiles, cancer, dysentery, diarrhea, bleeding	DPPH assay, 5- lipoxygenase assay and luminal/xanthine oxidase system (Chemiluminescence assay) showed a strong antioxidant activity	Surveswaran et al. (2006), Ricci and Giamperi (2006)
Rauwolfia serpentina Sarpgandha Apocynaceae	Stem	High blood pressure, nervous disorders, anxiety, excitement	Radical scavenging and antioxidant activity	Surveswaran et al. (2006)

(continued)

Table 13.1 (continued)

Name of plant/common name/family	Parts used	Medicinal uses/properties	Mechanism of action	Reference
Rubia cordifolia Indian madder Rubiaceae	Roots	Paralysis, jaundice, fever, obstruction; treatment of menstrual disorder	Neuroprotective properties via preventing the depletion and increasing GSH (glutathione) levels by inducing c-glutamylcysteine ligase expression, reducing oxidant levels by direct scavenging, and decreasing iNOS expression	Rawal et al. (2004)
Salvia officinalis Salvia Lamiaceae	Whole plant	Cough, cold, applied in wounds, chronic inflammation of skin, hemorrhoids, dysentery	Radical scavenging and antioxidant activity	Surveswaran et al. (2006)
Solanum tuberosum Alu, potato Solanaceae	Tuber	Ulcer, cough, weakness, diuretic, aperient, nutritive	Angiotensin converting enzyme inhibiting properties and radical scarenging properties	Pihlanto et al. (2008)
Sphaeranthus indicus Gorakmundi Asteraceae	Flowers	Insanity, tuberculosis, indigestion, bronchitis, spleen diseases, elephantiasis, anemia, pain in uterus, vagina, piles	The ethanolic extract scavenges radical cation, DPPH, SOD, and NO.	Shirwaikar and Prabhu (2006)
Striga orobanchoides – Scrophulariaceae	Whole plant	Antidiabetic, antiimplantation and estrogenic activity	Increase level of catalase, SOD and ascorbic acid, decrease level of thio-barbituric acid reactive substances; Flavonoids antioxidant including apigenin and luteolin	Raj and Shalini (1999), Badami et al. (2003)
Taxus baccata Himalayan yew, Thuner Taxaceae	Leaves, bark	Emmenogogue, sedative and anticancerous, used in asthma, bronchitis, hiccough, epilepsy, indigestion	Lignan derivative compounds showed antioxidant, anti-inflammatory, antinociceptive, anti-ulcerogenic, cytotoxic properties	Kucukboyaci et al. (2010)

(continued)

Table 13.1 (continued)

Name of plant/common name/family	Parts used	Medicinal uses/properties	Mechanism of action	Reference
Terminalia belerica Harad Combretaceae	Fruit	Asthma, sore throat, thirst, vomiting, hiccough, eye disorders, disease of heart and bladder, inflammation, piles, tumor	Inhibits lipid peroxidation in rat liver microsomes and free radical scavenging activity due to presence of tannins. also inhibits the development of duodenal ulcer and appeared to extract a cytoprotective effect on the gastric mucosa	Sahu and Kuttan (2009)
Thymus vulgaris Van ajwayan Lamiaceae	Whole plant	Digestive disorders	Radical scavenging and antioxidant reducing activity	Surveswaran et al. (2006)
Trigonella foenum-graecum Fenugreek, methi Fabaceae	Leaf, seed	Cancers, bronchitis, inflammation, rheumatism	Phenolics and ascorbic acid posses strong radical scavenging and antioxidant properties	Wojdylo et al. (2007), Gupta and Prakash (2009)
Valeriana jatamansi Tagar Valerianaceae	Roots	Ulcer, wounds, epilepsy, dyspepsia, flatulation, colic constipation, jaundice, dry cough, asthma, seminal weakness, neurological disorders	Antioxidant and radical scavenging properties	Kalim et al. (2010), Bhatt et al. (2012)
Zingiber officinale Adrak, ginger sunth Zingiberaceae	Rhizome	Swelling, tumors, colic, diabetes, heart disease	Antioxidant, anticancer, antipyretic, analgesic, anti-inflammatory, astringent	Surveswaran et al. (2006), Ghasemzadeh et al. (2010), Ramadan et al. (2011)

preparation of Chyawanprash, traditional ayurvedic mixture, which is known for strong antioxidant properties. Furanoid diterpene hedychenone isolated from the rhizome of the species have shown to possess antiinflammatory activities. Labdane diterpenes isolated from the rhizomes of species showed cytotoxicity against various types of cancer cells. Essential oil of the species and its extract possess strong reducing properties and free radical scavenging properties (Joshi et al. 2008; Rawat et al. 2011a). Total phenolic content varied from 2.84 to 4.69 GAE/g dw in different populations of the species (average value 3.80 mg/g dw). Antioxidant activities of the extract are ascribed due to the presence of phenolic content (Rawat et al. 2011a). The species is considered to be rich antioxidant phytochemicals (Bhatt et al. 2008).

13.4.2 Myrica Esculenta

Myrica esculenta Buch–Ham ex. D. Don, family Myricaceae, commonly known as 'Kaphal', is one of the wild edible and medicinally important plant growing between 900 and 2,100 m asl in IHR. The species is widely accepted among local peoples for its delicious fruits and its processed products. Species contain some antioxidant phenolic compounds, such as gallic acid, catechins, hydroxybenxoic acid, and coumaric acid. Total phenolic content varied from 1.78 to 2.51 GAE/g dw in different populations of the species (average value 2.12 mg/g dw) and total flavonoids from 1.31 to 1.59 QE among different populations. Also, the fruits of the species possess strong reducing properties and free radical scavenging properties with ABTS and DPPH assay which showed a significant relationship with phenolic and flavonoids content (Rawat et al. 2011b).

13.4.3 Habenaria Edgeworthii

Habenaria edgeworthii, (Family-Orchidaceae) common name Virdhi, grows in open grassy land in mountainous region in IHR with an altitudinal range of 1,500–3,000 m asl. Species is traditionally used in burning sensation, hyperdypsia, fever, cold, asthma, anemia, insanity, cataplexy, leprosy, skin diseases, anorexia, Helminthiasis, emaciation, haematemesis, gout, and general debility. The tubers are sweet, refrigerant, emollient, intellect promising, aphrodisiac, depurative, appetiser, antihelminthic, rejuvenating, and tonic. It is one of the components of 'Astvarga', which is mainly used in preparation of 'Chyavanprash' and used to cure cough, cold, calcium deficiency, and anemia. The main properties of Chyavanprash are protects against strain and stress, restores youth and vitality and gives strength and stamina. These properties are attributed to the presence of phenolic compounds in the species (Govindrajan et al. 2007). The total phenolic content of the species varied between 3.78 and 8.57 mg gallic acid equivalent/g dry weight rhizome among the different localities of Uttarakhand. Total phenolic content contributed a strong share in ferric reducing radical scavenging properties by DPPH and ABTS assay (Rawat et al. 2012).

13.4.4 Valeriana Jatamansi

Valeriana jatamansi Jones syn *V. wallichi* (Family Valerianaceae) commonly known as 'Tagar or Indian valerian, is a wild herb commonly distributed in sub-tropical and temperate Himalaya. The species grow in temperate zone of the western Himalaya at an altitudinal range of 1,200–3,300 m asl. The species is used as an aromatic, stimulant, carminative, and antispasmodic in Ayurvedic medicine especially in the preparation of Sudarshan churan, Darsan gaylep, Papalyasava, etc. The plant is widely known for its use in anxiety, insomnia, epilepsy, failing reflexes, hysteria, neurosis, sciatica, tranquiliser, emmenagogue, diuretic, and hepatoprotective (Baquar 1989). Plant extract of *V. jatamansi* has been reported to attenuate stress, anxiety, and symptoms of depression (Bhattacharya et al. 2007). The species has found beneficial for cerebrospinal system, hypochondriasis, insomnia, migraines, nervous unrest, nervous tension, neuralgia, and neuroasthemia. Valerian is reported for depressant action on the central nervous system and antispasmodic activity (Cionga 1961). Pharmacological screening of valerenal and some other components of *Valeriana* showed that the sedative action can be attributed to the essential oil and valepotriates fractions (Wagner et al. 1980). Valerenic acid inhibits the enzyme system responsible for central catabolism of GABA and the valerian extract releases [3H] GABA by reversal of the GABA carrier, which is Na (+) dependent and Ca (2+)-independent (Santos et al. 1994). The valepotriates (valtrate and didrovaltrate) of the species have been reported to exert a spasmolytic effect (Wagner et al. 1980). *V. jatamansi* essential oil exhibited antimicrobial activity against large number of pathogenic bacteria and antifungal activity against fungal pathogens (Thind et al. 1979). Antiinflammatory activity of the species in both methanolic and ethanolic extract are reported and are known to inhibit inflammation mediators, such as histamine, serotonin, prostaglandins, and bradykinins, etc. (Subhan et al. 2007). The species showed strong antioxidant activity in different types of system models, such as, scavenging of DPPH, ABTS+, nitric oxide, hydroxyl radical, peroxinitrile, nonenzymatic superoxide radicals, and prevent oxidative DNA damage (Kalim et al. 2010; Bhatt et al.2012).

13.4.5 Acorus Calamus

Acorus calamus Linn., commonly known as sweet flag or 'Bach' in India, is found growing wild in abundance ascending to 2,200 m in the Himalaya. Species is widely used for the treatment of epilepsy, chronic diarrhea, dysentery, bronchial catarrh and abdominal tumors and as analgesic for the relief of toothache or headache, for oral hygiene to cleanse and disinfect the teeth, relief the effects of exhaustion or fatigue. Methanolic extract showed that during exposure of noisy environment ROS generation led to increase in corticosterone, lipid peroxidase, and sulphoxide dismutase, but decrease in catalase, glutathione peroxidase, glutathione, protein thiols, vitamins C and E levels. Both the ethyl acetate and methanolic extract of *A. calamus* protected most of the changes in the rat brain induced by noise-stress (Manikanandan et al. 2005). Species showed presence of phenolic content ranged between 6.85 and 9.97 gallic acid equivalents (GAE) mg/g in acetone extract among the different populations. Antioxidant activity measured by ABTS assay, scavenging properties by DPPH assay, and reducing properties by FRAP assay showed a significant relationship with total phenolic content. The species showed potent against fish pathogen *Aeromonas hydrophila* (Bhuvaneswari and Balasundaram 2009). The antifungal activity of crude methanolic extract of *A. calamus* was reported (Ghosh 2006; Singh et al. 2010). The essential oil of *A. calmus* exhibited antibacterial activity against phytopathogenic bacteria and antioxidant activity of crude methanol extract of rhizome and leaf extract of *A. calamus* is reported (Phongpaichit et al. 2005; Asha Devi and Ganjewala 2009).

13.4.6 Roscoea Procera

Roscoea procera, family Zingiberaceae, is one of the important Himalayan medicinal plant distributed from Himachal Pradesh to Arunachal Pradesh between 1,800 and 3,000 m asl. The species is traditionally used as a tonic in seminal debility, malaria, and in many other folk medicines. The species is one of the ingredients of a polyherbal formulation 'Ashtavarga' which is used in the preparation of Ayurvedic formulation 'Chyavanprash'. The Chyavanprash is categorized into Rasayana group of drugs which is a rich source of antioxidant, good hepatoprotective and immunomodulating agent with nutritive, antiaging, and many other medicinal properties. Total phenolic and flavonoids content in the species were found 2.21–3.58 mg gallic acid equivalent/gram and 3.87–5.52 mg quercetin equivalent/gram fresh weight, respectively. Also, species showed ferric reducing antioxidant properties and free radical scavenging properties with ABTS and DPPH assay which showed a significant relationship with phenolic and flavonoids content (Rawat et al. 2012).

13.4.7 Berberis Asiatica

Berberis asiatica, family Berberidaceae is used for traditional system of medicine since historical time. In modern system of medicine, species is being used for preparation of drugs to cure various diseases, including eye-related disorders, intermittent fevers, as well as malaria, promoting the flow of bile, jaundice, inflammation of the gall-bladder, improving appetite, digestion, and assimilation. In addition, the fruits of species are well known for edibility value. Various antioxidant phytochemicals, such as xanthophylls, α-carotene, β- carotene, ascorbic acid, and phenolic content has been reported from the species (Andola et al. 2008). Extract of *Berberis species* (root) has reported strong potential to regulate glucose homeostasis through decreased gluconeogenesis and oxidative stress (Singh and Kakkar 2009).

13.4.8 Curcuma Longa

Curcuma longa, family Zingiberaceae is used as a dietary spice, coloring agent in foods and textiles. The major functional compound is curcumin which has been reported to reduce blood cholesterol, prevents Low Density Lipoprotein oxidation, inhibits platelet aggregation, suppresses thrombosis, myocardial infarction, symptoms associated with type II diabetes, rheumatoid arthritis, multiple sclerosis (MS), Alzheimer's disease, inhibits human immunodeficiency virus (HIV) replication, enhances wound healing, protects from liver injury, increases bile secretion, protects from cataract formation, pulmonary toxicity, and fibrosis (Aggarwal et al. 2007). *Curcuma* species are largely being used in a variety of food products due to their antioxidant properties. The species contain essential oil, including terpenes, alcohols, ketones, flavonoids, carotenoids, and phytoestrogens (Habash et al. 2000).

13.4.9 Zingiber Officinale

Ginger is the rhizome of the plant *Zingiber officinale*, family Zingiberaceae, consumed as a delicacy, medicine, or spice. 6-Zingirols is the major compound of *Zingiber* species and has reported to have high antioxidant activity, known to enhance high density lipoprotein in diabetic rats (Bhandari et al. 2005), reduced lipid peroxidation tissues (Bernd et al. 1997), and possess scavenging and high chelating capacity. Moreover, *Zingiber* extract is used in curry powder, sauces, ginger bread, carbonated drink, and in preparation of dietaries due to its high antioxidant activity (Stoilova et al. 2007). *Zingiber officinale* contains phenolic compounds (gingerol) which have antioxidant activity that is even greater than α-tocopherol.

13.4.10 Morus Alba

Morus alba, family Moraceae commonly known as white mulberry, is a short-lived, fast-growing, small to medium-sized tree. The species is native to northern

China and is widely cultivated and naturalized elsewhere. Phenols and flavonoids rich fraction of *Morus alba* showed beneficial effect against lipid and lipoproteins and delayed the onsets of arthrosclerosis. Roots of *Morus alba* is one of the constituent of Chinese drug named "Sohaku-hi" which reduces the plasma sugar level in mice. Some of the flavonoids obtained from leaves of *Morus alba*, such as quercetin 3-(6-malonyglucoside) attenuate the arthrosclerosis lesion development in LDL receptor deficient mice through enhancement of LDL resistance to oxidation modification (Enkhmaa et al. 2005).

13.4.11 Podophyllum Hexandrum

Podophyllum hexandrum commonly known as Himalayan mayapple or Indian may apple is native to Himalaya and found between an altitude of 2,800–3,500 m asl. The plant contains podophyllotoxin, which has an antimiotic effect (interferes with cell division and thus prevent the growth of cells). The roots contain several important anticancer lignans, including podophyllin and berberine. The roots are also antirheumatic. Radioprotective and antitumour properties in *Podophyllum hexandrum* extract treated animals was reported (Mittal et al. 2001). Aqueous extract of the species is reported to protect kidney and lung tissue against CCl_4-induced oxidative stress (Ganie et al. 2011). Studies of the expression patterns of various proteins associated with inhibition of apoptosis in the spleen of male Swiss albino strain 'A' mice by immunoblotting, has been reported (Kumar et al. 2011). In vitro studies using human hepato carcinoma cell lines revealed its ability to stabilize the state of mitochondrial oxidative burst, decreased TBARS, time-dependent inhibition of gamma radiation-induced leakage of electrons in the mitochondrial electron transport chain (ETC.) via reduction in ROS and NO generation and simultaneous enhancement in the thiol status via neosynthesis (Mittal et al. 2001; Chawla et al. 2006).

13.4.12 Picrorhiza Kurroa

Picrorhiza kurroa, family Scrophulariaceae, is an important medicinal herb in Ayurvedic medicines. The roots are rich source of various chemical compounds, such as picrorhiza, kutkin, D-mannitol, glycosides, cucurbitacin, kutkisterol, steroids, and vanillic acid. The species is used in the treatment of liver cirrhosis, ascites, treatment of jaundice, constipation, dyspepsia, and promotes stomach actions and provides strength. The root extract was found cytotoxic and were able to target cells toward apoptosis (Rajkumar et al. 2011). Roots are also reported to be useful in therapeutic action on gastric ulcer (Ray et al. 2002). Cardioprotective effect of *P. kurroa* against adriamycin-induced cardiomyopathy and low dose of combined methanolic extract of *P. kurroa* possess good hepatoprotective activity against paracetamol induce liver damage (Rajaprabhu et al. 2007).

13.4.13 Bergenia Ciliata

Bergenia ciliata is a perennial herb found in the Himalayan region between 900 to 3,000 m asl. The species is used as a poultice, treating boils, curing diarrhea and vomiting, treatment of fever, cough, menorrhagia, excessive uterine hemorrhage and pulmonary infections, and kidney stone. The marketed composite herbal formulations, Cystone (Himalaya Drug Company, India), Calcuri (Charak Pharmaceuticals, Bombay, India), and Chandraprabha Vati (Baidyanath, India) have widely been used clinically to dissolve urinary calculi in the kidney and urinary bladder (Prasad et al. 2007). Rhizome extracts of the species were found to possess antioxidant activity, including reducing power, free radical scavenging activity, and lipid peroxidation inhibition potential as well as DNA protection (Rajkumar et al. 2010).

13.4.14 Asparagus Racemosus

Asparagus racemosus is commonly found in India with a tuberous roots possesses a number of valuable medicinal properties, including immunostimulation, uterine relaxation, anticancer, antiulcer, antimicrobial, etc. Tuber extract contains flavonoids, polyphenols, and vitamin-C. The species is used as a bitter, sweet, emollient, cooling, nervine tonic, constipating, galactogogue, diuretic, carminative, appetizer, stomachic, antispasmodic, and tonic. Also used in nervous disorders, dyspepsia, diarrhea, dysentery, tumors, inflammation, burning sensation, hyperdipsia, nephropathy, agalactia, and general debility. The antioxidant effects of the species against membrane damage induced by the free radicals generated through gamma radiation were examined in rat liver mitochondria (Kamat et al. 2000). Also, the extract of the species is reported to inhibit lipid peroxidation and protein oxidation in mitochondrial membranes of rat liver (Kamat et al. 2000).

13.5 Conclusion

The present review clearly indicated that antioxidants play a vital role in preventing the oxidative stress and also certain degenerative diseases. In this context, medicinal plants have been viewed as a potential player. As such, it has been estimated that 70–80 % of the world's population cannot afford modern medicine, use of medicinal plants can be an important source of natural antioxidants. The review reveals that most of the plant-based extract/compounds are beneficial in suppressing multiple diseases, more research, however, is needed to achieve the full therapeutic potential. Also, the isolation of bioactive compounds from the medicinal plants and their impact on various health improvement/control of free radical-mediated diseases through in vivo studies is needed. Such information could subsequently be exploited as cost-effective measure for the health of well-being.

References

Aggarwal BB, Ichikawa H, Garodia P, Weerasinghe P, Sethi G, Bhatt ID, Pandey MK, Shishodia S, Nair MG (2006) From traditional ayurvedic medicine to modern medicine: identification of therapeutic targets for suppression of inflammation and cancer. Expert Opin Ther Tar 10(1):87–118

Aggarwal BB, Bhatt ID, Ichikawa H, Ahn KS, Sethi G, Sandur SS, Sundaram C, Seerum N, Shishodia S (2007) Curcumin: biological and medicinal properties. Turmeric; the genus *Curcuma*. Taylor & Francis, New York, pp 297–368

Ahmed F, Chandra JNNS, Urooj A, Rangappa KS (2009) In vitro antioxidant and anticholinesterase activity of *Acorus calamus* and *Nardostachys jatamansi* rhizomes. J Pharm Res 2:830–833

Akhondzadeh S, Noroozian M, Mohammadi M, Ohadinia S (2003) *Melissa officinalis* extract in the treatment of patients with mild to moderate Alzheimer's disease: a double-blind, randomised, placebo controlled trial. J Neurol Neurosurg Psychiatr 74:863–866

Ali SS, Kasoju N, Luthra A, Singh A, Sharanabasava H, Sahu A, Bora U (2008) Indian medicinal herbs as sources of antioxidants. Food Res Int 41:1–15

Amin IY, Norazaidah KI, Hainida E (2006) Antioxidant activity and phenolic content of raw and blanched *Amaranthus* species. Food Chem 94:47–52

Andola HC, Rawal RS, Bhatt ID (2008) Antioxidants in fruits and roots of *Berberis asiatica* Rox. ex.DC.: a highly valued Himalayan plant. Natl Acad Sci Lett 31(11–12): 337–340

Andola HC, Rawal RS, Bhatt ID (2011) Comparative studies on the nutritive and anti-nutritive properties of fruits in selected *Berberis* species of West Himalaya, India. Food Res Int 44:2352–2356

Aqil F, Ahmad I, Mahmood Z (2006) Antioxidant and free radical scavenging properties of twelve traditionally used medicinal plants. Tub Tak 30:177–183

Artemis PS, Gopalan C (2003) Plants in human health and nutrition policy. Karger Publishers, Basel, Switzerland

Arumugam P, Ramesh A (2009) Antigenotoxic and antioxidant potential of aqueous fraction of ethanol extract of *Mentha spicata* (L.) against 4-nitroquinoline-1-oxide-induced chromosome damage in mice. Drug Chem Toxicol 32:411–416

Asha Devi S, Ganjewala D (2009) Antimicrobial activity of *Acorus calamus* (L.) rhizome and leaf extract. Acta Biol Szegediensis 53(1):45–49

Auddy B, Ferreira M, Blasina F, Lafon L, Arredondo F, Dajas F, Tripathi PC, Seal T, Murkerjee B (2003) Screening of antioxidant activity of three Indian medicinal plants, traditionally used for the management of neurodegenerative diseases. J Ethnopharmacol 84:131–138

Badami S, Gupta MK, Suresh B (2003) Antioxidant activity of the ethanolic extract of *Striga orobanchioides*. J Ethnopharmacol 85:227–230

Baquar SR (1989) Medicinal and poisonous plants of Pakistan. Printas, Karachi, p 461

Baynes JW, Thorpe SR (1997) The role of oxidative stress in diabetic complications. Curr Opin Endocrinol 3:277–284

Bernd A, Theilig C, Kippenberger S, Ramirez-Bosca A, Podda M, Diaz J, Miquel J, Kaufmann R (1997) An extract of *Curcuma longa* exerts anti-oxidative, antiinflammatory and antiproliferative effects on human keratinocytes in vitro. J Investig Dermatol 109:460

Bhandari U, Kanojia R, Pillai KK (2005) Effect of ethanolic extract of *Zingiber officinale* on dyslipidaemia in diabetic rats. J Ethnopharmacol 97:227–230

Bhatt ID, Prasad K, Rawat S, Rawal RS (2008) Evaluation of antioxidant phytochemical diversity in *Hedychium spicatum*: a high value medicinal plant of Himalaya. Phcog Mag 4(16):202–206

Bhatt ID, Dauthal P, Rawat S, Gaira KS, Jugran A, Rawal RS, Dhar U (2012) Characterization of essential oil composition, phenolic content, and antioxidant properties in wild and planted individuals of *Valeriana jatamansi Jones*. Sci Hort 136:61–68

Bhattacharya D, Jana U, Debnath PK, Sur TK (2007) Initial exploratory observational pharmacology of *Valeriana wallichii* on stress management: a clinical report. Ind J Exp Biol 45:764–769

Bhuvaneswari R, Balasundaram C (2009) Anti-bacterial activity of *Acorus calamus* and some of its derivates against fish pathogen *Aeromonas hydrophila*. J Med Plants Res 3(7):538–547

Brewer MS (2011) Natural antioxidants: sources, compounds, mechanisms of action and potential applications. Comp Rev Food Food Saf 10:221–246

Campos KE, Diniz YS, Cataneo AC, Faine LA, Alves MJ, Novelli EL (2003) Hypoglycaemic and antioxidant effects of onion, *Allium cepa*: dietary onion addition, antioxidant activity and hypoglycaemic effects on diabetic rats. Int J Food Sci Nutr 54(3):241–246

Chatterjee M, Sarkar K, Parames C (2006) Sil herbal (*Phyllanthus niruri*) protein isolate protects liver from nimesulide induced oxidative stress. Pathophysiology 13:95–102

Chattopadhyay RR, Chattopadhyay RN, Nandy AK, Poddar G, Maitra SK (1987) Preliminary report on anti hyperglycemic effect of fraction of fresh leaves of *Azadiracta indica* (beng neem). Bull Calcutta Sch Trop Med 35:29–33

Chauhan S, Nath N, Tule V (2008) Antidiabetic and antioxidant effects of *Picrorhiza kurrooa* rhizome extract in diabetic rats. Ind J Clin Biochem 23:238–242

Chawla R, Arora R, Singh S, Sagar RK, Sharma RK, Kumar R (2006) *Podophyllum hexandrum* offers radioprotection by modulating free radical flux: role of aryl-tetralin lignans. ECAM 3:503–511

Chludil HD, Corbino GB, Leicach RC (2008) Soil quality effects on *Chenopodium album* flavonoid content and antioxidant potential. J Agric Food Chem 56:5050–5056

Cionga E (1961) Considerations on the root of valerian. Pharmazie 16:43–44

Dahanukar SA, Kulkarni RA, Rege NN (2000) Pharmacology of medicinal plants and natural products. Ind J Pharmacol 32:81–118

Davis RH, Maro NP (1998) *Aloe vera* and gibberellins, anti-inflammatory activity in diabetes. J Am Pediatr Med Assoc 79:24–26

Desai VR, Ramkrishnan R, Chintalwar GJ, Sainis KB (2007) G1-4A, an immunomodulatory polysaccharide from *Tinospora cordifolia*, modulates macrophage responses and protects mice against lipopolysaccharide induced endotoxic shock. Int Immunopharmacol 7:1375–1386

Devasangayam TPA, Tilak JC, Boloor KK, Sane KS, Ghaskadbi SS, Lele RD (2004) Free radicals and antioxidant in human health: current status and future prospects. JAPI 52:794–804

Dewanjee S, Maiti A, Sahu R, Duam TK, Mandal V (2009) Effective control of type-2 diabetes through antioxidant defense by edible fruits of *Diospyros peregrine*. ECAM. doi:10.1093/ecam/nep080

Emerit JM, Edeas FB (2004) Neurodegenerative diseases and oxidative stress. Biomed Pharmacother 58:39–46

Enkhmaa B, Shiwaku K, Katsube T, Kitajima K, Anuurad E, Yamasaki M, Yamane Y (2005) Mulberry (*Morus alba* L.) leaves and their major flavonol quercetin 3-(6-malonylglucoside) attenuate atherosclerotic lesion development in LDL receptor-deficient mice. J Nutr 135:729–734

Esposito E, Ratilio D, Di Matteo V, Di Glulia C, Cicchino M, Algeri S (2002) A review on specific dietary antioxidant and the effects on biochemical mechanism related to neurodegenerative processes. Neuribiol Aging 23:719–735

Ferrary CKB (2000) Free radical, lipid peroxidation and antioxidants in apoptosis: implications in cancer, cardiovascular and neurological diseases. Biologia 55:581–590

Ganie SA, Haq E, Hamid A, Qurishi Y, Mahmood Z, Zargar BA, Masood A, Zargar MA (2011) Carbon tetrachloride induced kidney and lung tissue damages and antioxidant activities of the aqueous rhizome extract of *Podophyllum hexandrum*. BMC Complem Altern Med 28(11):17

Gaster B, Holroyd J (2000) St. John's wort for depression: a systematic review. Arch Intern Med 160:152–156

Ghasemzadeh A, Jaafar HZE, Rahmat A (2010) Antioxidant activities, total phenolic content in two varieties of Malaysia young ginger (*Zingiber officinale* Roscoe). Molecules 15:4324–4333

Ghosh M (2006) Antifungal properties of haem peroxidase from *Acorus calamus*. Ann Bot 98(6):1145–1153
Govindarajan R, Singh DP, Rawat AKS (2007) High performance liquid chromatographic method for the quantification of phenolics in 'Chyavanprash', a potent Ayurvedic drug. J Pharm Biomed Anal 43:527–532
Gunju L, Karan D, Chanda S, Srivastava KK, Sawhney RC, Selvamurthy W (2003) Immunomodulatory effects of agents of plant origin. Biomed Pharmacother 57:296–300
Gupta S, Prakash J (2009) Studies on Indian leafy vegetables for their antioxidant activity. Plant Foods Hum Nutr 64:39–45
Gupta S, Mediratta PK, Singh S, Sharma KK, Shukla R (2006) Antidibetic, antihypercholestrolaemic and antioxidant effects of *Ocimum sanctum* seed oil. Ind J Exp Biol 44:300–304
Habash M, Amran M, Mackeen MM, Lajis NH, Kikuzaki H, Nakatani H, Rahman A, Ghafar A, Ali AM (2000) Screening of zingiberaceae extract for antimicrobial and antioxidant activities. J Ethanopharmacol 92:403–410
Halliwell B (1994) Free radicals, antioxidants and human diseases: curiosity, cause and consequences. Lancet 344:721–724
Halliwell B (2007) Flavonoids: a re-run of the carotenoids story? Dietary supplements and health. Novartis Foundation Symposium Chichester, Willey, pp 93–104
Hu Y, Hu J, Hu Q (2003) Evaluation of antioxidant potential of *Aloe vera* (*Aloe barbadensis* Miller) extracts. J Agric Food Chem 51:7788–7791
Ismail S, Asad M (2009) Immunomodulatory activities of *Acacia catechu*. Ind J Physiol Pharmacol 53:25–33
Jain AK, Sikarwar MK, Dubey SK, Jain AK (2007) Antioxidant activity of ethanolic extract of fruits of *Juglans regia* (L.). Int J Plant Sci 2:55–56
Jain A, Soni M, Deb A, Jain A, Rout SP, Gupta VB, Krishna KL (2008) Antioxidant and hepatoprotective activity of ethanolic and aqueous extracts of *Momordica dioica* Roxb. Leaves J Ethnopharmacol 115:61–66
Jayashree G, Muraleedhara GK, Sudarslal S, Jacob VB (2003) Anti-oxidant activity of *Centella asiatica* on lymphoma-bearing mice. Fitoterapia 74:431–434
Joshi S, Chanotiya CS, Agarwal G, Prakesh O, Pant AK, Methela CS (2008) Terpenoide compositions and antioxidant and antimicrobial properties of the rhizome essential oil of different *Hedychium* species. Chem Biodivers 5:299–309
Kaleen M, Asif M, Ahmed QU, Bano B (2006) Antidiabetic and antioxidant activity of *Annona squamosa* extract in streptorotocin indused diabetic rats. Singap Med J 47:670–675
Kalim MD, Bhattacharyya D, Banerjee A, Chattopadhyay S (2010) Oxidative DNA damage preventive activity and antioxidant potential of plants used in unani system of medicine. BMC Complem Altern Med 10:77
Kamalakkannan N, Prince P (2004) Antidiabetic and antioxidant activity of *Aegle marmelos* extract in streptozotocin-induced diabetic rats. Pharma Biol 42:125–130
Kamat JP, Boloor KK, Devasagayam TP, Venkatachalam SR (2000) Antioxidant properties of *Asparagus racemosus* against damage induced by gamma-radiation in rat liver mitochondria. J Ethnopharmacol 7:1425–1435
Kamboj VP (2000) Herbal medicine. Curr Sci 78(1):35–38
Kanowski S, Hoerr R (1997) Proof of the efficacy of the Gingko biloba special extract egb761 in outpatients suffering from mild to moderate primary degenerative dementia of the Alzheimer type of multi-infarct dementia. Phytomed 4:215–222
Karlsen A, Retterstøl L, Laake P et al (2007) Anthocyanins inhibit nuclear factor-kappa B activation in monocytes and reduce plasma concentrations of pro-inflammatory mediators in healthy adults. J Nutr 137(8):1951–1954
Kartik R, Rao CV, Pushpangadan P, Trivedi SP, Reddy GD (2010) Exploring the protective effects of *Abrus precatorius* in Hep G2 and N-nitrosodiethylamine-induced hepatocellular carcinoma in swiss albino rats. Ir J Pharm Sci 6:99–114
Karunanayake EH, Welihinda J, Sirimanne SR, Sinnadorai G (1984) Oral hypoglycemic activity of some medicinal plants of Sri Lanka. J Ethnopharmacol 11:223–231

Khopde SM, Priyadarshni KI, Mohan H, Gawandi VB, Satav JV, Banavaliker MM, Biyani MK, Mittal JP (2001) Characterizing the antioxidant activity of amla (*Phyllanthus emblica*) extract. Curr Sci 81:85–91

Kilani S, Ammar RB, Bouhlel I, Abdelwahed A, Hayder N, Mahmoud A, Ghedira K, Chekir-Ghedira L (2005) Investigation of extracts from (Tunisian) *Cyperus rotundus* as antimutagens and radical scavengers. Environ Toxicol Pharmacol 20:478–484

Kim HL, Streltzer J, Goebert D (1999) St. John's wort for depression: a meta-analysis of well-defined clinical trials. J Nerv Ment Dis 187:532–539

Kubish HM, Vang J, Bray TM, Phillips JP (1997) Targeted over expression of Cu/Zn superoxide dismutase protects pancreatic beta cells against oxidative stress. Diabetes 46:1563–1566

Kucukboyaci N, Orhan I, Ener BS, Nawaz SA, Iqbal MC (2010) Assessment of enzyme inhibitory and antioxidant activities of lignans from *Taxus baccata*. L Z Naturforsch 65c:187–194

Kumar R, Singh PK, Arora R, Chawala R, Sharma RK (2011) Radioprotective activities of *Podophyllum hexandrum*: current knowledge of the molecular mechanisms. Trees Life J 4:1–9

Kumaran A, Karunakaran RJ (2007) In vitro antioxidant activities of methanol extracts of five *Phyllanthus* species from India. LWT (Food Sci Technol) 40:344–352

Kusznierewicza B, Bartoszekb A, Wolskaa L, Drzewieckic J, Gorinsteind S, Namiesinik J (2008) Partial characterization of white cabbages (*Brassica oleracea* var. *capitata f. alba*) from different regions by glucosinolates, bioactive compoundsbioactive compounds, total antioxidant activities and proteins. LWT Food Sci Technol 41:1–9

Mahae N, Chaiseri S (2009) Antioxidant activities and antioxidative components in extracts of *Alpinia galanga* (L.) Sw. Kasetsart J Nat Sci 43:358–369

Mandana B, Russly AR, Ali G, Farah ST (2011) Antioxidant activity of spearmint (*Mentha spicata* L.) leaves extracts by supercritical carbon dioxide (SC-CO_2) extraction. Int Food Res J 18:543–547

Manikanandan S, Srikumar S, Yeya PN, Devi RS (2005) Protective effect of *Acorus calamus* Linn on free radical scavengers and lipid peroxidation in discrete regions of brain against noise stress exposed rat. Biol Pharm Bull 28:2327–2330

Mark PM (2004) Metal-catalyzed disruption of membrane protein and lipid signaling in the pathogenesis of neurodegenerative disorders. Ann NY Acad Sci 1012:37–50

Mashour NH, George IL, William HF (1998) Herbal medicine for the treatment of cardiovascular disease. Arch Intern Med 158:2225–2234

Materska M, Perucka I (2005) Antioxidant activity of the main phenolic compounds isolated from hot pepper fruit (*Capsicum annuum* L.). J Agric Food Chem 53:1750–1756

Matteucci E, Giampietro O (2000) Oxidative stress in families of type 1 diabetic patients. Diabetes Care 23:1182–1186

Mazumdar UK, Gupta M, Maiti S, Mukhejee D (1997) Antitumor activity of *Hygrophila spinosa* in ehrlich ascites carcinoma and sarcoma-180 induced mice. Ind J Exp Biol 35:473–477

McGregor GP, Biesalski HK (2006) Rationale and impact of vitamin C in clinical nutrition. Curr Opin Clin Nutr Mctabol Care 9(6):697–703

Mehrotra S, Mishra KP, Maurya R, Srimal RC, Yadav VS, Pandey R, Singh VK (2003) Anticellular and immunosuppressive properties of ethanolic extract of *Acorus calamus* rhizome. Int Immunopharmacol 3:53–61

Miguel MG (2010) Antioxidant activity of medicinal and aromatic plants. A Rev Flavour Fragr J 25:291–312

Mittal A, Pathania V, Agrawala PK, Prasad J, Singh S, Goel HC (2001) Influence of *Podophyllum hexandrum* on endogenous antioxidant defence system in mice: possible role in radioprotection. J Ethnopharmacol 76:253–262

Modak M, Dixit P, Lonthe J, Ghaskabdi A, Devasagayam TPA (2007) Indian herbs and herbal drugs used for the treatment of diabetes. Recent Adv Ind Herb Drug Res 40:163–173

Mondal SK, Chakraborty G, Gupta M (2005) In vitro antioxidant activity of *Diospyros malabarica* Kostal bark. Ind J Exp Biol 44:39–44

Mongkolsilp S, Pongbupakit I, Sae-Lee N, Sitthithaworn W (2004) Radical scavenging activity and total phenolics content of medicinal plants used in primary health care. SWU J Pharm Sci 9:32–35

Mungantiwar AA, Nair AM, Shinde UA, Dixishit VJ, Saraf MN, Thakur VS, Sainis KB (1999) Studies on the immunomodulatory effects of *Boerhavia diffusa* alkaloidal fraction. J Ethnopharmacol 65:125–133

Murthy KN, Rajasekaran T, Giridhar P, Ravishankar GV (2006) Antioxidant property of *Decalepis hamiltonii* Wight and Arn. Ind J Exp Biol 44:832–837

Naik GH, Priydarshni KI, Satav JG, Banavalikar MM, Soholi DP, Biyani MK, Mohan H (2003) Comparative antioxidant activity of individual herbal components used in ayurvedic medicine. Phytochemistry 63:97–104

Naziroglu M, Cay M (2001) Protective role of intraperitoneally administered vitamin E and selenium on the oxidative defense mechanisms in rats with diabetes induced by streptozotocin. Biol Stress Elem Res 47:475–488

Nijveldt RJ, van Nood E, van Hoorn DEC, Boelens PG, van Norren K, van Leeuwen PAM (2001) Flavonoids: a review of probable mechanisms of action and potential applications. Am J Clin Nutr 74:418–425

Oberlay LW (1988) Free radicals and diabetes. Free Rad Biol Med 5:113–124

Pal RS, Ariharasivakumar G, Girhepunje K, Upadhyay A (2009) In vitro antioxidant Activity of phenolic and flavonoids compounds extracted from seeds of *Abrus precatorius*. Int J Pharm Pharm Sci 1:136–140

Peter PZ, Anthon JC, Khachaturian AS, Stone SV, Deborah G, JoAnn T, Tschanz MC, Norton KA, Welsh-Bohmer, Breitner JCS (2004) Reduced risk of Alzheimer disease in users of antioxidant vitamin supplements-the cache county study. Arch Neurol 61:82–88

Phongpaichit S, Pujenjob N, Rukachaisirikul V, Ongsakul M (2005) Antimicrobial activities of the crude methanol extract of *Acorus calamus* Linn. M J Sci Technol 26:517–523

Pihlanto A, Akkanen S, Korhonen HJ (2008) ACE inhibitory and antioxidant properties of potato (*Solanum tuberosum*). Food Chem 108:104–112

Poltanov EA, Shikov AN, Dorman HJD, Pozharitskaya ON, Makarow VG, Tikhonow VP, Hiltunen (2009) Chenical and antioxidant ecaluation of Indian Gooseberry (*Embelica officinale* Gaertn., syn *Phyllanthus enblica* L.) suplimants. Phytother Res 23:1309–1315

Prasad KVSRG, Sujatha D, Bharathi K (2007) Herbal drugs in urolithiasis—a review. Pharmacog Rev 1:176–179

Prochazkova D, Bousova I, Wilhelmova N (2011) Antioxidant and prooxidant properties of flavonoids. Fitoterapia 82(4):513–523

Punitha ISR, Rajendran K, Shirwaikar A, Shirwaikar A (2005) Alcoholic stem extract of *Coscinium fenestratum* regulates carbohydrate metabolism and improves antioxidant status in streptozotocin nicotinamide induced diabetic rats. ECAM 2:375–381

Raffo A, Malfa GL, Fogliano V, Maiania G, Quaglia G (2006) Seasonal variations in antioxidant components of cherry tomatoes (*Lycopersicon esculentum* cv. Naomi F1). J Food Compos Anal 19:11–19

Rai P, Wahile A, Mukhargee K, Saha B, Mukhargee RK (2006) Antioxidant activity of *Nelumbo nucifera* (sacred lotus) seed. J Ethanopharmacol 104:322–327

Raj KJ, Shalini K (1999) Flavonoids—a review of biological activities. Ind Drugs 36:668–676

Rajaprabhu D, Rajesh R, Jayakumar R, Buddhan S, Ganesan B, Anandan R (2007) Protective effects of *Picrorhiza kurroa* on antioxidant defense status in adrimyci-induced cardimyopathy in rats. J Med Plant Res 1:080–085

Rajkumar V, Guha G, Kumar RK, Mathew L (2010) Evaluation of antioxidant activity of *Berginia ciliata* rhizome. Rec Nat Prod 4:38–48

Rajkumar V, Guha G, Kumar RA (2011) Antioxidant and anti-neoplastic activities of *Picrorhiza kurro*a extracts. Food Chem Toxicol 49:363–369

Ramadan G, Al-Kahtani MA, El-Sayed WM (2011) Anti-inflammatory and anti-oxidant properties of *Curcuma longa* (turmeric) versus *Zingiber officinale* (ginger) rhizomes in rat adjuvant-induced arthritis. Inflammation 34:291–301

Ramakrishnappa K (2002) Impact of cultivation and gathering of medicinal plants on biodiversity: case studies from India. Biodiversity and the ecosystem approach in agriculture, forestry and fisheries. FAO, Rome

Ramalakshmi K, Rahath KI, Rao LJM (2007) Antioxidant potential of low-grade coffee beans. J Food Sci 41:96–103

Ranjith MS, Ranjitsingh AJA, Shankar SG, Vijayalaksmi GS, Deepa K, Sidhu HS (2008) Enhanced phagocytosis and antibody production by *Tinospora cordifolia*: a new dimension in immunomodulation. Afr J Biotechnol 7:81–85

Rates SMK (2001) Plants as source of drugs. Toxicon 39:603–613

Rawal A, Muddeshwar M, Biswas S (2004) Effect of *Rubia cordifolia, Fagonia cretica linn,* and *Tinospora cordifolia* on free radical generation and lipid peroxidation during oxygen–glucose deprivation in rat hippocampal slices. Biochem Biophys Res Commun 324:588–596

Rawat S, Bhatt ID, Rawal RS (2011a) Variation in total phenolic compounds and antioxidant potential of *Hedychium spicatum* Buch. Ham. ex D. Don in West Himalaya. India J Food Compos Anal 24:574–579

Rawat S, Jugran A, Giri L, Bhatt ID, Rawal RS (2011b) Assessment of antioxidant properties in fruits of *myrica esculenta*: a popular wild Edible species in Indian Himalayan region. ECAM. doi:10.1093/ecam/neq055

Rawat S, Andola H, Giri L, Dhyani P, Jugran A, Bhatt ID, Rawal RS (2012) Assessment of nutritional and antioxidant potential of selected vitality strengtheningHimalayan medicinal plants. Int J Food Prop (in press)

Ray R, Chaudhary SR, Majumdar B, Bandhopadhyay SK (2002) Antioxidant activity of ethanolic extract of *Picrorhiza kurroa* on indomethacin induces gastric ulcer during healing. Ind J Clin Biochem 17:44–51

Ricci D, Giamperi L (2006) Antioxidant activity of *Punica granatum* fruits. Fitoterapia 77:310–312

Sahu MC, Kuttan R (2009) Antidiabetic and antioxidant properties of *Terminalia belerica.* Ind J Exp Biol 47:270–275

Saini SA, Muruganandam AV, Agrawal S (2007) Antioxidant activities of some essential oils. Ind Drugs 44:236–238

Santos MS, Ferreira F, Faro C, Pires E, Carvalho AP, Cunha AP, Macedo T (1994) The amount of GABA present in aqueous extracts of valerian is sufficient to account for [^{3}H] GABA relaese in synaptosomes. Planta Med 60:475–476

Satheesh MA, Pari L (2004) Antioxidant effect of *Boerhavia diffusa* Linn. In tissues of alloxan induced diabetic rats. Ind J Exp Biol 42:989–992

Satoh K, Sakagami H (1997) Effect of metal ions on radical intensity and cytotoxic activity of ascorbate. Anticancer Res 17(2A):1125–1129

Sbartai H, Djebar MR, Rouabhi R, Sbartai I, Berrebbah H (2011) Antioxidative response in tomato plants *Lycopersicon esculentum* L. Roots and leaves to zinc. Am Eur J Toxicol Sci 3:41–46

Serrano J, Puupponen–Pimia R, Dauer A, Aura AM, Saura-Calixto F (2009) Tannins: Current knowledge of food sources, intake bioavailability and biological effects. Mol Nutr Food Res 53:1–20

Shanmugasundaram KR, Sujatha R, Shanmugasundaram ERB (1994) Amrita Bindu—a salt spice herbal health food supplement for the prevention of nitrosamine induced depletion of antioxidants. J Ethnopharmacol 42:83–93

Sheela CG, Augusti KT (1992) Antidiabetic effects of S-allyl cysteine sulphoxide isolated from garlic *Allium sativum* Linn. Indian Exp Biol 30:523–526

Shirwaikar A, Prabhu KS (2006) In vitro antioxidant studies of *Sphaeranthus indicus* Linn. Ind J Exp Biol 44:993–996

Shirwaikar A, Somashekar AP, Udupa AL, Udupa SL, Somashekar S (2003) Wound healing studies of *Aristolochia bracteolata* Lam. with supportive action of antioxidant enzymes. Phytomed 10:558–562

Sies H (1997) Oxidative stress: oxidants and antioxidants. Exp Physiol 82(2):291–295

Singh A, Handa SS (1995) Hepatoprotective activity of *Apium graveolens* and *Hygrophila auriculata* against paracetamol and thioacetamide intoxication in rats. J Ethnopharmacol 49:119–126

Singh J, Kakkar P (2009) Anti hyperglycemic and antioxidant effect of *Berberis aristata* root extract and its role in regulating carbohydrate metabolism in diabetic rats. J Ethnopharmacol 123(1):22–26

Singh S, Srivastava R, Choudhary S (2010) Antifungal and HPLC analysis of the crude extracts of *Acorus calamus*, *Tinospora cordifolia* and *Celestrus paniculatus*. J Agri Tech 6:149–158

Sivakumar V, Niranjali DS (2006) Protective effect of *Plumbago zeylanica* against cyclophosphamide-induced genotoxicity and oxidative stress in Swiss albino mice. Drug Chem Toxicol 29(3):279–288

Stahelin HB (1999) The impact of antioxidants on chronic diseases in aging and in old age. Int J Vitam Nutri Res 69:149

Stoilova I, Krastanov A, Stoyanova A, Denev P, Gargova S (2007) Antioxidant activity of a ginger extract (*Zingiber officinale*). Food Chem 102:764–770

Subhan F, Karim N, Ibrar M (2007) Anti- inflammatory activity of methanolic and aquous extracts of *Valeriana wallichi* DC rhizome. Pak J Plant Sci 13:103–108

Sun T, Xu Z, Wu CT, Janes M, Prinyawiwatkul W, No HK (2007) Antioxidant activities of different colored sweet bell peppers (*Capsicum annuum* L.). J Food Chem 72:S98–S102

Sundaram R, Mitra SK (2007) Anyioxidant activity of ethyl acetate soluble fraction of *Acacia arabica* bark in rats. Ind J Pharmacol 39:33–38

Surveswaran S, Cai Y, Corke H, Sun M (2006) Systemtic evaluation of natural phenolics antioxidants from 133 Indian medicinal plants. Food Chem 102:938–953

Szatmari SZ, Whitehouse PJ (2003) Vinpocetine for cognitive impairment and dementia. Cochrane Database Sys Rev 1:CD003119

Thind TS, Suri RK (1979) In vitro antifungal efficacy of four essential oils. Indian Perfumer 23:138–140

Thirugnanasampandan R, Mahendran G, Narmatha Bai V (2008) Antioxidant properties of some medicinal aristolochiaceae species. Afr J Biotechnol 7(4):357–361

Tyler VE (1994) Herbs of choice: the therapeutic use of phytomedicinals. Pharmaceutical Product Press, New York

Veerapur VP, Prabhakar RK, Parihar VK, Kandadi MC, Ramakrishana S, Mishra V, Rao BSS, Srinivasan KK, Priyadarsini KI, Unnikrishnan MK (2009) *Ficus racemosa* stem bark extract: a potent antioxidant and a probable natural radioprotector. ECAM doi.10.1093./ecam/nem119

Venkateswaran S, Pari L (2002a) Effect of *Coccinia indica* on blood glucose, insulin and hepatic key enzymes in experimental diabetes. Pharm Biol 40:165–170

Venkateswaran S, Pari L (2002b) Antioxidant effect of *Phaseolus vulgaris* in streptozotocin-induced diabetic rats. Asia Pacific J Clin Nutr 11:206–209

Wadkar KA, Magdum CA, Patil SS, Naikwade NS (2008) Anti-diabetic potential and Indian medicinal plants. J Herb Med Toxicol 2(1):45–50

Wadood A, Wadood N, Shah SA (1989) Effects of *Acacia arabica* and *Caralluma edulis* on blood glucose levels on normal and alloxan diabetic rabbits. J Pak Med Asso 39:208–212

Wagner H, Jurcic K, Schaette R (1980) Comparative studies on the sedative action of *Valeriana* extracts, valepotriates and their degradation products. Planta Med 39:358–365

Wang SY, Jiao H (2000) Scavenging capacity of berry crops on superoxide radicals, hydrogen peroxide, hydroxyl radicals, and singlet oxygen. J Agric Food Chem 48(11):5677–5684

Wojdylo A, Oszmianki J, Czemerys R (2007) Antioxidant activity and phenolic compounds in 32 selected herbs. Food Chem 105:940–949

Wong LF, Lim YY, Omar M (2009) Antioxidant and antimicrobial activities of some *Alpinia* species. J Food Biochem 33:835–851

Yogisha S, Raveesha KA (2009) In vitro antibacterial effect of selected medicinal plant extracts. J Nat Prod 2:64–69

Zhang D, Hamauzu Y (2004) Phenolics, ascorbic acid, carotenoids and antioxidant activity of broccoli and their changes during conventional and microwave cooking. Food Chem 88:503–509

Chapter 14
Metabolic Engineering and Synthetic Biology for the Production of Isoquinoline Alkaloids

Yit-Lai Chow and Fumihiko Sato

14.1 Introduction

Nature has bestowed upon us a myriad of plants, which have been a rich source of food and medicine since ancient times. According to a survey on sources of new drug substances from 1981 to 2006, almost half of the approved drugs since 1994 are based on natural products derived from plants and microbes (Newman and Cragg 2007). Furthermore, nutraceuticals, which are defined as functional food ingredients with human health benefits beyond basic nutrition, are part of a growing trend. The global market for the nutraceuticals industry could reach nearly $22 billion by 2013, according to a report published by the Freedonia Group (2009). Multinational pharmaceutical and food companies are diversifying their product lines. As a result, they have propelled the fields of synthetic chemistry, metabolic engineering, and synthetic biology to develop valuable bioactive molecules as well as to increase production efficiency and yield.

Unlike primary metabolites, the target compounds, so-called secondary metabolites, are not always necessary for plant growth and development. Especially, the physiological roles of alkaloids, which are naturally occurring, low molecular weight, nitrogenous secondary metabolites found in about 20 % of flowering plant species (Croteau et al. 2000), are still not clearly understood, except that they help to protect against pathogens, insects, and herbivores (Hartmann 2004). These highly bioactive alkaloids can have therapeutic effects in other organisms, including human beings. For example, the anti-tumor compound paclitaxel is a diterpene alkaloid that was first isolated from the bark of the Pacific

Y.-L. Chow · F. Sato (✉)
Division of Integrated Life Science, Graduate School of Biostudies,
Kyoto University, Main Bldg. of Faculty of Agriculture N252,
Oiwake, Kitashirakawa, Sakyo, Kyoto 606-8502, Japan
e-mail: fsato@lif.kyoto-u.ac.jp

S. Chandra et al. (eds.), *Biotechnology for Medicinal Plants*,
DOI: 10.1007/978-3-642-29974-2_14, © Springer-Verlag Berlin Heidelberg 2013

yew tree, *Taxus brevifolia,* and is a toxic product of this plant species. Other well-known alkaloids, including atropine, caffeine, vincristine, quinine, cocaine, and morphine, are also toxic when they are misused. On the other hand, other major groups of plant secondary metabolites, phenylpropanoids, and terpenoids are less toxic and commonly used for nutraceutical purposes.

In this chapter, we focus on the production of alkaloids, especially isoquinoline alkaloids, since they have been most intensively studied at the molecular level and include many useful pharmaceutical compounds. While the cultivation of some isoquinoline alkaloid-producing plant species, such as opium poppy, is well established, many isoquinoline alkaloid-producing plant species are difficult to cultivate and there are unavoidable problems with naturally harvested products: limited availability of their natural sources, concerns about the conservation of rare or near-extinct plants, unavoidable contamination of plant extracts with misidentified materials and/or environmental pollutants, seasonal variations in plant harvesting, poor standardization of the final product due to variable conditions for plant growth, and the difficulty of extracting secondary metabolites from the parts of the grown plant (Korkina and Kostyuk 2012).

14.2 Three Alternative Methods for the Production of Secondary Metabolites

14.2.1 Synthetic Chemistry

Synthetic chemistry is a powerful approach for overcoming the limitations associated with plant metabolite-derived chemicals. Over the past several decades, numerous synthetic endeavors have led to the development of new methodologies and strategies regarding principles of reactivity and selectivity. Total syntheses have been established for many complicated chemicals. Ancistrocladidine, a rare 7,3′-linked naphthylisoquinoline alkaloid, has been produced, and the key feature of its synthesis is the formation of an extremely hindered biaryl linkage by ortho-arylation of a naphthol with an aryllead triacetate (Fig. 14.1; Bungard and Morris 2006). The asymmetric total synthesis of (−)-emetine and (+)-emetine has also been achieved in a completely stereoselective manner (Fig. 14.2; Itoh et al. 2006; Tietze et al. 2003). Recently, the total synthesis of benzo[*c*]phenanthridine alkaloids has been achieved via palladium-catalyzed ring-opening coupling-cyclization sequence reactions, which enabled the construction of benzo[*c*]phenanthridines with more diverse substituents (Fig. 14.3; Lv et al. 2011).

The idea that an "ideal synthesis is one which creates a complex molecule…in a sequence of only construction reactions involving no intermediary refunctionalizations, and leading directly to the target, not only its skeleton but also its correctly placed functionality" was first stated by Hendrickson (1975) and continues to reflect the ultimate goal of organic chemists in synthesizing chemicals

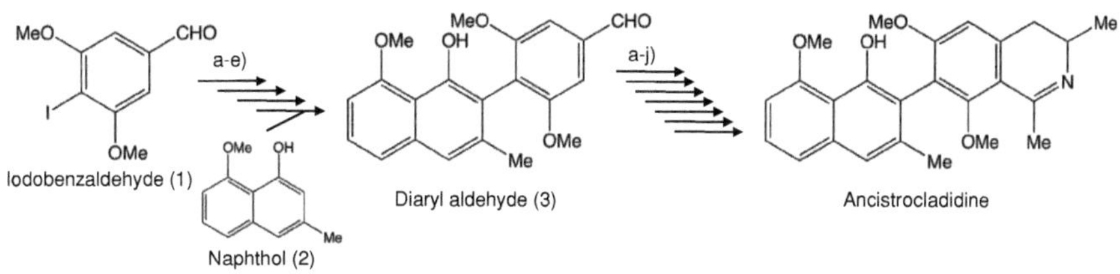

Fig. 14.1 Synthesis of ancistrocladidine (Bungard and Morris 2006). Ancistrocladidine was synthesized via 10 reaction steps in 26 % overall yield from diaryl aldehyde (3), which was prepared via 5 steps from iodobenzaldehyde (1) in 67 % yield from naphthol (2)

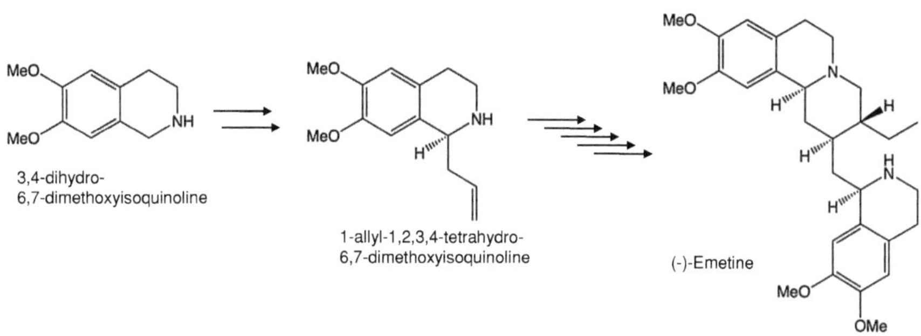

Fig. 14.2 Stereoselective synthesis of (−)-emetine (Itoh et al. 2006). Catalytic asymmetric allylation of 3,4-dihydro-6,7-dimethoxyisoquinoline was carried out in 11 total reaction steps using allyltrimethoxysilane in the presence of Cu(I) and tol-BINAP. The allyl adduct thus obtained was transformed to a chiral synthetic intermediate for (−)-emetine in an overall yield of 8.5 % from 3,4-dihydro-6,7-dimethoxyisoquinoline

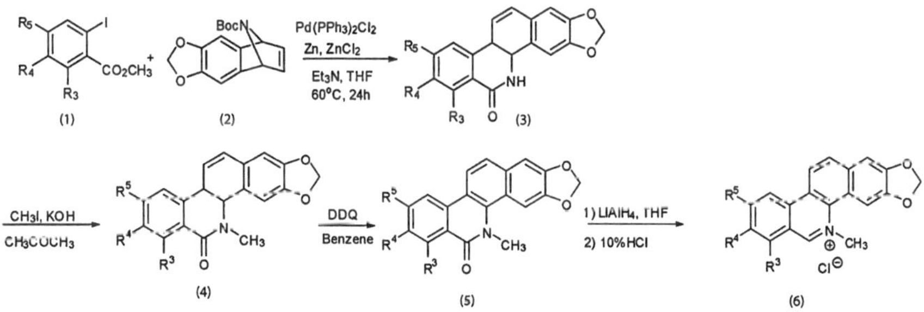

Fig. 14.3 Synthesis of benzo[*c*] phenanthridine alkaloids (Lv et al. 2011). Palladium-catalyzed ring-opening coupling of 1,3-benzodioxole-azabicycle (2) with an *o*-iodobenzoate (1) in THF gave the key intermediate, cis-dihydrobenzo[*c*] phenanthridinones (3), in 90 % yield. Further *N*-methylation (4) and oxidation with DDQ produced oxybenzo[*c*]phenanthridine alkaloids (5), and finally, reduction of (5) with $LiAlH_4$ followed by treatment with HCl gave the desired quaternary benzo[*c*]phenanthridine chlorides (6) in yields of about 97 % (chelerythrine, nitidine, and avicine) to 98 % (sanguinarine)

derived from nature. Although there has been rapid growth in the production of nonnatural alkaloids as a result of bioorganic research, many natural complex molecules still cannot be efficiently synthesized in the laboratory. The field of organic synthesis has reached an awe inspiring level, and many observers have proclaimed that synthesis matured in 2010. Nonetheless, significant challenges remain, and one of the greatest challenges is the total synthesis of large quantities of complex natural products with a minimal amount of labor and material (Gaich and Baran 2010).

14.2.2 Metabolic Engineering of Plants

14.2.2.1 Cell and Organ Culture

Due to technological difficulties regarding the cultivation and relatively long culture periods of medicinal plants, they have commonly been harvested from the wild. Plant cell and tissue culture are feasible ways to produce secondary metabolites under a controlled environment, independent of climatic changes and soil conditions, and free of microbe and insect contamination (Sato and Yamada 2008; Verpoorte et al. 1991). In some cases, a further improvement in yield by the selection of high-yield lines and treatment with an elicitor such as methyl jasmonate (MeJA) has been achieved for industrial production. For example, high shikonin-producing *Lithospermum erythrorhizon* cells have been established for the commercial production of shikonin (Fujita 1988/2007). A remarkable improvement in paclitaxel production (up to 0.5 % of dry weight compared to 0.01 %) has been achieved in Pacific yew cell cultures through treatment with MeJA (Yukimune et al. 1996). Although some technological breakthroughs are being investigated to establish high-yield culture (Lee et al. 2010), there are still some difficulties to achieve both high productivity and stability for industrial application.

While there are still challenges to establish high metabolite-producing cells from totipotent cells/tissues, the development of hairy roots by the transformation of plant cells with *Agrobacterium rhizogenes*, a Gram-negative soil bacterium, provides a solution with unlimited growth potential due to autonomous production of the growth hormone auxin and the morphological differentiation of roots (Shanks and Morgan 1999). Hairy roots also have high potential for the mass production of valuable secondary metabolites (see Chap. 10). They often exhibit about the same or greater biosynthetic capacity for secondary metabolite production as their parent plants (Kim et al. 2002), and can also produce a spectrum of secondary metabolites that are not present in the parent plant (Veerasham 2004). Additional treatment of hairy roots with an elicitor might be a more effective method for the production of metabolites and the characterization of plant secondary metabolite pathways (Zhou et al. 2011).

Other attractive features of hairy root cultures are their high genetic stability, relatively fast growth rates and physiological strength for fermentation. In vitro cultures of hairy roots derived from *Catharanthus roseus* via *A. rhizogenes* infection exhibited greater genetic stability than cell suspension cultures. They accumulate higher levels of terpenoid indole alkaloids including catharanthine, which has been widely used in many laboratories (Zhou et al. 2009, 2010). Similarly, Le Flem-Bonhomme et al. (2004) developed an efficient transformation system for opium poppy (*Papaver somniferum*) using *A. rhizogenes* LBA9402, which led to the production of *P. somniferum* hairy root cultures that produced morphinan alkaloids and sanguinarine. The total alkaloid content was higher in transformed roots, which accumulated three times more codeine than intact roots.

Hairy roots are also good hosts for metabolic engineering. Transgenic California poppy (*Eschscholzia californica*) roots expressing berberine bridge enzyme (BBE) from *P. somniferum* were found to have higher levels of BBE mRNA and increased accumulation of benzophenanthridine alkaloids compared to control roots transformed with a beta-glucuronidase gene, while roots transformed with an antisense-BBE construct had lower levels of BBE mRNA and reduced benzophenanthridine alkaloid accumulation relative to controls (Park et al. 2003). Similarly, *Hyoscyamus niger* hairy root harboring putrescine *N*-methyltransferase (PMT) was found to produce more methylputrescine than a control hairy root line (Zhang et al. 2007). The details of metabolic engineering will be discussed in the next section.

14.2.2.2 Metabolic Pathway Engineering

Increasing Molecular Tools

Advances in molecular biology have paved the way for improving the productivity of bioactive metabolites through metabolic engineering, in terms of the quality and quantity of metabolites. The rational design of pathways by the trimming of undesired pathways, the enhancement of rate-limiting steps, and the introduction of new pathways to produce novel compounds is an important approach. Such metabolic engineering could turn plant cells into "green chemical factories" to produce the desired products. The identification of many biosynthetic genes and characterization of the spatial and developmental regulation of their expression have clarified their importance in the biosynthesis of secondary metabolites and revealed bottlenecks for their production in cell culture (Sato et al. 2007a, b).

In the metabolic engineering of alkaloids, isoquinoline alkaloid (IQA) biosynthesis has been most intensively studied and modified through the use of metabolic engineering. The biosynthetic pathway for IQAs starts with L-tyrosine. All of the enzyme genes of the total nine enzymatic reaction steps from norcoclaurine to berberine have been characterized at the DNA level (Fig. 14.4): a norcoclaurine synthase (NCS; an entry enzyme in IQA biosynthesis, proteins in the PR10 family such as TfNCS from *Thalictrum flavum* (Lee and Facchini 2010; Samanani et al. 2004) and a novel dioxygenase-like protein, CjNCS1, from *Coptis japonica*

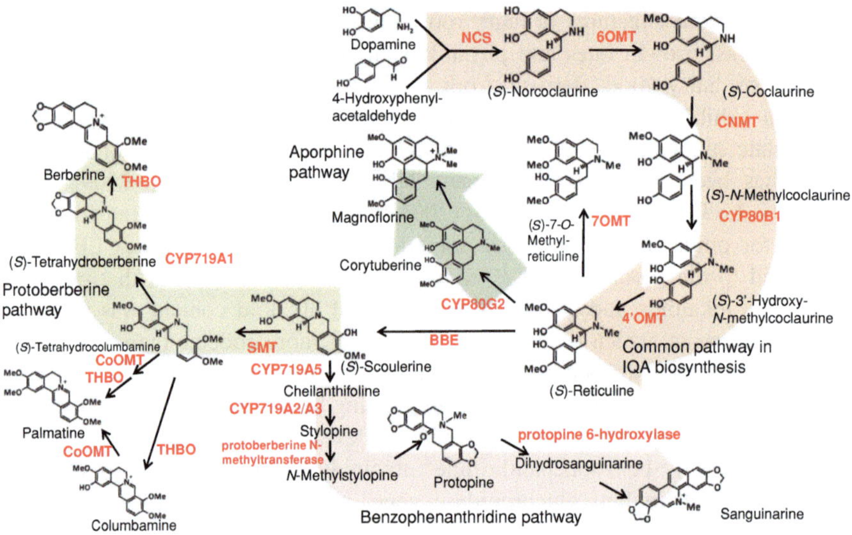

Fig. 14.4 Biosynthetic pathway for aporphine-, benzophenanthridine- and protoberberine-type isoquinoline alkaloids. Major isoquinoline alkaloids are synthesized through a common pathway derived from norcoclaurine to reticuline. Aporphine alkaloids are synthesized from (*S*)-reticuline by the reaction of corytuberine synthase (CYP80G2). Protoberberine and benzophenanthridine alkaloids are synthesized from (*S*)-scoulerine produced from (*S*)-reticuline by the reaction of berberine bridge enzyme (BBE). The names of the biosynthetic enzymes for which cDNA has been isolated and characterized are shown in red. Abbreviations for biosynthetic enzymes are defined in the text, except CoOMT (columbamine *O*-methyltransferase) and 7OMT(reticuline 7-*O*-methyltransferase)

(Minami et al. 2007)), an *N*-methyltransferase (coclaurine *N*-methyltransferase; CNMT, Choi et al. 2002), three *O*-methyltransferases (OMTs; norcoclaurine 6OMT, 3′-hydroxy-*N*-methyl-coclaurine 4′OMT (Morishige et al. 2000), (*S*)-scoulerine 9-*O*-methyltransferase (SMT) (Takeshita et al. 1995)), a hydroxylase (CYP80B1) (Pauli and Kutchan 1998), a BBE (Dittrich and Kutchan 1991), a methylenedioxy ring-forming enzyme (canadine synthase; CDS or CYP719A1) (Ikezawa et al. 2003), and a tetrahydroprotoberberine oxidase (Gesell et al. 2011; Minami et al. unpublished data). The number of available enzyme genes is still growing rapidly, such as the genes of corytuberine synthase for aporphine-type (CYP82G2) (Ikezawa et al. 2008), those of salutaridine synthase (CYP719B1) (Gesell et al. 2009), salutaridinol-7-*O*-acetyltransferase (SAT) (Grothe et al. 2001), thebaine 6-*O*-demethylase, and codeine *O*-demethylase (Hagel and Facchini 2010), and NADPH-dependent codeinone reductase (COR) (Unterlinner et al. 1999) for morphinan-type, those of chelanthifoline synthase (CYP719A5) (Ikezawa et al. 2009), stylopine synthase (CYP719A2/A3) (Ikezawa et al. 2007), protopine 6-hydroxylase (Takemura et al. 2012), and sanguinarine reductase (Vogel et al. 2010) for benzophenanthridine-type, those of berbamunine synthase (CYP80A1) (Kraus and Kutchan 1995) for bis-benzylisoquinoline alkaloids, and those of *O*-methyltransferases for emetine biosynthesis (Cheong et al. 2011; Nomura and Kutchan 2010).

Pathway Engineering with Overexpression of the Rate-Limiting Enzyme, Trimming of Pathways, and the Introduction of New Branch Pathways

Molecular characterization of biosynthetic enzyme genes has enabled the molecular design of the metabolic flow of plant cells (Fig. 14.5). Overexpression of the gene for the enzyme in the rate-limiting step is commonly used to improve the yield. For example, the overexpression of golden thread (*Coptis japonica*) SMT in *C. japonica* cells increased berberine production (Sato et al. 2001). However, overexpression of an endogeneous gene often induces cosuppression of the gene and decreased production of the end metabolite (Sato et al. unpublished results). Thus, heterologous expression of a rate-limiting enzyme, such as *C. japonica* 6OMT in California poppy cells, has been recommended to overcome the rate-limiting step (Inui et al. 2007). The bottleneck step(s) will differ among plant species and isolated cell lines due to the variation in the expression level of genes. In California poppy cells, the overexpression of 6-OMT increased alkaloid production more effectively than that of 4′-OMT (Inui et al. 2007). On the other hand, the overexpression of salutaridinol 7-*O*-acetyltransferase (salAT) in opium poppy affected the accumulated levels of alkaloid products (Allen et al. 2008). Changes in the bottleneck step may also occur after modification of the enzyme expression level through metabolic engineering.

Since there may be multiple bottleneck steps, the overall regulation of the expression levels of biosynthetic genes in the pathway would be more effective, and further studies are seeking to isolate general transcription factors. Recently, Sato's group identified CjWRKY1 and CjbHLH1 as possible general regulators in IQA biosynthesis using a transient RNAi screening system (Dubouzet et al. 2005; Kato et al. 2007; Yamada et al. 2011). The successful application of the overexpression of heterologous AtWRKY transcription factor in opium poppy to increase morphinan alkaloid content has proven that it is possible to improve the yield of IQAs through the regulation of transcription (Apuya et al. 2008).

Whereas overexpression of the enzyme in the rate-limiting step is a useful method for improving productivity, modification of the metabolic flow at the branching point by reducing enzymes for undesired pathways is another promising approach to increase desired metabolites while reducing unwanted byproducts. When this approach is combined with the introduction of a new pathway, transgenic cells can produce novel products. Among the methods used to reduce gene expression (e.g., antisense, RNAi, knockout with T-DNA insertion, mutation etc.), RNA silencing with double-stranded RNA (RNA interference, RNAi with siRNA or miRNA) is the most powerful and reliable.

The versatility of RNAi for controlling the multiple genes responsible for metabolite production and across various tissues and developmental steps has been well recognized (Sato 2005; Borgio 2009). For example, Fujii et al. (2007) reported the successful accumulation of a direct substrate of BBE, reticuline, by RNAi of BBE in California poppy cells, whereas, little accumulation of reticuline was reported by Park et al. (2002) using the antisense method. However, we should note that transgenic California poppy cells with RNAi of BBE also produced a

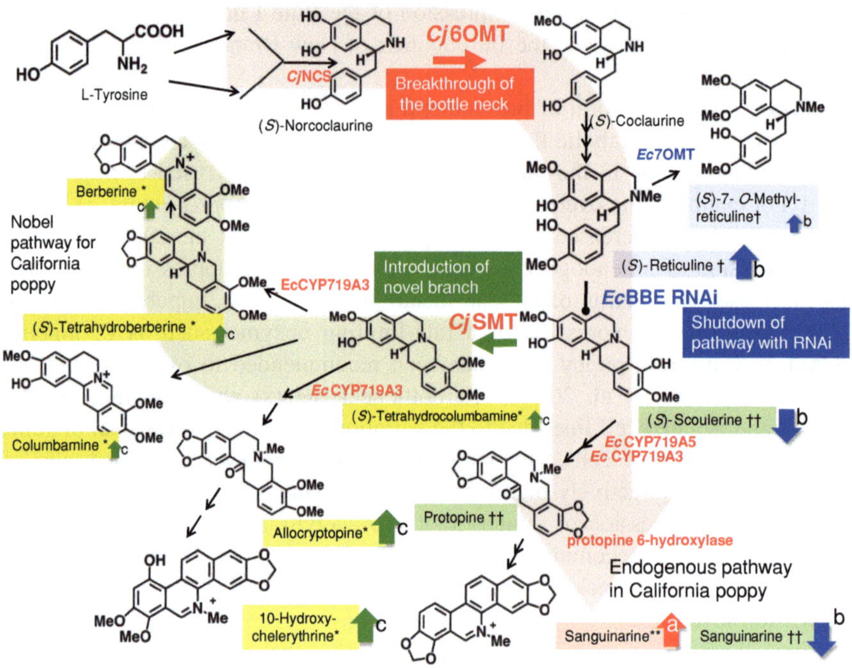

Fig. 14.5 Examples of metabolic engineering of IQA in California poppy cells. Ectopic expression of *C. japonica* (*Cj*) 6-OMT overcame the rate limitation at this enzyme step and increased endogenous products such as sanguinarine, as shown by the upward pointing red arrow (Inui et al. 2007, label a). Knockdown of BBE expression with double-stranded RNA strongly reduced endogenous alkaloid accumulation and increased the accumulation of precursor, reticuline, and some metabolites, such as 7-O-methylreticuline, as shown with downward and upward pointing blue arrows (Fujii et al. 2007, label b). The introduction of a novel branch pathway in California poppy induced the production of uncommon protoberberine-type alkaloids, whereas, major products were further metabolized to dimethoxy-type benzophenanthridine alkaloids by endogenous enzymes, as shown with upward pointing green arrows (Takemura et al. 2010, label c)

methylated derivative of reticuline and laudanine, which could scarcely be detected in control cells. In addition, Allen et al. (2004) reported that RNAi of COR genes in opium poppy resulted in the accumulation of a far-upstream precursor, (*S*)-reticuline, and might interfere with the formation of an enzyme complex in morphinan alkaloid biosynthesis, whereas, there is no direct evidence of an enzyme complex in IQA biosynthesis. The development of a supercomplex is a challenging field in metabolic engineering, since only limited supercomplex formation has been documented in the photosynthetic apparatus, such as photosynthetic NAD(P)H dehydrogenase (Endo et al. 2008), and the active principle of supercomplex formation is not known.

As mentioned above, the combination of enhanced expression of a desired pathway and reduced expression of an unwanted branch is a more promising strategy for improving the quantity and quality of metabolites. Two successful examples in secondary metabolite engineering are discussed here. The content of artemisinin, an effective anti-malarial drug isolated from *Artemisia annua L.* was increased when the expression of squalene synthase, which is a key enzyme in the sterol pathway, was suppressed by means of a hairpin-RNA-mediated RNAi technique (Zhang et al. 2009). The artemisinin content of some transgenic plants was significantly increased; as much as 3.14-fold greater than that in untransformed control plants. Blue rose is another example of a successfully commercialized product that was developed by the ectopic expression of a novel branch pathway enzyme gene for blue pigment (i.e., flavonoid 3′,5′-hydoxylase) and the RNAi-mediated downregulation of endogeneous dihydroflavonol 4-reductase in an undesired flavonoid biosynthetic pathway in plants (Katsumoto et al. 2007).

14.2.3 Synthetic Biological Approaches

When our molecular biological tools were limited, the one-step bioconversion of substrate was commonly used to improve metabolite quality. In contrast, the reconstitution of an entire metabolic pathway is now possible in microorganisms (Ajikumar et al. 2010; Minami et al. 2008; Ro et al. 2006). Due to their smaller genome size, microorganisms have a simpler internal structure than plants, which simplifies metabolite transport between enzymatic steps (Leonard et al. 2009). The reconstitution of a secondary metabolite pathway in an engineered microbial host offers several advantages over methods directed toward native plant hosts such as the isolated production of key intermediate molecules, rapid biomass accumulation, ease of purification, and the availability of genetic tools for forward engineering and pathway optimization (Hawkins and Smolke 2008). A combination of microbial enzyme genes and plant-derived genes offers some advantages for the establishment of efficient and productive systems for the entry compound (Minami et al. 2008; Sato et al. 2007b). On the other hand, microbial systems also have disadvantages, such as a limited substrate availability for plant metabolites, although this limitation is currently being challenged (see below).

Synthetic biological approaches may use a single host system such as *Escherichia coli* or *Saccharomyces cerevisiae* alone, or a combination of both. In the first reconstitution of IQA biosynthesis in a microbe, Minami et al. (2008) used a two-step system with *E. coli* and *S. cerevisiae* (Fig. 14.6). First, an artificial pathway to produce reticuline, which is a key intermediate in benzylisoquinoline alkaloid (BIA) biosynthesis, was assembled in *E. coli* by the introduction of genes for monoamine oxidase (MAO) together with *C. japonica* NCS, 6OMT, CNMT and 4′-OMT in plasmid-based expression systems. The use of microbial MAO made it possible to incorporate the hydroxyl group early in the reticuline pathway through the synthesis of 3,4-dihydroxyphenylacetaldehyde from dopamine, thereby skipping the need to express the plant cytochrome P450 hydroxylase (CYP80B1) in the bacterium, which is often problematic in microbial expression. Upon the induction of enzyme expression and dopamine supplementation, the final

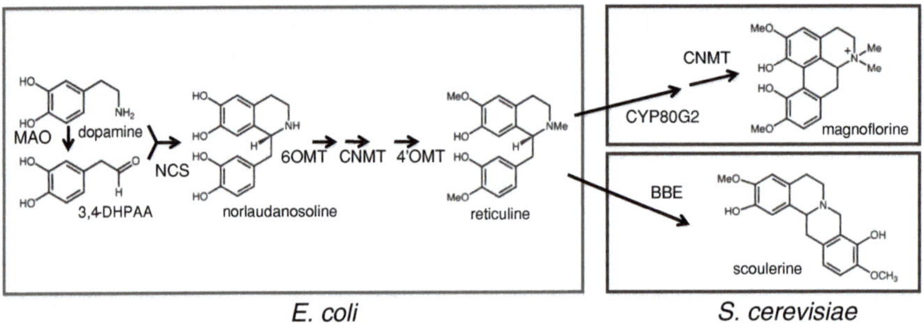

Fig. 14.6 Two-step synthesis of alkaloids with *E. coli* and *S. cerevisiae* (Minami et al. 2008). Transgenic *E. coli* and *S. cerevisiae* can be combined to produce the desired compounds. MAO: monoamine oxidase, 6-OMT: norcoclaurine 6-*O*-methyltransferase, CNMT: coclaurine *N*-methyltransferase, 4′-OMT: 3′-hydroxy *N*-methylcoclaurine 4′-O-methyltransferase, CYP80 G2: corytuberine synthase, BBE: berberine bridge enzyme

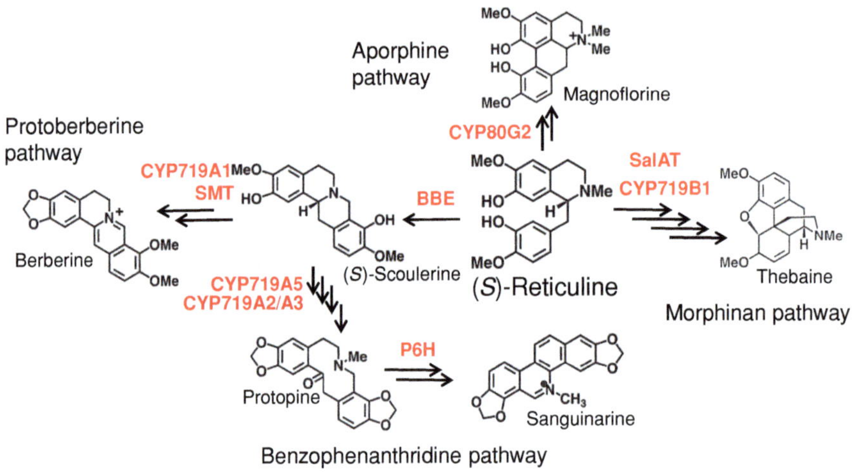

Fig. 14.7 Various benzylisoquinoline alkaloids can be synthesized from reticuline. The syntheses of magnoflorine in the aporphine pathway, tetrahydrocolumbamine in the protoberberine pathway, and salutaridine in the morphinan pathway have been reported in a combined system of *E. coli* and *S. cerevisiae* or single *S. cerevisiae* culture

yield of (*S*)-reticuline synthesized in vitro by the crude enzymes from recombinant *E. coli* was 55 mg l^{-1} within 1 h. When reticuline-producing *E. coli* was cocultured with *S. cerevisiae* expressing *C. japonica* BBE or CYP80G2 in the presence of dopamine, (*S*)-scoulerine and magnoflorine were detected at 8.3 mg l^{-1} and 7.2 mg l^{-1}, respectively, after incubation for 48–72 h (Fig. 14.7). The use of two microbial systems for pathway construction should be useful for establishing different biosynthetic modules, whereas, it may have reduced the efficiency of alkaloid synthesis due to the need for metabolite transport between the cell membranes of the two microorganisms.

On the other hand, Hawkins and Smolke (2008) used *S. cerevisiae* as a sole host organism for the assembly of artificial BIA pathways; they combined three enzymes from plant sources (6-OMT, CNMT and 4′-OMT derived from either *Thalictrum flavum* or *P. somniferum*) and a human P450 enzyme, and used (*R*, *S*)-norlaudanosoline as a substrate. After the stable insertion of genes into the yeast genome under a reduced-strength promoter variant (TEF7), heterologous gene expression resulted in the creation of an artificial plant pathway with reduced transcriptional activities, while maintaining high catalytic activities. Plasmid-based expression of *P. somniferum* BBE together with *T. flavum* (S)-scoulerine 9-*O*-methyltransferase (SMT) in (*R, S*)-reticuline–producing yeasts yielded $\sim$60 mg l^{-1} (*S*)-tetrahydrocolumbamine from $\sim$1 g l^{-1} of (*R*, *S*)-norlaudanosoline in 48 h (Fig. 14.7). Expression of the *C. japonica* P450 enzyme CYP719A1 and integration of the *A. thaliana* P450 redox partner protein ATR1 in the genome resulted in the accumulation of $\sim$30 mg l^{-1} (*S*)-canadine, the direct precursor of berberine. The synthesis of reticuline by the engineered yeast strain also enabled the synthesis of salutaridine, an intermediate in the branch pathway of morphine, through a shorter path by the expression of a human CYP2D6 involved in morphine metabolism together with human CPR1 reductase in the reticuline-producing yeasts. Approximately 20 mg l^{-1} salutaridine was synthesized from (*R*, *S*)-norlaudanosoline. Stable integration of transgene into the genome would be advantageous for stable expression, but a continuous supply of IQA precursor remains a limiting factor, as mentioned above.

The reconstitution of alkaloid biosynthesis in microbial systems has shown promising results, as described above. However, the existing need to exogenously supply expensive intermediate precursors limits its commercial application. Our incomplete understanding of the complex alkaloid biosynthetic networks, especially the early steps, hinders the synthesis of alkaloids from simple precursors. In addition, the cytotoxicity of alkaloids in yeast may also pose a similar problem in other microbial systems. Therefore, practical and effective strategies to minimize toxicity are needed to create alkaloid-overproducing microbes, such as the engineering of transcription factors to increase microbial tolerance toward toxic metabolites (Alper et al. 2006). Further attempts have enabled the successful production of (*S*)-reticuline from simple carbon sources without the need for additional substrate in an *E. coli* fermentation system (Fig. 14.8; Nakagawa et al. 2011). Nakagawa et al. (2011) produced tyrosine-overproducing *E. coli* with the overexpression of biosynthetic enzymes (TKT; transketolase, PEPS; phosphoenolpyruvate synthetase, fbr-DAHPS; feedback-inhibition-resistant 3-deoxy-d-arabino-heptulosonate-7-phosphate synthase, fbr-CM/PD; feedback-inhibition-resistant chorismate mutase/prephenate dehydrogenase and deletion of repressor of aromatic amino acid biosynthesis; *tyrR*), and then modified the pathway to produce dopamine from tyrosine with the introduction of pathway-specific tyrosinase (TYR) of *Ralstonia solanacearum* and DOPA decarboxylase (DODC) of *Pseudomonas*. Such modification combined with a previously developed alkaloid production module enabled reticuline production from glucose or glycerol. This achievement may lead to industrial-scale alkaloid production in the near future.

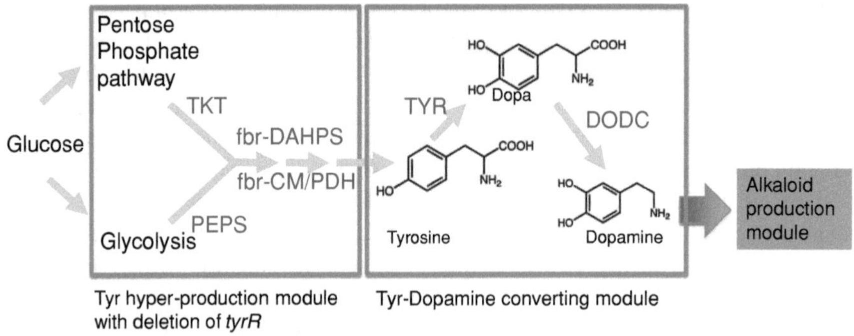

Fig. 14.8 Fermentation production of benzylisoquinoline alkaloids from a simple precursor. (1) Construction of a tyrosine hyper-production module with the overexpression of biosynthetic enzymes (TKT; transketolase, PEPS; phosphoenolpyruvate synthetase, fbr-DAHPS; feedback-inhibition-resistant 3-deoxy-d-arabino-heptulosonate-7-phosphate synthase, fbr-CM/PD; feedback-inhibition-resistant chorismate mutase/prephenate dehydrogenase) and deletion of repressor of aromatic amino acid biosynthesis; *tyrR*), (2) construction of a tyrosine-dopamine-converting module with the introduction of pathway-specific tyrosinase (TYR) of *Streptomyces castaneoglobisporus* and DOPA decarboxylase (DODC) of *Pseudomonas* would be a useful platform for the production of various secondary metabolites including alkaloids

14.3 Perspectives

While various approaches to the production of useful secondary metabolites are being studied, each has strengths and limitations with regard to its large-scale capability. Researchers will have to consider all possibilities to generate value-added compounds from natural as well as nonnatural sources. A multidisciplinary approach may offer the best solution to this problem. For instance, while both plant cell or tissue culture and microbial systems enable alkaloid production on a relatively large scale, metabolic engineering is still needed to improve productivity, product tolerance, and stability for practical commercialization.

The establishment of transcriptome and proteome databases for the investigation of natural product metabolism in non-model plant systems has provided useful information for metabolic engineering. Desgagné-Penix et al. (2010) showed that profiling of the more abundant proteins in elicitor-treated opium poppy cell cultures revealed several uncharacterized enzymes that potentially catalyze steps in sanguinarine biosynthesis.

In addition, the emergence of synthetic biology offers the possibility of improving the production of useful products through the design and engineering of complex biological systems. Computer-assisted molecular design software is being developed to facilitate the synthesis of novel compounds with drug-like properties and desired activities (Hartenfeller and Schneider 2011). Recently, the use of information science for the reconstruction of metabolic pathways based on the identification and analysis of biosynthetic enzymes and pathways found in nature is gaining interest as a useful tool in metabolic engineering. Ideally, this data-guided "retrobiosynthetic" system works by enabling an end-user (for example, a metabolic engineering researcher) to sketch a desired natural product analog in a software environment. The software then lists potential routes from central cellular metabolites to the target, putative enzymes to carry out each catalytic step, the catalytic efficiency and the potential compatibility of different components of the pathway, thus allowing the user to select the combination of enzymes with the highest probability of success. Finally, the software would provide a DNA synthesis and assembly scheme for experimental testing of the pathway design. This approach has the potential to reduce the manpower and resources used in conventional attempts to develop synthetic metabolic pathways based on trial-and-error (Bayer 2010). Singla et al. (2010) developed a database that provides comprehensive information about benzylisoquinoline alkaloids with tools for visualization of the 2D and 3D structures, hyperlinks to related databases (http://crdd.osdd.net/raghava/biadb/), and tools for searching structurally similar molecules. This database is integrated with Drugpedia and thus enables the use of BIA information in computer-aided drug design. This database will undoubtedly require continuous effort to incorporate synthesis pathways of the BIA and components to create shorter or more efficient processes and to increase yields. Nevertheless, such an integrated database should be useful for researchers in the field of natural products drug discovery, including those involved in metabolic engineering and synthetic biology.

14.4 Conclusion

In summary, a concerted effort among various disciplines will be needed to hasten the development of an efficient and robust system for industrial-scale metabolite production in the near future. Isoquinoline alkaloid biosynthesis is at the forefront of metabolic engineering and the synthetic biological reconstitution of pathways with most abundantly characterized and isolated genes. Technological advances, including next-generation sequencing, suggest that there is a bright future for metabolite production of not only isoquinoline alkaloids but also other alkaloids.

References

Ajikumar PK, Xiao W-H, Tyo KEJ, Wang Y, Simeon F, Leonard E, Mucha O, Phon TH, Pfeifer B, Stephanopoulos G (2010) Isoprenoid pathway optimization for taxol precursor overproduction in *Escherichia coli*. Science 330:70–74

Allen RS, Millgate AG, Chitty JA, Thistleton J, Miller JAC, Fist AJ, Gerlach WL, Larkin PJ (2004) RNAi-mediated replacement of morphine with the non narcotic alkaloid reticuline in opium poppy. Nat Biotechnol 22:1559–1566

Allen RS, Miller JAC, Chitty JA, Fist AJ, Gerlach WL, Larkin PJ (2008) Metabolic engineering of morphinan alkaloids by overexpression and RNAi suppression of salutaridinol 7-*O*-acetyltransferase in opium poppy. Plant Biotechnol J 6:22–30

Alper H, Moxley J, Nevoigt E, Fink GR, Stephanopoulos G (2006) Engineering yeast transcription machinery for improved ethanol tolerance and production. Science 314: 1565–1568

Apuya NR, Park JH, Zhang L, Ahyow M, Davidow P, Van Fleet J et al (2008) Enhancement of alkaloid production in opium and California poppy by transactivation using heterologous regulatory factors. Plant Biotechnol J 6(2):160–175

Bayer T (2010) Transforming biosynthesis into an information science. Nat Chem Biol 6:859–861

Borgio J (2009) RNA interference (RNAi) technology: a promising tool for medicinal plant research. J Med Plant Res 3:1176–1183

Bungard C, Morris J (2006) Total synthesis of the 7,3′-linked naphthylisoquinoline alkaloid ancistrocladidine. J Org Chem 71(19):7354–7363

Cheong BE, Takemura T, Yoshimatsu K, Sato F (2011) Molecular cloning of an *O*-methyltransferase from adventitious roots of *Carapichea ipecacuanha*. Biosci Biotechnol Biochem 75(1):107–113

Choi KB, Morishige T, Shitan N, Yazaki K, Sato F (2002) Molecular cloning and characterization of coclaurine *N*-methyltransferase from cultured cells of *Coptis japonica.* J Biol Chem 277:830–835

Croteau R, Kutchan TM, Lewis NG (2000) Natural products (Secondary metabolites). In: Buchanan BB, Gruissem W, Jones RL (eds) Biochemistry and molecular biology of plants. American Society of Plant Physiologists, Maryland

Desgagné-Penix I, Khan MF, Schriemer DC, Cram D, Nowak J, Facchini PJ (2010) Integration of deep transcriptome and proteome analyses reveals the components of alkaloid metabolism in opium poppyopium poppy cell cultures. BMC Plant Biol 10:252–268

Dittrich H, Kutchan TM (1991) Molecular cloning, expression, and induction of berberine bridge enzyme, an enzyme essential to the formation of benzophenanthridine alkaloids in the response of plants to pathogenic attack. Proc Natl Acad Sci U S A 88:9969–9973

Dubouzet JG, Morishige T, Fujii N, An CI, Fukusaki E, Ifuku K et al (2005) Transient RNA silencing of scoulerine 9-*O*-methyltransferase expression by double stranded RNA in *Coptis japonica* protoplasts. Biosci Biotechnol Biochem 69:63–70

Endo T, Ishida S, Ishikawa N, Sato F (2008) Chloroplastic NAD(P)H dehydrogenase complex and cyclic electron transport around photosystem I. Mol Cells 25(2):1–6

Freedonia Group (2009) Study #2565, World Nutraceutical Ingredients Industry Study with Forecasts for 2012 & 2018. The Freedonia Group, USA. http://www.freedoniagroup.com/brochure/25xx/2565smwe.pdf

Fujii N, Inui T, Iwasa K, Morishige T, Sato F (2007) Knockdown of berberine bridge enzyme by RNAi accumulates (*S*)-reticuline and activates a silent pathway in cultured California poppy cells. Transgenic Res 16:363–375

Fujita Y (1988/2007) Industrial production of shikonin and berberine, in Ciba Foundation Symposium 137. In: Bock G, Marsh J (eds) Applications of plant cell and tissue culture (Ciba Foundation Symposium 137), Wiley, Chichester, pp 228–238

Gaich T, Baran P (2010) Aiming for the ideal synthesis. J Org Chem 75:4657–4673

Gesell A, Rolf M, Ziegler J, Luisa Díaz Chávez ML, Huang F-C, Kutchan TM (2009) CYP719B1 is salutaridine synthase, the C–C phenol-coupling enzyme of morphine biosynthesis in opium poppy. J Biol Chem 284:24432–24442

Gesell A, Díaz Chávez ML, Kramell R, Piotrowski M, Macheroux P, Kutchan TM (2011) Heterologous expression of two FAD-dependent oxidases with (*S*)-tetrahydroprotoberberine oxidase activity from *Argemone mexicana* and *Berberis wilsoniae* in insect cells. Planta 233:1185–1197

Grothe T, Lenz R, Kutchan TM (2001) Molecular characterization of the salutaridinol-7-*O*-acetyltransferase involved in morphine biosynthesis in opium poppy *Papaver somniferum.* J Biol Chem 276:30717–30723

Hagel JM, Facchini PJ (2010) Dioxygenases catalyze the *O*-demethylation steps of morphine biosynthesis in opium poppy. Nat Chem Biol 6(4):273–275

Hartenfeller M, Schneider G (2011) De novo drug design. Methods Mol Biol 672:299–323

Hartmann T (2004) Plant-derived secondary metabolites as defensive chemicals in herbivorous insects: a case study in chemical ecology. Planta 219:1–4

Hawkins K, Smolke C (2008) Production of benzylisoquinoline alkaloids in *Saccharomyces cerevisiae*. Nat Chem Biol 4:564–573

Hendrickson J (1975) Systematic synthesis design. IV. Numerical codification of construction reactions. J Am Chem Soc 97(20):5784–5800

Ikezawa N, Tanaka M, Nagayoshi M, Shinkyo R, Sakaki T, Inouye K, Sato F (2003) Molecular cloning and characterization of CYP719, a methylenedioxy bridge-forming enzyme that belongs to a novel P450 family, from cultured *Coptis japonica* cells. J Biol Chem 278: 38557–38565

Ikezawa N, Iwasa K, Sato F (2007) Molecular cloning and characterization of methylenedioxy bridge-forming enzymes involved in stylopine biosynthesis in *Eschscholzia californica*. FEBS J 274:1019–1035

Ikezawa N, Iwasa K, Sato F (2008) Molecular cloning and characterization of CYP80G2, a cytochrome P450 that catalyzes an intramolecular C–C phenol coupling of (*S*)-reticuline in magnoflorine biosynthesis, from cultured *Coptis japonica* cells. J Biol Chem 283:8810–8821

Ikezawa N, Iwasa K, Sato F (2009) CYP719A subfamily of cytochrome p450 oxygenase and isoquinoline alkaloid biosynthesis in *Eschscholzia californica*. Plant Cell Rep 28(1):123–133

Inui T, Tamura K, Fujii N, Morishige T, Sato F (2007) Overexpression of *Coptis japonica* norcoclaurine 6-*O*-methyltransferase overcomes the rate-limiting step in benzylisoquinoline alkaloid biosynthesis in cultured *Eschscholzia californica*. Plant Cell Physiol 48:252–262

Itoh T, Miyazaki M, Fukuoka H, Nagata K, Ohsawa A (2006) Formal total synthesis of (-)-emetine using catalytic asymmetric allylation of cyclic imines as a key step. Org Lett 8: 1295–1297

Kato N, Dubouzet E, Kokabu Y, Yoshida S, Taniguchi Y, Dubouzet JG et al (2007) Identification of a WRKY protein as a transcriptional regulator of benzylisoquinoline alkaloid biosynthesis in *Coptis japonica*. Plant Cell Physiol 48:8–18

Katsumoto Y, Fukuchi-Mizutani M, Fukui Y, Brugliera F, Holton TA, Karan M et al (2007) Engineering of the rose flavonoid biosynthetic pathway successfully generated blue-hued flowers accumulating delphinidin. Plant Cell Physiol 48:1589–1600

Kim Y, Wyslouzil B, Weathers P (2002) Secondary metabolism of hairy root cultures in bioreactors. In Vitro Cell Dev Biol Plant 38:1–10

Korkina L, Kostyuk V (2012) Biotechnologically produced secondary plant metabolites for cancer treatment and prevention. Curr Pharm Biotechnol 13:265–275

Kraus PF, Kutchan TM (1995) Molecular-cloning and heterologous expression of a cDNA-encoding berbamunine synthase, a C-O-phenol-coupling cytochrome-P450 from the higher-plant *Berberis stolonifera*. Proc Natl Acad Sci U S A 92(6):2071–2075

Le Flem-Bonhomme V, Laurain-Mattar D, Fliniaux MA (2004) Hairy root induction of *Papaver somniferum* var. album, a difficult-to-transform plant, by *A. rhizogenes* LBA 9402. Planta 218(5):890–893

Lee EJ, Facchini P (2010) Norcoclauine synthase is a member of the pathogenesis-related 10/Bet v1 protein family. Plant Cell 22:3489–3503

Lee E-K, Jin Y-W, Park JH, Yoo YM, Hong SM, Amir R et al (2010) Cultured cambial meristematic cells as a source of plant natural products. Nat Biotechnol 28:1213–1217

Leonard E, Runguphan W, O'Connor S, Prather KJ (2009) Opportunities in metabolic engineering to facilitate scalable alkaloid production. Nat Chem Biol 5:291–300

Lv P, Huang K, Xie L, Xu X (2011) Palladium-catalyzed tandem reaction to construct benzo[*c*]phenanthridine: application to the total synthesis of benzo[*c*]phenanthridine alkaloids. Org Biomol Chem 9:3133–3135

Minami H, Dubouzet E, Iwasa K, Sato F (2007) Functional analysis of norcoclaurine synthase in *Coptis japonica*. J Biol Chem 282(9):6274–6282

Minami H, Kim J-S, Ikezawa N, Takemura T, Katayama T, Kumagai H, Sato F (2008) Microbial production of plant benzylisoquinoline alkaloids. Proc Natl Acad Sci U S A 105:7393–7398

Morishige T, Tsujita T, Yamada Y, Sato F (2000) Molecular characterization of the *S*-adenosyl-L-methionine:3′-hydroxy-*N*-methylcoclaurine 4′-*O*-methyltransferase involved in isoquinoline alkaloid biosynthesis in *Coptis japonica*. J Biol Chem 275:23398–23405

Nakagawa A, Minami H, Kim JS, Koyanagi T, Katayama T, Sato F, Kumagai H (2011) A bacterial platform for fermentative production of plant alkaloids. Nat Commun 2, Article number: 326

Newman D, Cragg G (2007) Natural products as sources of new drugs over the last 25 years. J Nat Prod 70:461–477

Nomura T, Kutchan TM (2010) Three new *O*-methyltransferases are sufficient for all O-methylation reactions of Ipecac alkaloid biosynthesis in root culture of *Psychotria ipecacuanha*. J Biol Chem 285:7722–7738

Park S, Yu M, Facchini P (2002) Antisense RNA-mediated suppression of benzophenanthridine alkaloid biosynthesis in transgenic cell cultures of California poppy. Plant Physiol 128:696–706

Park S, Yu M, Facchini P (2003) Modulation of berberine bridge enzyme levels in transgenic root cultures of California poppy alters the accumulation of benzophenanthridine alkaloids. Plant Mol Biol 51(2):153–164

Pauli H, Kutchan TM (1998) Molecular cloning and functional heterologous expression of two alleles encoding (*S*)-*N*-methylcoclaurine 3′-hydroxylase (CYP80B1), a new methyl jasmonate-inducible cytochrome P-450-dependent mono-oxygenase of benzylisoquinoline alkaloid biosynthesis. Plant J 13:793–801

Ro DK, Paradise EM, Ouellet M, Fisher KJ, Newman KL, Ndungu JM et al (2006) Production of the antimalarial drug precursor artemisinic acid in engineered yeast. Nature 440:940–943

Samanani N, Liscombe DK, Facchini PJ (2004) Molecular cloning and characterization of norcoclaurine synthase, an enzyme catalyzing the first committed step in benzylisoquinoline alkaloid biosynthesis. Plant J 40:302–313

Sato F (2005) RNAi and functional genomics. Plant Biotech 22:431–442

Sato F, Yamada Y (2008) Engineering formation of medicinal compounds in cell cultures. In: Bohnert HJ, Nguyen H, Lewis NG (eds) Advances in plant biochemistry and molecular biology, vol 1. Elsevier Ltd., Amsterdam, pp 311–345

Sato F, Hashimoto T, Hachiya A, Tamura K-i, Choi K-B, Morishige T, Fumimoto H, Yamada Y (2001) Metabolic engineering of plant alkaloid biosynthesis. Proc Natl Acad Sci U S A 98:367–372

Sato F, Inai K, Hashimoto T (2007a) Metabolic engineering in alkaloid biosynthesis: case study in tyrosine- and putrescine-derived alkaloids. In: Verpoorte R et al (eds) Applications of plant metabolic engineering. Springer, The Netherlands

Sato F, Inui T, Takemura T (2007b) Metabolic engineering in isoquinoline alkaloid biosynthesis. Curr Pharm Biotechnol 8:211–218

Shanks J, Morgan J (1999) Plant ‘hairy root’ culture. Curr Opin Biotechnol 10:151–155

Singla D, Sharma A, Kaur J, Panwar B, Raghava GP (2010) BIAdb: a curated database of benzylisoquinoline alkaloids. BMC Pharmacol 10:4

Takemura T, Ikezawa N, Iwasa K, Sato F (2010) Metabolic diversification of benzylisoquinoline alkaloid biosynthesis through the introduction of a branch pathway in *Eschscholzia californica*. Plant Cell Physiol 51:949–959

Takemura T, Ikezawa N, Iwasa K, Sato F (2012) Molecular cloning and characterization of a cytochrome P450 in sanguinarine biosynthesis from *Eschscholzia californica* cells. Phytochem doi:10.1016/j.phytochem.2012.02.013

Takeshita N, Fujiwara H, Mimura H, Fitchen J, Yamada Y, Sato F (1995) Molecular cloning and characterization of *S*-adenosyl-L-methionine: scoulerine-9-*O*-methyltransferase from cultured cells of *Coptis japonica*. Plant Cell Physiol 36:29–36

Tietze L, Rackelmann N, Sekar G (2003) Catalyst-controlled stereoselective combinatorial synthesis. Angew Chem Int Ed 42:4254–4257

Unterlinner B, Lenz R, Kutchan TM (1999) Molecular cloning and functional expression of codeinone reductase: the penultimate enzyme in morphine biosynthesis in the opium poppy *Papaver somniferum*. Plant J 18(5):465–475

Veerasham C (2004) Medicinal plant biotechnology. CBS, New Delhi, pp 377–419

Verpoorte R, van Der Heijden R, van Gulik WM, Ten Hoopen HJG (1991) Plant biotechnology for the production of alkaloids: present status and prospects. In: Brossi A (ed) The Alkaloids: chemistry and pharmacology, vol 40. Academic, New York, pp 1–187

Vogel M, Lawson M, Sippl W, Conarad U, Roos W (2010) Structure and mechanism of sanguinarine reductase, an enzyme of alkaloid detoxification. J Biol Chem 285:18397–18406

Yamada Y, Kokabu Y, Chaki K, Yoshimoto T, Ohgaki M, Yoshida S, Kato N, Koyama T, Sato F (2011) Isoquinoline alkaloid biosynthesis is regulated by a unique bHLH-type transcription factor in *Coptis japonica*. Plant Cell Physiol 52:1131–1141

Yukimune Y, Tabata H, Higashi Y, Hara Y (1996) Methyl jasmonate-induced overproduction of paclitaxel and baccatin III in taxus cell suspension cultures. Nat Biotechnol 14:1129–1132

Zhang L, Yang B, Lu BB, Kai GY, Wang ZN, Xia Y et al (2007) Tropane alkaloids production in transgenic *Hyoscyamus niger* hairy root cultures over-expressing putrescine *N*-methyltransferase is methyl jasmonate-dependent. Planta 225:887–896

Zhang L, Jing F, Li F, Li M, Wang Y, Wang G, Sun X, Tang K (2009) Development of transgenic *Artemisia annua* (Chinese wormwood) plants with an enhanced content of artemisinin, an effective antimalarial drug, by hairpin-RNA mediated gene silencing. Biotechnol Appl Biochem 52:199–207

Zhou M, Zhu ZM, Shao JR, Tang YX, Wu YM (2011) Production and metabolic engineering of bioactive substances in plant hairy root culture. Appl Microbiol Biotechnol 90:1229–1239

Zhou M, Shao JR, Tang YX (2010) Transcriptional response of the catharanthine biosynthesis pathway to methyl jasmonate/nitric oxide elicitation in *Catharanthus roseus* hairy root culture. Appl Microbiol Biotechnol 88:737–750

Zhou M, Shao J, Tang Y (2009) Production and metabolic engineering of terpenoid indole alkaloids in cell cultures of the medicinal plant *Catharanthus roseus* (L.) G. Don (Madagascar periwinkle). Biotechnol Appl Biochem 52:313–323

Chapter 15
Jasmonate-Responsive Transcription Factors: New Tools for Metabolic Engineering and Gene Discovery

Tsubasa Shoji and Takashi Hashimoto

15.1 Introduction

Concerted gene action of a multistep pathway is essential for maintaining appropriate metabolic flux in developmental and environmental contexts. Transcriptional regulation of metabolic genes plays central roles in such coordination and is generally governed by specific transcription factors, which positively or negatively modulate the expression of target genes through sequence-specific DNA binding in promoter regions. The pioneering discovery of maize flavonoid regulators COLOELESS1 (C1) and RED (R) and subsequent characterization of related proteins from various species have demonstrated the central regulatory roles of these factors in flavonoid and anthocyanin pathways (Hichri et al. 2011). Ectopic overexpression of C1- and R-family transcription factors induces increased accumulation of flavonoids in various transgenic plants (Lloyd et al. 1992; Grotewold et al. 1998; Borevitz et al. 2000). Recent advances in genomic approaches have fueled the identification of a number of transcriptional regulators responsible for various secondary products (van der Fits and Memelink 2000; Verdonk et al. 2005; Hirai et al. 2007; Kannangara et al. 2007; Shoji et al. 2010), which have emerged as potential tools for metabolic engineering. Because transcription factors simultaneously control multiple steps or even a whole pathway of particular secondary products, they may have significant impact on the entire flux of the target pathway, thus enabling increased end product accumulation (Broun 2004; Grotewold 2008).

T. Shoji (✉) · T. Hashimoto (✉)
Graduate School of Biological Sciences,
Nara Institute of Science and Technology,
Takayama 8916-5, Ikoma, Nara 630-0101, Japan
e-mail: t-shouji@bs.naist.jp

T. Hashimoto
e-mail: hasimoto@bs.naist.jp

S. Chandra et al. (eds.), *Biotechnology for Medicinal Plants*,
DOI: 10.1007/978-3-642-29974-2_15, © Springer-Verlag Berlin Heidelberg 2013

15.2 Jasmonates Up-Regulate Defensive Secondary Metabolism

Jasmonates (JAs), fatty acid derivatives synthesized via the octadecanoid pathway, play critical regulatory roles in plant defense responses, especially against herbivores and necrotrophic pathogens (Browse 2009). JA signaling induces the production of a wide range of defensive secondary products, conferring adaptive advantages to plants in nature (Baldwin 1998). Elicitors inducing JA production and JA itself have been widely used to improve the production of useful phytochemicals in plant cell and tissue cultures (Gundlach et al. 1992; Menke et al. 1999a, Blechert et al. 1995; Yukimune et al. 1996; Caretto et al. 2011). However, still little is known of how JA signaling regulates the gene expression related to secondary metabolism.

Core signaling modules from JA perception to transcriptional activation have been identified in *Arabidopsis* based on molecular genetic studies (Fig. 15.1; Browse 2009; Fonseca et al. 2009). The bioactive form of JA, a JA-isoleucine conjugate (JA-Ile) (Fonseca et al. 2008), is perceived by a JA receptor complex (Sheard et al. 2010) composed of CORONATINE INSENSTIVE1 (COI1), which is an F-box protein of the SCF-type E3-ubiqutin ligase complex (SCF^{COI1}), and JASMONATE ZIM-domain (JAZ) proteins. JA perception is believed to induce the ubiquitination of JAZ proteins, followed by their 26S proteasome-dependent degradation. There are 12 functionally redundant JAZs in *Arabidopsis*, whose primary structures are relatively divergent except for the Jas and TIFY motifs (Chini et al. 2007; Thines et al. 2007). These conserved motifs are important interfaces for protein–protein interactions; JAZs form homo and heterodimers by interacting at the TIFY motif, whereas the Jas motif mediates COI1-JAZ interaction (Melotto et al. 2008; Chini et al. 2009). JAZs target jasmonate-responsive genes by interacting with transcription factors, and recruit a protein complex containing Groucho/Tup1-type corepressors (Pauwels et al. 2010), which repress transcription of the target genes, possibly through chromosomal remodeling. JA-dependent removal of JAZs liberates the JAZ-interacting transcription factors from the repressor complex, leading to the JA-responsive activation of downstream genes under the control of the liberated factors. Related basic helix-loop-helix (bHLH)-family transcription factors, MYC2, MYC3, and MYC4, interact with multiple JAZ proteins and control wide spectra of JA-responsive genes, including those involved in the defense against bacterial pathogens and insect herbivory (Dombrecht et al. 2007; Chini et al. 2007; Thines et al. 2007; Fernández-Calvo et al. 2011). Other transcription factors, such as MYB-family proteins, MYB21, and MYB24, are required for stamen development (Song et al. 2011), and bHLH-and R2R3 MYB-family members mediating anthocyanin accumulation and trichome initiation (Qi et al. 2011), were recently found to interact with JAZs, and are involved in subsets of JA-regulated metabolic and morphological responses.

In the following sections, we will focus on the JA-responsive and regulatory transcription factors for alkaloid biosynthesis in medicinal plants, particularly the

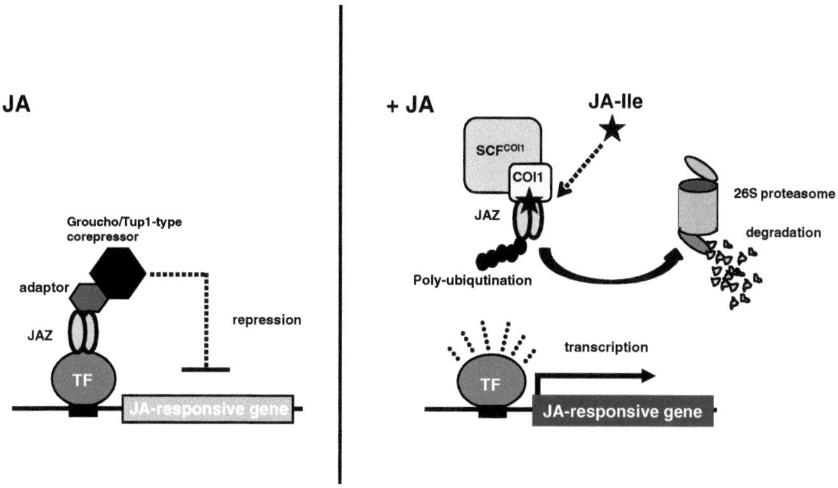

Fig. 15.1 Jasmonate signal transduction in plants JA. Signaling mechanism mainly elucidated in *Arabidopsis* is schematically depicted from the perception of JA-Ile by the COI1-JAZ co-receptor to the gene activation targeted by transcription factors (TFs), such as bHLH-family MYC2, whose activities are suppressed by a co-repressor complex recruited by JAZs in the absence of JA

terpenoid indole alkaloids (TIAs) in *Catharanthus roseus* and nicotine alkaloids in tobacco.

15.3 Regulators of Terpenoid Indole Alkaloid Biosynthesis in *Catharanthus roseus*

TIAs are produced mainly in Apocynaceae, Logniaceae, and Rubiaceae species, and include pharmaceutically important metabolites, such as antineoplastic vinblastine and vincristine, and antihypertensive ajmalicine. Madagascar periwinkle (*Catharanthus roseus*) has been used as a commercial source of vinblastine and vincristine, and to characterize TIA biosynthetic pathways (Fig. 15.2a; De Luca 2011). Tryptamine and secologanin are condensed by strictosidine synthase (STR) (Kutchan et al. 1988; McKnight et al. 1990) to generate strictosidine, a key intermediate common to all TIAs. Tryptamine is derived from tryptophan by decarboxylation catalyzed by tryptophan decarboxylase (TDC) (De Luca et al. 1989). A series of enzymatic steps from tabersonine to vindoline have been characterized (O'Connor and Maresh 2006). Biosynthetic enzymes of the TIA pathway are distributed in several subcellular compartments and in different cell types, implying inter and intracellular trafficking of metabolic intermediates and end products (De Luca 2011), although the transporters involved have not been identified.

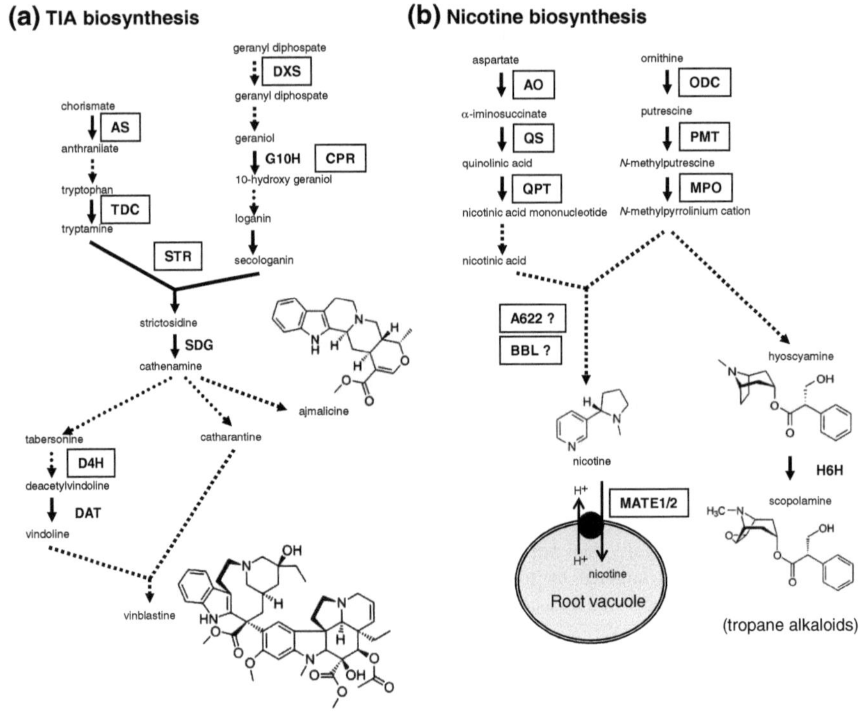

Fig. 15.2 Alkaloid biosynthetic pathways. Solid arrows indicate defined biosynthetic steps catalyzed by known enzymes, whereas dashed arrows indicate multiple steps or ill-defined reactions. Abbreviations for the genes are shown at relevant steps. **a** Terpenoid indole alkaloid (TIA) pathway in *C. roseus*. Abbreviations of the structural genes induced by *ORCA3* overexpression in *C. roseus* cultured cells are boxed. Cytotochrome P450 reductase (CPR) acts with a P450 enzyme, geraniol 10-hydroxylase (G10H). AS, anthranilate synthase; DXS, D-1-deoxyxylulose 5-phosphate synthase; TDC, tryptophan decarboxylase; STR, strictosidine β-D-glucosidase; D4H, desacetoxyvindoline 4-hydroxylase; DAT, acetyl-CoA:4-O-deacetylvindoline 4-O-acetyltransferase. **b** Nicotine alkaloid pathway in tobacco. Abbreviations of the structural genes regulated by *NIC2*-locus *ERF* genes are boxed. A622, a PIP-family oxidoreductase (De Boer et al. 2009; Kajikawa et al. 2009), and BBL, a berberine bridge enzyme-like protein (Kajikawa et al. 2011), are involved in late steps. MATE1 and MATE2 (MATE1/2) are multidrug and toxic compound extrusion transporters implicated in vacuolar sequestration of nicotine in tobacco root (Shoji et al. 2009). Tropane alkaloids, hyoscyamine, and scopolamine, produced in some Solanaceae species, are also derived from *N*-methylpyrrolinum cation; hyoscyamine 6-β-hydroxylase (H6H) catalyzes a two-step epoxidation of hyoscyamine to scopolamine. AO, aspartate oxidase; MPO, *N*-methylpurescine oxidase; ODC, ornithine decarboxylase; PMT, putrescine *N*-methyltransferase; QPT, qunolinate phosphoribosyltransferase; QS, qunolinate synthase

In *C. roseus* cell cultures, fungal elicitors induce TIA biosynthesis, and JA functions as an essential secondary messenger for elicitor-induced *STR* and *TDC* expression (Menke et al. 1999a). A JA- and elicitor-responsive element (JERE) was identified in the *STR* promoter, and was then used as bait in yeast one-hybrid screening to isolate a transcription factor, ORCA2 (octadecanoid-responsive *Catharanthus* AP2-domain protein) (Menke et al. 1999b). ORCA2 belongs to the subgroup IXa of the APETARA2 (AP2)/ETHYLENE RESPONSIVE FACTOR (ERF) family (Nakano et al. 2006). *ORCA2* transcription is rapidly induced by JA and elicitor, and when transiently expressed in *C. roseus* cells, *ORCA2* transactivates the *STR* promoter by binding around a GCC box within JERE (Menke et al. 1999b). In addition to ORCA2, other transcription factors have been isolated based on their ability to bind with non JERE elements in the *STR* or *TDC* promoter (Pré et al. 2000, van der Fits et al. 2000, Sibéril et al. 2001, Pauw et al. 2004), but *in planta* roles of these factors have yet to be defined.

ORCA3, which is closely related to ORCA2 and belongs to the IXa subgroup, was identified by a T-DNA activation tagging approach which utilized the TDC activity to detoxify 4-methyltryptophan, a cytotoxic artificial substrate of TDC, as a selectable marker in cultured *C. roseus* cells (van der Fits and Memelink 2001). Like ORCA2 transcription, ORCA3 transcription is induced by JA, and ORCA3 binds and transactivates the promoters of the *TDC*, *STR*, and *cytochrome P450 reductase* (*CPR*) genes. CPR and geraniol 10-hydroxylase (G10H) are involved in secologanin formation. Overexpression of *ORCA3* in cultured *C. roseus* cells up-regulated the expression of *TDC*, *STR*, *CPR*, *desacetoxyvindoline 4-hydroxylase* (*D4H*), and two primary metabolic genes that supply precursors for TIAs, *anthranilate synthase* (*AS*) and *D-1-deoxyxylulose 5-phosphate synthase* (*DXS*). *ORCA3* overexpression did not induce the expression of *G10H* or the acetyl CoA:deacetylvindoline 4-O-acetyltransferase gene (*DAT*), and increased the accumulation of tryptophan and tryptamine, but not of TIAs, indicating that the terpenoid branch of the TIA pathway remained limiting. When the *ORCA3*-overexpressing cells were fed with a terpenoid intermediate, loganin (van der Fits and Memelink 2001), or when *G10H* was overexpressed together with *ORCA3* (Wang et al. 2010), TIA formation was considerably increased. Therefore, ORCA3 (and possibly ORCA2 as well) is a transcriptional activator of some, but not all, steps in the TIA pathway.

JA-responsive *ORCA3* transcription is directly regulated by *C. roseus* MYC2 (CrMYC2) which binds to a G-box in the *ORCA3* promoter (Endt et al. 2007; Zhang et al. 2011). Downregulation of *CrMYC2* expression strongly decreases the levels of JA-responsive *ORCA3* and *ORCA2* expression in cultured *C. roseus* cells. When a short *ORCA3* promoter fragment containing a G-box, which was recognized by *Arabidopsis* MYC2 in vitro, was introduced into *Arabidopsis* plants, it showed *At-MYC2*-dependent activation by JA (Montiel et al. 2011), indicating that CrMYC2 and *Arabidopsis* MYC2 function similarly. Although *CrMYC2* was initially isolated based on its activity to bind with a G-box in the *STR* promoter by yeast one-hybrid screening (Pré et al. 2000), the *STR* promoter is not regulated by *CrMYC2* (Zhang et al. 2011).

15.4 Regulators of Nicotine Biosynthesis in Tobacco

Nicotine is a predominant alkaloid in tobacco (*Nicotiana tabacum*) and other *Nicotiana* species, and responsible for addiction to smoking. Two heterocyclic rings of nicotine, pyridine and pyrrolidine, are derived from aspartate and ornithine, respectively (Fig. 15.2b; Shoji and Hashimoto 2011a). While early steps of the nicotine pathway leading to the formation of nicotinic acid and *N*-methylpyrrolium cation intermediates have been defined at the enzyme and molecular levels (Imanishi et al. 1998; Hibi et al. 1994; Katoh et al. 2006, 2007), later steps involving coupling of the two rings have remained unclear. Two different families of oxidoreductases have been shown to be necessary for the late steps of nicotine biosynthesis, but their exact enzymatic reactions are not clear (De Boer et al. 2009; Kajikawa et al. 2009, 2011). Nicotine is sequestrated into vacuoles for storage by Multidrug And Toxic compound Extrusion (MATE) transporters (Morita et al. 2009; Shoji et al. 2009). Nicotine is produced exclusively in roots and translocated to aerial portions via the xylem. Tropane alkaloids, including anticholigenic hyoscyamine and scopolamine, are also derived from *N*-methylpyrrolinium cation, and are produced in some Solanaceae species (Fig. 15.2b).

Nicotine is produced constitutively at a basal level, but insect herbivory and wounding of leaves substantially increase its production in roots (Baldwin 1989). Although the JA content of the roots increases after leaf damage, it is not clear whether JA is the mobile signal transmitted from the leaf to the root (Baldwin 1989; Baldwin et al. 1994). JA coordinately activates nearly an entire set of genes for nicotine biosynthesis in roots and cultured cells of tobacco (Imanishi et al. 1998; Shoji et al. 2000; Goossens et al. 2003). Tobacco has functional counterparts of *Arabidopsis* COI1, JAZs, and MYC2, which are all required for JA-responsive induction of nicotine biosynthesis (see below for NtMYC2, Fig. 15.3b). In transgenic tobacco plants and hairy roots, RNAi-mediated silencing of *NtCOI1* or expression of a dominant-negative form of NtJAZs severely compromised JA-inducible nicotine production, as well as other JA-mediated responses (Paschold et al. 2007; Shoji et al. 2008).

JA-mediated induction of the genes encoding putrescine *N*-methyltranasferase (PMT) (Hibi et al. 1994) and quinolinate phosphoribosyltransferase (QPT) (Sinclair et al. 2000) requires two distinct *cis*-elements, G-box, and GCC-box, located in the proximal promoter regions (Fig. 15.3b; Shoji et al. 2000, 2010; Xu and Tinko 2004; Oki and Hashimoto 2004; De Boer et al. 2011, Shoji and Hashimoto 2011b, c). The G-box and GCC-box *cis*-elements are recognized by NtMYC2 and the *NIC2*-locus ERF transcription factors, respectively (Todd et al. 2010; Shoji et al. 2010; De Boer et al. 2011; Shoji and Hashimoto 2011b, c, see above). The *NIC2*-locus ERFs belong to the subgroup IXa, which includes *C. roseus* ORCAs. *NtMYC2* and the *NIC2*-locus *ERF*s are induced by JA (Shoji et al. 2010; De Boer et al. 2011; Shoji and Hashimoto 2011b), and cooperatively transactivate *PMT* and *QPT* promoters when expressed transiently in tobacco cells (De Sutter et al. 2005; Shoji et al. 2010; De Boer et al. 2011; Shoji and Hashimoto 2011b, c). Moreover, in transgenic *Nicotiana* plants and hairy roots, suppression of these transcription factor genes decreased the expression of genes involved in nicotine biosynthesis, and consequently inhibited accumulation of tobacco alkaloids (Todd et al. 2010; Shoji et al. 2010; Shoji and Hashimoto 2011b). While NtMYC2 apparently controls a wide spectrum of JA-responsive defensive genes and is a target of NtJAZ repressors, the *NIC2*-locus ERFs specifically regulate nicotine pathway genes and

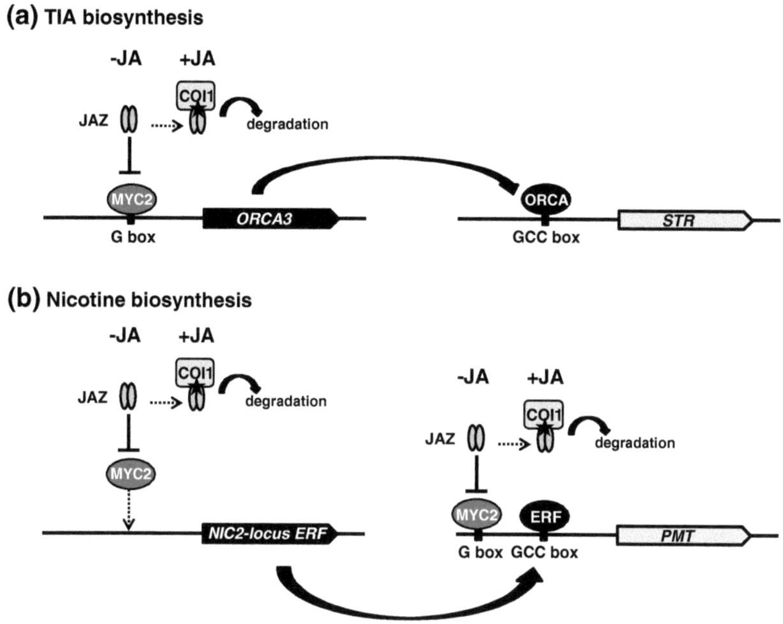

Fig. 15.3 Model of jasmonate signaling leading to expression of alkaloid biosynthetic genes. ORCA/NIC2-type ERF transcription factors regulate biosynthetic genes, such as *STR* for the TIA pathway (**a**) and *PMT* for the nicotine pathway (**b**), by binding at GCC boxes in their promoters, while the JA-responsive transcription of the ERF factors is mediated by a bHLH-family MYC2 transcription factor, a direct target of upstream JAZ repressors (see Fig. 15.1). In nicotine biosynthesis (**b**), MYC2 also regulates some of the structural genes directly by binding at G boxes in the promoters

do not interact with NtJAZs in yeast two-hybrid assays (Shoji et al. 2010; De Boer et al. 2011; Shoji and Hashimoto 2011b).

Low-nicotine regulatory loci have been used to breed commercial tobacco varieties for low-nicotine cigarette production (Legg and Collins 1971). Semi-dominant mutant alleles at two unlinked *NIC1* and *NIC2* loci synergistically decrease nicotine concentrations in tobacco leaves (Legg and Collins 1971), and downregulate the expression of multiple structural genes specific to the nicotine pathway (Hibi et al. 1994; Katoh et al. 2007; Shoji et al. 2009; Shoji et al. 2010; Kajikawa et al. 2011). Indeed, several structural genes involved in nicotine accumulation have been isolated after differential screening of tobacco genes specifically downregulated in mutant roots, compared to wild-type roots, by cDNA subtraction (Hibi et al. 1994), differential display (Shoji et al. 2009), and cDNA microarray (Katoh et al. 2007; Shoji et al. 2010; Kajikawa et al. 2011). The microarray analysis also provided a clue to the molecular identity of the *NIC2* locus; we found the expression of a distinct set of tobacco *ERF* genes to be suppressed in the *nic2* mutant. At least seven *ERF* genes of the subgroup IXa are clustered at the *NIC2* locus, therefore referred as the *NIC2*-locus *ERFs*, and are found to be deleted altogether in the *nic2* mutant genome (Shoji et al. 2010). The deleted *ERF* genes all originate from the *N. tomentosiformis* ancestor, while the corresponding *ERF* genes of *N. sylvestris* origin appear to be intact in the *nic2* genome (Shoji et al. 2010). Tobacco is a natural alloteraploid that originated by hybridisation between two diploid ancestral species closely related to *N. tomentosiformis* and *N. sylvestris* (Clarkson et al. 2005). The presence of functionally redundant *ERF* genes at the *N. sylvestris*-derived locus likely explains why the extensive genomic deletion of the *NIC2*-locus *ERF* genes gives only mild low-nicotine phenotypes with the *nic2* mutant allele.

While both NtMYC2 and *NIC2*-locus ERFs bind their cognate *cis*-elements in the promoter regions of nicotine biosynthesis genes, thereby activating them, is there hierarchical control between these two distinct transcription factors? When *NtMYC2* was silenced by RNA interference in tobacco hairy roots, all of the *NIC2*-locus *ERF*s, as well as nicotine biosynthesis and transporter genes, were down-regulated (Shoji and Hashimoto 2011b). In contrast, the *NtMYC2* expression level was not altered either in the *nic2* mutant or in transgenic tobacco hairy roots expressing a dominant-repressive form of a *NIC2*-locus ERF (Shoji and Hashimoto 2011b). These results suggest that *NtMYC2* regulates the *NIC2*-locus *ERF* genes, but not vice versa (Fig. 15.3b). Since NtMYC2-binding G-boxes are predicted in the proximal promoter regions of some *NIC2*-locus *ERF* genes (our unpublished data), NtMYC2 may directly activate these *ERF* genes. Some *NIC2*-locus *ERF* genes are immediately induced by JA, while others are gradually induced at later time points (Shoji et al. 2010). Future studies should address how individual members of the *NIC2*-locus *ERF* genes are regulated by NtMYC2 and possibly by particular ERFs at the *NIC2* locus.

15.5 Biotechnological Applications

Overexpression of the gene for an enzyme in a particular alkaloid pathway in alkaloid-producing plants may considerably boost levels of the pathway's end products, alkaloids, if the enzymatic activity constitutes a metabolic bottleneck of pathway flux (Yun et al. 1992). However, since multiple steps are frequently rate-limiting in secondary metabolism, overexpression of one or a few enzymes in a long biosynthetic pathway likely has very limited effects on the accumulation of end products. A promising strategy would be to constitutively (over) express transcription factors that specifically control all or many of the enzymes in the target pathway. Specific regulators, such as the *NIC2*-locus ERFs of nicotine biosynthesis, should be more useful than general regulators, such as MYC2. If transcription factors are post-translationally modified for activation, for example by phosphorylation, mutated factors with constitutive activation are ideal for metabolic engineering.

Overexpression of *ERF189*, a *NIC2*-locus *ERF*, in tobacco hairy roots increased levels of tobacco alkaloids two- to threefold both in uninduced and JA-induced conditions (Shoji et al. 2010). Similarly, when *ORC1/ERF221*, another of the *NIC2*-locus *ERF*s, was overexpressed in tobacco plants or in *N. glauca* hairy roots, nicotine contents are increased two- to four-fold (De Boer et al. 2011). In the case of TIA biosynthesis, overexpression of *ORCA3* in cultured cells or hairy roots of *C. roseus* induced the expression of several genes for enzymes in the pathway, but not others, including *G10H* and *DAT* (van der Fits and Memelink 2000; Peebles et al. 2009). Thus, cultures overexpressing *ORCA3* accumulated increased amounts of tryptamine, but not later products of the TIA pathway. Feeding the terpenoid precursor loganin to the cultures was necessary to circumvent the metabolic bottlenecks. Apparently, other yet-to-be-identified transcription factors are required to activate the *ORCA3*-nonresponsive genes in the pathway. Simultaneous overexpression of *ORCA3* and such missing regulators in *C. roseus* may overcome difficulties encountered in metabolic engineering of TIA biosynthesis.

Specific transcriptional regulators are useful tools for to identifying their target genes which encode biosynthetic enzymes and transporters required for alkaloid biosynthesis and accumulation. Natural mutants, and transgenic cultured cells, tissues, or plants, in which specific transcription factor genes are suppressed or overexpressed, have been comprehensively analyzed for a broad range of transcriptomes, and are compared for control transcriptomes, using various techniques, such as PCR-based differential screening, differential display, cDNA, or oligonucleotide microarrays, and next generation sequencing technologies. When target transcription factor is encoded in several functionally redundant genes, overexpression of a dominant negative form which acts as an engineered repressor by the function of a fused repressor domain should be useful (Shoji et al. 2010). Successful examples can be found for identifying candidate tobacco genes involved in nicotine biosynthesis and transport (see above).

While transcription factor-based approaches provide a list of candidate genes, identifying gene functions of orphan proteins is not straightforward. If enzymatic reactions have been reported previously, recombinant candidate proteins can be tested as to whether they catalyze the predicted reactions. However, when exact enzymatic reactions are not known and when possible substrates are not available, recombinant proteins cannot be readily used for in vitro enzyme assays. Metabolomic profiling of the transgenic plant materials in which a candidate gene is suppressed may identify pathway intermediates that accumulate in the absence of the gene's function. Knockdown of *A622*, encoding an orphan oxidoreductase of the PIP family, in tobacco hairy roots induced a considerable accumulation of *N*-methyl-pyrrolinium cation and nicotinic acid *β*-*N*-glucoside (Kajikawa et al. 2009). However, the glucoside is not a substrate for A622, and is likely a detoxification metabolite of nicotinic acid. When *BBL*, encoding an uncharacterized FAD-containing oxidoreductase of the berberine bridge enzyme family, was suppressed in tobacco hairy roots and tobacco plants, a novel alkaloid, dihydromethanicotine, accumulated (Kajikawa et al. 2011). Again, dehydromethanicotine does not appear to be a true substrate of BBL, and may be formed from an unstable metabolite that accumulates when the BBL reaction is blocked. These examples illustrate the considerable difficulties in assigning exact functions to orphan enzymes. Activity-based metabolomic profiling may be a promising approach that allows unbiased discovery of enzymatic activities of uncharacterized proteins (De Carvalho et al. 2010). With this approach, a purified target recombinant enzyme and potential cofactors are incubated with partially purified cell extracts from transgenic plant samples in which a target enzyme gene has been suppressed. Metabolomic profiling of the reaction mixtures by liquid-chromatography mass spectrometry-based methods and subsequent statistical analysis may identify unique metabolites which decrease during the enzymatic reactions.

It is quite interesting that closely related ERF transcription factors of the IX subgroup (periwinkle ORCAs and tobacco *NIC2*-locus ERFs) specifically control the JA-responsive biosynthesis of unrelated bioactive alkaloids (TIA and nicotine) in two evolutionally distant species, periwinkle (Apocynaceae) and tobacco (Solanaceae). An intriguing possibility is that homologs of ORCA/NIC2 ERFs may be involved in the regulation of JA-inducible secondary metabolite biosynthesis in commercially valuable medicinal plants. Biotechnological applications of these transcription factors are eagerly anticipated.

References

Baldwin IT (1989) Mechanism of damage-induced alkaloid production in wild tobacco. J Chem Ecol 15:1661–1680

Baldwin IT (1998) Jasmonate-induced responses are costly but benefit plants under attacks in native populations. Proc Natl Acad Sci USA 95:8113–8118

Baldwin IT, Scmelz EA, Ohnmeiss TE (1994) Wound-induced changes in root and shoot jasmonic acid pools correlated with induced nicotine synthesis in *Nicotiana sylvestris* Spegazzini and Comes. J Chem Eco 20:2139–2157

Blechert S, Brodschelm W, Hölder S, Kammerer L et al (1995) The octadecanoic pathway: signal molecules for the regulation of secondary pathways. Proc Natl Acad Sci USA 92:4099–4105

Borevitz JO, Xia Y, Blount J, Dixon RA et al (2000) Activation tagging identifies a conserved MYB regulator of phenylpropanoid biosynthesis. Plant Cell 12:2383–2394

Broun P (2004) Transcription factors as tools for metabolic engineering in plants. Curr Opin Plant Biol 7:202–209

Browse J (2009) Jasmonate passes muster: a receptor and targets for the defense hormone. Ann Rev Plant Biol 60:183–205

Caretto S, Quarta A, Durante M, Nisi R et al (2011) Methyl jasmonate and miconazole differently affect arteminisin production and gene expression in *Artemisia annua* suspension cultures. Plant Biol (Stuttg) 13:51–58

Chini A, Fonseca S, Chico JM, Fernández-Calvo P et al (2009) The ZIM domain mediate homo- and heteromeric interactions between *Arabidopsis* JAZ proteins. Plant J 59:77–78

Chini A, Fonseca S, Fernández G, Adie B et al (2007) The JAZ family of repressors is the missing link in jasmonate signaling. Nature 448:666–671

Clarkson JJ, Lim KY, Kovarik A, Chase MW et al (2005) Long-term genome diploidization in allopolyploid *Nicotiana* section Repandae (Solanaceae). New Phytol 168:241–252

De Boer K, Lye JC, Aiken CD, Su AK et al (2009) The *A622* gene in *Nicotiana glauca* (tree tobacco): Evidence for a functional role in pyridine alkaloid synthesis. Plant Mol Biol 69:299–312

De Boer K, Tileman S, Pauwels L, Bossche RV et al (2011) APETALA2/ETHYENE RESPONSE FACTOR and basic helix-loop-helix transcription factors cooperatively mediate jasmonate-elicited nicotine biosynthesis. Plant J 66:1053–1065

De Carvalho LPS, Zhao H, Dickinson CE, Arango NM et al (2010) Activity-based metabolomic profiling of enzymatic function: identification of Rv1248c as a mycobacterial 2-hydroxy-3-oxoadipate synthase. Chem Biol 17:323–332

De Luca V (2011) Monoterpenoid indole alkaloid biosynthesis. In: Ashihara H, Crozier A, Komamine A (eds) Plant metabolism and biotechnology. Wiley, New York, pp 263–291

De Luca V, Marineau C, Brisson N (1989) Molecular cloning and analysis of cDNA encoding a plant tryptophan decarboxylase. Proc Natl Acad Sci USA 86:2582–2586

De Sutter V, Vanderhaeghen R, Tilleman S, Lammertyn F et al (2005) Exploration of jasmonate signalling via automated and standardized transient expression assays in tobacco cells. Plant J 44:1065–1076

Dombrecht B, Xue GP, Sprague SJ, Kirkegaard JA et al (2007) MYC2 differentially modulates diverse jasmonate-dependent functions in *Arabidopsis*. Plant Cell 19:2225–2245

Endt DV, Silva MS, Kijne JW, Pasquali G et al (2007) Identification of a bipartite jasmonate-responsive promoter element in the *Catharanthus roseus ORCA3* transcription factor gene that interacts specifically with AT-hook DNA-binding proteins. Plant Physiol 144:1680–1689

Fernández-Calvo P, Chini A, Fern ández-Barbero G, Chico J.M et al (2011) The *Arabidopsis* bHLH transcription factors MYC3 and MYC4 are targets of JAZ repressors and act additively with MYC2 in the activation of jasmonate responses. Plant Cell 23:701–715

Fonseca S, Chini A, Hamberg M, Adie B et al (2008) (+)-7-*iso*-Jasmonoyl-L-isoleucine is the endogenous bioactive jasmonate. Nat Chem Biol 5:344–350

Fonseca S, Chico JM, Solano R (2009) The jasmonate pathway: the ligand, the receptor and the core signaling module. Curr Opin Plant Biol 12:539–547

Goossens A, Häkkinen ST, Laakso I, Seppänen-Laakso T et al (2003) A functional genomics approach toward the understanding of secondary metabolism in plant cells. Proc Natl Acad Sci USA 100:8595–8600

Grotewold E (2008) Transcription factors for predictive plant metabolic engineering: are we there yet? Curr Opin Biotechnol 19:138–144

Grotewold E, Chamberlin M, Snook M, Siame B et al (1998) Engineering secondary metabolism in maize cells by ectopic expression of transcription factors. Plant Cell 10:721–740

Gundlach H, Müller MJ, Kutchan TM, Zenk MH (1992) Jasmonic acid is a signal transducer in elicitor-induced plant cell cultures. Proc Natl Acad Sci USA 89:2389–2393

Hibi N, Higashiguchi S, Yamada Y, Hashimoto T (1994) Gene expression in tobacco low-nicotine mutant. Plant Cell 6:723–735

Hichri I, Barrieu F, Bogs J, Kappel C et al (2011) Recent advances in the transcriptional regulation of the flavonoid biosynthetic pathway. J Exp Bot 62:2465–2483

Hirai MY, Sugiyama K, Sawada Y, Tohge T et al (2007) Omics-based identification of *Arabidopsis* Myb transcription factors regulating aliphatic glucosinolate biosynthesis. Proc Natl Acad Sci USA 104:6478–6483

Imanishi S, Hashizume K, Nakakita M, Kojima H et al (1998) Differential induction of methyl jasmonate of genes encoding ornithine decarboxylase and other enzymes involved in nicotine biosynthesis in tobacco cell cultures. Plant Mol Biol 38:1101–1111

Kajikawa M, Hirai N, Hashimoto T (2009) A PIP-family protein is required for biosynthesis of tobacco alkaloids. Plant Mol Biol 69:287–298

Kajikawa M, Shoji T, Katoh A, Hashimoto T (2011) Vacuole-localized berberine bridge enzyme-like proteins are required for a late step of nicotine biosynthesis in tobacco. Plant Physiol 155:2010–2022

Kannangara R, Branigan C, Liu Y, Penfield T et al (2007) The transcription factor WIN1/SHN1 regulates cutin biosynthesis in *Arabidopsis thaliana*. Plant Cell 19:1278–1294

Katoh A, Shoji T, Hashimoto T (2007) Molecular cloning of *N*-methylputrescine oxidase from tobacco. Plant Cell Physiol 48:550–554

Katoh A, Uenohara K, Akita M, Hashimoto T (2006) Early steps in the biosynthesis of NAD in *Arabidopsis* start with aspartate and occur in the plastid. Plant Physiol 141:851–857

Kutchan TM, Hampp N, Lottspeich F, Beyreuther K et al (1988) The cDNA clone for strictosidine synthase from *Rauvolfia serpentine*. DNA sequence determination and expression in *Esherichia coli*. FEBS Lett 237:40–44

Legg PG, Collins GB (1971) Inheritance of percent total alkaloids in *Nicotiana tabacum* L. II. genetic effects of two loci in Burley21 X LA Burley21 populations. Can J Genet Cytol 13:287–291

Lloyd AM, Walbot V, Davis RW (1992) *Arabidopsis* and *Nicotiana* antocyanin production activated by maize regulator R and C1. Science 258: 1773–1775

McKnight TD, Roessner CA, Devagupta R, Scott AI et al (1990) Nucleotide sequence of a cDNA encoding the vacuolar protein strictosidine synthase from *Catharanthus roseus*. Nucleic Acids Res 18:4939

Melotto M, Mecey C, Niu Y, Chung HS et al (2008) A critical role of two positively charged amino acids in the Jas motif of *Arabidopsis* JAZ proteins in mediating coronatine- and jasmonoyl isoleucine-dependent interactions with the COI1 F-box protein. Plant J 55:979–988

Menke FLH, Parchmann S, Müller MJ, Kijne JW et al (1999a) Involvement of the octadecanoid pathway and protein phosphorylation in fungal elicitor-induced expression of terpenoid indole alkaloid biosynthetic genes in *Catharanthus roseus*. Plant Physiol 119:1289–1296

Menke FLH, Champion A, Kijne JW, Memelink J (1999b) A novel jasmonate- and elicitor-responsive element in the periwinkle secondary metabolite biosynthetic gene Str interacts with a jasmonate- and elicitor-inducible AP2-domain transcription factor, ORCA2. Embo J 18:4455–4463

Montiel G, Zarei A, Körbes AP, Memelink J (2011) The jasmonate-responsive element from the ORCA3 promoter from *Catharanthus roseus* is active in *Arabidopsis* and is controlled by the transcription factor AtMYC2. Plant Cell Physiol 52:578–587

Morita M, Shitan N, Sawada K, Van Montagu MCE et al (2009) Vacuolar transport of nicotine is mediated by a multidrug and toxic compound extrusion (MATE) transporter in *Nicotiana tabacum*. Proc Natl Acad Sci USA 106:2447–2452

Nakano T, Suzuki K, Fujimura T, Shinshi H (2006) Genome-wide analysis of the ERF gene family in *Arabidopsis* and rice. Plant Physiol 140:411–432

O'Connor SE, Maresh JJ (2006) Chemistry and biology of monoterpene indole alkaloid biosynthesis. Nat Prod Rep 23:532–547

Oki H, Hashimoto T (2004) Jasmonate-responsive regions in a *Nicotiana sylvestris PMT* gene involved in nicotine biosynthesis. Plant Biotechnol 21:269–274

Paschold A, Hailtschke R, Baldwin IT (2007) Co(i)-ordinating defenses: NaCOI1 mediates herbivore-induced resistance in *Nicotiana attenuate* and reveals the role of herbivore movement in avoiding defenses. Plant J 51:79–91

Pauw B, Hilliou FAO, Martin VS, Chatel G et al (2004) Zinc finger protein act as transcriptional repressors of alkaloid biosynthesis genes in *Catharanthus roseus*. J Biol Chem 279: 52940–52948

Pauwels L, Barbero GF, Geerinck J, Tilleman S et al (2010) NINJA connects the co-repressor TOPLESS to jasmonate signaling. Nature 464:788–791

Peebles CAM, Hughes EH, Shanks JV, San KY (2009) Transcriptional response of the terpenoid alkaloid pathway to the overexpression of *ORCA3* along with jasmonic acid elicitation of *Catharanthus roseus* hairy roots over time. Metabolic Eng 11:76–86

Pré M, Sibéril Y, Memelink J, Champion A et al (2000) Isolation by the yeast one-hybrid system of cDNAs encoding transcription factors that bind to the G-box element of the *strictosidine synthase* gene promoter from *Catharanthus roseus*. Int J Biochrom 5:224–229

Qi T, Song S, Ren Q, Wu D et al (2011) The Jasmonate-ZIM-domain proteins interact with the WD-repeat/bHLH/MYB complexes to regulate jasmonate-mediated anthocyanin accumulation and trichome initiation in *Arabidopsis thaliana*. Plant Cell 23:1795–1814

Sheard JB, Tan X, Mao H, Withers J et al (2010) Jasmonate perception by inositol-phosphate-potentiated COI1-JAZ co-receptor. Nature 468:400–405

Shoji T, Hashimoto T (2011a) Nicotine biosynthesis. In: Ashihara H, Crozier A, Komamine A (eds) Plant metabolism and biotechnology. Wiley, New York, pp 191–216

Shoji T, Hashimoto T (2011b) Tobacco MYC2 regulates jasmonate-inducible nicotine biosynthesis genes directly and by way of the *NIC2*-locus *ERF* genes. Plant Cell Physiol 52: 1117–1130

Shoji T, Hashimoto T (2011c) Recruitment of a duplicated primary metabolism gene into the nicotine biosynthesis regulon in tobacco. Plant J 67:949–959
Shoji T, Inai K, Yazaki Y, Sato Y et al (2009) Multidrug and toxic compound extrusion-type transporters implicated in vacuolar sequestration of nicotine in tobacco roots. Plant Physiol 149:708–718
Shoji T, Kajikawa M, Hashimoto T (2010) Clustered transcription factor genes regulate nicotine biosynthesis in tobacco. Plant Cell 22:3390–3409
Shoji T, Ogawa T, Hashimoto T (2008) Jasmonate-induced nicotine formation in tobacco is mediated by tobacco *COI1* and *JAZ* genes. Plant Cell Physiol 49:1003–1012
Shoji T, Yamada Y, Hashimoto T (2000) Jasmonate induction of putrescine *N*-methyltransferase genes in the root of *Nicotiana sylvestris*. Plant Cell Physiol 41:831–839
Sibéril Y, Benhamron S, Memelink J, Giglioli-Guivarc'h N et al (2001) *Catharanthus roseus* G-box binding factors 1 and 2 act as repressors of strictosidine synthase gene expression in cell culture. Plant Mol Biol 45:427–440
Sinclair SJ, Murphy PJ, Birch CD, Hamill D (2000) Molecular characterization of quinolinate phosphoribosyltransferase (QPRTase) in *Nicotiana*. Plant Mol Biol 44:603–617
Song S, Qi T, Huang H, Ren Q et al (2011) The Jasmonate-ZIM domain proteins interact with the R2R3-MYB transcription factors MYB21 and MYB24 to affect jasmonate-regulated stamen development. Plant Cell 23:1000–1013
Thines B, Katsir L, Melotto M, Niu Y et al (2007) JAZ repressor proteins are targets of the SCF^{COI1} complex during jasmonate signalling. Nature 448:661–665
Todd AT, Liu E, Polvi SL, Pammett RT et al (2010) A functional genomic screen identifies diverse transcription factors that regulate alkaloid biosynthesisin *Nicotiana benthamiana*. Plant J 62:589–600
van der Fits L, Memelink J (2000) ORCA3, a jasmonate-responsive transcriptional regulator of plant primary and secondary metabolism. Science 289:295–297
van der Fits L, Memelink J (2001) The jasmonate-inducible AP2/ERF-domain transcription factor ORCA3 activates gene expression via interaction with a jasmonate-responsive promoter element. Plant J 25:43–53
van der Fits L, Zhang H, Menke FLH, Deneka M et al (2000) A *Catharanthus roseus* BPF-1 homologue interact an elicitor-responsive region of the secondary metabolite biosynthetic gene *Str* and is induced by elicitor via a JA-independent signal transduction pathway. Plant Mol Biol 44:675–685
Verdonk JC, Haring MA, van Tunen AJ, Schuurink RC (2005) ODORANT1 regulates fragrance biosynthesis in petunia. Plant Cell 17:1612–1624
Wang CT, Liu H, Gao XS, Zhang HX (2010) Overexpression of *G10H* and *ORCA3* in the hairy roots of *Catharanthus roseus* improves catharanthine production. Plant Cell Rep 29:887–894
Xu B, Tinko MP (2004) Methyl jasmonate induced expression of the tobacco putrescine *N*-methyltransferase genes requires both G-box and GCC-motif elements. Plant Mol Biol 55:743–761
Yukimune Y, Tabata H, Higashi Y, Hara Y (1996) Methyl jasmonate-induced overproduction of paclitaxel and baccatin III in *Taxus* cell suspension cultures. Nat Biotechnol 14:1129–1132
Yun DJ, Hashimoto T, Yamada Y (1992) Metabolic engineering of medicinal plants: Transgenic *Atropa belladonna* with an improved alkaloid composition. Proc Natl Acad Sci USA 89: 11799–11803
Zhang H, Hedhili S, Montiel G, Zhang Y et al (2011) The basic helix-loop-helix transcription factor CrMYC2 controls the jasmonate-responsive expression of the *ORCA* genes regulating alkaloid biosynthesis in *Catharanthus roseus*. Plant J 67:61–71

Chapter 16
Metabolic Engineering of Plant Cellular Metabolism: Methodologies, Advances, and Future Directions

Rafael Zárate, Nabil el Jaber-Vazdekis and Robert Verpoorte

16.1 Introduction

Plants produce a myriad of secondary metabolites with an immense number of applications for man (fragrances, phytochemicals, dyes, drugs, insecticides, toxins, etc.). For the producing plants these compounds are important for their interaction with the environment and their survival (symbiosis, pollinator attractants, defense against herbivores, and phytopathogens, etc.). Structures of an estimated 200,000 natural products have been elucidated (Dixon and Steele 2002) and each year approximately 4,000–5,000 novel compounds are characterized. At the same time, the determination, characterization, and sequence of genes involved in the biosynthesis of secondary metabolites are progressing at an increasing pace (Oksman-Caldentey and Barz 2002; Verpoorte and Alfermann 2000).

Metabolic engineering constitutes an important tool to improve the plant cell factory directed for the production of desired phytochemicals (Verpoorte and Alfermann 2000; Zárate and Yeoman 2001; Barampuram and Zhang 2011). In theory, any biosynthetic pathway is amenable to metabolic engineering, although in practice different positive outcomes have been reported although unsuccessful experiments are difficult to get published and thus might not be accounted for.

R. Zárate (✉) · N. el Jaber-Vazdekis
Instituto Canario de Investigación del Cáncer, Hospital Universitario La Candelaria, Carr. El Rosario 45, 38010, Santa Cruz de Tenerife, Spain
e-mail: rzarate@icic.es

N. el Jaber-Vazdekis
e-mail: njvazdekis@icic.es

R. Verpoorte
Natural Products Laboratory, Institute of Biology Leiden, University of Leiden, P.O. Box 9502, 2300 RA, Leiden, The Netherlands
e-mail: verpoort@chem.leidenuniv.nl

S. Chandra et al. (eds.), *Biotechnology for Medicinal Plants*,
DOI: 10.1007/978-3-642-29974-2_16,

Metabolic engineering is possible. But what are the targets, why and what should one like to alter in the metabolism of plants? The following goals can be considered:

(a) Improved quality for producer (farmer)

1. Improved yield
2. Improved resistance against pests and diseases
3. Improved cold, salt, and drought resistance
4. Nitrogen fixation

(b) Improved quality for processing (industry)

1. Storage, increased shelf life
2. Water content
3. Unwanted by-products (e.g. amylopectine in starch production, lignin for cellulose production)
4. Higher level of speciality chemicals; e.g. medicines, biocides, flavors, fragrances, cosmetics, dyes
5. Novel compounds for drug development

(c) Improved quality for consumer

1. Taste food
2. Color food, flowers
3. Increased level of health improving compounds
4. Lower level of undesired compounds

Looking at this list of possibilities, one can clearly distinguish that the applications concern in part changes in primary metabolism and in part changes in secondary metabolism. Secondary metabolism is per definition species-specific, it serves the producing organism to survive within its ecosystem. In plants, that means that among others it is involved in defence against pests and diseases, and in attracting pollinators. The defence compounds are of different character, some are constitutively expressed (phytoanticipins), others are only biosynthesized after wounding or following infection (phytoalexins). That means that the regulation of secondary metabolism is, in part, developmentally regulated, and in part dependent on external (stress) signals.

The length and intricacy of secondary metabolite pathways differs considerably. For instance, the biosynthesis of a compound like the phytoalexin resveratrol consists of only a single step, starting from primary metabolism, catalyzed by one single enzyme, encoded by one single gene (Hain et al. 1990). Whereas the biosynthesis of a terpenoid indole alkaloid like vinblastine, includes at least 30 different steps, at least three different cell types and in each type of cell three different cellular compartments, and thus is also regulated by transport (Van der Heijden et al. 2004; Roytrakul and Verpoorte 2007). Because of the fact that secondary metabolism is plant species-specific, the knowledge about most pathways is limited and few pathways have been fully elucidated to all levels of intermediates, enzymes and genes.

These aspects have to be kept in mind in considering metabolic engineering of plant secondary metabolism. Another consideration is whether one wants to increase or decrease the flux through a pathway. The latter could, for example, be of interest in case of undesired (toxic) compounds, or to switch off certain pathways that compete with the pathway of interest. Also, catabolic pathways might be of interest to cut, in order to increase the level of a desired compound. The aim in all these cases is to reduce the level of the protein of interest by an antisense or RNAi approach, or by overexpressing an antibody of the protein in the plant.

In case that one wants to increase the level of a compound, it has to be kept in mind that in most cases not all genes of the biosynthetic pathway are known. Engineering long pathways thus first needs quite some work to elucidate the pathway. In case of inducible pathways, one might also consider the engineering of the signal transduction pathway leading to the induction of the pathway of interest, e.g. overexpression of a regulatory gene such as a transcription factor.

To produce new compounds for a plant, one can add extra steps to an already existing route, or introduce a new enzyme for the plant that catalyzes an early step in a route, e.g. a terpene synthase, resulting in the formation of a novel molecular skeleton for that plant, which might be further processed by enzymes naturally occurring in that plant to yield a novel product for the plant, or maybe even a totally novel compound. Such a recombinatorial biochemistry approach aiming at production of novel compounds would be of interest for developing novel leads for drug development. It might also be of interest for increasing the resistance of the plant against pests and diseases; however, from the point of view of safety, it might need quite some work to prove that the new compound is not toxic for the consumer. Producing a known compound in another plant might be of interest for several reasons. The most obvious is the introduction of health promoting compounds in food plants, e.g. the vitamin A in rice (Ye et al. 2000). Another reason could be that the target plant has better properties for producing the compound than others. The *Atropa belladonna* producing scopolamine is such an example (Yun et al. 1992).

Anyway in principle there are infinite possibilities, but in practice they are limited by: public acceptance, safety issues, lack of knowledge of biosynthetic pathways, or viability of plants with altered metabolism.

The major area for application on the short term is thus the production of medicinal compounds, either known or new, in plants or plant cell or tissue cultures. At present, the only commercial application is in altered flower colors.

In this chapter, the more common plant transformation techniques are reviewed; furthermore, the different genetic approaches and advances made on the genetic engineering of biosynthetic pathways are also presented, as well as various new points that should be addressed for a more predictable outcome of the genetic engineering of plant species. Finally, under future directions, references are made to the expected impacts of new or more comprehensive approaches, such as, functional genomics, or systems biology, to be likely applied in future genetic engineering programs of target plant species.

16.2 Methodology

Several methods for stable or transient genetic transformation of plants or plant cells have been developed (Potrykus 1991; Hansen and Wright 1999; Lorence and Verpoorte 2004). These include: *Agrobacterium*-mediated transformation, particle bombardment, floral dip transformation, agrodrench, viral vectors, protoplast transformation, and ultrasound. These are the main techniques for the genetic transformation of plants, and many of them have also been applied for the transformation of secondary metabolite pathways in an attempt to alter the metabolic pathways of target compounds (Zárate 2010).

16.2.1 Agrobacterium-Mediated Transformation

Two different *Agrobacterium* species i.e. *A. tumefaciens* and *A. rhizogenes* are routinely used for genetic transformation of plants. These belong to the *Rhizobiaceae* family and are the causal agents of the plant diseases crown gall and hairy root, respectively. Diseases caused by the presence of bacterial DNA, the transferred DNA (T-DNA) within the plant cells, which are involved in the production of plant growth regulators, auxin and cytokinin type, which participate in the induction of tumor or hairy roots (Hooykaas 2000). These bacteria have been considered natural metabolic engineers for their capacity to transfer genes into the target plant cells, thus showing their ability to genetically cross kingdoms (Chilton et al. 1977; Tepfter 1990). Moreover, these bacteria have been reported to transform mainly dicotyledonous plants, although there are records on their use to also transform monocotyledonous plants (Ishida et al. 1996), and in some instances even yeast and animals (Bundock et al. 1995; Michielse et al. 2005; Kunik et al. 2001).

Agrobacterium-mediated transformation occurs after bacterial infection of the plant cells or tissues. Upon infection, the T-DNA, which can be engineered to harbor the genes of interest, gets inserted into the plant nuclear DNA. Similarly, other components of the bacterial plasmids, the Ti-plasmid from *A. tumefaciens* (tumour inducer) or the Ri-plasmid from *A. rhizogenes* (root inducer) are also involved in this process. Both plasmids show large functional homologies and appear to have evolved from a common ancestor (Sinkar et al. 1987). These plasmids also possess a virulence region, with various silent *vir* genes that do not enter the plant genome but are essential for the T-DNA transfer. These genes are activated by lignin precursors and acetosyringone, wound tissue metabolites, which may explain why tissue wounding is necessary for efficient infection (Melchers et al. 1989). On the other hand, the *onc*-genes and *rol*-genes encode enzymes involved in the production of plant growth regulators, as well as other opine synthase genes that activate the synthesis of different classes of opines, as well as their catabolism. These are unique natural metabolites, pseudoaminoacids, which serve as a nutrient source of carbon and nitrogen for the pathogenic bacteria (Dessaux et al. 1993).

Agrobacterium transformation requires an acidic pH (5–6), the presence of phenolic inducers produced by wounded tissue, and more recently it has been reported that light also enhances *Agrobacterium* transformation (Zambre et al. 2003). Nevertheless, how the T-DNA integrates into the plant nuclear genome is not totally understood, but it seems to resemble illegitimate recombination, and it is conducted in a similar way in dicot and monocot plants (Hooykaas 2000). Integration of the T-DNA occurs at random positions in the genome, although it shows preferences for transcriptionally active regions. Once integrated, the T-DNA is maintained stably. Contrary to other gene insertion techniques, the plant transgenic lines produced via *Agrobacterium*, often contain one copy or a low copy number of the T-DNA, although cell lines with multiple T-DNA copies can also be found.

Recently, a novel transient assay has been developed and efficiently established, based on the cocultivation of young *Arabidopsis thaliana* seedlings with *A. tumefaciens* in the presence of a surfactant. This new method does not require any dedicated equipment and can be carried out within 1 week from sowing seeds to protein analysis. This Fast Agro-mediated Seedling Transformation (FAST) was used successfully to express a wide variety of constructs driven by different promoters in *Arabidopsis* seedling cotyledons in diverse genetic backgrounds (Li et al. 2009). The FAST assay offers particular advantages: (i) only routine techniques and reagents are used which breaks the constraint of specialized devices, such as a particle gun reducing the overall cost of the experiments; (ii) unlike particle bombardment and protoplast transfection where high quality plasmid DNA has to be prepared each time, *A. tumefaciens* cells used in this assay can be stored indefinitely and can be repeatedly generated from the glycerol stock before use; (iii) this transient assay could achieve higher cotransformation efficiency for two constructs when they are simultaneously carried in the same agrobacterium cell; (iv) small-scale assay produces sufficient proteins for downstream analysis (e.g. Western blot), and the protein production can be easily scaled up for other applications; (v) the use of 4-day-old *Arabidopsis* seedlings instead of mature plants allows for rapid screening with minimal manipulations, which could even be adapted for a 96-well plate format.

Analogously, important crops, such as cereals have been modified using *Agrobacterium*-mediated transformation employing cocultivation of highly embryogenic callus tissue or healthy immature embryos with *A. tumefaciens* being considered one of the critical factors in successful genetic transformation of crop plants (Shrawat and Good 2011).

16.2.2 Particle Bombardment

This is a direct DNA delivery technique developed by Sandford et al. (Klein et al. 1987; Sandford et al. 1987; Sandford 1988), also referred to as biolistics (biology + ballistics). This technique employs high velocity particles or microprojectiles coated with DNA to deliver exogenous genetic material into the target cell or tissue, which is then cultured in vitro and regenerated to produce mature transformed plants.

The particles, of gold or tungsten, are of small size (0.5–5 μm) but large enough to have the necessary mass to be accelerated and able to penetrate the target cell or tissue carrying the coated DNA on their surface. The microprojectiles are DNA coated following for instance the $CaCl_2$ protocol, with the addition of spermidine to protect the DNA (Klein et al. 1987). Nevertheless, a recent report describes the use of *Agrobacterium* as coating material which has been used successfully (Cordero-Mesa et al. 2000). The microprojectiles are propelled under partial vacuum, using helium or CO_2 pressure to produce the necessary blast to propel the coated particles. Particle penetration may be controlled by altering different parameters, and instrument setup. These include the particle size, the distance between sample holder and target, pressure applied to thrust the particles, the presence of a retaining screen employed to disperse the particle before hitting the target, as well as the biological stage of the cells or tissues to be transformed.

Most of the coated particles are either degraded or inactivated, and many do not reach the nucleus. The DNA can be expressed after reaching the nucleus and getting stably integrated in the genome. Gene insertion using biolistic technique does not show a preference for insertion sites and when landing in a transcriptionally active region, it may be expressed at a high rate; whereas, if it integrates in a nonactive region, gene expression may be reduced or absent. In contrast to *Agrobacterium*-mediated transformation where DNA tends to be inserted in transcriptionally active sites (Ingelbrecht et al. 1991), biolistic technique generates higher copy numbers of the inserted DNA which often results in gene silencing at higher frequencies.

Biolistics offer unique advantages over conventional techniques such as *Agrobacterium* (Leech et al. 2000). For instance, (1) the ability to transfer foreign DNA directly into regenerable cells, tissues or organs; (2) the instrument allows fine tuning and control, permitting precise targeting of DNA coated particles to specific cells or tissue areas; (3) it does not show host specificity, hence it offers the potential to transform any plant species, and it has been successfully applied for the transformation of recalcitrant species such as many monocots and some dicots; (4) it permits a simultaneous bombardment of different gene constructs, finding accounts whereby even 12 different gene constructs were bombarded and expressed in plant cells (Hadi et al. 1996); (5) the input DNA size is not a limitation and it is possible to bombard almost any plasmid, although larger plasmids tend to disintegrate after bombardment resulting in a poor transformation efficiency; (6) this technique has been successfully applied for the transformation of not only plant cells but even animal, bacteria, yeast and fungi (Smith et al. 1992; Wang et al. 2001; O'Brien and Lummis 2004; Olmedo-Monfil et al. 2004). This technique has been successfully used to easily deliver plasmid DNA into the epidermis of *Arabidopsis* leaves comparing with *Agrobacterium* infiltration, a method that does not work efficiently in *Arabidopsis* leaf tissue. This approach was characterized by its high efficiency, reproducibility, and suitability for the robust transient expression of a variety of functional *Arabidopsis* proteins with diverse biological activities and subcellular localization specificities. Furthermore, this procedure also allows for the coexpression of numerous proteins, tagged and/or untagged, by bombardment with multigene expression constructs (Ueki et al. 2009).

16.2.3 Vacuum Infiltration and Floral Dip

Both methods can be considered to be simple techniques for the genetic transformation of *Arabidopsis thaliana*, although other plant species have also been successfully transformed following these methods (Curtis and Nam 2001; Curtis 2005). Species with large number of flowers together with a higher number of set seeds following infiltration are the best candidates. The major advantage of these methods is that they make it possible to transform these species without necessitating the labor-intensive plant tissue culture or regeneration steps required with other techniques such as *Agrobacterium*-mediated transformation or particle bombardment.

Vacuum infiltration was first developed (Bechtold and Pelletier 1998) and involves the growth of *Arabidopsis thaliana* to an early flowering stage (prior to anthesis). The plants are uprooted, and submerged in cloned *A. tumefaciens* liquid cultures, which after vacuum infiltration acquires the bacterium thus carrying the transgene(s) of interest. Plants are replanted and further grown, and seeds collected for assessment of transformants employing appropriate selection agents. Following this procedure, it has been demonstrated that the ovules of *A. thaliana* are the target for *A. tumefaciens* in transformed plants (Ye et al. 1999).

Similarly, floral dip is a variation of the vacuum infiltration method, developed as a simplified method for *Agrobacterium*-mediated transformation of *A. thaliana* (Clough and Bent 1998). It also employs *A. tumefaciens* as the DNA vector carrier, here floral dipping into the liquid bacterial culture is performed avoiding the vacuum infiltration process. Again, plants are susceptible to transformation when flowers are still immature; the attached flower buds are immersed in a solution containing a mixture of *A. tumefaciens* cells, sucrose and a surfactant. The appropriate amounts and ratio of the latter ingredients are important for a successful transformation. This method allows repeated dipping in the transformation mixture after a few days; besides, covering the plants with a plastic dome after infection to maintain humidity was also reported to increase transformation by 2-fold (Clough and Bent 1998). Transformants could be obtained at a high rate, 0.5–3% of all progeny seed, and for multiple *Arabidopsis* ecotypes. The success of the floral dip method implies that the germ-line tissues to be transformed should be exposed and accessible to the *Agrobacterium*. It was reported that female tissues such as developing ovules within the gynoecium of young flowers are the primary target of *Agrobacterium*-mediated floral-dip transformation of *Arabidopsis* (Desfeux et al. 2000). Besides, it has also been reported that *A. thaliana* transgenic plants obtained following floral dip transformation presents genetic stability (Labra et al. 2004).

A further modification of the above methods has been the application of *A. tumefaciens* by spraying. The results indicate that the floral spray method of *Agrobacterium* can achieve rates of *in planta* transformation comparable to the vacuum-infiltration and floral dip methods. The floral spray method opens up the possibility of *in planta* transformation of plant species which are too large for dipping or vacuum infiltration (Chung et al. 2000).

Plant transformation using these techniques produces genetically uniform progeny and more importantly obviates the somaclonal variation often encountered with tissue culture and regeneration. In addition, practicing these methodologies does not require costly or precise equipment being approachable by many laboratories. Nevertheless, the major drawback is that few species appear susceptible to transformation following these methods, although further research may provide new attractive possibilities, for example, transformation of banana meristematic tissue after vacuum infiltration (Acereto-Escoffié et al. 2005), and other potential plant species such as the medicinal plants *Catharanthus roseus* or *C. pusillus* which produce large number of flowers and many seeds per follicles (Zárate et al. 2001).

16.2.4 Agrodrench

This is a newly devised technique developed as an agroinoculation method for virus-induced gene silencing (VIGS) (Ryu et al. 2004). It is known that virus-induced gene silencing is an RNA-mediated posttranscriptional gene-silencing mechanism that protects plants against foreign gene invasion (Baulcombe 1999), and it represents a powerful tool, for instance, for plant functional genomics which allows knocking out gene expression of target plant genes in some plants, following an RNA silencing mechanism via small interference RNA. One advantage of VIGS is that it does not alter the gene itself, but transiently suppresses the expression of the gene via degradation of mRNA transcripts.

It has been developed employing tobacco rattle virus (TRV)-derived vectors, in conjunction with *A. tumefaciens* which harbors the designed TRV-VIGS constructs. For agrodrenching, the cloned bacterium is applied as a solution to the soil adjacent to the rhizosphere of previously grown 3–4-week-old seedlings, after infection gene silencing was demonstrated. Furthermore, it was established that an avirulent *A. tumefaciens,* and *Agrobacterium*-mediated transformation is required for agrodrench method of agroinoculation, and that agrodrench was more effective than the leaf infiltration method of agroinoculation (Ryu et al. 2004).

Furthermore, the agrodrench technique was employed to carry out a virus-induced gene silencing using Tobacco rattle virus (TRV)-based VIGS vector. This method was used to demonstrate that the duration of gene silencing by VIGS can occur for more than 2 years and that TRV is necessary for longer duration VIGS.

Also, inoculation of TRV-VIGS constructs by both agrodrench and leaf infiltration greatly increased the effectiveness and duration of VIGS (Senthil-Kumar and Mysore 2011). These results demonstrated the transmittance of VIGS to progeny seedlings via seeds. Moreover, the transmittance of VIGS to the progeny will be useful in studying the effect of gene silencing in young seedlings.

This method proved also to be a better choice for high-throughput gene silencing and for plants which appear difficult to infiltrate. On the other hand, this technique shuts down transcript levels to a greater degree in roots than leaf infiltration does, indicating advantages when roots are to be engineered.

Although this technique has been recently described for virus-induced gene silencing in primarily Solanaceous species, and for functional genomic analyzes, its employment might soon be broadened to other plant species and likely for the genetic modification of metabolic pathways by gene silencing.

16.2.5 Viral Vectors

Another approach to genetically transform plants has been the employment of viral vectors, thus different plant viruses have been used for the insertion of genes into a variety of plant species. The majority of plant viruses have an RNA genome, and just a few are of single or double stranded DNA. The main drawback is that viruses can propagate easily and spread to other plant parts and even to other neighboring plants. Nevertheless, their use offers interesting advantages such as being highly multiplicative, production of large number of copies of the transgene, as well as its expression (Porta and Lomonosaff 2002).

One of the early plant viruses used was the cauliflower mosaic virus (CaMV), a double stranded DNA virus that replicates through an RNA intermediate (Gronenborn et al. 1981). Using this virus extra DNA insertions are difficult to achieve, and often molecular recombinations rapidly eliminate the inserted transgene. In addition, replication of this virus requires an RNA step which can introduce errors because the inverse transcriptase does not have a proof-reading activity.

Similarly, regarding RNA viruses, two viruses, i.e. the tobacco mosaic virus (TMV) and the cowpea mosaic virus (CMV) have been successfully employed (Wilson and Davies 1992).

Two different strategies have been followed to clone and insert foreign gene(s) using plant viruses; either to replace the coat protein of the virus by a transgene or to insert the gene(s) next to the initiation codon of the coat protein that will then express the product of the inserted gene and the coat protein. Nonetheless, it has been observed that the inserted gene(s) can be removed by the virus, thus returning to its original form.

Several drawbacks have been reported when using plant viruses as vectors. These include: their instability, manifested by the loss of infectivity when compared to intact viruses and elimination of the transgene when this is larger than

1 kb. It appears that the ease of insert elimination might be an evolutionary strategy of the virus to avoid the accumulation of unwanted genomic material which does not offer any advantage to the host.

16.2.6 Protoplast Transformation

Protoplasts are osmotically fragile plant cells deprived of their cell wall, also referred to as naked cells. Upon removal of the cell wall, protoplast can behave like animal cells which by nature have no cell wall barrier. Protoplasts have the potential ability to redifferentiate into whole plants due to their totipotency, and this has been exploited for genetic modification of plants; however, there exist many recalcitrant plant species which have not been able to be transformed by this means, either because of failure to isolate viable protoplast or impossibility of regeneration.

Generally, removal of the cell wall is achieved by treatment of plant material (leaves, tissue cultures, suspended cells, etc.) with a designed cocktail of enzymes, such as cellulases, pectinases, and/or hemicellulases, in an incubation medium of the right osmolarity to avoid cell burst. In addition, the physiological status of the source tissue also influences the release of viable protoplasts, and several other factors influence protoplast release, including the extent of thickening of cell walls, temperature, duration of enzyme incubation, pH of the enzyme solution, gentle agitation, and nature of the osmoticum (Sinha et al. 2003).

Insertion of transgenes into protoplast can be achieved either by fusion of different protoplasts from different species, or DNA insertion into the protoplast through the plasma membrane. Several approaches have been devised for this gene insertion; these include: (a) *chemical techniques*, such as the use of PEG (polyethylene glycol), Ca^{++}-DNA coprecipitation or use of liposomes; (b) *electrical techniques*, where electrical pulses are applied to the protoplasts resulting in an increase in the protoplast permeability which facilitates DNA transfer; (c) *microinjection*, whereby DNA is physically injected into the protoplast nucleus or cytoplasm, although the presence of vacuole (comprising 90% of total cell volume) can produce cell death if they become damaged releasing hydrolases and toxic metabolites.

Although other genetic transformation techniques have shown increasing success, the use of protoplasts has also demonstrated to be effective in genetically transforming both monocot and dicot species. Furthermore, it has also proved suitable for other applications such as to investigate plant cell responses to growth regulators, genetic basis of developmental regulation and specificity, regulation of osmotic water transport across cell membranes, intracellular responses to drought and salinity stress, transient gene expression, and proteomics. A comprehensive review of the current status of plant protoplasts in genetic engineering has been recently published (Davey et al. 2005).

This technique has also been used for the transformation of fungal species, i.e. *Ozonium* sp. a producer of taxol and taxol-related compounds (Zhou et al. 2008). The authors were able to obtain protoplasts using a cocktail of commercially available enzymes, which were then successfully transformed using PEG and regenerated into mature fungus, thus establishing a model to efficiently transform via protoplast transformation different fungal species.

16.2.7 Ultrasound (Sonoporation)

Ultrasound is a sound whose frequency is higher than that perceptible to the human ear, although many animal species rely on ultrasound for their survival. It shows multiple industrial, medical, chemical, and biotechnological applications. Nevertheless, ultrasound can destroy microbial and other cells and this well-known effect has perhaps discouraged research on beneficial effects of ultrasound on many biological applications. Potentially, finely tuned ultrasound can enhance diffusive transport both within and outside a cell to influence rates of reactions and yields of metabolites (Christi 2003), and has been frequently used as a means for genetic transformation in animal cells and tissues. Nonetheless, there are few reports about the use of ultrasound in plant cells or tissues transformation. It seems that the cell wall of plant cells is the big obstacle for gene transfer using this technology (Liu et al. 2005).

Gene transfer by ultrasonication employs the same simple procedure irrespective of the nature of the plant material to be transformed. The protoplasts, suspension cells or small pieces of tissue are suspended in a few millilitres of sonication medium in a microcentrifuge tube. Plasmid DNA (and possibly carrier DNA) is then added, and after rapid mixing the samples are ready for sonication. The pulses of ultrasound of selected intensity and duration are delivered by ordinary machines used for homogenization of various tissues. Finally, the cells are transferred to fresh growth medium (Liu et al. 2005). Success on the insertion of plasmid DNA has been reported for sugar beet and tobacco protoplasts after a brief exposure to 20 kHz ultrasound (Joersbo and Brunstedt 1990).

Currently most studies are focused on sonication-assisted *Agrobacterium*-mediated transformation (SAAT) in plant cells or tissues (Weber et al. 2003). SAAT is a new technology which involves subjecting the plant tissue to brief periods of ultrasound in the presence of cloned *A. tumefaciens*. It has been demonstrated that for all tissues tested, the SAAT treatment greatly increased the levels of transient expression. In addition, it was established that SAAT treatment tremendously improved the efficiency of *Agrobacterium* infection by introducing large numbers of micro-wounds into the target plant cells or tissues which facilitated gene insertion (Trick and Finner 1997; Tang 2003).

16.3 Advances on Metabolic Engineering of Secondary Metabolite Pathways

Metabolic engineering is the redirection or modification of one or more biosynthetic steps (enzymes) for either the production of new compounds, enhancement of existing metabolites or halting the accumulation of a specific product, either by knocking out specific genes or by degradation of that metabolite.

On the other hand, the scientific advances made by a combination of biochemistry, genetics, and molecular biology (i.e. recombinant DNA technology) on biosynthetic pathway studies of target secondary metabolites have allowed the identification, function determination, cloning, and expression of many genes, which has shed some light on the complexity of secondary metabolism networks and their regulation. This knowledge has permitted to engineer metabolic pathways to yield tailored compounds, but has also resulted in successes, as well as in unexpected disappointments (Zárate and Yeoman 2001; DellaPenna 2001). Nonetheless, the future to elucidate and manipulate metabolic pathways is very challenging and further success is clearly foreseen.

Under the following subheadings, an attempt has been made to review the major strategies conducted to tailor secondary metabolism, highlighting some of the most current and interesting results, avoiding repeating the biotechnology of those products referred in this book. Furthermore, references to several important aspects which are often overlooked but should be addressed for successful metabolic engineering are also presented.

16.3.1 Gene Silencing

Although the first undesired results of gene silencing were considered disappointing, later it was realized its great potential to specifically block the expression of endogenous genes, and it is now considered a powerful tool to knockout genes in a given pathway (Waterhouse et al. 2001a, b; Lessard et al. 2002).

Downregulation or blockage of specific genes forms another approach to manipulate secondary metabolic pathways to either avoid formation of competitive intermediates or to completely turn off metabolic pathway branches or undesired metabolite synthesis. An enzymatic step in a pathway can be knocked out or downregulated by reducing the level of the corresponding mRNA either via antisense, cosuppression or employing RNAi (RNA interference), or by activating the synthesis of an antibody against the enzyme.

The antisense gene approach, whereby a gene sequence (cDNA), is fused in the inverted orientation to a plasmid, containing an upstream promoter and a downstream terminator sequence was the initial approach in the 1980s. The resulting mRNA becomes double stranded, and depending on the size of the inserted DNA, the double stranded mRNA could be partial or along all its length. Thus, the ribosome finds it difficult to process mRNA and none or very little protein is produced. Consequently, a pioneering work on the genetic manipulation of the flavonoid flower pigments in *Petunia hybrida* was reported by inserting an antisense chalcone synthase gene (a key enzyme for anthocyanin and flavonoid biosynthesis), which resulted in pigmentation changes of the transgenic plant flowers (Van der Krol et al. 1988). Further on this strategy but altering two genes, antisense insertion of chalcone synthase and dihydroflavonol-4-reductase genes in torenia plants resulted also in an altered flower pigmentation (Aida et al. 2000a, b). Following a similar approach, engineering the 7-xanthosine methyl transferase gene in antisense orientation via *A. tumefaciens* transformation for the production of caffeine free coffee plants has also been attempted (Moisyadi et al. 1998). More recently, antisense-mediated downregulation of putrescine *N*–methyltransferase (PMT) in transgenic *Nicotiana tabacum* plants and hairy root cultures resulted in elevated levels of anatabine, and a clear decrease in nicotine (Chintapakorn and Hamill 2003). In *Catharanthus roseus* antisense has also been used in an attempt to elucidate the participation of the enzyme isochorismate synthase (ICS) in the formation of 2,3-dihydroxybenzoic acid, indicating that in *C. roseus* the biosynthesis of this metabolite probably involves a multienzyme pathway (Rodríguez-Talou et al. 2001; Zárate and Verpoorte 2007).

RNA interference (RNAi) is a newer strategy for gene silencing, displaying a posttranscriptional gene silencing mechanism. However, it represents the oldest and most ubiquitous antiviral system in plants (Sharp 2001). RNAi refers to the introduction of homologous double stranded RNA (dsRNA) to specifically silence a gene. The presence of dsRNA is responsible for producing the interfering activity, the dsRNA is cleaved into 21–23 bp short interfering RNA (siRNA) by an enzyme called Dicer, thus producing multiple molecules from the original dsRNA. These form a RISC complex (RNA-induced silencing complex) that after becoming activated targets the homologous transcript by base pairing interactions and cleaves the mRNA. The readers are referred to comprehensive reviews on RNA silencing in plants for further information on this topic (Waterhouse et al. 2001a, b; Baulcombe 2004; McGinnis 2010).

RNAi offers important advantages compared to the antisense approach for being highly specific, remarkably potent since only a few dsRNA molecules per cell are required for effective interference and dsRNA can cause interference in cells and tissues away from the site of introduction and its effects are maintained through the progeny.

The chalcone synthase gene has also been silenced following the RNAi technique, thus the flower colour of the garden plant *Torenia hybrida* was successfully modulated. These results demonstrate that RNAi is a powerful technique for flower colour modulations of commercially important garden plants (Fukusaki et al. 2004).

An elegant piece of research worth quoting has been the genetic engineering of the opium poppy whereby morphine was replaced by the non-narcotic alkaloid reticuline by RNAi (Allen et al. 2004, Page 2005). Silencing of the codeinone reductase (COR) in *Papaver somniferum* plants was achieved employing a chimeric hairpin RNA construct designed to silence all members of the multigene

COR family. The precursor alkaloid (*S*)-reticuline, 7 enzymatic steps upstream of codeinone, accumulated at the expense of morphine, codeine, oripavine and thebaine. (*S*)-reticuline is a potential substrate for the synthesis of various bioactive compounds (antimalarial or anticancer) but its availability is limited, though not in these silenced poppy plants which show (*S*)-reticuline as the major metabolite. The eight enzyme branch leading to morphine can be downregulated in response to the loss of the penultimate enzyme, COR. This represents the most dramatic example of gene silencing-induced feedback in secondary metabolism ever reported (Allen et al. 2004). These authors suggest three possible processes responsible for the results observed, even though the full chain of events is still unelucidated, i.e. (a) the build up of COR substrates–codeinone and neopinone–might switch negative feedback on earlier enzyme or transport step(s); (b) or might inhibit transcription of those genes; and (c) loss of COR enzyme from a larger enzyme complex might disable other enzyme reactions associated with this complex. The initial expectation of this research was the accumulation of thebaine and oripavine following silencing of COR. The unexpected accumulation of (*S*)-reticuline demonstrates that the morphinan-alkaloid pathway can be coordinately regulated independently of the benzylisoquinoline pathway.

Similarly, RNAi has also been employed to further elucidate the metabolic pathway of caffeine synthesis, which has been proposed to involve three distinct *N*-methyltransferases, xanthosine methyltransferase (XMT), 7-N-methylxanthine methyltransferase (MXMT; theobromine synthase), and 3,7-dimethylxanthine methyltransferase (DXMT; caffeine synthase). Silencing of CaMXMT1 expression in *Coffea arabica* and *C. canephora* plants by the double stranded RNAi method showed that not only transcripts for CaMXMT1 but also for CaXMT1 and CaDXMT1 were reduced, indicating that indeed caffeine was synthesized in vitro by the combination of their three gene products although the major pathway of caffeine synthesis is mediated by the theobromine synthase, CaMXMT1. The resulting transgenic tissues and plantlets exhibited a marked reduction of theobromine and caffeine contents, indicating that RNAi can be attractively used for low or free caffeine plants (Ogita et al. 2004). Furthermore, in comparison with the above-mentioned antisense approach for caffeine-free plants, the construction of transgenic coffee plants in which the expression of the gene encoding theobromine synthase (CAMXMT1) was repressed by RNAi has been attempted (Ogita et al. 2003). This resulted in a major caffeine reduction of up to 70%, indicating that it should be feasible to produce coffee beans that are intrinsically deficient in caffeine following this strategy. Similarly, in *Camellia sinensis* an *Agrobacterium*-mediated silencing of caffeine synthesis through root transformation was carried out to suppress the expression of the caffeine synthase gene, by means of a RNAi construct, resulting in a marked reduction in both caffeine and theobromine contents in young shoots of tea seedlings after root transformation through *Agrobacterium* infiltration (Mohanpuria et al. 2010).

In the same fashion, Carotenoid Cleavage Dioxygenase 4 (CCD4) activity was studied in potato (*Solanum tuberosum*) in order to check whether the carotenoid content would be affected after manipulation. The expression level of the potato

CCD4 gene was downregulated using an RNA interference (RNAi), resulting in a 2- to 5-fold increased carotenoid content (Campbell et al. 2010).

More recently, a novel gene-silencing system using the chimeric transcriptional repressor, which is a designated chimeric repressor gene-silencing technology (CRES-T) has been developed. This system can induce a phenotype that was not expressed by antisense or gene knockout lines by suppressing expression of target enzyme genes in the pathway through modification of a metabolic regulator (Hiratsu et al. 2003). The CRES-T system can convert a transcriptional activator into a dominant repressor simply by fusion of the EARrepression domain with the activator, referring as chimeric repressor (Shikata and Ohme-Takagi 2008 and references therein). An outstanding advantage of the CRES-T system to other gene-silencing technologies, namely antisense, RNAi, and T-DNA insertion, is that the repressive activity of the chimeric repressor can overcome those of the endogenous and functionally redundant transcription factors, resulting in the induction of a dominant negative phenotype.

16.3.2 Homologous and Heterologous Overexpression of Biosynthetic Encoding Genes

Progress on enzyme purification, characterisation, and encoding genes determination of a myriad of secondary metabolism pathways has allowed the genetic engineering of biosynthetic networks. In an attempt to tailor and boost secondary metabolite yields, the overexpression of a biosynthetic enzyme encoding gene was the first approach conducted, which was further developed by the overexpression of multiple, even heterologous genes.

Thus, the manipulation of flavonoid and anthocyanin biosynthesis was the first example of genetic engineering, since the pathway was well characterized and the results could easily be observed by flower colour changes. Several genes have been overexpressed aiming at obtaining different flower colour; for instance, the dihydroflavonol reductase gene from maize or *Gerbera* into *Petunia*, as well as the introduction of alfalfa chalcone reductase gene into *Petunia* resulted in clear flower colour changes (Dixon and Strack 2003 and references therein). Moreover, based on the antioxidant activity of these metabolites, attempts have also been made to increase their yields in tomato. Thus, the overexpression of *Petunia* chalcone isomerase into tomato led to a 78-fold increase of flavonol levels in tomato peel, and 21-fold increase in tomato paste; demonstrating that it is feasible to enhance the yield of health beneficial metabolites by overexpression of genes (Muir et al. 2001). Tomatoes have also been metabolically engineered to enhance the aroma and flavor compound *S*-linalool by heterologous overexpression of *Clarkia breweri* *S*-linalool synthase (LIS) gene, under the control of the tomato late-ripening-specific E8 promoter. The accumulation of *S*-linalool and 8-hydroxylinalool was increased in ripening fruits. Apart from the difference in volatiles, no other phenotypic alterations were noted, including the levels of other terpenoids, such as γ- and α-tocopherols, lycopene, β-carotene, and lutein (Lewinsohn et al. 2001).

Dietary carotenoids offer important health benefits (protection against prostate cancer, age-related macular degeneration and cardiovascular diseases), and attempts for the metabolic engineering of potato tubers (*Solanum tuberosum* and *S. phureja*) containing enhanced levels of β-carotene and lutein have been undertaken (Ducreux et al. 2005 and references therein]. A phytoene synthase gene (*crtB*) from the bacterium *Erwinia uredovora* cloned using a tuber-specific patatin promoter in conjunction with a plastid targeting sequence, has been overexpressed in potato tubers. This resulted in an enhanced accumulation of β-carotene and lutein with total carotenoid levels increasing 6.25-fold and 4-fold in *Solanum tuberosum* and *S. phureja,* respectively. Similarly, attempts to increase isoprenoids (carotenoids and phytosterols) have been successfully performed in tomato by metabolic engineering of the mevalonate (MVA) and nonmevalonate (MEP) isopentenyl diphosphate-forming pathways. Independent overexpression of the two genes 3-hydroxymethylglutaryl CoA (*hmgr-1*) from *A. thaliana*, and 1-deoxy-D-xylulose-5-phosphate synthase (*dxs*) from *Escherichia coli* resulted in an increased carotenoid content with phytoene and β-carotene exhibiting the greatest increases (Enfissi et al. 2005). Further information can be found in a recent comprehensive review, which presents the current state of the molecular tools for the metabolic engineering of carotenoid in plants (Sauret-Güeto et al. 2003).

Another group of health promoting metabolites which has received immense attention for their metabolic engineering is one of the fatty acids. Particularly long-chain polyunsaturated fatty acids (PUFA) among which eicosapentaenoic (EPA) and docosahexaenoic acids (DHA) are of major importance due to their multiple important roles in human health and nutrition. These provide protection from cardiovascular diseases, obesity, diabetes (type II) as well as in an important number of mental illnesses such as schizophrenia and depression; they also participate in retinal development, and nerve and brain development of growing fetuses and infants, antiinflammatory responses, and against some types of cancer (Browning 2003; Simonopoulus 1991; Horrobin 1999; Simonsen et al. 1998; Trautwein 2001).

Different oilseed crops have been genetic engineered in an attempt to alter the fatty acid content aiming mainly to enhance n-3 and/or n-6 PUFA yield. This has been achieved either by overexpressing or suppressing single genes (i.e. desaturases, elongases), and in most cases, seed-specific promoters have been used to restrict the changes to the storage oils of seeds (Thelen and Ohlorgge 2002). Linseed plants have been engineered by the heterologous expression of three genes encoding a Δ6-desaturase, a Δ6-elongase, and a Δ5-desaturase which resulted in enhanced levels of arachidonic and eicosapentaenoic acids (Abbadi et al. 2004). Likewise, canola plants have been engineered to alter their PUFA content. Transgenic canola lines that expressed in seeds the Δ6 and Δ12-desaturases isolated from the fungus *Mortierella alpina*, and the Δ15-desaturase from canola (*Brassica napus*) were generated. Seed oil accumulated stearidonic acid (SDA), the highest yield was up to 23% of the lipids F1 seed. The total omega-3 content in the seed lipids (α-linolenic acid and SDA) exceeded 55% of the seed lipids, whereas the total omega-6 fatty acid content of the seed lipids (γ-linolenic and linoleic acids) was 22% (Ursin 2003).

Furthermore, the expression of the borage Δ6-desaturase in transgenic nonoleaginous plants i.e., tobacco and *Arabidopsis* resulted in the production of gamma linolenic and stearidonic acids (Sayanova and Napier 2004; Sayanova et al. 1997). *Arabidopsis* was further engineered for the production of PUFA (Qi et al. 2004). Here, sequential transformation of this plant with genes encoding a Δ9-specific elongating activity from *Isochrysis galbana*, a Δ8-desaturase from *Euglena gracilis,* and a Δ5-desaturase from *Mortierella alpina* resulted in the production of substantial quantities of arachidonic and eicosapentaenoic acids in this higher plant, suggesting the construction of an alternative sustainable source of fish oils. On the other hand, cotton has also been engineered by hpRNA-mediated silencing (hair pin RNA) of the *ghFAD2-1* Δ12-desaturase which raised the oleic acid content from 12 to 38%. Analogously, silencing of the *ghSAD-1* Δ9-desaturase gene, substantially increased stearidonic acid from the normal 2 to 40%. Intercrossing of these two plants resulted in a wide range of unique intermediate combinations of palmitic, stearic, oleic, and linoleic acids (Liu et al. 2002).

In another example, the endogenous *FAD2* gene was downregulated in *Brassica carinata* plants resulting in the desaturation of oleic acid to enhance substrate availability for the biosynthesis of erucic acid, via elongation with a significant increase in the relative proportions of erucic acid (12–27%) and other very long-chain fatty acids (Jadhav et al. 2005).

Similarly, using *Agrobacterium rhizogenes* mediated transformation, hairy root cultures of *Echium acanthocarpum* were established as the first example of a *Boraginaceae* species transformation for fatty acid studies and production. This system is capable to accumulate a large range of polyunsaturated fatty acids including stearidonic acid and gamma-linolenic acid, which are the precursors of important fatty acids, i.e. EPA and DHA (Cequier-Sánchez et al. 2011). This stable system, capable to produce economically important fatty acid, has been genetically manipulated overexpressing a Δ6-desaturase gene resulting in a clear increase of SDA amounts (unpublished).

The heterologous coexpression of the nematode *Caenorhabditis elegans* genes Δ6-elongase together with the Δ6- and Δ5-desaturases has been attempted using yeast. Following feeding with linoleic and alfa linolenic acid of the transgenic yeast, small but significant accumulation of arachidonic and eicosapentaenoic acids was recorded (Beauduin et al. 2000). Further information is available in recent comprehensive reviews dealing with the production of PUFA in transgenic plants, which demonstrate the feasibility of establishing transgenic plants for the supply of fish oils (Cahoon and Kinney 2005; Napier et al. 2004; Jaworski and Cahoon 2003).

16.3.3 Multiple Expressions of Transgenes

The possibility to overexpress homologous or heterologous genes in different plant systems has been highlighted and presented above. Progress has been further achieved whereby entire biosynthetic pathways have been engineered, resulting in the accumulation of the target metabolites, and attempts are being undertaken to further extend on the manipulation of entire pathways confirming that multigene transfer allows researchers to face challenges previously impossible (Naqvi et al. 2010).

One of the initial successes in genetic engineering an entire pathway was the production of β-carotene in rice endosperm (Ye et al. 2000), also known colloquially as the golden rice. The entire β-carotene biosynthetic pathway, vitamin A precursor, was introduced and expressed into rice endosperm in a single transformation effort with three vectors harboring four transgenes, i.e. *psy* plant phytoene synthase; *crt-1* bacterial phytoene desaturase; *lcy* lycopene β-cyclase and *tp* transient peptide. The transformed rice grains were yellow colored indicating the accumulation of carotenoids, and in some lines β-carotene was the only carotenoid detected.

Verberne et al. (2000) introduced isochorismate synthase (*entC* from *E. coli*) and isochorismate pyruvate lyase (*pmsB* from *Pseudomonas fluorescens*) with a chloroplast targeting signal into tobacco. This results in plants constitutively producing salicylate. The transformed plants were shown to be more resistant against viral, microbial, and fungal infections.

Hallard et al. (2000) reported the overexpression of the strictosidine synthase and tryptophan decarboxylase genes from *Catharanthus roseus* in hairy roots of *Weigela styriaca*. This plant makes secologanin, the precursor for the terpenoid indole alkaloids, and due to the presence of the two mentioned genes, the hairy roots made small amounts of ajmalicine and serpentine, two terpenoid indole alkaloids which are the major alkaloids of *C. roseus* roots, showing that it is feasible to introduce the production of alkaloids in nonalkaloid plants from a different family.

Transgenic *A. thaliana* plants expressing the entire biosynthetic pathway for the tyrosine-derived cyanogenic glucoside dhurrin has been achieved following the insertion of the cytochrome P450 enzymes CYP79A1, CYP71E1, and the glucosyltransferase UGT85B1 genes from *Sorghum bicolor* (Tattersall et al. 2001; Kristensen et al. 2005). Dhurrin is a plant cyanogenic glucoside that plays a role in defence against herbivore attack, and it is absent in *A. thaliana* plants but highly abundant in *S. bicolor*. Only in transgenic *A. thaliana* plants expressing the three transgenes, dhurrin was the single metabolite to accumulate in high abundance, together with minor amounts of p-hydroxybenzylglucosinolate. The presence of this metabolite also provided a clear deterrent effect against adults and larvae of flea beetles. In addition, the high accumulation of dhurrin did neither appear to provoke any physiological problem for the transgenic *A. thaliana* nor significant changes to the transcriptome and metabolome. Nonetheless, when incomplete pathways were introduced, metabolic crosstalk or detoxification reactions were found to induce significant changes in plant morphology, the transcriptome and metabolome. These results demonstrate that insertion of a full pathway is feasible in *A. thaliana* resulting in the production of the cyanogenic glucoside dhurrin with marginal inadvertent effects on the transcriptome and metabolome.

16.3.4 Transcription Factors

Currently a different and novel approach to engineering the expression of biosynthetic pathway genes is the employment of regulatory genes, i.e. transcription factors, that control multiple genes in a pathway and are able to regulate or induce various genes concurrently, representing the most important strategy for studying secondary metabolic pathways mainly for their ability to activate multiple genes simultaneously (Grotewold 2008; Iwase et al. 2009; Hussain et al. 2011).

Transcription factors are regulatory proteins that modulate the expression of specific set of genes through sequence-specific DNA binding to motifs usually in gene promoters and protein–protein interactions, showing homologies with mammalian transcription factors (i.e. protein encoded by the proto-oncogene *c-MYB* and basic-Helix-Loop-Helix protein encoded by the proto-oncogene *c-MYC*) (Davies and Schwinn 2003; Broun 2004). They can function as repressors (less frequent) or activators (more abundant) of gene expression, provoking a decrease or increase in the mRNA synthesis by RNA polymerase II. These proteins regulate gene transcription depending on tissue type and/or in response to internal signals; moreover, external signals (UV light, microbial elicitors, etc.) may also induce production of internal signals (Vom Endt et al. 2002).

The first pioneering discovery in plants was the determination of the transcription factors C1 and R of the maize flavonoid-anthocyanin pathways which were shown to regulate this biosynthetic pathway (Dooner et al. 1991). The anthocyanin pathway has been extensively studied and other transcription factors have been determined in *Anthirrinum*, factor *DELILA* (Goodrich et al. 1992), and in *Petunia*, factors *ANTHOCYANIN2* (*AN2*) and *JAF13* (Quattrocchio et al. 1998; Quattrocchio et al. 1999). Similarly, in *Arabidopsis* two transcription factors, *PRODUCTION OF ANTHOCYANIN PIGMENT1* (*PAP1*) and *PAP2* have been identified (Borevitz et al. 2000). Recently, also in *A. thaliana* in the proanthocyanidin (condensed tannins) pathway the presence of the gene regulators *Transparent testa 2* (*TT2*) and *TT8* has been determined (Nesi et al. 2000; Nesi et al. 2001). Moreover, with the completion of the *Arabidopsis* genome sequence, and due to the well-conserved DNA-binding domains, the identities of the complete set of *Arabidopsis* transcription factors can be characterized in the near future. On the other hand, in a different metabolic pathway, i.e. the biosynthesis of the medicinally important terpenoid indole alkaloids (TIAs) vincristine and vinblastine in *Catharanthus roseus* the determination of various transcription factors such as *ORCA2* and *ORCA3* (octadecanoid responsive *Catharanthus roseus* AP2) have been reported (Memelink et al. 2001a, b).

The maize transcription factors *LC* and *C1* have been specifically expressed in the fruits of tomato plants to increase flavonoid levels. Expression of both genes was needed and sufficient to upregulate the flavonoid pathway in the fruit flesh, a tissue that normally does not produce any flavonoids. High levels of kaempferol and, to a lesser extent, naringenin were recorded, and all flavonoids detected were present as glycosides. Anthocyanins were present in *LC/C1* tomato leaves but absent in ripe *LC/C1* fruits. Furthermore, RNA analysis of ripening fruits revealed that, except the chalcone isomerase, all of the structural genes required for the production of kaempferol-type flavonols and pelargonidin-type anthocyanins were strongly induced by the LC/C1 transcription factors (Bovy et al. 2002). In another case, several transcription factors from monocot and dicot species, i.e. the maize *C1* and *B-PERU* genes (MYC family of transcription factors) and *DELILA* gene from *Antirrhinum* as well as *MYB-PH2* from *Petunia hybrida* (MYB family of transcription factors) were overexpressed in white clover (*Trifolium repens*). The transformed plants exhibited enhanced anthocyanin accumulation in a range of tissues. In particular, one plant transformed with the *B-PERU* gene, displayed a unique pattern of anthocyanin accumulation in the leaf. The accumulation of anthocyanin in this plant was closely associated with the crescent of leaves, which is normally white. The red pigmentation declined in intensity in the oldest leaf stage. The *B-PERU* message was detected in all leaf stages of this white clover plant. Furthermore, this anthocyanin pattern was shown to be heritable (de Maijnik et al. 2000).

Regarding the TIAs biosynthetic pathway in *C. roseus*, overexpression of *ORCA3* in suspension cultures was sufficient to induce several genes controlling the biosynthesis of indole precursors resulting in enhanced expression of several biosynthetic genes (*Tdc*, *Str*, *Cpr*, *Sgd*, *D4 h*) and, consequently, in increased accumulation of TIAs, after feeding loganin. Nevertheless, the biosynthetic genes *G10 h* and *Dat* were not induced, suggesting that these genes are not controlled by *ORCA3*. Besides, two genes involved in primary metabolism leading to TIA precursor synthesis (*As* α and *Dxs*) were also induced by *ORCA3*. These results show that *ORCA3* is indeed a regulator of primary as well as secondary metabolite biosynthetic genes involved in TIAs biosynthesis (Zárate and Verpoorte 2007; Memelink et al. 2001a; van der Fits and Memelink 2000).

These data show that expression of specific transcription factors can more efficiently redirect the metabolism of plant cells by acting simultaneously and coordinately on different metabolic events. The manipulation of transcription factors have also shown that they are involved not just in secondary metabolism, but in cell differentiation as well as growth and development (Montiel et al. 2003). Furthermore, one undesired effect of the overexpression of transcription factors is that it can negatively affect plant growth and development. This may be due to the sensitivity of certain cell types to the simultaneous overexpression of genes, which not only operate on the target metabolic step but may divert to unwanted points. This may be controllable by restricting transcription factor activity to limit the potential deleterious effects. This could be achieved by an appropriate choice of the promoter, which would then impede the occurrence of undesirable effects.

In another attempt, transcription factors have been widely used for engineering stress tolerance in plants (Hussain et al. 2011) and even different floral traits (Shikata and Ohme-Takagi 2008 and references therein). The manipulation of native plant regulatory networks therefore represents a new era for genetically modified crops.

16.3.5 Further Considerations in Metabolic Engineering

So far we discussed the methods needed for transforming plant cells, and the successful experiments in this field by silencing or overexpressing structural genes encoding proteins that catalyze certain reactions in a biosynthetic pathway or overexpressing regulatory genes. However, it is also clear from the results so far, that just overexpression of one or more structural genes, not always result in the desired effect for increasing productivity. This is due to the complexity of most biosynthetic pathways (Verpoorte et al. 2000). A number of reasons can be mentioned: Stability, Viability cells, Pathway architecture, Enzyme regulation, Metabolic channels, Substrate availability, Cofactor availability, Sub-cellular, and cellular compartmentation and Transport.

16.3.5.1 Stability

In plant cell cultures a continuous selection pressure exists for fast growing cells. Overexpressing an energy-consuming secondary metabolite pathway is a negative trait in such an environment. Thus, long-term stability of transgenic cell lines might be a problem. For example, *Catharanthus roseus* cell lines overexpressing some structural genes were found to maintain increased levels of the overexpressed proteins, but the initially increased levels of alkaloids, eventually went down to much lower levels, comparable with the wild-type cells (Whitmer et al. 2003 and unpublished results).

16.3.5.2 Viability

The production of high levels of certain compounds may lead to toxic effects. Many times changes in the production of secondary metabolites in plants lead to impaired growth, or plants which are more susceptible for pests and diseases. Overexpression of the isochorismate lyase gene, for example, in tobacco plants leads to impaired growth, probably due to lack of vitamin K production in the plant (Verberne et al. 2000 and unpublished results). Other examples have been reported about reduced growth rates after genetic engineering metabolic pathways that may compete with essential primary pathways.

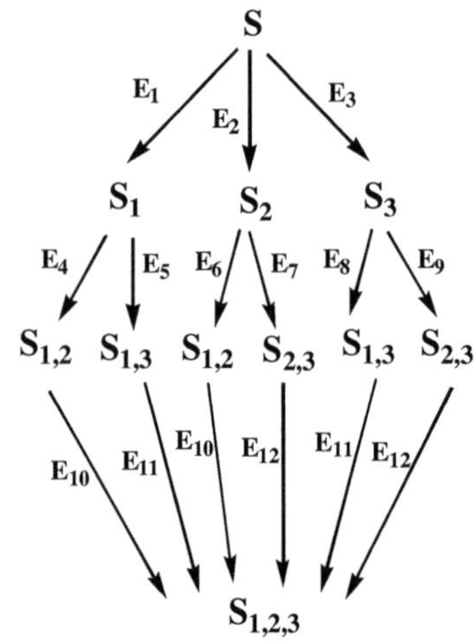

Fig. 16.1 Schematic biosynthetic network. S is a basic skeleton to which the functional groups 1, 2, and 3 are added. E_{1-12} are enzyme catalyzed steps. With high enzyme substrate specificity, 12 different enzymes are required to produce $S_{1,2,3}$, and with very low substrate specificity 3 enzyme would suffice (taken from Verpoorte and Alferman (2000) with permission from the authors and publisher)

16.3.5.3 Pathway Architecture

In a simple pathway consisting of a series of consecutive steps, it is easy to determine a possible rate limiting step and overexpress the gene concerned. However, in more complex networks, as it is most common in plants, an extensive knowledge about the specificity of every enzyme involved is needed to decide for the approach to be taken to improve the production of a certain desired compound (Verpoorte et al. 2000). Like in Fig. 16.1, the specificity of the enzymes is an important factor, is each step catalyzed by a separate enzyme, or are three enzymes with less specificity involved? For example, for making more of compound S2,3 in the latter case, only one single enzyme (E1) needs to be knockedout, whereas in case of a number of different enzymes being involved, several candidates needs to be considered (E1, E3 and E12). Moreover, competitive pathways may exist, or catabolism of the products of interest may be an important factor.

16.3.5.4 Enzyme Regulation

Regulation is not only on the level of the genes, but also on the level of the enzymes, e.g. activation of the enzyme via phosphorylation, feedback mechanisms, feed-forward mechanisms. The simple overexpression of a protein is thus not necessarily sufficient to come to an active enzyme. In case of feedback and feed-forward mechanisms, one might choose for genes encoding proteins that lack these properties, e.g. from other organisms or by protein engineering. For instance, to increase tryptophan levels in plants, overexpression of the first committed enzyme in the tryptophan biosynthesis, anthranilate synthase (ASA), requires a gene which is not sensitive for allosteric inhibition by tryptophan. However, still the effect might be limited as tryptophan induces the first enzyme in the

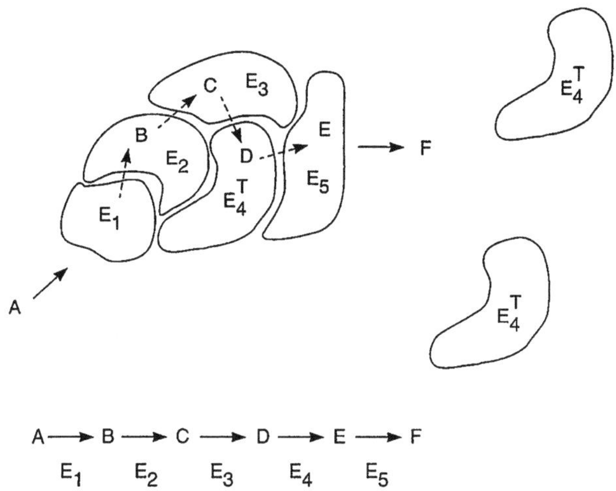

Fig. 16.2 Enzyme aggregates may play an important role in the regulation of pathways (taken from Verpoorte et al. 1999 with permission from the authors and publisher)

phenylalanine biosynthesis, also requiring chorismate, thus increasing the flux into that competitive pathway.

16.3.5.5 Metabolic Channels

Often enzymes occur in aggregates with other enzymes (metabolons, or metabolic channels). The cartoon of Fig. 16.2 illustrates this point. The aggregate problem is the same as concluding that there is more than one limiting step in a pathway, and consequently, by overcoming one, immediately the next one will be observed. Only by overexpressing all limiting enzymes involved, an increase can be achieved. Recently Choi et al. (2011) postulated that besides water and lipids a third phase is present in cells. This consists of Natural Deep Eutectic Solvents (NADES) that are made up from certain combinations of sugars, sugar alcohols, organic acids, amino acids, and choline derivatives. Such liquids have a clear structure like a liquid crystal and are excellent solvents for proteins as well as all kind of medium to low polar metabolites, up to several orders of magnitude better than in water. In fact, metabolons or metabolic channels could be such NADES in which both enzymes and substrates are freely moving, thus explaining the synthesis of complex water insoluble compounds. Such NADES could be in the ER and be attached to various membranes. The presence or absence of such a medium might also be a restricting factor in biosynthetic pathways. For studying biosynthesis the consequence is that *in-planta* enzymes might behave differently in such an environment than in water in which one is used to determine enzyme characteristics.

16.3.5.6 Substrate Availability

In the terpenoid indole alkaloid biosynthesis it has been found that the terpenoid part of the pathway is a major limiting factor. Overexpression of later steps in the pathway has only limited effect. By feeding the terpenoid precursor loganin, for example, an enormous production could be achieved in transgenic cell cultures of *C. roseus* overexpressing TDC and/or STR (Whitmer et al. 2003; Verpoorte et al. 1999; Whitmer et al. 1998; Whitmer et al. 2002a, b, c). But also substrates as nitrogen and methyl donors may be limiting in certain steps.

16.3.5.7 Cofactor Availability

Many enzymes require cofactors such as NAD(P)H, H_2O_2, ATP, also here the plant cell has the possibility to control the production of compounds on the level of the enzymes.

16.3.5.8 Subcellular and Cellular Compartmentation

All the factors mentioned above concern enzyme catalysed chemical reactions in the pathway itself, but also other regulatory mechanisms do play an important role. Intra and intercellular compartmentation are such regulatory mechanisms. The biosynthesis of terpenoid indole alkaloids in *C. roseus* is an excellent example of such compartmentation in which different steps of the pathway occur in different cellular compartments (plastids, chloroplasts, cytosol, vacuoles) and some of the later steps are even in different cells (Van der Heijden et al. 2004; St. Pierre et al. 1999).

16.3.5.9 Transport

Inter and intracellular compartmentation means that transport through cellular membranes and through membranes of organelles play a major role in controlling the flux through a pathway. This transport can be through physical phenomena (concentration or pH, driven diffusion), and/or the participation of active transporters that can translocate substrates and products of enzyme reactions to the right compartment (Yazaki 2005). Roytrakul et al. showed that in *C. roseus* vacuoles active ABC-transporters of the MDR-(efflux) and MRP-type (influx) are involved in selective bidirectional transport of various indole alkaloids through vacuolar membranes (Roytrakul and Verpoorte 2007). All alkaloids and iridoids tested showed a different accumulation in the vacuoles caused by different affinity of the various transport systems for the individual alkaloids. Accumulation in vacuoles is thus dependent on the activity of transporter proteins. This can be illustrated by the results of feeding tryptophan and secologanin to tobacco cells overexpressing two indole alkaloid biosynthetic genes (tryptophan decarboxylase and strictosidine synthase) which resulted in the production of alkaloids, but instead of being stored in vacuoles, they were excreted from the cells into the medium, showing the importance of transport (Hallard et al. 1997). The overexpression of a specific berberine ABC-transporter gene from *Coptis japonica* (Shitan et al. 2003) in *Catharanthus roseus* (Pomahacova et al. 2009) surprisingly showed improved uptake by the transformed cells of ajmalicine, but no effect at all on berberine uptake, showing the complexity of the transport systems.

Considering all these possibilities the conclusion is that the overexpression of a single structural gene may not lead to the desired effect. To overcome all the above-mentioned problems regulatory genes have been overexpressed. Indeed, it was shown that this approach may lead to the upregulation of a series of structural genes, but not necessarily to an increased flux through the pathway.

The final conclusion is thus that to come to successful projects for the upregulation of the production of a certain desired compound, detailed information is required on all steps in the metabolic pathways concerned. Adding a single step to a pathway may result in the formation of the product of the enzyme, in fact that has so far been the most successful approach.

16.3.6 Unraveling Biosynthetic Pathways

It is well known that the synthesis of plant secondary metabolites is a complex and highly regulated process that involves the participation of multiple factors such as subcellular compartments (vacuoles, endoplasmic reticulum, cytosol, plastids, nucleus, etc.); self regulation by endogenous feed-back and/or feed-forward mechanisms; transport of intermediates and end-products within cells, tissues and even organs; storage; catabolism, and chemical modifications such as acylation and glycosylation just to mention a few. Therefore, attempting to genetic engineering these complex networks requires a broad knowledge on each of the factors and control points participating in the biosynthesis of a target metabolite. In the previous sections, the general approaches conducted for the genetic engineering of secondary metabolite pathways have been outlined, as well as relevant published results; nevertheless, further considerations and attention to important parameters cannot be overlooked if a successful genetic manipulation is desired.

Indeed, it is not always possible to foresee the effect(s) of the knockout and/or overexpression of a biosynthetic enzyme(s) on an entire pathway, although large successes have been reported when overexpression and/or downregulation of single or multiple genes have been performed. Thus, it is evident that knowing the basic network of metabolite intermediates and enzymes involved in a pathway is crucial in unravelling secondary metabolism. However, further steps are required to keep advancing at an attractive pace. Particularly, it appears interesting to study the diverse regulatory mechanisms that control gene expression and transcriptional regulators (Memelink et al. 2001b) as well as identifying and characterizing the diverse vital transport mechanisms and proteins governing many secondary metabolite routes which have been demonstrated to be of crucial importance (Yazaki 2005). It is also essential to look at the various postbiosynthetic events and mechanisms involved, occurring in plant secondary metabolites (Zhang et al. 2002) for a more controllable genetic engineering approach.

With the fast advancement of biological research and DNA technology, the entire genome of important plants species has been already sequenced as is the case of arabidopsis, rice, maize, poplar, soybean, tomato. Other genome sequencing programs are in progress, i.e., cotton, soya, sunflower, etc. Access to genomics information permits to characterize gene expression under different conditions, thus establishing with precision gene function, a research field referred to as functional genomics. Functional genomics aims at quantitatively determining the spatial and temporal accumulation of specific mRNA, proteins, and metabolites using high-throughput technologies (Oliver et al. 2002).

Thus, new approaches are being applied to unravel plant secondary metabolism on all levels from genes to metabolites, from genome via transcriptome and proteome to the metabolome. Linking data from the different levels should enable the identification of genes involved in the production of certain metabolites. Transcriptomics is nowadays a well-developed technology, the same applies for proteomics. However, using plants from which no sequences are known, studies of the proteome may end up with a large number of unknown peptide sequences, for which it will be difficult to determine the function of the encoding genes. For example, Jacobs et al. (2005) found about 100 proteins to be connected with the production of alkaloids in *C. roseus* cell cultures, but only 60 of these gave some peptide sequences that match known proteins, and of these, just 3 had direct connection with the terpenoid indole pathway. On the other hand, 40 unknown sequences were obtained. Many known proteins from the indole alkaloid biosynthesis were not detected, which might be due to the fact that they are membrane bound, or appear at too low levels to be detected in the presence of the primary metabolism proteins. Therefore, when working with nonsequenced plants other approaches need to be applied. These include for instance metabolite profiling (i.e. metabolomics) which intends to qualitatively and quantitatively determine all metabolites in an organism and in that way have a complete view of the metabolism as a network (Trethewey 2004). Metabolomics has emerged in the past years as a very useful technology platform in different fields, such as drug discovery, clinical diagnostics, plant biochemistry, and quality control of food and botanicals. It requires highly specific state of the art analytical equipment, such as LC–MS, GC–MS, MS–MS, and NMR to trace the metabolic changes following genetic manipulation of the target plant or after physiological or environmental alterations. By using biostatistical methods, metabolomic data can be correlated with transcriptomics and proteomics data, thus allowing a holistic view on all correlations between these data sets. Thus, for example allowing to find links between secondary metabolites biosynthesis and genes.

Such an integrated approach, which should be considered when developing strategies for metabolic engineering, is now known as systems biology. Systems biology can be defined as the study of the mechanisms implicated in complex biological processes as integrated systems of many, varied, interacting components (DNA, RNA, proteins and cells). This involves (1) collection of large sets of experimental data; (2) proposal of mathematical models that might account for at least some significant aspects of this data set, (3) accurate computer solution of the mathematical equations to obtain numerical predictions, and (4) assessment of the quality of the model by comparing numerical simulations with the experimental data. However, full understanding of all collected data and design of the mathematical models are demanding tasks, but these can prove highly informative and attractive for a better design for a successful metabolic engineering of a target biosynthetic network (Kell 2004; Oksman-Cladentey et al. 2004). The major constraint in fact is what we call the transcriptome, proteome and metabolome of an organism, in fact is the sum of these "omes" of all individual cells that means cells with totally different functions, and consequently display a specific "ome" for that function. So to understand a cell and the organism we have to deconvolute the "omes" in the four dimensions (3 of space and 1 of time) which will be a major challenge for the coming years. Measuring fluxes rather than taking the snapshot approach of the present "omics" and single cell analysis will be key technologies for unravelling plant metabolism.

16.4 Future Directions

The early strategies used to boost secondary metabolite yields in the plant cell factory included cell line selection, nutrient regime, choice of culture system and conditions, level of plant growth regulators, precursor feeding, elicitation, removal of end-product. These were followed by the application of recombinant DNA technology and other biotechnological tools, resulting in spectacular advances.

During the past few years, great progresses in the metabolic and enzyme engineering for the production of pharmaceutical products using plant systems have been witnessed. Nonetheless, our knowledge on how secondary metabolites are synthesised and which genes govern and conduct these processes is far from complete, although more biosynthetic pathways are well understood and elucidated. Furthermore, in the near future new genes will become available, increasing the possibilities for further metabolic engineering of many new biosynthetic networks. However, plant metabolic engineering currently proceeds more by trial and error than by intelligent system design, a reality that need to be addressed with the application of the new tools and approaches such as those considered in subheading 16.3.6 of this chapter.

On the other hand, for positive metabolic engineering to be realised, natural product biosynthesis must be considered as a system of many interacting parts: promoters, transcription factors, enzymes, transporters, and intracellular structures, such as vesicles and membranes; ingredients which need particular attention when attempting to manipulate biosynthetic routes. In fact, it has been presented under subheading 16.3.4 and can now be emphasized, the powerful use of transcription factors for the regulation of biosynthetic pathways. Consequently, the quest for new transcription factors in medicinal plant species appear as an attractive goal for the subsequent genetic manipulation.

Interdisciplinary cooperation between several research fields, such as biology, chemistry, instrumentation, computer science, physics, and mathematics are also required for a more foreseeable genetic engineering outcome, thus, avoiding many unwanted effects following the current strategies for the genetic modification of a target biosynthetic network.

It is known that the current knowledge of secondary metabolite pathways is limited, and this represents the major barrier for the successful genetic manipulation of pathways. Thus, improving this level of knowledge is a challenging goal and would provide precise and exquisite information to further understand the control mechanisms and processes involved in the functioning of a pathway. Nonetheless, despite these obvious obstacles, we believe that many stimulating progresses and tailor-made designs generating appealing results on the metabolic engineering of medicinal compounds biosynthetic routes are to be expected in the near future, and inevitable progress will continue. Furthermore, the enormous potential for the genetic engineering of plant secondary metabolism has been clearly demonstrated, and the possibility to obtain plants able to accumulate new products useful as specialty chemical is an attainable objective.

The metaphor of the cell factory is in fact quite interesting, if one compares the cell with a car factory, one may easily understand that one extra piece of equipment in the total production line will not lead to a higher output. Only when the complete infrastructure is functioning and all parts are on the right time on the right place and the energy supply is functioning, such a factory will have its maximum output. The same applies for the cell, all logistics must be functioning properly to enable the cell to perform optimally. And all cells must function in a coordinated way for an optimal performance of the organism. So engineering such a factory to perform better is a real challenge!

Acknowledgments The authors would like to thank Dr. Suman Chandra for the invitation to contribute to this book. RZ and NJV acknowledge Instituto Canario de Investigación del Cáncer and Agencia Canaria de Investigación Innovación y Sociedad de la Información, Spain for financial support.

References

Abbadi A, Domergue F, Bauer J, Napier JA, Welti R, Zähringer U, Cirpus P, Heinz E (2004) Biosynthesis of very-long-chain polyunsaturated fatty acids in transgenic oilseeds: constrains on their accumulation. Plant Cell 16:2734–2748

Acereto-Escoffié POM, Chi-Manzanero BH, Echeverría-Echeverría S, Grijalva R, James Kay A, González-Estrada T, Castaño E, Rodríguez-Zapata LC (2005) *Agrobacterium*-mediated transformation of *Musa acuminata* cv. "Grand Nain" scalps by vacuum infiltration. Sci Horticult 105:359–371

Aida R, Kishimoto S, Tanaka Y, Shibata M (2000a) Modification of flower color in torenia (*Torenia fournieri* Lind.) by genetic transformation. Plant Sci 153:33–42

Aida R, Yoshida K, Kondo T, Kishimoto S, Shibata M (2000b) Copigmentation gives bluer flowers on transgenic torenia plants with the antisense dihydroflavonol-4-reductase gene. Plant Sci 160:49–56

Allen RS, Millgate AG, Chitty JA, Thisleton J, Miller JAC, Fist AJ, Gerlach WL, Larkin PJ (2004) RNAi-mediated replacement of morphine with the nonnarcotic alkaloid reticuline in opium poppy. Nat Biotechnol 22:1559–1566

Barampura S, Zhang AJ (2011) Recent advances in plant transformation. Methods Mol Biol 701:1–35

Baulcombe D (2004) RNA silencing in plants. Nature 431:356–363

Baulcombe DC (1999) Viruses and gene silencing in plants. Arch Virol Suppl 15:189–201

Beaudoin F, Michaelson LV, Hey SJ, Lewis MJ, Shewry PR, Sayanova O, Napier JA (2000) Heterologous reconstitution in yeast of the polyunsaturated fatty acid biosynthetic pathway. Proc Natl Acad Sci USA 97:6421–6426

Bechtold N, Pelletier G (1998) In planta *Agrobacterium*-mediated transformation of adult *Arabidopsis thaliana* plants by vacuum infiltration. Meth Mol Biol 82:259–266

Borevitz JO, Xia Y, Blount J, Dixon RA, Lamb C (2000) Activation tagging identifies a conserved MYB regulator of phenylpropanoid biosynthesis. Plant Cell 12:2383–2394

Bovy A, de Vos R, Kemper M, Schijlen E, Almenar-Pertejo M, Muir S, Colllins G, Robinson S, Verhoeyen M, Hughes S, Santos-Buelga C, van Tunen A (2002) High-flavonol tomatoes resulting from the heterologous expression of the maize transcription factor genes *LC* and *R1*. Plant Cell 14:2509–2526

Broun P (2004) Transcription factors as a tool for metabolic engineering in plants. Curr Op Plant Biol 7:202–209

Browning LM (2003) *n*-3 polyunsaturated fatty acids, inflammation and obesity-related disease. Proc Nutr Soc 62:447–453

Bundock P, den Dulk-Ras A, Beijersbergen A, Hooykaas PJJ (1995) Trans-kingdom T-DNA transfer from *Agrobacterium tumefaciens* to *Saccharomyces cerevisiae*. EMBO J 14:3206–3214

Cahoon EB, Kinney AJ (2005) The production of vegetable oils with novel properties: using genomic tools to probe and manipulate plant fatty acid metabolism. Eur J Lipid Sci Technol 107:239–243

Campbell R, Ducreux LJ, Morris WL, Morris JA, Suttle JC, Ramsay G, Bryan GJ, Hedley PE, Taylor MA (2010) The metabolic and developmental roles of carotenoid cleavage dioxygenase 4 from potato. Plant Phisiol 154(2):656–664

Cequier-Sánchez E, Rodríguez C, Dorta-Guerra R, Ravelo AG, Zárate R (2011) *Echium acanthocarpum* hairy root cultures, a suitable system for polyunsaturated fatty acid studies and production. BMC Biotechnol 11:42

Chilton MD, Drummond MH, Merlo DJ, Sciaky D, Montoya AL, Gordon MP, Nester EW (1977) Stable incorporation of plasmid DNA into higher plant cells: the molecular basis of crown gall tumorogenesis. Cell 11:263–271

Chintapakorn Y, Hamill JD (2003) Antisense-mediated down-regulation of putrescine *N*-methyltransferase activity in transgenic *Nicotiana tabacum* L. can lead to elevated levels of anatabine at the expense of nicotine. Plant Mol Biol 53:87–105

Choi YH, van Spronsen J, Dai Y, Verberne M, Hollmann F, Arends IWCE, Witkamp GJ, Verpoorte R (2011) Are natural deep eutectic solvents the missing link in understanding cellular metabolism and physiology? Plant Physiol 156:1701–1705

Christi Y (2003) Ultrasound the power of a silent gong. Biotechnol Adv 21:1

Chung M-H, Chen M-K, Pan S-M (2000) Floral spray transformation can efficiently generate *Arabidopsis* transgenic plants. Transgen Res 9:471–476

Clough SJ, Bent AF (1998) Floral dip: a simplified method for *Agrobacterium*-mediated transformation of *Arabidopsis thaliana*. Plant J 16:735–743

Cordero-Mesa M, Jiménez-Bermúdez S, Pliego-Alfaro F, Quesada MA, Mercado JA (2000) *Agrobacterium* cells as microprojectile coating: a novel approach to enhance stable transformation rates in strawberry. Aust J Plant Physiol 27:1093–1100

Curtis IS (2005) Production of transgenic crops by the floral-dip method. Meth Mol Biol 286 (Transgenic Plants): 103–109

Curtis IS, Nam HG (2001) Transgenic radish (*Raphanus sativus* L. *longipinnatus* Bailey) by floral-dip method - plant development and surfactant are important in optimizing transformation efficiency. Transgen Res 10:363–371

Davey MR, Anthony P, Power BJ, Lowe KC (2005) Plant protoplasts: status and biotechnological perspectives. Biotechnol Adv 23:131–171

Davies KM, Schwinn KE (2003) Transcriptional regulation of secondary metabolism. Funct Plant Biol 30:913–925

de Majnik J, Weinman JJ, Djordjevic MA, Rolfe BG, Tanner GJ, Joseph RG, Larkin PJ (2000) Anthocyanin regulatory gene expression in transgenic white clover can result in an altered pattern of pigmentation. Aust J Plant Physiol 27:659–667

DellaPenna D (2001) Plant metabolic engineering. Plant Physiol 125:160–163

Desfeux C, Clough SJ, Bent AF (2000) Female reproductive tissues are the primary target of *Agrobacterium*-mediated transformation by the *Arabidopsis* floral-dip method. Plant Physiol 123:895–904

Dessaux Y, Petit A, Tempé J (1993) Chemistry and biochemistry of opines, chemical mediators of parasitism. Phytochemistry 34:31–38

Dixon RA, Steele CL (2002) Flavonoids and isoflavonoids—a gold mine for metabolic engineering. Trends Plant Sci 4:394–400

Dixon RA, Strack D (2003) Phytochemistry meets genome analysis, and beyond. Phytochemistry 62:815–816

Dooner HK, Robbins TP, Jorgensen RA (1991) Genetic and developmental control of anthocyanin biosynthesis. Annu Rev Genet 25:173–199

Ducreux LJM, Morris WL, Hedley PE, Shepherd T, Davies HV, Millam S, Taylor MS (2005) Metabolic engineering of high carotenoid potato tubers containing enhanced levels of β-carotene and lutein. J Exp Bot 56:81–89

Enfissi EMA, Fraser PD, Lois LM, Boronat A, Schuch W, Bramley PM (2005) Metabolic engineering of the mevalonate and non-mevalonate isopentenyl diphosphate-forming pathways for the production of health-promoting isoprenoids in tomato. Plant Biotechnol J 3:17–27

Fukusaki E, Kawasaki K, Kajiyama S, An C-I, Suzuki K, Tanaka Y, Kobayashi A (2004) Flower color modulations of *Torenia hybrida* by downregulation of chalcone synthase genes with RNA interference. J Biotechnol 111:229–240

Goodrich J, Carpenter R, Coen ES (1992) A common gene regulates pigmentation pattern in diverse plant species. Cell 68:955–964

Gronenborn B, Gardner RC, Schaefer S, Shepherd RJ (1981) Propagation of foreign DNA in plants using cauliflower mosaic virus as vector. Nature 294:773–776

Grotewold E (2008) Transcription factors for predictive plant metabolic engineering: are we there yet? Curr Opin Biotechnol 10(2):138–144

Hadi MZ, McMullen MD, Finner JJ (1996) Transformation of 12 different plasmids into soybean via particle bombardment. Plant Cell Rep 15:500–505

Hain R, Bieseler B, Kindl H, Schroeder G, Stocker R (1990) Expression of a stilbene synthase gene in *Nicotiana tabacum* results in synthesis of the phytoalexin resveratrol. Plant Mol Biol 15:325–335

Hallard DAC, van der Heijden R, Verpoorte R, Lopez-Cardoso I, Pasquali G, Memelink J, Hoge JHC (1997) Suspension cultured transgenic cells of *Nicotiana tabacum* expressing tryptophan decarboxylase and strictosidine synthase cDNAs from *Catharanthus roseus* produce strictosidine upon feeding of secologanin. Plant Cell Rep 17:50–54

Hallard DAC, Palacios N, van der Heijden R, Memelink J, Verpoorte R (2000) Production of ajmalicine in transgenic *Weigela styriaca* tissue cultures expressing *tdc* and *str* cDNAs from

Catharanthus roseus. Chapter 6. Transgenic plant cells for the production of indole alkaloids, Hallard DAC, PhD thesis, Leiden, University, pp 69–85

Hansen G, Wright MS (1999) Recent advances in the transformation of plants. Trends Plant Sci 4:226–231

Hiratsu K, Matsui K, Koyama T, Ohme-Takagi M (2003) Dominant repression of target genes by chimeric repressors that include the EAR motif, a repression domain, in *Arabidopsis*. Plant J 34:733–739

Hooykaas PPJ (2000) *Agrobacterium*, a natural metabolic engineer of plants. In: Verpoorte R, Alfermann AW (eds) Metabolic Engineering of Plant Secondary Metabolism. Kluwer Academic Publishers, The Netherlands

Horrobin DF (1999) Lipid metabolism, human evolution and schizophrenia. Prostaglandins Leukot Essent Fatty Acids 60:431–437

Hussain SS, Kayani MA, Amjad M (2011) Transcription factors as tools to engineer enhanced drought stress tolerance in plants. Biotechnology Prog 27(2):297–306

Ingelbrecht I, Breyne P, Vancompernetlle K, Jacobs A, van Montagu M, Depicker A (1991) Transcriptional interference in transgenic plants. Gene 109:239–242

Ishida Y, Saito H, Ohta S, Hiei Y, Komari T, Kumashiro T (1996) High frequency transformation of maize (*Zea mays*) mediated by *Agrobacterium tumefaciens*. Nat Biotechnol 14:745–750

Iwase A, Matsui K, Ohme-Takagi M (2009) Manipulation of plant metabolic pathways by transcription factors. Plant Biotechnol 26:29–38

Jacobs DI, Gaspari M, van der Greef J, van der Heijden R, Verpoorte R (2005) Proteome analysis of the medicinal plant *Catharanthus roseus*. Planta 221:690–704

Jadhav A, Katavic V, Marillia EF, Giblin EM, Barton DL, Kumar A, Sonntag C, Babic V, Keller WA, Taylor DC (2005) Increased levels of erucic acid in *Brassica carinata* by co-suppression and antisense repression of the endogenous *FAD2* gene. Metab Eng 7:215–220

Jaworski J, Cahoon EB (2003) Industrial oils from transgenic plants. Curr Opin Plant Biol 6: 178–184

Joersbo M, Brunstedt J (1990) Direct gene transfer to plant protoplasts by mild sonication. Plant Cell Rep 9:207–210

Kell DB (2004) Metabolomics and systems biology: making sense of the soup. Curr Opin Microbiol 7:296–307

Klein TM, Wolf ED, Wu R, Sanford JC (1987) High-velocity microprojectiles for delivering nucleic acids into living cells. Nature 327:70–73

Kristensen C, Morant M, Olsen CE, Ekstrom CT, Galbraith DW, Moller BL, Bak S (2005) Metabolic engineering of dhurrin in transgenic Arabidopsis plants with marginal inadvertent effects on the metabolome and transcriptome. Proc Natl Acad Sci USA 102:1779–1784

Kunik T, Tzfira T, Kapulnik Y, Gafni Y, Dingwall C, Citovsky V (2001) Genetic transformation of HeLa cells by *Agrobacterium*. Proc Natl Acad Sci USA 98:1871–1876

Labra M, Vannini C, Grassi F, Bracale M, Balsemin M, Basso B, Sala F (2004) Genomic stability in *Arabidopsis thaliana* transgenic plants obtained by floral dip. Theor Appl Genet 109:1512–1518

Leech MJ, Burtin D, Hallard D, Hillou F, Kemp B, Palacios N, Rocha P, O'Callaghan D, Verpoorte R, Christou P (2000) Particle gun methodology as a tool in metabolic engineering. In: Verpoorte R, Alfermann AW (eds) Metabolic Engineering of Plant Secondary Metabolism. Kluwer Academic Publishers, The Netherlands

Lessard PA, Kulaveerasingam H, York GM, Strong A, Sinskey AJ (2002) Manipulating gene expression for the metabolic engineering of plants. Metab Eng 4:67–79

Lewinsohn E, Schalechet F, Wilkinson J, Matsui K, Tadmor Y, Nam K-H, Amar O, Lastochkin E, Larkov O, Ravid U, Hiatt W, Gepstein S, Pichersky E (2001) Enhanced levels of the aroma and flavor compound *S*-linalool by metabolic engineering of the terpenoid pathway in tomato fruits. Plant Physiol 127:1256–1265

Li J-F, Park E, von Arnim AG, Nebenführ A (2009) The FAST technique: a simplified *Agrobacterium*-based transformation method for transient gene expression analysis in seedlings of *Arabidopsis* and other plant species. Plant Methods 5(6):1–15

Liu Q, Singh S, Green A (2002) High-oleic and high-stearic cottonseed oils: nutritionally improved cooking oils developed using gene silencing. J Am Coll Nutr 21:205S–211S

Liu Y, Yang H, Sakanishi A (2005) Ultrasound: mechanical gene transfer into plant cells by sonoporation. Biotechnol Adv 24(1):1–16

Lorence A, Verpoorte R (2004) Gene transfer and expression in plants. In: Balbás P, Lorence A (eds) Methods in Molecular Biology, Recombinant Gene Expression. Reviews and Protocols. Humana Press Inc, Totowa

McGinnis KM (2010) RNAi for functional genomics in plants. Brief Funct Genomics 9(2):111–117

Melchers LS, Regensburg-Tuink AJG, Schilperoort RA, Hooykaas PJJ (1989) Specificity of signal molecules on the activation of *Agrobacterium* virulence gene expression. Mol Microbiol 3:969–977

Memelink J, Kijne JW, van der Heijden R, Verpoorte R (2001a) Genetic modification of plant secondary metabolite pathways using transcriptional regulators. Adv Biochem Engineering/ Biotechnol 72:103–125

Memelink J, Verpoorte R, Kijne JW (2001b) ORCAnisation of jasmonate-responsive gene expression in alkaloid metabolism. Trends Plant Sci 6:212–219

Michielse CB, Hooykaas PJ, Van der Hondel CA, Ram AF (2005) *Agrobacterium*-mediated transformation as a tool for functional genomics in fungi. Curr Genetics 48:1–17

Mohanpuria P, Kumar V, Ahuja PS, Yadav SK (2010) *Agrobacterium*-Mediated Silencing of Caffeine Synthesis through Root Transformation in *Camellia sinensis* L. Mol Biotechnol doi:10.1007/s12033-010-9364-4

Moisyadi S, Neupane KR, Stiles JI (1998) Cloning and characterization of a cDNA encoding xanthosine-N7-methyltransferase from coffee (*Coffea arabica*). Acta Horticult 461:367–377

Montiel G, Breton C, Doireau P, Jay-Allemand C, Gantet P (2003) Transcription factors that regulate secondary metabolism biosynthesis pathways: key actors for plant eco-physiology and ontogeny. Rec Res Devel Plant Cell Physiol 1:83–98

Muir SR, Collins GJ, Robinson S, Hughes S, Bovy A, de Vos CHR, van Tunen AJ, Verhoeyen ME (2001) Overexpression of petunia chalcone isomerase in tomato results in fruits containing increased levels of flavonols. Nat Biotechnol 19:470–474

Napier JA, Sayanova O, Qi B, Lazarus CM (2004) Progress toward the production of long-chain polyunsaturated fatty acids in transgenic plants. Lipids 39:1067–1075

Naqvi S, Farré G, Sanahuja G, Capell T, Zhu C, Christou P (2010) When more is better: multigene engineering in plants. Trends Plant Sci 15(1):48–56

Nesi N, Debeaujon I, Jond C, Pelletier G, Caboche M, Lepiniec L (2000) The TT8 gene encodes a basic helix-loop-helix domain protein required for expression of DFR and BAN genes in *Arabidopsis thaliana*. Plant Cell 12:1863–1878

Nesi N, Jond C, Debeaujon I, Caboche M, Lepiniec L (2001) The *Arabidopsis TT2* gene encodes an R2R3 MYB domain protein that acts as a key determinant for proanthocyanidin accumulation in developing seed. Plant Cell 13:2099–2114

O'Brien J, Lummis SCR (2004) Biolistic and diolistic transfection: using the gene gun to deliver DNA and lipophilic dyes into mammalian cells. Methods 33:121–125

Ogita S, Uefuji H, Morimoto M, Sano H (2004) Application of RNAi to confirm theobromine as the major intermediate for caffeine biosynthesis in coffee plants with potential for construction of decaffeinated varieties. Plant Mol Biol 54:931–941

Ogita S, Uefuji H, Yamaguchi Y, Koizumi N, Sano H (2003) RNA interference: producing decaffeinated coffee plants. Nature 423:823

Oksman-Caldentey K-M, Barz WH (2002) Plant Biotechnology and Transgenic Plants. In: Oksman-Caldentey K-M, Barz WH (ed) C.H.I.P.S. Texas, USA

Oksman-Caldentey K-M, Inzé D, Oresic M (2004) Connecting genes to metabolites by a systems biology approach. Proc Natl Acad Sci USA 101:9949–9950

Oliver DJ, Nikolau B, Syrkin-Wurtele E (2002) Functional genomics: high throughput mRNA, protein and metabolite analyses. Metab Eng 4:98–106

Olmedo-Monfil V, Cortes-Penagos C, Herrera-Estrella A (2004) Three decades of fungal transformation: key concepts and applications. Meth Mol Biol 267:297–313

Page JE (2005) Silencing nature's narcotics: metabolic engineering of the opium poppy. Trends Biotechnol 23:331–333

Pomahacová B, Dusek J, Dusková J, Yazaki K, Roytrakul S, Verpoorte R (2009) Improved accumulation of ajmalicine and tetrahydroalstonine in *Catharanthus* cells expressing an ABC transporter. J Plant Physiol 166:1405–1412

Porta C, Lomonossoff GP (2002) Viruses as vectors for the expression of foreign sequences in plants. Biotechnol Gen Eng Rev 19:245–291

Potrykus I (1991) Gene transfer to plants: assessment of published approaches and results. Annu Rev Plant Physiol Plant Mol Biol 42:205–225

Qi B, Fraser T, Mugford S, Dobson G, Sayanova O, Butler J, Napier JA, Stobart AK, Lazarus CM (2004) The production of very long chain polyunsaturated omega-3 and omega-6 fatty acids in transgenic plants. Nat Biotechnol 22:739–745

Quattrocchio F, Wing J, van der Woude K, Mol JN, Koes R (1998) Analysis of bHLH and MYB domain proteins: species-specific regulatory differences are caused by divergent evolution of target anthocyanin genes. Plant J 13:475–488

Quattrocchio F, Wing J, van der Woude K, Souer E, de Vetten N, Mol JN, Koes R (1999) Molecular analysis of the anthocyanin2 gene of petunia and its role in the evolution of flower color. Plant Cell 11:1433–1444

Rodríguez-Talou J, Verbene MC, Budi Muljono RA, van Tegelen LJP, Gonsalvez-Bernal B, Linthorst HJM, Wullems GJ, Bol JF, Verpoorte R (2001) Isochorismate synthase transgenic expression in *Catharanthus roseus* cell suspensions. Plant Physiol Biochem 39:595–602

Roytrakul S, Verpoorte R (2007) Role of vacuolar transporter proteins in plant secondary metabolism: Catharanthus roseus cell culture. Phytochemistry Rev 6:383–396

Ryu C-M, Anand A, Kang L, Mysore KS (2004) Agrodrench: a novel and effective agroinoculation method for virus-induced gene silencing in roots and diverse Solanaceous species. Plant J 40:322–331

Sanford JC (1988) The biolistic process. Trends Biotechnol 6:299–302

Sanford JC, Klein TM, Wolf ED, Allen N (1987) Delivery of substances into cells and tissue using a particle bombardment process. Trends Biotechnol 6:299–302

Sauret-Güeto S, Botella-Pavia P, Rodríguez-Concepción M (2003) Molecular tools for the metabolic engineering of carotenoids biosynthesis in plants. Recent Res Devel Plant Mol Biol 1:339–363

Sayanova OV, Napier JA (2004) Eicosapentaenoic acid: biosynthetic routes and the potential for synthesis in transgenic plants. Phytochemistry 65:147–158

Sayanova OV, Smith MA, Lapinskas P, Stobart AK, Dobson G, Christie WW, Shewry PR, Napier JA (1997) Expression of a borage cDNA containing an N-terminal cytochrome b_5 domain results in the accumulation of high levels of Δ6-desaturated fatty acids in transgenic tobacco. Proc Natl Acad Sci USA 94:4211–4216

Senthil-Kumar M, Mysore KS (2011) Virus-induced gene silencing can persist for more than 2 years and also be transmitted to progeny seedlings in *Nicotiana benthamiana* and tomato. Plant Biotechnol J 9:797–806

Sharp PA (2001) RNA interference. Genes Dev 15:485–490

Shikata M, Ohme-Takagi M (2008) The utility of transcription factors for manipulation of floral traits. Plant Biotechnol 25:31–36

Shitan N, Bazin I, Dan K, Obata K, Kigawa K, Ueda K, Sato F, Forestier C, Yazaki K (2003) Involvement of CjMDR1, a plant multidrug-resistance-type ATP-binding cassette protein, in alkaloid transport in *Coptis japonica*. Proc Natl Acad Sci USA 100:751–756

Shrawat AK, Good AG (2011) *Agrobacterium* tumefaciens-mediated genetic transformation of cereals using immature embryos. Methods Mol Biol 710:355–372

Simonopoulus AP (1991) Omega-3 fatty acids in health and disease and in growth and development. Am J Clin Nutr 54:438–463

Simonsen S, van't Veer P, Strain JJ (1998) Adipose tissue omega-3 and omega-6 fatty acid content and breast cancer in the EURAMIC study. Am J Epidemiol 147:342–352

Sinha A, Wetten AC, Caligari PDS (2003) Effect of biotic factors on the isolation of *Lupinus albus* protoplasts. Aust J Bot 51:103–109

Sinkar VP, White FF, Gordon MP (1987) Molecular biology of Ri-plasmid. A rev J Biosci 11:47–57

Smith FD, Harpending PR, Sanford JC (1992) Biolistic transformation of prokaryotes: factors that affect biolistic transformation of very small cells. J Gen Microbiol 138:239–248

St-Pierre B, Vazquez-Flota F, De Luca V (1999) Multicellular compartmentation of *Catharanthus roseus* alkaloid biosynthesis predicts intercellular translocation of a pathway intermediate. Plant Cell 11:887–900

Tang W (2003) Additional virulence genes and sonication enhance *Agrobacterium tumefaciens*-mediated loblolly pine transformation. Plant Cell Rep 21:555–562

Tattersall DB, Bak S, Jones PR, Olsen CE, Nielsen JK, Hansen ML, Hoj PB, Moller BL (2001) Resistance to an herbivore through engineered cyanogenic glucoside synthesis. Science 293:1826–1828

Tepfter D (1990) Genetic transformation using *Agrobacterium rhizogenes*. Physiol Plant 79: 140–146

Thelen JJ, Ohlorgge JB (2002) Metabolic engineering of fatty acid biosynthesis in plants. Metab Eng 4:12–21

Trautwein EA (2001) n-3 Fatty acids–physiological and technical aspects for their use in food. Eur J Lipid Sci Technol 103:45–55

Trethewey RN (2004) Metabolite profiling as an aid to metabolic engineering in plants. Curr Opin Plant Biol 7:196–201

Trick HR, Finner JJ (1997) SAAT: sonication-assisted *Agrobacterium*-mediated transformation. Transgen Res 6:329–336

Ueki S, Lacroix B, Krichevsky A, Lazarowitz SG, Citovsky V (2009) Functional transient genetic transformation of Arabidopsis leaves by biolistic bombardment. Nat Protoc 4(1):71–77

Ursin VA (2003) Modification of plant lipids for human health: development of functional land-based omega-3 fatty acids. Symposium: improving human nutrition through genomics, proteomics and biotechnologies. American Society Nutrition science, pp 4271–4276

Van der Fits L, Memelink J (2000) ORCA3, a jasmonate-responsive transcriptional regulator of plant primary and secondary metabolism. Science 289:295–297

Van der Heijden R, Jacobs DI, Snoeijer W, Hallard D, Verpoorte R (2004) The *Catharanthus* alkaloids: pharmacognosy and biotechnology. Curr Med Chem 11:1241–1253

Van der Krol AR, Lenting PE, Veenstra J, Van der Meer IM, Koes RE, Gerats AGM, Mol JNM, Stuitje AR (1988) An anti-sense chalcone synthase gene in transgenic plants inhibits flower pigmentation. Nature 333:866–869

Verberne M, Verpoorte R, Bol J, Mercado-Blanco J, Linthorst HJM (2000) Overproduction of salicylic acid in plants by bacterial transgenes enhances pathogen resistance. Nat Biotechnol 18:779–783

Verpoorte R, Alfermann AW (2000) Metabolic engineering of plant secondary metabolism. In: Verpoorte R, Alfermann AW (ed) Kluwer Academic Publishers, Dordrecht, The Netherlands

Verpoorte R, van der Heijden R, Memelink J (2000) Engineering the plant cell factory for secondary metabolite production. Transgen Res 9:323–343

Verpoorte R, van der Heijden R, Ten Hoopen HJG, Memelink J (1999) Metabolic engineering of plant secondary metabolite pathways from the production of fine chemicals. Biotechnol Lett 21:467–479

Vom Endt D, Kijne JW, Memelink J (2002) Transcription factors controlling plant secondary metabolism: what regulates the regulators? Phytochemistry 61:107–114

Wang TT, Choi YJ, Lee BH (2001) Transformation systems of non-*Saccharomyces* yeasts. Crit Rev Biotechnol 21:177–218

Waterhouse PM, Wang M, Finnegan EJ (2001a) Role of short RNAs in gene silencing. Trends Plant Sci 6:297–301

Waterhouse PM, Wang M, Lough T (2001b) Gene silencing as an adaptive defence against viruses. Nature 411:834–842

Weber S, Friedt W, Landes N, Molinier J, Himber C, Rousselin P, Hahne G, Horn R (2003) Improved *Agrobacterium*-mediated transformation of sunflower (*Helianthus annuus* L.): assessment of macerating enzymes and sonication. Plant Cell Rep 21:475–482

Whitmer S, Canel C, Hallard D, Goncalves C, Verpoorte R (1998) Influence of precursor availability on alkaloid accumulation by transgenic cell lines of *Catharanthus roseus*. Plant Physiol 116:853–857

Whitmer S, Canel C, van der Heijden R, Verpoorte R (2003) Long-term instability of alkaloid production by stably transformed cell lines of *Catharanthus roseus*. Plant Cell Tiss Org Cult 74:73–80

Whitmer S, van der Heijden R, Verpoorte R (2002a) Effect of precursor feeding on alkaloid accumulation by a tryptophan decarboxylase overexpressing transgenic cell line T22 of *Catharanthus roseus*. J Biotechnol 96:193–203

Whitmer S, van der Heijden R, Verpoorte R (2002b) Effect of precursor feeding on alkaloid accumulation by a strictosidine synthase over-expressing transgenic cell line S1 of *Catharanthus roseus*. Plant Cell Tiss Org Cult 69:85–93

Whitmer S, van der Heijden R, Verpoorte R (2002c) Genetic engineering of the plant cell factory for secondary metabolite production. Indole Alkaloid production in *Catharanthus roseus* as a model. In: Oksman-Caldentey K-M, Barz WH (ed) Plant biotechnology and transgenic plants. Scitech publishing, New York

Wilson TMA, Davies JW (1992) Genetic engineering with plant viruses. CRC Press, Boca Raton, Florida

Yazaki K (2005) Transporters of secondary metabolites. Curr Opin Plant Biol 8:301–307

Ye GN, Stone D, Pang SZ, Creely W, Gonzalez K, Hinchee M (1999) *Arabidopsis* ovule is the target for *Agrobacterium in planta* vacuum infiltration transformation. Plant J 19:249–257

Ye X, Al-Babili S, Klöti A, Zhang J, Lucca P, Beyer P, Potrykus I (2000) Engineering the provitamin A (*β*-carotene) biosynthetic pathway into (carotenoid free) rice endosperm. Science 287:303–305

Yun DJ, Hashimoto T, Yamada Y (1992) Metabolic engineering of medicinal plants: transgenic *Atropa belladonna* with an improved alkaloid composition. Proc Natl Acad Sci USA 89:11799–11803

Zambre M, Terryn N, de Clercq J, de Buck S, Dillen W, van Montagu M, van der Straeten D, Angenon G (2003) Light strongly promotes gene transfer from *Agrobacterium tumefaciens* to plant cells. Planta 216:580–586

Zárate R (2010) Plant secondary metabolism engineering: methods, strategies, advances, and omics. In. Mander L, Lui H-W (ed) Comprehensive natural products II chemistry and biology, vol. 3. Elsevier, Oxford, pp 629–668

Zárate R, Dirks C, van der Heijden R, Verpoorte R (2001) Terpenoid indole alkaloid profile changes in *Catharanthus pusillus* during development. Plant Sci 160:971–977

Zárate R, Verpoorte R (2007) Strategies for the genetic modification of the medicinal plant *Catharanthus roseus* (L.) G. Phytochem Rev 6(2–3):475–491

Zárate R, Yeoman MM (2001) Application of recombinant DNA technology to studies on plant secondary metabolism. In: Bender L, Kumar A (eds) From Soil to Cell—A Broach Approach to Plant Life. Giessen Electronic Library, Germany

Zhang W, Curtin C, Franco C (2002) Towards manipulation of post-biosynthetic events in secondary metabolism of plant cell cultures. Enzyme Microbiol Technol 30:688–696

Zhou X, Wei Y, Zhu H, Wang Z, Lin J, Liu L, Tang K (2008) Protoplast formation, regeneration and transformation from the taxol-producing fungus *Ozonium* sp. African J Biotechnol 7(12):2017–2024

Chapter 17
Use of Metabolomics and Transcriptomics to Gain Insights into the Regulation and Biosynthesis of Medicinal Compounds: *Hypericum* as a Model

Matthew C. Crispin and Eve Syrkin Wurtele

17.1 Introduction

Herbal remedies have been used across cultures for thousands of years, and despite the availability of many pharmaceuticals, recently there has been a renaissance toward using herbal supplements. Although many of these products are not approved by the Food and Drug Administration, more than 15 million people in the United States use herbal supplements (Tachjian et al. 2010), grossing amounts above $34 billion per year (Eisenberg et al. 1998; Tachjian et al. 2010). According to Nalawade and Tsay (2004) this boom in popularity has created an overexploitation of medicinal plants, and threatened the survival of many medicinal species. Consequently, there has been an increased need for propagation. Combined basic and applied research has aimed to better propagate plants and to increase bioactive compound yields in species such as: *Aloe Vera* (Liao et al. 2004), *Atropa belladonna* (Yang et al. 2011), *Catharanthus roseus* (Murata et al. 2008), *Digitalis purpurea* (Einbond et al. 2010), *Echinacea purpurea* (Perry et al. 1997), and *Hypericum* (Karppinen et al. 2006). Yet, current research strategies are not always effective in informing optimal propagation methods, discovering biosynthetic pathways for specialized plant compounds, or elucidating their effects. In this chapter, incorporation of additional strategies is evaluated. We discuss several research approaches

M. C. Crispin
Department of Genetics, Developmental, and Cell Biology,
Iowa State University, 433 Bessey Hall, Ames, IA, USA
e-mail: crispy11@iastate.edu

E. S. Wurtele (✉)
Department of Genetics, Developmental, and Cell Biology,
Iowa State University, 441 Bessey Hall, Ames, IA 50010, USA
e-mail: mash@iastate.edu

S. Chandra et al. (eds.), *Biotechnology for Medicinal Plants*,
DOI: 10.1007/978-3-642-29974-2_17, © Springer-Verlag Berlin Heidelberg 2013

enabled by recent technological advances in metabolomics and sequencing, in part using *Hypericum* as a case study. These are: (1) strategies for localizing the sites of synthesis and accumulation of bioactive compounds; (2) combined metabolomic and transcriptomic approach as a method of novel gene discovery; and (3) combined metabolomic and transcriptomic approach to determine the positive and negative effects of botanicals in humans. The knowledge gleaned from such studies can provide clues to development of propagation strategies (Fig. 17.3).

17.2 Overview

This chapter discusses strategies to gain a better understanding of medicinal plants. First, to determine the location of bioactive compound synthesis and accumulation, and evaluate the genetic and environmental factors influencing the synthesis and accumulation of these bioactive compounds. Second, to leverage this knowledge to identify the genes that contribute to the metabolism and storage of the bioactive compounds, and to establish molecular "fingerprints" that can be used in breeding and research. Third, to better understand the intricacies of the effects of bioactive compounds on humans. These data can ultimately inform the propagation, breeding, and genetic engineering of plants such that they produce the compounds of interest at levels that are desired by the consumer/breeder/researcher/company.

17.3 Detailing the Sites/Conditions of Accumulation of Bioactive Compounds in the Plant

17.3.1 Importance

Understanding the factors that impact accumulation of bioactive compounds at the agronomic, physiological, and molecular levels is greatly facilitated by a detailed understanding of where/under what conditions the compounds occur. The term "bioactive compounds" is defined as the specialized metabolites present in particular plant clades/species, for which there is either direct evident of a biological effect on other biological systems, or a structural similarity to known bioactive compounds. In most cases, the specific bioactive compounds or combination of bioactive compounds that induce particular effects in humans are not well understood. And many bioactive compounds in plants, even in currently medicinally used species, remain to be identified.

Here, we focus on the use of this understanding as a tool to identify the genes involved in bioactive compound metabolism: if synthesis and storage of bioactive compounds co-occurs, samples with high levels of bioactive compounds could be used to dissect the biosynthetic and regulatory genes. Thus, one major advantage of knowing the discrete regions and/or conditions of accumulation of bioactive compounds is that it empowers a comparative transcriptomics approach. In a transcriptomic approach, comparing plant samples that are synthesizing high levels of a given bioactive compound(s) to other samples not synthesizing these compounds can be used to tease out the transcripts/genes involved in their synthesis.

A goal in selecting samples for sequencing is to minimize differences in transcriptomes between samples *except for* those transcripts involved with the metabolism of the compounds of interest. For example, comparing *Echinacea* flower petals at a stage of development in which alkamides are not present with a slightly later stage in which alkamides synthesis is rapid provides a sensitive approach to identifying alkamides biosynthetic genes. As a counterexample, if comparing transcripts (expressed genes) of an *Echinacea* root that accumulates alkamides, with transcripts of an *Echinacea* leaf that does not accumulate these compounds, identification of the genes involved in metabolism of alkamides will be confounded by the whole range of other genes expressed in the leaf but not in the root that have nothing to do with metabolism of the alkamides. Likewise, extracts made using broad classes of organs, such as "flowers", present a challenge because a flower contains a different metabolic "fingerprint" in the petals, stamen, carpals, and sepals, and each of these organs has a distinct pattern of development and pattern of synthesis of bioactive compounds.

17.3.2 Technologies Used for Metabolomics Analysis

Multiple methods are available and emerging for metabolomic analysis (Emmett et al. 1998; Ganzera et al. 2002; Johnson et al. 2008; Cha et al. 2008; Blakney et al. 2011). Techniques for metabolomic analysis of bioactive compounds must be tailored to the compounds of interest (Nikolau and Wurtele 2007), for example, fatty acids and essential oils can be analyzed using GC–MS (Cakir et al. 2004). However; for the cyclic polyketides, LC–MS (or UV) are preferable because these compounds decompose at high temperatures and elute as multiple peaks in GC instead of only one peak. Spatial detection is emerging as a powerful method for evaluating distribution of metabolites; one interesting approach (Cha et al. 2008) describes the use of graphite-assisted laser desorption ionization MS (GALDI-MS) to profile the spatial accumulation of flavonoids in cryo-sectioned *Arabidopsis thaliana* flowers and sepals. This strategy has the distinct advantage that it gives metabolite localization and levels simultaneously and requires no microdissection, however, it also requires specialized equipment and expertise.

17.3.3 Comprehensive Metabolomics to Identify Known and Putative Bioactive Compounds

In most medicinal species, there are a plethora of related chemicals, many of which may have bioactivities in animals, and others of which have not been identified. As an example of the complements of metabolites that occur in a medicinal plant species and its relatives, we focus on the genus *Hypericum*, which contains over 450 species and is within the Clusiaceae family. The popular medicinal plant *Hypericum perforatum* (St. John's Wort) accumulates compounds with putative anti-viral (Richman 1991; Axarlis et al. 1998), anti-inflammatory (Hammer et al. 2007), anti-depressive (Mennini and Gobbi 2004; Butterweck and Schmidt 2007; Kasper et al. 2008), and anti-bacterial (Franklin et al. 2009; Saddiqe et al. 2010) bioactivities. In *Hypericum* there are three major classes of specialized compounds. The first is dianthrones, which include hypericins (hypericin, pseudohypericin, protopseudohypericin, and protohypericin) (Tatsis et al. 2007). These chemicals are present in *H. perforatum*, but not in all *Hypericum* species (e.g., *Hypericum gentianoides*, Hillwig et al. 2008). A second class of compounds is flavonoids, including quercetin and the quercetin glycosides (quercetin, rutin, hyperoside, isoquercitrin, quercitrin) (Ganzera et al. 2002). A third class is acyl-phloroglucinols, which consist of hyperforin and adhyperforin in *H. perforatum* (Kirakosyan et al. 2004), and a wide variety of other compounds (Ishiguro et al. 1987), including nine structurally similar acyl-phloroglucinols in *H. gentianoides* (Hillwig et al. 2008). Despite the popularity of extracts from *H. perforatum* as medicinal, only two genes that may be involved in hypericin synthesis (Hyp-1 and HpPKS2) have been identified, but their roles are under debate (Bais et al. 2003; Karppinen et al. 2008; Košuth et al. 2011).

17.3.4 Spatial Differences in the Same Organ

A strategy that employs dissection of organs can be useful for obtaining samples that vary in specialized metabolites, particularly if there are distinct structures or morphologies that might correspond to the accumulation of a compound of interest. For example, in most species of *Hypericum*, leaves have translucent glands (Fig. 17.1a shows translucent glands in *H. gentianoides*), which appear to the eye as light spots. It has been hypothesized that hyperforins accumulate in translucent glands in *H. perforatum* (Ernst 2003). If this is the case, dissection of leaves to separate the regions with the light glands from the regions without, could provide a set of samples that vary in the concentration (and hopefully the synthesis) of bioactive compounds (for example, hyperforins or other acyl-phloroglucinols).

Fig. 17.1 *Hypericum gentianoides*, accession Ames 27729 (http://www.ars-grin.gov/cgi-bin/npgs/acc/display.pl?1668188), germinated and grown in greenhouse conditions. **a** The adaxial side of a leaf from the vegetative phase of a 3-month-old plant. Its many translucent glands appear as light spots on the leaf. **b** The vegetative phase of an 8.5-month-old plant. **c** The reproductive phase of an 11-month-old plant

17.3.5 Developmental Stage

Collection of distinct developmental stages can provide an approach for identifying samples that vary in specialized metabolites (Brown et al. 2003). A discrete and easily recognizable example is *H. gentianoides*, which has two morphologically distinct phases; vegetative and reproductive (Fig. 17.1b, c). In this species, the vegetative phase accumulates higher levels of acyl-phloroglucinols than the reproductive phase (Fig. 17.2). Depending on the population being studied, it is possible to collect both reproductive and vegetative phases from the same plants, ensuring that these parts have been grown under the same conditions.

17.3.6 Different Populations

Due to genotypic differences, different populations/accessions have different levels of certain traits that can be used for examining specialized compounds or phenotypes. For example, in *H. perforatum*, the accumulation of hyperforin varied

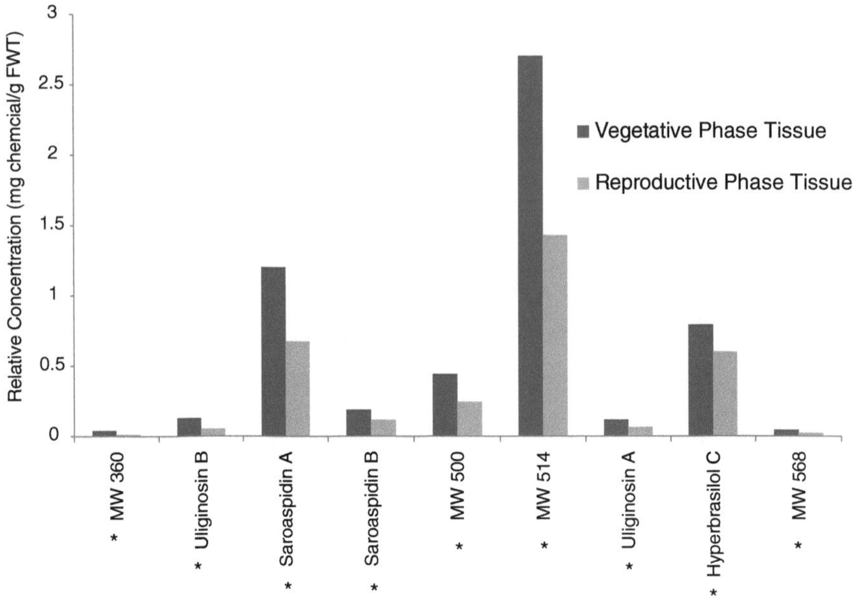

Fig. 17.2 Accumulation pattern of acyl-phloroglucinol polyketides in the vegetative and reproductive phases of *H. gentianoides*. *Babka (2009)

between two genotypes, with one genotype having low but steady levels of hyperforin, whereas in the other, hyperforin levels were initially higher but decreased during development (Bölter and Bölter 2002).

17.3.7 Induction by Stress or Other Environmental Perturbation

Identifying a treatment that induces bioactive compound synthesis can also facilitate obtaining samples that have many expressed genes in common, but vary in specialized metabolites (Brechner et al. 2011). For example, in *H. perforatum*, UV-B, tryptophan, methyl jasmonate, tridiazuron (a plant defoliant) and 2,3-dihydroxypropyl jasmonate each induces accumulation of bioactive compounds (Liu et al. 2007a, b; Germ et al. 2010). This provides the added advantage that if the accumulation of the bioactive compounds is induced, the biosynthetic transcripts will likely be accumulated also, since induction typically requires expression of the genes required to make the compounds. Timing is important, if samples are collected following a stressor, the lag time between the induction and increase in a transcript and the induction of the corresponding metabolite needs to be considered.

17.4 Use of Transcriptomics and Metabolomics to Discover Putative Biosynthetic Genes

17.4.1 Importance

Patterns of transcript and metabolite accumulation across multiple data sets can be used to provide fingerprints of the status of medicinal plants. In addition, as discussed below, comparisons of transcripts expressed in samples synthesizing the bioactive compounds versus transcripts in samples not synthesizing the bioactive compounds provide a major criteria for identifying the putative metabolic and regulatory genes.

17.4.2 Technologies for Transcriptomic Analysis

The power of nucleic acid sequencing has vastly impacted all of biology (Schuster 2008). Previously, Sanger sequencing was used for sequencing nucleic acids; however next generation sequencing (NGS) has immerged as a far better option. Unlike Sanger, NGS technologies do not use traditional PCR techniques, which can cause a loss of some sequences; more importantly, NGS platforms have provided researchers with the ability to sequence an entire genome or transcriptome for much cheaper than possible with Sanger (Metzker 2010), and the ability to detect alternative splicing and gene fusions (Ozsolak and Milos 2011). Methods for sequencing and its computational analysis are rapidly evolving (Jiang and Wong 2009; Trapnell et al. 2009). For transcriptomic sequencing, Illumina Genomic Analyzers are currently common; these have advantages over Roche 454 because 454 can get hundreds of thousands of reads, whereas Illumina normally gets tens of millions. Illumina has lower costs for reagents, and are far better at variant discovery (Mardis 2008; Metzker 2010). Even longer reads/more depth is becoming possible with, e.g., Pacific Bioscience real-time sequencing platform (Eid et al. 2009).

17.4.3 Assumptions and Experimental Procedures

To identify biosynthetic transcripts for bioactive compounds by comparing plant samples with high accumulation of these compounds with those of no/low accumulation, two criteria must be met. First, the compounds must be synthesized in the same place that they are accumulated. For example in *H. perforatum*, hypericin accumulates in the black glandular nodules on the leaves (Ciccarelli et al. 2001). If hypericin is not synthesized in the leaves, but rather transported from the stems, despite hypericin being present in leaves there would be no biosynthetic transcripts, because the biosynthetic machinery is located elsewhere. The second criterion is that the compound(s) of interest must be actively being

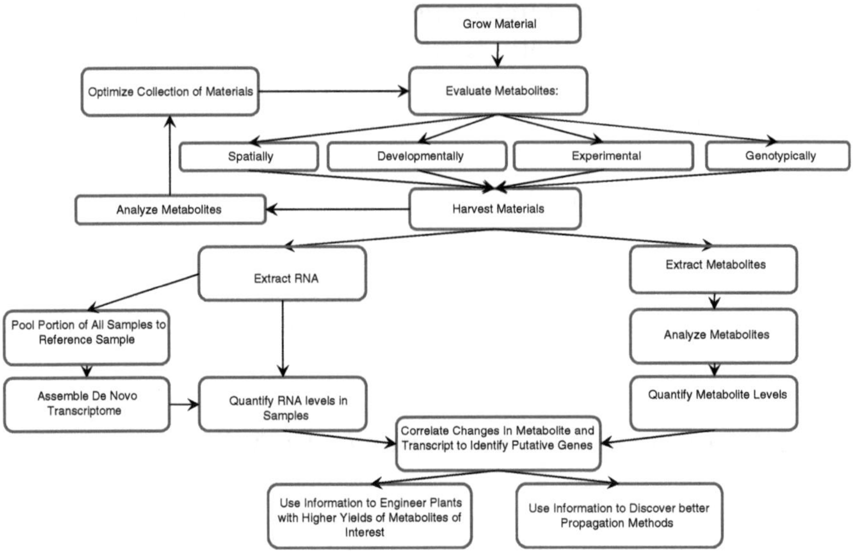

Fig. 17.3 Flowchart illustrating a strategy to co-evaluate the metabolome and transcriptome of a previously unsequenced medicinal plant species. The strategy is designed to facilitate the identification of the enzyme-coding and regulatory transcripts involved in the synthesis, accumulation and catabolism of bioactive compounds. It also provides a means to identify a fingerprint of transcripts that characterize particular stages of the plant

synthesized at the time the plant material is collected. If not, despite high concentrations of the compounds, there would be no corresponding transcripts.

Here, localization becomes very critical. With an idea of which tissues to harvest, this approach can be used and material can be extracted for RNA and metabolites.

Figure 17.3 shows a flowchart of a possible protocol. This approach uses different aliquots from the same tissue samples for both transcript and the metabolite analysis. For next-generation sequencing of a previously unsequenced species, a reference sample should be sequenced in order to assemble what is sometimes referred to as a "de novo transcriptome" or "artificial genome" (Meyer et al. 2009). In order to most thoroughly represent the possible transcriptome space, this reference sample should consist of a pooled collection of material from a combination of tissues, developmental stages, and stresses. After the de novo transcriptome is assembled, the individual samples can be sequenced, contigs are assembled and quantified according to fragments per kilobase per million reads (Trapnell et al. 2010). Concurrently, metabolites can be extracted and analyzed from these samples (Sects. 17.3.4–17.3.7).

17.4.4 Advantages

Detecting differentially expressed genes by sequencing, rather than a microarray-based technology, provides higher accuracy, quantitative analysis, and much more depth

(Trapnell et al. 2010). Because there is no longer a requirement to use an organism with a sequenced genome or annotated genome, any organism of interest, such as coral (Meyer et al. 2009), can be sequenced and the transcriptome of individual samples can be determined. Large-scale sequencing efforts focused on medicinal plants have accumulated significant sequence data for 28 medicinal species (http://medplant transcriptomes.uic.edu for the Medicinal Plant/Human Health Consortium, and http://medicinalplantgenomics.msu.edu/ for the Medicinal Plant Consortium).

This approach is very advantageous when samples can be obtained in which the transcriptomic profile should be very similar except for the transcripts of interest, for example in regions of an organ with differential accumulation of bioactive compounds (e.g., Sect. 17.3.4). The advantage of mass spectrometry is that if precursors are unknown, but are observed to correlate to a compound of interest, further identification of that precursor can be done using MS/MS.

17.4.5 Disadvantages

As is with most techniques, there are some disadvantages. For example, as discussed, the bioactive compounds of interest must be actively synthesized in a subset of the samples. Also, most medicinal species are not well characterized genetically, and different accessions/genotypes of a species will have different transcript sequences, making comparisons challenging. In addition, plant contamination (such as *Escherichia coli*, thrips, or an unknown pathogen) can confound sequence data. A researcher needs to consider that metabolomics and transcriptomics are rapidly changing technologies, and both the instrumentation and the analytic tools are evolving and becoming more powerful.

17.5 Effects of Medicinal Plants in Humans

The interplay for survival and reproduction among each plant clade and the pathogens, competitors, symbionts, and pollinators of that clade has been evolving for hundreds of thousands of years. As a consequence, plants have complements of compounds with complex bioactivities toward many organisms, and some of these compounds also have bioactivities in humans. Because of potential variation among genetic lines and growth and extraction conditions, and the vast array of unusual bioactive compounds in most medicinal plants, bioactivity must be evaluated in conjunction with the metabolomics studies of the specific extracts used in each study; this is often only partially done or not done at all. In this section, bioactive effects of a single genus, *Hypericum*, will be discussed as an example. Table 17.1 lists some of the medicinal properties observed for *Hypericum*. Transcriptomics studies evaluating the effects of low levels of *Hypericum* extracts on human cells show that the transcriptome of different cell types are complexly and differently affected (Holtrup et al. 2011).

17.5.1 Positive Effects

17.5.1.1 Effect of *Hypericum* on Human Immunodeficiency Virus Infectivity

Considerable evidence indicates that the putative bioactive compounds of *Hypericum* have many beneficial effects in humans (Table 17.1). *H. perforatum* extracts added to human immunodeficiency virus (HIV)-infected cells decreases the infectivity of this virus (Richman 1991). Possibilities suggested for the targets of the extract include virus adsorption, virus-cell fusion, reverse transcription, integration, and translation (De Clercq 2000). However, *H. perforatum* extracts interact with human CYP3A4 (Moore et al. 2000), and alter transcription in mammalian cells (Hammer et al. 2010; Holtrup et al. 2011; Jungke et al. 2011), hence it is possible the extract interaction with human cells also impacts the virus.

17.5.1.2 *Hypericum* as an Anti-depressant

Besides anti-retroviral properties, one of the commonly studied effects of *Hypericum* is anti-depressive. According to Brenner et al. (2000), the Hamilton Rating Score for Depression (HAM-D) and Clinical Global Impression Scale were significantly decreased by *Hypericum* treatments and by the selective serotonin reuptake inhibitor (SSRI) sertraline. Of the patients in the study who had a clinical response (as characterized by a 50% decrease in HAM-D), the difference between *Hypericum* and the SSRI was not statistically significant, demonstrating for this small group that the two treatments were effective in treating mild to moderate depression. This study does not identify the specific compound(s) causing this effect. Other studies, such as Mennini and Gobbi (2004), indicated that hyperforin is the active compound in the anti-depressive response.

17.5.2 Negative Effects

17.5.2.1 Cytotoxicity of *Hypericum*

Several *Hypericum* compounds show significant cytotoxicity at high concentrations. Of these, the *H. perforatum* dianthrones, pseudohypericin, and hypericin, are most toxic, causing unpigmented regions of skin in mammals to become photosensitive (Fields et al. 1990; Traynor et al. 2005), and death of cells in culture (Schmitt et al. 2006).

Table 17.1 Effects of *Hypericum* extracts on mammalian cells and mammals

Effect	Investigated in	References
Anti-viral	Friend murine leukemia virus in human[1]; human diploid embryonic lung fibroblasts (MRC-5) (Biomerieux-France) against human cytomegalovirus[2]; HIV in mice[3]	Richman (1991)[1], Axarlis et al. (1998)[2] and Meruelo et al. (1988)[3]
Anti-depressive	Humans with mild to moderate depression[1]; and in mice[2]	Brenner et al. (2000)[1], Kasper et al. (2008)[1], Mennini and Gobbi (2004)[2], Butterweck and Schmidt (2007)[2] and Crupi et al. (2011)[2]
Anti-inflammatory	264.7 mouse macrophage cells[1]; and in mice[2]	Hammer et al. (2007)[1], Hammer et al. (2010)[1] and Galati et al. (2008)[2]
Anti-bacterial	Many species of gram positive and gram negative bacteria incubated on plates and inoculated with extracts	Rabanal et al. (2002), Peeva-Naumovska et al. (2010) and Saddiqe et al. (2010)
Anti-fungal	Fungal species *Collectotrichum acutatum*, *Collectotrichum fragariae*, *Collectotrichum gloeosporioides*[1]	Crockett et al. (2010)[1]
Anti-cancer	Human patients with reoccurring malignant gliomas[1]; DA3 breast carcinoma-derived lung metastases[2]; SQ2 murine squamous cell carcinoma-derived metastases[2]; cell cultures of human leukemic lymphoma cell[3]; human prostate cells in vivo and implanted in mice[4]; autologous MT-450 breast carcinoma in immunocompetent Wistar rats[5]	Couldwell et al. (2011)[1]*, Blank et al. (2004)[2], Schempp et al. (2002b)[3], Martarelli et al. (2004)[4] and Schempp et al. (2002a)[5]
Anti-malarial	A multidrug-resistant W2mef laboratory strain, and a field isolate (SHF4) of *Plasmodium falciparum*[1]	Zofou et al. (2011)[1]*
Treatment of mild to moderate atopic dermititis	Humans using skin cream with *Hypericum*[1]	Schempp et al. (2003)[1]
Cytotoxicity in cells	HaCaT human keratinocytes[1]; A431 tumor cells[2]; LLC-MK2 monkey kidney epithelial cells[3]	Schmitt et al. (2006)[1], Vandenbogaerde et al. (1998)[2] and Zofou et al. (2011)[3]*
Photosensitivity	Mammals[1]; HaCaT human keratinocytes supplemented with UV-A[2]	Fields et al. (1990)[1] and Traynor et al. (2005)[2]
Lowers anti-retroviral drug concentration	Human patients taking anti-retroviral drugs for HIV[1]	Adam et al. (2002)[1]
Increases blood pressure and heart rate	Humans on monoamine oxidase inhibitors[1,2]	Tachijan et al. (2010)[1] and Schroeder et al. (2004)[2]*
Lowers serum concentration of many drugs	Human patients using several categories of heart disease medications[1]	Tachijan et al. (2010)[1]

*No statistically significant effect observed
Numbers in superscripts connect "investigated in" column to "references" column

17.5.2.2 *Hypericum perforatum* Decreases the Effect of Anti-retroviral Drugs

Previous studies have shown that *H. perforatum* can act as a double-edged sword in the treatment of retrovirus, most specifically in HIV, because although the extracts have inhibitory effects on the virus, they also cause a decrease in the effectiveness of anti-retroviral drugs. One explanation behind this is the lipophilic nature of the *Hypericum* polyketides. The polyketides elicit an increase in cytochrome P450s that remove the polyketides and the anti-retroviral drugs (Adam et al. 2002); it is unclear whether there are additional interacting factors.

17.5.2.3 Other Interactions of Drugs with *H. perforatum*

Anti-retroviral drugs are not the only targets for inhibition by *H. perforatum* (De Clercq 2000). Other compounds whose effectiveness is reduced or altered by *Hypericum* include digoxin, clopidogrel, warfarin, simvastatin, paroxetine, class IA and III antarrhythmic agents, cyclosporine, and theophylline. In clopidogrel, however the activity increases giving some level of evidence to suggest that there are multiple factors interacting with these drugs and it is not always the cytochrome P450s that are moving compounds out of the serum.

17.5.2.4 *Hypericum perforatum* Can Cause Interactions in People with Cardiovascular Diseases

People with cardiovascular diseases are at risk for negative effects. According to Tachjian et al. (2010), *H. perforatum* extracts can cause increased heart rate and blood pressure when combined with monoamine oxidase inhibitors, and decrease the concentration of digoxin. It is interesting and ironic that heart disease is quite prevalent in the United States, yet more and more people are using extracts that may exacerbate cardiovascular disease.

17.5.3 Medicinal Supplement Use

Although there may be numerous positive effects of medicinal plants, there can be negative properties as well. As with any other pharmaceutical, people should do their best to evaluate specific risk factors with respect to their own health before they use medicinal plant extracts on a regular basis. Personalized medicine will provide additional information. Because medicinal species contain a complement of bioactive compounds, it is important to provide well-characterized and consistent extracts to the public.

17.6 Conclusions

Medicinal plants have been critical in medicine since civilization and probably earlier (Fox et al. 2001). Despite many studies on medicinal species, much remains to be understood about how the bioactive compounds are synthesized, what drives their accumulation, and mechanistically how the extracts and their components affect humans. The revolution in transcriptomics and metabolomics technologies provides new strategies to gain a further understanding, as well as to further characterize plant material. With these strategies, more mysteries associated with the accumulation and bioactivities of bioactive compounds in these species can be unraveled.

Acknowledgments This publication was made possible in part by grant 9P50AT004155-06 from National Center for Complementary and Alternative Medicine and Office of Dietary Supplements, and grant NIGM-99511 from National Institute of General Medical Sciences, National Institutes of Health, its contents are solely the responsibility of the authors; and by the Genetics Program, Iowa State University (to MCC). We thank Dr. Heather Babka, Iowa State University, for training MC in *H. gentianoides* biology, Dr. Robin Buell and Dr. Elsa Gongora Castillo, Michigan State University, for insightful discussions into RNA-seq technologies. We are grateful to Dr. Mark Wiederlichtner for helpful discussions and for defined accession seeds of *H. gentianoides* (Ames 27729 (http://www.ars-grin.gov/cgi-bin/npgs/acc/display.pl?1668188)).

References

Adam P, Arigoni D, Bacher A, Eisenreich W (2002) Biosynthesis of Hyperforin in *Hypericum perforatum*. J Med Chem 45:4786–4793

Axarlis S, Mentis A, Demetzos C, Mitaku S, Skaltsounis AL, Marselos M, Malamas M (1998) Antiviral in vitro activity of *Hypericum perforatum* L. extract on the human cytomegalovirus (HCMV). Phytother Res 12:507–511

Babka HL (2009) Acetyl-CoA in plant biology. Iowa State University, Iowa

Bais HP, Vepachedu R, Lawrence CB, Stermitz FR, Vivanco JM (2003) Molecular and biochemical characterization of an enzyme responsible for the formation of hypericin in St. John's Wort (*Hypericum perforatum* L.). J Biol Chem 278(34):32413–32422

Blakney GT, Hendrickson CL, Marshall AG (2011) Predator data station: a fast data acquisition system for advanced FT-ICR MS experiments. Int J Mass Spectrom 306:246–252

Blank M, Lavie G, Mandel M, Hazan S, Orenstein A, Meruelo D, Keisari Y (2004) Antimetastatic activity of the photodynamic agent hypericin in the dark. Int J Cancer 111:596–603

Brechner ML, Albright LD, Weston LA (2011) Effects of UV-B on secondary metabolites of St. John's Wort (*Hypericum perforatum* L.) grown in controlled environments. Photochem Photobiol 87:680–684

Brenner R, Azbel V, Madhusoodanan S, Pawlowska M (2000) Comparison of an extract of *Hypericum* (LI 160) and sertraline in the treatment of depression: a double-blind, randomized pilot study. Clin Ther 22(4):411–419

Brown PD, Tokuhisa JG, Reichelt M, Gershenzon J (2003) Variation of glucosinolate accumulation among different organs and developmental stages of *Arabidopsis thaliana*. Phytochemistry 62(3):471–481

Bölter KB, Bölter B (2002) Ontogenetic variation regarding hypericin and hyperforin levels in four accessions of *Hypericum perforatum* L. J Herbs Spices Med Plants 9(2&3):95–100
Butterweck V, Schmidt M (2007) St John's Wort: role of active compounds for its mechanism of action and efficacy. Wien Med Wochenschr 157:356–361
Cakir A, Kordali S, Zengin H, Izumi S, Hirata T (2004) Composition and antifungal activity of essential oils isolated from *Hypericum hyssopifolium* and *Hypericum heterophyllum*. Flavour Frag J 19:62–68
Cha S, Zhang H, Ilarslan HI, Wurtele ES, Libuse B, Nikolau BJ, Yeung ES (2008) Direct profiling and imagining of plant metabolites in intact tissues by using colloidal graphite-assisted laser desorption ionization mass spectrometry. Plant J 55:348–360
Ciccarelli D, Andreucci AC, Pagni AM (2001) The "black nodules" of *Hypericum perforatum* L. subsp. *perforatum*: morphological, anatomical, and histochemical studies during the course of ontogenesis. Israel J Plant Sci 49:33–40
Couldwell WT, Surnock AA, Tobia AJ, Cabana BE, Stillerman WT, Forsyth PA, Appley AJ, Spence AM, Hinton DR, Chen TC (2011) A phase ½ study of orally administered synthetic hypericin for treatment of recurrent malignant gliomas. Cancer 117:4905–4915
Crockett SL, Poller B, Tabanca N, Pferschy-Wenzig E, Kunert O, Wedge DE, Bucar F (2010) Bioactive xanthones from the roots of *Hypericum perforatum* (common St. John's Wort). J Sci Food Agric 91:428–434
Crupi R, Mazzon E, Marino A, La Spada G, Bramanti P, Battaglia F, Cuzzocrea S, Spina E (2011) *Hypericum perforatum* treatment: effect on behaviour and neurogenesis in a chronic stress model in mice. BMC Complement Altern Med 11(7):1–10
De Clercq E (2000) Current lead natural products for the chemotherapy of human immunodeficiency virus (HIV) infection. Med Res Rev 20:323–349
Eid J, Fehr A, Gray J, Luong K, Lyle J, Otto G, Peluso P, Rank D, Baybayan P, Bettman B, Bibillo A, Bjornson K, Chaudhuri B, Christians F, Cicero R, Clark S, Dalal R, DeWinter A, Dixon J, Foquet M, Gaertner A, Hardenbol P, Heiner C, Hester K, Solden D, Kearns G, Kong X, Kuse R, Lacroix Y, Lin S, Lundquist P, Ma C, Marks P, Maxham M, Murphy D, Park I, Pham T, Phillips M, Roy J, Sebra R, Shen G, Sorenson J, Tomaney A, Travers K, Trulson M, Vieceli J, Wegener J, Wu D, Yang A, Zaccarin D, Zhoa P, Zhong F, Korlach J, Turner S (2009) Real-time DNA sequencing from single polymerase molecules. Science 323(5910):133–138
Einbond LS, Wu H, Su T, Chang T, Panjikaran M, Wang X, Goldberry S (2010) Digitoxin activates EGR1 and synergizes with paclitaxel on human breast cancer cells. J Carcinog 9:10–23
Eisenberg DM, Davis RB, Ettner SL (1998) Trends in alternative medicine use in the United States, 1990–1997: results of a follow-up national survey. J Am Med Assoc 280:1569–1575
Emmett MR, White FM, Hendrickson CL, Shi SDH, Marshall AG (1998) Application of micro-electrospray liquid chromatography techniques to FT-ICR MS. J Am Soc Mass Spectrom 9(4):333–340
Ernst E (2003) *Hypericum*: the genus *Hypericum*. Taylor & Francis CRC Press, London
Fields P, Arnason J, Fucher R (1990) The spectral properties of *Hypericum perforatum* leaves: the implications for its photoactivated defences. Can J Bot 68:1166–1170
Fox E, Murphy RF, McCully CL, Adamson PC (2001) Plasma pharmacokinetics and cerebrospinal fluid penetration of hypericin in nonhuman primates. Cancer Chemother 47(1):41–44
Franklin G, Conceição LFR, Kombrink E, Dias ACP (2009) Xanthone biosynthesis in *Hypericum perforatum* cells provides antioxidant and antimicrobial protection upon biotic stress. Phytochemistry 70:60–68
Galati EM, Contartese G, Miceli N, Taviano MF, Sdrafkakis V, Couladis M, Tzakou O, Lanuzza F (2008) Antiinflammatory and antioxidant activity of *Hypericum rumeliacum* Boiss. Subsp. *apollinis* (Boiss. & Heldr.) Robson & Strid methanol extract. Phytother Res 22(6):766–771
Ganzera M, Zhao J, Khan IA (2002) *Hypericum perforatum*—chemical profiling and quantitative results of St. John's Wort products by an improved high-performance liquid chromatography method. J Pharm Sci 91(3):623–630

Germ M, Stibilj V, Kreft S, Gaberščik A, Kreft I (2010) Flavonoid, tannin, and hypericin concentrations in the leaves of St. John's Wort (*Hypericum perforatum* L.) are affected by UV-B radiation levels. Food Chem 122:471–474

Hammer KDP, Hillwig ML, Solco AKS, Dixon PM, Delate K, Murphy PA, Wurtele ES, Birt DF (2007) Inhibition of prostaglandin E2 production by anti-inflammatory *Hypericum perforatum* extracts and constituents in RAW264.7 mouse macrophage cells. J Agric Food Chem 55:7323–7331

Hammer KDP, Yum M, Dixon PM, Birt DF (2010) Identification of JAK-STAT pathways as important for the anti-inflammatory activity of a *Hypericum perforatum* fraction and bioactive constituents in RAW 264.7 mouse macrophages. Phytochemistry 71:716–725

Hillwig ML, Hammer KD, Birt DF, Wurtele ES (2008) Characterizing the metabolite fingerprint and anti-inflammatory activity of *Hypericum gentianoides*. J Agric Food Chem 56(12):4359–4366

Holtrup F, Bauer A, Fellenberg K, Hilger RA, Wink M, Hoheisel J (2011) Microarray analysis of nemorosone-induced cytotoxic effects on pancreatic cancer cells reveals activation of the unfolding protein response (UPR). Br J Pharmacol 162(5):1045–1059

Ishiguro K, Yamaki M, Kashihara M, Takagi S (1987) Saroaspidin A, B, and C: additional antibiotic compounds from *Hypericum japonicum*. Planta Med 53(5):415–417

Jiang H, Wong WH (2009) Statistical inferences for isoform expression in RNA-Seq. Bioinformatics 25(8):1026–1032

Johnson H, Wong SCC, Simpson DM, Beynon RJ, Gaskell SJ (2008) Protein quantification by selective isolation and fragmentation of isotopic pairs using FT-ICR MS. J Am Soc Mass Spectrom 19:973–977

Jungke P, Ostrow G, Li JL, Norton S, Nieber K, Kelber O, Butterweck V (2011) Profiling of hypothalamic and hippocampal gene expression in chronically stressed rats treated with St. John's Wort extract (STW 3-VI) and fluoxetine. Psychopharmacology 213(4):757–772

Karppinen K, György Z, Kauppinen M, Tolonen A, Jalonen J, Neubauer P, Hohtola A, Häggman H (2006) In vitro propagation of *Hypericum perforatum* L. and accumulation of hypericins, pseudohypericins, and phloroglucinols. Propag Ornam Plants 6(4):170–179

Karppinen K, Hokkanen J, Mattila S, Neubauer P, Hohtola A (2008) Octaketide-producing type III polyketide synthase from *Hypericum perforatum* is expressed in dark glands accumulating hypericins. FEBS J 275(17):4329–4342

Kasper S, Volz HP, Moller HJ, Dienel A, Kieser M (2008) Continuation and long-term maintenance treatment with hypericum extract WS5570 after recovery from an acute episode of moderate depression—a double-blind, randomized, placebo controlled long-term trial. Eur Neuropsychopharmacol 18:803–813

Kirakosyan A, Sirvent TM, Gibson DM, Kaufman PB (2004) The production of hypericins and hyperforin by in vitro cultures of St. John's Wort (*Hypericum perforatum*). Biotechnol Appl Biochem 39:71–81

Košuth J, Smelcerovic A, Borsch T, Zuehlke S, Karppinen K, Spiteller M, Hohtola A, Čellárová E (2011) The Hyp-1 gene is not a limiting factor for hypericin biosynthesis in the genus *Hypericum*. Funct Plant Biol 38:35–43

Liao Z, Chen M, Tan F, Sun X, Tang K (2004) Micropropagation of endangered Chinese aloe. Plant Cell Tissue Org 76(1):83–86

Liu XN, Zhang XQ, Zhang SX, Sun JS (2007a) Regulation of metabolite production by precursors and elicitors in liquid cultures of *Hypericum perforatum*. Plant Cell Tissue Organ Cult 91:1–7

Liu XN, Zhang XQ, Sun JS (2007b) Effects of cytokinins and elicitors on the production of hypericins and hyperforin metabolites in *Hypericum sampsonii* and *Hypericum perforatum*. Plant Growth Regul 53:207–214

Mardis ER (2008) Next-generation DNA sequencing methods. Annu Rev Genomics Human Genet 9:387–402

Martarelli D, Martarelli B, Pediconi D, Nabissi MI, Perfumi M, Pompei P (2004) *Hypericum perforatum* methanolic extract inhibits growth of human prostatic carcinoma cell line orthotopically implanted in nude mice. Cancer Lett 210(1):27–33

Mennini T, Gobbi M (2004) The antidepressant mechanism of *Hypericum perforatum*. Life Sci 75:1021–1027

Meruelo D, Lavie G, Lavie D (1988) Therapeutic agents with dramatic antiretroviral activity and little toxicity at effective doses: aromatic polycyclic diones hypericin and pseudohypericin. Proc Natl Acad Sci USA 85(14):5230–5234

Metzker ML (2010) Sequencing technologies—the next generation. Nat Rev Genet 11(1):31–46

Meyer E, Aglyamova GV, Wang S, Buchanan-Carter J, Abrego D, Colbourne JK, Willis BL, Matz MV (2009) Sequencing and de novo analysis of a coral larval transcriptome using 454 GSFIx. BMC Genomics 10:219–238

Moore LB, Goodwin B, Jones SA, Wisely GB, Serabjit-Singh CJ, Willson TM, Collins J, Kliewer SA (2000) St. John's Wort induces hepatic drug metabolism through activation of the pregnane X receptor. Proc Natl Acad Sci USA 97(13):7500–7502

Murata J, Roepke J, Gordon H, De Luca V (2008) The leaf epidermome of *Catharanthus roseus* reveals its biochemical specialization. Plant Cell 20:524–542

Nalawade SM, Tsay HS (2004) In vitro propagation of some important Chinese medicinal plants and their sustainable usage. In Vitro Cell Dev Biol Plant 40(2):143–154

Nikolau BJ, Wurtele ES (2007) Concepts in plant metabolomics. Springer, Heidelberg, pp 145–158

Ozsolak F, Milos PM (2011) RNA sequencing: advances, challenges and opportunities. Nat Rev Genet 12(2):87–98

Peeva-Naumovska V, Panovski N, Grdanovska T, Fredro-Kumbaradzi E (2010) Formulations of St. John's Wort oil ointment and evaluation of its antibacterial effect. AMAPSEEC. www.amapseec.org/cmapseec.1/Papers/papp067.htm. Accessed 15 June 2011

Perry NB, Van Klink JW, Burgess EJ, Parmenter GA (1997) Alkamide levels in *Echinacea purpurea*: a rapid analytical method revealing differences among roots, rhizomes, stems, leaves and flowers. Planta Med 63(1):58–62

Rabanal RM, Arias A, Prado B, Hernández-Pérez M, Sánchez-Mateo CC (2002) Antimicrobial studies on three species of *Hypericum* from the Canary Islands. J Ethnopharmacol 81:287–292

Richman DD (1991) Antiviral therapy of HIV infection. Annu Rev Med 42:69–90

Saddiqe Z, Naeem I, Maimoona A (2010) A review of the antibacterial activity of *Hypericum perforatum* L. J Ethnopharmacol 131:511–521

Schempp CM, Kirkin V, Simon-Haarhaus B, Kersten A, Kiss J, Termeer CC, Glib B, Kaufmann T, Borner C, Sleeman JP, Simon JC (2002a) Inhibition of tumour cell growth by hyperforin, a novel anticancer drug from St. John's Wort that acts by induction of apoptosis. Oncogene 21(8):1242–1250

Schempp CM, Simon-Haarhaus B, Simon JC (2002b) Phototoxic and apoptosis-inducing capacity of pseudohypericin. Planta Med 68(2):171–173

Schempp CM, Windeck T, Hezel S, Simon JC (2003) Topical treatment of atopic dermatitis with St. John's Wort cream—a randomized, placebo controlled, double blind half-side comparison. Phytomed Suppl 4:31–37

Schmitt LA, Liu Y, Murphy PA, Birt DF (2006) Evaluation of the light-sensitive cytotoxicity of *Hypericum perforatum* extracts, fractions, and pure compounds. J Agric Food Chem 54: 2681–2890

Schroeder C, Tank J, Goldstein DS, Stoeter M, Haertter S, Luft FC, Jordan J (2004) Influence of St. John's Wort on catecholamine turnover and cardiovascular regulation in humans. Clin Pharmacol Ther 76(5):480–489

Schuster SC (2008) Next-generation sequencing transforms today's biology. Nat Methods 5(1):16–18

Tachjian A, Maria V, Jahangir A (2010) Use of herbal products and potential interactions in patients with cardiovascular diseases. J Am Coll Cordiol 55(6):515–525

Tatsis EC, Boeren S, Exarchou V, Troganis AN, Vervoort J, Gerothanassis IP (2007) Identification of the major constituents of *Hypericum perforatum* by LC/SPE/NMR and/or LC/MS. Phytochemistry 68:383–393

Trapnell C, Pachter L, Salzberg SL (2009) TopHat: discovering splice junctions with RNA-Seq. Bioinformatics 25(9):1105–1111

Trapnell C, Williams BA, Pertea G, Mortazavi A, Kwan G, Van Baren MJ, Salzberg SL, Wold BJ, Pachter L (2010) Transcript assembly and quantification by RNA-Seq reveals unannotated transcripts and isoform switching during cell differentiation. Nat Biotechnol 28:511–515

Traynor NJ, Beattie PE, Ibbotson SH, Moseley H, Ferguson J, Woods JA (2005) Photogenotoxicity of hypericin in HaCaT keratinocytes: implications for St. John's Wort supplements and high dose UVA-1 therapy. Toxicol Lett 158:220–224

Vanderbogaerde AL, Kamuhabwa A, Delaey E, Himpens BE, Merlevede WJ, DeWitte PA (1998) Photocytotoxic effect of pseudohypericin versus hypericin. J Photochem Photobiol B Biol 45(2–3):87–94

Yang C, Chen M, Zeng L, Zhang L, Liu X, Lan X, Tang K, Liao Z (2011) Improvement of tropane alkaloids production in hairy root cultures of *Atropa belladonna* by overexpressing pmt and h6h genes. Plant Omics 4(1):29–33

Zofou D, Kowa TK, Wabo HK, Ngemenya MN, Tane P, Titanji VPK (2011) *Hypericum lanceolatum* (Hypericaceae) as a potential source of new anti-malarial agents: a bioassay-guided fractionation of the stem bark. Malar J 10:167

Chapter 18
Multivariate Analysis of Analytical Chemistry Data and Utility of the KNApSAcK Family Database to Understand Metabolic Diversity in Medicinal Plants

Taketo Okada, Farit Mochamad Afendi, Akira Katoh, Aki Hirai and Shigehiko Kanaya

18.1 Introduction

Plants produce a vast number of metabolites; the estimated number of plant metabolites ranges between 200,000 and 1,000,000 (Pichersky and Gang 2000; Saito and Matsuda 2010), which is more than 25 times the number of human metabolites. [Currently the estimate for human metabolites is around 7,900 according to human metabolome database, HMDB (http://www.hmdb.ca/) (Wishart et al. 2009).] Plant metabolites are produced in primary and secondary metabolisms. Although primary metabolism is essential for the life of the plants themselves and found in almost all plant species (e.g., photosynthesis and respiration), metabolite diversity in plants is generated by secondary metabolism. Secondary metabolites are necessary for self-protection in nature and adaptation to

T. Okada (✉)
Faculty of Pharmaceutical Sciences at Kagawa Campus, Tokushima Bunri University, Shido 1314-1, Sanuki-City 769-2193, Kagawa, Japan
e-mail: okadat@kph.bunri-u.ac.jp

F. Mochamad Afendi · A. Hirai · S. Kanaya
Graduate School of Information Science, Nara Institute of Science and Technology, Takayama 8916-5, Ikoma, Nara, Japan
e-mail: ampuun@yahoo.com

A. Hirai
e-mail: hirai@gtc.naist.jp

S. Kanaya
e-mail: skanaya@gtc.naist.jp

A. Katoh
Graduate School of Biological Sciences, Nara Institute of Science and Technology, Takayama 8916-5, Ikoma, Nara, Japan
e-mail: akira-katoh@bs.naist.jp

S. Chandra et al. (eds.), *Biotechnology for Medicinal Plants*,
DOI: 10.1007/978-3-642-29974-2_18,

the environment of each plant and are produced by plant species-specific biosynthetic pathways (e.g., phenylpropanoid pathway and mevalonate pathway). For humans and animals, many secondary metabolites isolated from plants have been used to treat disease and to maintain and improve health. In fact, many modern medicines, including chemical and synthetic drugs, have been manufactured and created from secondary metabolites. Moreover, the pharmacological effects of herbal medicines, including crude drugs and traditional medicines [e.g., traditional Chinese medicine (TCM), Kampo medicines in Japan, and Jamu medicines in Indonesia], depend on both major pharmacological constituents and other minor constituents to promote biological actions. In addition, major pharmacological constituents in herbal medicines may be minor constituents in the plant. Therefore, comprehensive and global studies to chemically analyze metabolites in plants, to determine the role in each metabolite in the metabolic pathways of plants, and to elucidate the biological association of metabolites with pharmacological effects of herbal medicines are essential to comprehend plant function and properties.

For comprehensive and simultaneous analysis of metabolites in plants, metabolomics has been applied to evaluate and discriminate analyzed samples. Metabolomics aims to comprehensively and globally analyze the metabolome, which is defined as all metabolites produced in a cell or organism. The innovative methodologies and techniques associated with metabolomics have advanced metabolism research on plants, humans, animals, and bacteria. Metabolomics typically employs two principal analyses: analytical chemistry and numerical computational analyses of multivariate analysis and bioinformatics. These techniques are necessary to process and arrange massive amounts of chemical analysis data (Fig. 18.1).

In metabolomic studies, chemical analysis is used to profile metabolites contained in the sample via its comprehensive, simultaneous, high throughput, and accurate metabolite analysis (Fig. 18.1). Chromatographic separation analyses, such as liquid chromatography (LC) and gas chromatography (GC), are often used in conjunction with highly sensitive and accurate detection methods such as mass spectrometry (MS). Photodetectors such as a UV detector and a photodiode array (PDA) detector have also been used for metabolite detection in LC analysis. In addition, capillary electrophoresis (CE)/MS has been used to analyze charged, neutral, polar, and hydrophobic compounds in metabolome analysis. Furthermore, nuclear magnetic resonance spectroscopy (NMR) analysis and Fourier transform infrared (FT-IR) spectroscopy have been frequently employed in metabolomic studies as these approaches can efficiently acquire chemical structure data. Then these raw metabolite fingerprint data are subjected to multivariate analysis.

Multivariate analysis is a computational method to statistically and mathematically process massive quantities of analytical chemistry data (Fig. 18.1). Several methods are used to reduce analytical chemistry data: principal component analysis (PCA), hierarchical cluster analysis (HCA), discrimination map analysis [e.g., batch-learning self-organizing map (BL-SOM) analysis], discriminant analysis [e.g., partial least-squares projection to latent structures-discriminant

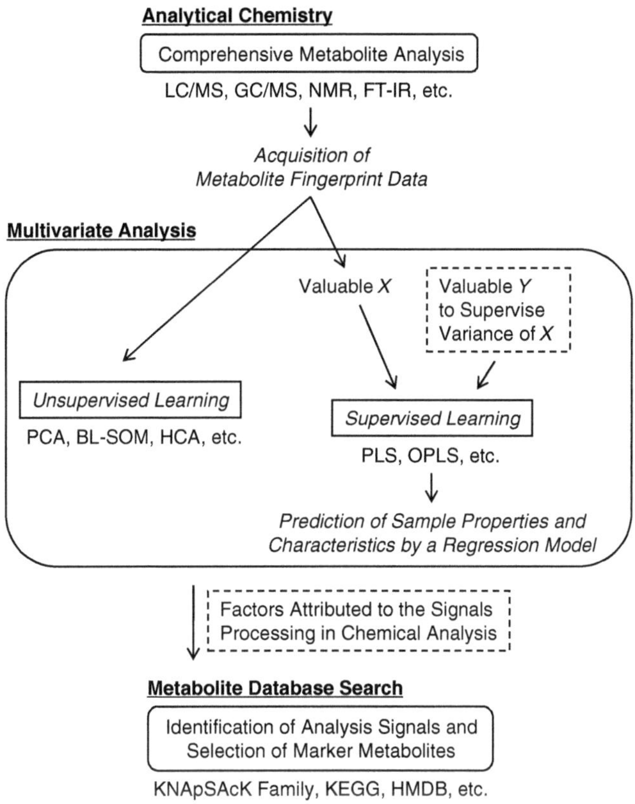

Fig. 18.1 Typical scheme of metabolome analysis via a combination of analytical chemistry, multivariate analysis, and a metabolite database search

analysis (PLS-DA) and orthogonal projections to latent structures-discriminant analysis (OPLS DA)], as well as prediction analysis by a regression model constructed from PLS and OPLS algorithms. In metabolome analysis, PCA, HCA, and SOM analyses are generally demonstrated as unsupervised learning-multivariate analyses. In unsupervised learning, the variables processed by multivariate analysis are one data set acquired by chemical analysis or is composed from knowledge about the properties and characteristics of the tested samples. On the other hand, multivariate analyses based on PLS and OPLS models are supervised learning. In supervised learning, the data set consists of two independent variables, X and Y, where the variance of X is supervised by Y [e.g., the result of LC/MS analysis (X) and physiological property of tested sample (Y)].

The numerical data points are plotted in high-dimensional graphs, which are usually score plots, loading plots, dendrograms, discrimination maps, and linear regression models. Score and loading plots are generated by PCA and discriminant analyses. The numerical data of the analyzed samples are represented as vectors on the score plot, and these vectors can be regarded as the reduced form of the metabolite fingerprint of each sample tested. On the other hand, the loading plot shows the contribution of each signal on the score plot. Therefore, variations in metabolite fingerprints are visualized on score plots, and the marker signals attributed to the biomarkers, chemical markers, and marker metabolites are selected from the loading plots. The dendrogram is used to display the HCA results, and the distance among tested samples on the dendrogram corresponds to the variance among the metabolite fingerprints of the tested samples. Consequently, this approach effectively compares and visualizes metabolite fingerprints as well as score plots. Discrimination map analysis can also categorize and visualize the variance of the tested samples. In BL-SOM analysis, a PCA algorithm is incorporated to process raw chemical analysis data and the results are outputted on a colored feature map. Accordingly, the metabolite fingerprints of the analyzed samples can be visually compared, and the signals contributing to the variance among samples can be selected from a map. A linear regression model is constructed by supervised learning-multivariate analysis such as PLS and OPLS algorithms. This model is generally employed to predict sample properties and characteristics. Finally, the metabolite fingerprint data reduced by multivariate analysis is available to compare the properties and characteristics of the analyzed samples and to select marker metabolites, including biomarkers and chemical markers.

After multivariate analysis reduces and arranges the data, numerical bioinformatics analysis is applied to simultaneously identify and assign the analytical chemistry data to known metabolites and chemical structures by searching metabolite databases (DBs) (Fig. 18.1). Metabolite DBs such as Kyoto encyclopedia of genes and genomes (KEGG) (http://www.kegg.jp/) (Kanehisa et al. 2010, 2006; Kanehisa and Goto 2000) and HMDB have been updated as research progresses. Because metabolite DBs contain vast quantities of data, candidate metabolites are selected from the DBs based on the signal processing in chemical analysis signals obtained from comprehensive metabolite analyses such as *m/z* values of MS analysis and chemical shifts of NMR analysis. Herein, we describe the KNApSAcK Family DB (http://kanaya.naist.jp/KNApSAcK_Family/) (Okada et al. 2010; Shinbo et al. 2006; Takahashi et al. 2011), which is constructed by our group and is applicable to analyze medicinal plants.

18.2 Analytical Chemistry to Acquire Comprehensive Metabolite Analysis Data

Metabolites are biotic compounds, including salts, sugars, acids, bases, lipids, hormonal steroids, as well as other small and large molecules. The diversity of their chemical nature prevents a comprehensive understanding of all the metabolites in a biological sample. Applying high-throughput detectors such as MS in combination with high-throughput separation instruments like GC or LC instruments and employing multivariate data analysis to process the sheer volume of data may realize a comprehensive understanding of biotic compounds. GC/MS, LC/MS, NMR, and FT-IR are four of the most common analytical platforms to acquire high-throughout data in metabolome analysis of medicinal plants.

18.2.1 MS-Based Metabolome Analysis

Mass spectrometry is a sensitive analytical technique that measures the mass-to-charge ratio (*m/z*) of charged compounds via a spectrometer, which consists of three basic components: an ion source, mass analyzer, and detector. In the ion source, molecules induced from the separation instrument are ionized. In comprehensive metabolite analysis, an electron ionization (EI) system (in GC/MS) and electrospray ionization (ESI) system (in LC/MS), are common. In the EI system, gas-phase molecules from a GC instrument are ionized by interacting with an electron beam from the filament to the receptor (Chapman 1996; Vestal 2001). A sample solution pumped into the LC instrument is passed through a narrow-bore metal capillary needle holding electro-voltage and is sprayed into a warm drying gas chamber. The liquid droplets formed by the spray become charged and evaporate. Consequently, molecules in the droplets are ionized during the ESI process (Chapman 1996; Smith et al. 1994; Tomer 2001; Vestal 2001). Then the *m/z* of charged molecules are measured using a mass analyzer. The quadrupole (Q)-MS, time-of-flight (TOF), triple quadrupole (QQQ), ion trap-MS, and their combined system are popular in metabolomics research.

18.2.1.1 LC/MS

Liquid chromatography/mass spectrometry has been used to analyze a wide range of metabolite classes containing polar thermolabile molecules without a chemically derived volatilization process. Recently, an LC system using reversed-phase columns combined with MS equipments has emerged where molecules are separated by their affinities for the stationary and mobile phases in the column. The mobile phase typically has a set polarity. A molecule more attracted to the mobile phase will elute quickly, i.e., have an earlier retention time. High performance liquid chromatography (HPLC), ultra performance liquid chromatography (UPLC), and ultra high performance liquid chromatography (UHPLC) are commonly applied to LC/MS systems. The liquid phase is generally pumped up to 500 psi to strengthen the column in the HPLC system, and is up to 15,000 psi in UHPLC and UPLC systems. The high pressure induced by the small particle size of the stationary phase in a packed column gives a stable and rapid analysis performance.

18.2.1.2 GC/MS

Gas chromatography/mass spectrometry platforms combine MS and GC separation instruments. The GC instrument performs high resolution molecular separation using fused silica capillary columns. The sample mixture is separated based on the distribution of the analytes between a gaseous mobile phase and a liquid or solid stationary phase in the column. The separation results (retention times) depend on the column. The separation results are detected by MS equipment at the end of the column. In some cases, molecules are volatilized by chemical derivatization. The most common derivatization procedures are methoxy groups and trimethylsilyl (TMS) groups as well as molecules without polarity using methyloxylamine hydrochloride and *N*-methyl-*N*-(trimethylsilyl)trifluoroacetamide (MSTFA).

18.2.1.3 Metabolic Analysis of Medical Plants using GC/MS and LC/MS

Gas chromatography/mass spectrometry and LC/MS systems are advantageous for research requiring comprehensive, sensitive, qualitative, and quantitative molecular analysis. Herbal medicines contain diverse molecules and their contents provide competitive and synergistic effects on human physiology. Metabolic analysis using GC/MS and LC/MS is one approach to investigate the complicated effects of medicinal plants. Here, we focus on recent chemical analysis of medicinal plants.

Comprehensive analysis or concurrent multi-analysis for plural target molecules has been employed to assess the quality of medicinal plants and herbal medicines. Qualitative and quantitative analytical methods of bioactive flavonoids in floating aquatic plants have been established using LC/MS (Qiao et al. 2011). Polar molecules, effective chlorogenic acids, triterpene saponins in Asteraceae plants (Jaiswal et al. 2011; Qiao et al. 2010), prenylated flavonoids in Berberidaceae plant (Chen et al. 2011), flavonoids in Rosaceae plant (Wang et al. 2011), polar metabolite in *Angelica* roots (Tianniam et al. 2009) and green tea (Pongsuwan et al. 2008), polar metabolites or alkaloids in a hot water extract in TCM (Tilton et al. 2010; Yan et al. 2010) have been profiled using LC/MS. Additionally, GC/MS profiling of polar and nonpolar metabolites such as whole polar metabolites in *Angelica* roots (Tianniam et al. 2008), and essential oils in *Curcuma* plants (Xiang et al. 2011) have been reported. Furthermore, LC/MS and GC/MS are advantageous for pharmacokinetic studies based on chemical analysis of plasma samples due to their sensitivity and ability to simultaneously analyze plural targets (e.g., Liu et al. 2010; Tong et al. 2010; Zhu et al. 2010).

18.2.2 NMR Spectroscopy

NMR spectroscopy is used to analyze chemical structures. Typically, ^{1}H and ^{13}C NMR spectra are used to assign atoms and functional groups in a compound.

This technique can efficiently analyze metabolite fingerprints based on chemical structures of compounds and has been widely employed for metabolomic studies of various biosamples, including medicinal plants (Kim et al. 2011; Okada et al. 2010; van der Kooy et al. 2009; Weljie et al. 2006; Zhang et al. 2010). Additionally, NMR provides spin–spin coupling constants, which are used to assign chemical structures and yields additional information about the chemical shifts. NMR is highly reproducible because each functional group appears within a specific region of the spectrum. To acquire metabolomic data, NMR instruments over 400 MHz are generally employed.

Multivariate analysis is generally applied to analyze the values of chemical shifts and their corresponding spectral intensities (peak integrals) because these values are variables in the chemical fingerprint of the total metabolite. In one-dimensional (1D) ^{1}H NMR analysis, chemical shifts and spectral intensities are shown in a two-dimensional graph, and this method is a standard approach to acquire metabolite fingerprint data. Multivariate analysis often excludes spectral data from an internal standard such as tetradeutero-3-(trimethylsilyl)propionic acid (TSP) and water. ^{13}C NMR analysis is also an effective approach to acquire metabolome data, but its sensitivity is much lower than ^{1}H NMR analysis due to the low natural abundance of ^{13}C to ^{12}C (1.1%) and ^{13}C is a weaker gyromagnet than ^{1}H. On the other hand, two-dimensional (2D) NMR analysis is particularly effective, especially when overlapping and complex spectra are observed in 1D-NMR analysis. 2D-NMR spectroscopy has two approaches: ^{1}H homonuclear correlation spectral analysis [e.g., correlation spectroscopy (COSY) and total correlation spectroscopy (TOCSY)] and ^{1}H-^{13}C heteronuclear correlation spectral analysis [e.g., heteronuclear multiple-bond correlation (HMBC) and heteronuclear multiple quantum coherence (HMQC)]. These results are shown in three-dimensional space. Therefore, the chemical shifts of ^{1}H and ^{13}C, and the spectral intensities applied to multivariate analysis should be acquired by efficient methods capable of analyzing each sample.

NMR signals are assigned to the chemical structure of metabolites, and multivariate analysis can identify the signals selected from metabolite fingerprints. Marker metabolites, including biomarkers and chemical markers, are selected from specific NMR signals that contribute to the variance among tested samples. In this process, the loading plot analysis of PCA and discrimination analysis are generally used to select marker signals. In NMR analysis of metabolomics, a tested sample is analyzed as a metabolite mixture, but chemical structural identification of one compound is often performed using a pure chemical fraction. To elucidate this chemical complexity of a tested sample, statistical total correlation spectroscopy (STOCSY) analysis is useful to identify metabolites in NMR analysis (Cloarec et al. 2005). STOCSY is a statistical analysis based on the correlation of the changes in the spectral intensity across different samples and generates pseudo-2D-NMR spectra. Therefore, this approach can effectively identify marker metabolites in complex 1D-NMR spectra.

18.2.3 FT-IR Spectroscopy

Fourier transform infrared spectroscopies, including near-infrared (NIR), mid-infrared (MIR), and far-infrared (FIR), have been used to determine elements as well as bonding states of atoms. The infrared wavelength is converted into a wavenumber by a Fourier transform. Accordingly, a wavenumber and the corresponding absorption and transmittance are applied to multivariate analysis. Because this chemical analysis can be non-destructively performed on solids, liquids, and gases, it has been employed in metabolomic studies of medicinal plants to acquire chemical fingerprints of metabolome (Okada et al. 2010). Moreover, other FT-IR-based metabolomic studies have been demonstrated, including chemotyping of plant cell wall (Gorzsás et al. 2011), plant–pathogen interaction (Allwood et al. 2010), silage fermentation (Johnson et al. 2004), analysis of chiral compounds in microorganisms (Wharfe et al. 2010), phenotypic characterization of bacterium under different growth conditions (Wang et al. 2010), monitoring of metabolites during a biotransformation (Winder et al. 2011), and phenotypic response of viral cells exposed to a drug (Kim et al. 2010).

18.3 Multivariate Analysis of Analytical Chemistry Data

To process massive amounts of metabolome data obtained from chemical analysis, raw chemical analysis data are arranged and visualized using multivariate analysis. This process is almost equal to the reduction of metabolite fingerprints. The data used in multivariate analysis can be generally represented by x_{ij}, where x is the variable representing the quantity of the jth element for the ith object. For example, x_{ij} represents the quantity of the jth metabolite for the ith sample. Accordingly, the data set for multivariate analysis can be represented by a matrix such as X as shown in Eq. (18.1)

$$\mathbf{X} = \begin{pmatrix} x_{11} & x_{12} & \cdots & x_{1j} \\ x_{21} & x_{22} & \cdots & x_{2j} \\ \cdots & \cdots & \cdots & \cdots \\ x_{i1} & x_{i2} & \cdots & x_{ij} \end{pmatrix} \tag{18.1}$$

18.3.1 Unsupervised Learning

Unsupervised learning-multivariate analysis such as PCA, HCA, and BL-SOM employed in metabolomic studies is applicable to most chemical analysis data, except data to supervise raw chemical data such as pharmacological activities of metabolites and geographical origins of tested plants. Unsupervised learning-multivariate analysis has been applied to *m/z* values and mass spectral intensities from LC/MS and GC/MS analyses, peak intensities and retention times of chromatograms, and chemical shifts and spectral intensities. The data matrix is generally represented by X in Eq. (18.1), and the objects on the basis of similarity of j variables are classified. Therefore, the result from this approach depends mostly on the chemical fingerprints of tested samples.

18.3.1.1 PCA

Principal component analysis is an unsupervised learning-multivariate analysis to simply summarize the variance among multiple data points (Jackson 1991). This approach is recommended as a starting point and can rapidly provide an overview of the information hidden in the multivariate data as well as detect outliers. In recent medicinal plant studies, PCA is the most common chemometric tool to understand chemical data structures (Okada et al. 2010) (Table 18.1). In PCA, the original variables are linearly transformed into axes with the largest variance, called principal components (PCs). A score plot visualizes the similarity of individual metabolome data, and it can be interpreted by factor loadings from the loading plot. In general, a PCA projection is effective for a data set consisting of a small number of objects (less than 100 samples) but numerous variables. PCA is widely employed for quality control, and discrimination of geographic origin, plant species, cultivars, and crude drugs, metabolic fingerprints, and physiological properties of medicinal plants (Table 18.1). A score plot is especially effective in the visualization of variance of each metabolite fingerprint, and represents the reduced form of comprehensive metabolite analysis data. On the other hand, a loading plot shows the contribution of a factor attributed to the signal processing in chemical analysis to the variance of each tested sample on a score plot. Consequently, marker metabolite candidates can be selected via a loading plot.

18.3.1.2 HCA

Hierarchical cluster analysis represents the similarities and differences among tested samples by a dendrogram (Everitt et al. 2011; Ward 1963). This approach effectively clusters large data sets based on their variables. Similarities and differences among the samples, which are calculated based on chemical analysis values, are denoted by the distance on the dendrogram. This approach is performed as an unsupervised method in metabolome analysis of medicinal plants (Okada et al. 2010). Although HCA can be employed in metabolome analysis of medicinal plants alone [e.g., quality control of TCM, *Houttuynia cordata* injection (Lu et al. 2006)], this analysis is often employed alongside the PCA analysis. Because both PCA and HCA are generally applied to the same data set, clustering in an HCA dendrogram effectively visualizes the relationships among the tested samples with the variance suggested by PCA. For example, HCA and PCA are coupled in

Table 18.1 Examples of PCA-based metabolome analysis of medicinal plants

Purpose	Samples	Analysis methods	References
Quality control	Astragali radix	LC/MS	Tanaka et al. (2008c)
	Dihuang		Li et al. (2009)
	Raw/steamed *Panax notoginseng*		Chan et al. (2007b)
	Citrus spp.	HPLC	Yi et al. (2007)
	Ginkgo biloba		Liang et al. (2004
	Uncaria sinensis	HPLC and GC/MS	Tan et al. (2011)
	Angelica acutiloba	GC/MS	Tianniam et al. (2008)
	Chamomile Flowers	NMR	Wang et al. (2004)
	Cortex Phellodendri	FT-IR	Chan et al. (2007a)
Quality control and discrimination of geographic origins	*Atracylodes chinensis*	HPLC	Ni et al. (2008)
Quality control and discrimination of species	*Curcuma* spp.	GC/MS	Xiang et al. (2011)
Discrimination of geographic origins	*Panax* spp.	LC/MS	Xie et al. (2008)
	Podophyllum peltatum	LC/UV/MS	Avula et al. (2011)
	Scrophularia ningpoensis	HPLC and LC/MS	Yang et al. (2011)
	Origanum vulgare	GC and GC/MS	Verma et al. (2010)
	Notopterygium incisum	GC×GC/MS and FID (Flame Ionization Detector)	Qiu et al. (2007)
	Galphimia glauca	NMR	Cardoso-Taketa et al. (2008)
	Narcissus Bulbs		Lubbe et al. (2009)

(continued)

Table 18.1 (continued)

Purpose	Samples	Analysis methods	References
Discrimination of species	*Ephedra* spp.	LC/MS	Okada et al. (2009)
	Taxus spp.	HPLC and LC/MS	Ge et al. (2008)
	Bauhinia spp.	HPLC	Soares & Scarminio (2008)
	Papaver spp.	GC/MS	Choe et al. (2011)
	Echinacea spp.	NMR	Frédérich et al. (2010)
	Ephedra spp.		Kim et al. (2005)
	Epilobium and *Hypericum* spp.	FT-IR	Kokalj et al. (2011)
	Achyrocline satureioides and *A. venosa*	ICP-OES (Inductively coupled plasma optical emission spectrometry)	Cantarelli et al. (2010)
Discrimination of cultivars	*Piper betle*	DART (Direct analysis in real time) MS	Bajpai et al. (2010)
	Cannabis sativa	NMR	Wang et al. (2005)
			Choi et al. (2004)
Discrimination of crude drugs	*Citrus* spp.	HPLC	Qin et al. (2009)
Metabolic fingerprinting	*Panax notoginseng*	LC/MS	Dan et al. (2008)
	Cannabis sativa	GC	Fischedick et al. (2010)
	Rhodiola rosea	NMR	Ioset et al. (2011)
	Withania somnifera		Sidhu et al. (2011)
Positive intervention effects of Yin Chen Hao Tang	Urine of rats	LC/MS	Wang et al. (2008)
Monitoring of hydrocortisone-induced state		GC/MS	Chen et al. (2005)
Effects of ginkgo leaf extracts on rats		NMR	Wang et al. (2005)
Improvement of blood stasis/stagnant syndrome by Kampo medicine	Human plasma	Proteomics	Matsumoto et al. (2008)

studies on the discrimination of *Taxus* (Ge et al. 2008), *Bauhinia* (Soares and Scarminio 2008), *Achyrocline* (Cantarelli et al. 2010) and *Papaver* species (Choe et al. 2011), metabolic fingerprinting of *Cannabis* species (Fischedick et al. 2010), discrimination of geographic origin of *Podophyllum* species (Avula et al. 2011), and the diagnosis of '*Oketsu*' (blood stasis) in Kampo medicine (Matsumoto et al. 2008), which is Japanese traditional medicine based on TCM. Interestingly, in the studies of Ge et al. (2008), PCA is employed to reduce the data to effectively perform HCA.

18.3.1.3 SOM Analysis

SOM analysis is a neural-network algorithm that implements a characteristic nonlinear projection of high-dimensional input signals onto a low-dimensional array of weights (Kohonen 1990). This approach is useful not only to visualize metabolite fingerprints, but also to classify closely related data points. For example, SOM analysis has discriminated *Ephedra* plants based on metabolite analysis by FT-NIR spectroscopy (Fan et al. 2010) and predicted the cytotoxic activity of sesquiterpene lactones (Fernandes et al. 2008). BL-SOM analysis is an improved method of traditional SOM analysis because (1) the learning process of SOM analysis is independent of the order of input vectors and (2) representative model vectors are initialized based on PCA instead of using random vectors (Kanaya et al. 2001; Okada et al. 2010). In BL-SOM analysis, the data points of interest (e.g., the factors contributing to the variance among tested samples) are easily selected from the mapping results. Individual data points are classified into the closest representative vectors arranged in a two-dimensional lattice. Thus, the similarity of objects can be recognized if they are classified into closely related representative vectors. The representative vectors have identical dimensions as in the original data set. Thus, the features of the individual variables can assess the similarity of the original variables. In general, the resolving power to classify objects is much larger than that of PCA, and the similarity of a very large number of objects (1,000–100,000) can be visualized. In medicinal plant studies, BL-SOM analysis has been employed to visualize the metabolite fingerprints of *Ephedra* plants based on chemical analysis by UPLC/Q-TOF–MS (Okada et al. 2009).

18.3.2 Supervised Learning

Unlike unsupervised learning, supervised learning requests two variables X and Y. The data matrix for supervised learning generally consists of independent variables X_n ($n = 1, 2, \dots M$) and dependent variables Y, which is represented by Eq. (18.2)

$$\mathbf{y} = \begin{pmatrix} y_1 \\ y_2 \\ \cdots \\ y_N \end{pmatrix} \tag{18.2}$$

Typically the relation between independent variables X_n and dependent variable Y is represented by Eq. (18.3).

$$Y = f(X_1, X_2, \ldots, X_M) \tag{18.3}$$

Accordingly, the variance of variable X is supervised by Y.

18.3.2.1 PLS and OPLS Models

A *PLS* model shows linear relationships between variables X and Y (Wold et al. 2001). Generally in metabolomic studies, X represents the signal processing in chemical analysis obtained from comprehensive metabolite analysis (e.g., *m/z* values of MS analysis and chemical shifts of NMR analysis); Y is the metabolic and biological data, which is correlated with the property of the tested samples and metabolites contained in the samples (e.g., pharmacological effects and biological activities of tested samples or metabolites, and the geographic origin of plant species and cultivars). The OPLS model is an improvement of the PLS model (Trygg 2004; Trygg and Wold 2002). The OPLS model allows systematic variation from the input data set (X) not correlated to the response set (Y) to be removed. Therefore, the OPLS model reduces the complexity compared to the PLS model.

PLS and OPLS models are frequently employed in metabolomics studies and provides score and loading plots. Discriminant analyses based on these models (PLS-DA and OPLS-DA) are often employed in medicinal plant studies (Table 18.2). X is generally supervised by plant-specific properties such as geographic origin, while Y is the metabolite content targeted in each study. In a unique case, the sensory quality of crude drugs, which is determined by experts, is set as Y to supervise the data obtained by NMR analysis (Tarachiwin et al. 2008). To evaluate the physiological activity of medicinal plants, pharmacological and biological activities have been used as Y.

In addition to discriminant analyses by supervised learning, the regression model has been constructed to predict the properties and characteristics of the tested samples. The calibration line is generated based on the variables of X_{obs} and Y_{obs}, which are acquired from the analysis of known samples, and is generally called an "observed" plot. Using this calibration model, the variable Y_{pred} of an unknown sample is predicted by extrapolation of X_{pred}, which is acquired by chemical analysis of the unknown sample. For example, in medicinal plant studies, regression models to predict sample properties have been employed to evaluate *Angelica* plants (Tarachiwin et al. 2008) and the inhibitory activity of cyclooxygenase-2 (COX-2) of Curcuma Rhizome (Tanaka et al. 2008a), to quantify curcuminoids in Curcuma Rhizome (Tanaka et al. 2008b), and to discriminate

Table 18.2 Examples of PLS and OPLS models-based metabolome analysis of medicinal plants

Methods	Samples	Y	X	References
PLS	*Citrus* spp.	Geographic origin	HPLC	Yi et al. (2007)
	Curcuma spp.	Species	GC/MS	Xiang et al. (2011)
	Angelica acutiloba	Sensory quality	NMR	Tarachiwin et al. (2008)
	Galphimia glauca	Neuropharmacological effect		Cardoso-Taketa et al. (2008)
	Panax spp.	Geographic origin and sample discrimination		Lee et al. (2009)
	Cortex Phellodendri	Berberine and total alkaloid content	FT-IR	Chan et al. (2007a)
	Curcuma longa	Curcuminoid content		Tanaka et al. (2008b)
	Honghua oil	A-pinene, methyl salicylate and eugenol		Wu et al. (2008)
	Curcuma phaeocaulis	Inhibition of COX-2	Metabolic profile	Tanaka et al. (2008a)
	Radix *Puerariae lobatae*	Antioxidant activity		Chau et al. (2009)
	Zuo Gui Wan	Geographic origin		Liang et al. (2009)
OPLS	Dihuang	Raw/processed dihuang	LC/MS	Li et al. (2009)
	Scutellaria baicalensis, *Alisma orientale*, *Atractyloides japonica* and *Pueraria lobata*	Geographic origin	NMR	Kang et al. (2008)

Curcuma species (Xiang et al. 2011) by a PLS model and the geographic origin of Korean and Chinese herbal medicines by an OPLS model (Kang et al. 2008).

18.4 Medicinal Plant Databases Connected with Metabolites: KNApSAcK Family DB

One hurdle to understanding the effects of components in medicinal plants on human health is the integration of knowledge on the usage of medicinal plants and current information about omics, including metabolomics (Okada et al. 2010;

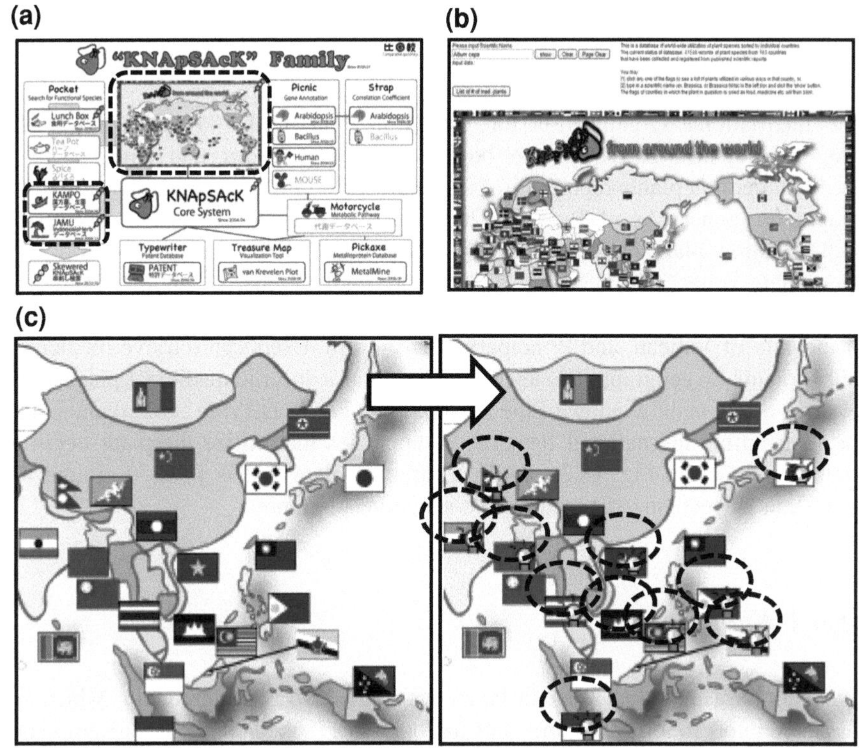

Fig. 18.2 KNApSAcK Family database. (**a**) Main window of KNApSAcK Family database. *Dotted lines* indicate icons that link databases of "KNApSAcK from around the world" (the world map at *top center*), traditional Japanese medicine "KAMPO", and traditional Indonesia medicine "JAMU". (**b**) Main window of "KNApSAcK from around the world". (**c**) Change in the windows of "KNApSAcK from around the world" by a database search. *Dotted lines* indicate the change by marking the national flags of the countries

Qiu 2007). The KNApSAcK Family DB (http://kanaya.naist.jp/KNApSAcK_Family/) allows metabolites from plant usage to be retrieved for multifaceted purposes (Okada et al. 2010; Shinbo et al. 2006; Takahashi et al. 2011). Figure 18.2a shows the main window of the KNApSAcK Family. The "Pocket" includes a search system for species and metabolites related to human life such as edible plants in Japan ("Lunch Box"), herbal teas ("Tea Pot", in progress), spices ("Spice", in progress), traditional Japanese medicine ("KAMPO"), and traditional Indonesian medicine ("JAMU"). "KNApSAcK from around the world" (at the top center denoted by the world map) includes medicinal and edible plants utilized by country. The dot circles in Fig. 18.2a indicate three DBs: KAMPO, JAMU, and KNApSAcK from around the world (KNApSAcK World).

18.4.1 KNApSAcK World DB "KNApSAcK from Around the World"

The International Organization for Standardization (ISO3166) has defined 251 geographic zones based on the borders between nations and small islands. Medicinal/edible plants reported in the scientific literature are classified into these geographic zones. As of June 22, 2011, KNApSAcK World DB comprises of 41,548 nation–plant pair entries with 15,240 medicinal/edible plants from 222 geographic zones, which covers 92% of the total number of nations excluding mini-states such as the Principality of Liechtenstein, Principality of Monaco, State of the City of Vatican, and Principality of Andorra. Users can retrieve medicinal/ edible plants by geographic zones. If a user inputs a specific medicinal plant in the text box of the main window of the KNApSAcK World DB (Fig. 18.2b) and clicks the List button, the national flags of the countries that utilize the plant begin to blink, as shown in Fig. 18.2c. Thus, a targeted plant that is utilized around the globe can be recognized based on the scientific literature.

18.4.2 Kampo DB

The National Health Insurance in Japan has approved 1,581 ingredients, which are classified into 336 formula names and are found in 278 medicinal plants. Kampo is a herbal medicine consisting of multiple raw herbs where the ratio of each herb in the Kampo formula is fixed. Because the KAMPO DB includes (i) groups of formulas for identical medicinal purposes and (ii) formula names and ingredients of plants, formula names from medicinal plants as well as plant ingredients can be retrieved.

18.4.3 Jamu DB

Almost every region in Indonesia has its own traditional way of formulating Jamu, which depends on the regional plant resources (Adnyana and Soemardji 2008; Sangat et al. 2000). To realize a health care system, one program of the Indonesian government has systematized the individual Jamu formulas. In addition to the traditional formulas, many Jamu are also produced commercially. These commercial Jamu must be registered at The National Agency for Drug and Food Control (NA-DFC) of Indonesia. The JAMU DB contains formulas from medicinal plants as well as medical plants and the formula names. There are over 7,000 Jamu formulas commercially registered at NA-DFC (http://www.pom.go.id/nonpublic/obat_tradisional/default.asp). Of these, the JAMU DB currently contains 5,310 formulas, which encompasses 550 medicinal plants and 12 morphological parts as well as formula names and plant ingredients.

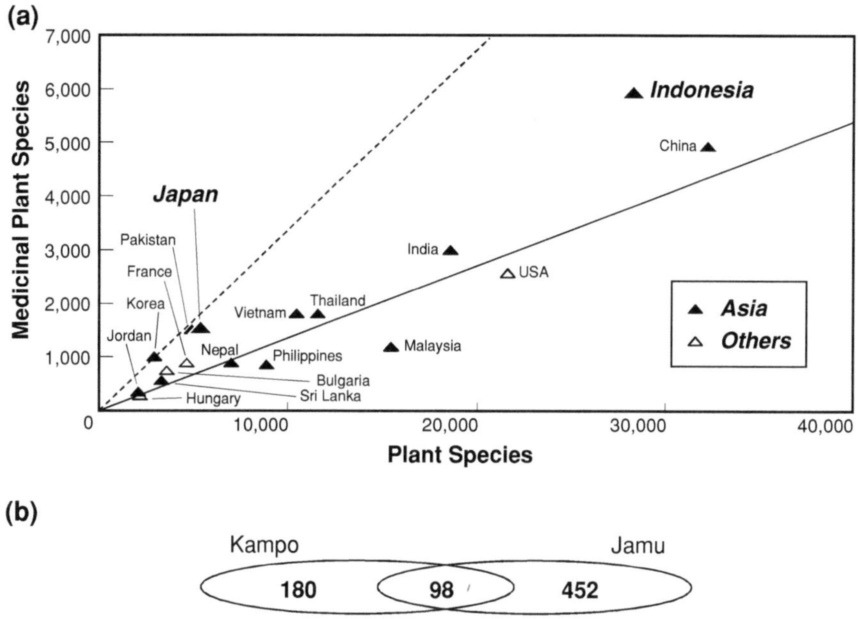

Fig. 18.3 Comparison of the amount of natural medicinals between Kampo and Jamu. (**a**) Amount of medicinal plant species and total plant species as well as their rate. *Solid line* represents the relationship between the number of medicinal plants and that of wild plants. *Dotted line* denotes the anticipated maximum ratio of that of medicinal plants to wild plants. (**b**) Overlap of medicinal plants used in Kampo and Jamu

18.4.4 Comparison of Kampo and Jamu Formula Based on Plant Usage

Indonesia is rich in natural resources. Although Indonesia's territory encompasses only 1.32% of the total world area, there are approximately 49 different types of ecosystems; it has 10% of plant species, 12% of mammals, 16% of reptiles and amphibians, 17% of birds, 25% of fish, and 15% of the insect species found on Earth. Thus, Indonesia is a country with "megadiverse" natural resources (Fig. 18.3a). The number of naturally occurring plant species in Japan is much smaller than that of Indonesia (5,000–6,000 vs. more than 30,000), but blended herbal medicines, Kampo and Jamu medicines, are very popular in their respective countries. Kampo and Jamu medicines are regulated by NA-DFC in Japan and by the National Agency of Drug and Food Control (NA-DFC) in the Republic of Indonesia, respectively.

To demonstrate the KNApSAcK Family DB, we compared the usage of medicinal plants between Kampo and Jamu. Figure 18.3b shows medicinal plants used in both Kampo and Jamu. A total of 35% of plants utilized in Kampo are also utilized in Jamu. However, only 17% of plants used in Jamu are also used in Kampo (Table 18.3). Moreover, the number of geographic zones and metabolites

Table 18.3 Plants utilized in both Kampo and Jamu medicines

The number of GZs	Plants	The number of metabolites reported
65	*Zingiber officinale*	105
52	*Foeniculum vulgare*	37
44	*Sesamum indicum*	19
38	*Cyperus rotundus*	81
36	*Curcuma longa*	19
33	*Citrus aurantium*	48
	Oryza sativa	129
32	*Hordeum vulgare*	103
26	*Coix lacryma-jobi*	0
21	*Syzygium aromaticum*	13
20	*Melia azedarach*	28
	Vitex trifolia	29
19	*Setaria italica*	12
18	*Areca catechu*	17
	Tribulus terrestris	51
17	*Tussilago farfara*	10
16	*Curcuma zedoaria*	80
	Mentha arvensis	8
	Morus alba	111
14	*Arctium lappa*	18
	Glycine max	175
13	*Nelumbo nucifera*	14
12	*Carthamus tinctorius*	33
	Cucumis melo	20
11	*Eugenia caryophyllata*	15
	Prunus persica	69
10	*Cinnamomum cassia*	24
	Eriobotrya japonica	66
	Gardenia jasminoides	43
	Terminalia chebula	6
9	*Allium tuberosum*	19
	Caesalpinia sappan	24
	Croton tiglium	6
	Pogostemon cablin	19
	Prunus armeniaca	12
	Rheum palmatum	5
8	*Allium fistulosum*	0
	Glycyrrhiza uralensis	172
7	*Amomum kravanh*	0
	Citrus reticulate	72
	Cnidium monnieri	16
	Rheum officinale	1

(continued)

Table 18.3 (continued)

The number of GZs	Plants	The number of metabolites reported
6	*Artemisia capillaris*	17
	Astragalus membranaceus	14
	Coptis chinensis	1
	Lonicera japonica	37
	Ophiopogon japonicus	27
	Platycodon grandiflorum	30
	Schisandra chinensis	27
	Thea sinensis	19
5	*Alpinia officinarum*	6
	Boswellia carteri	0
	Bupleurum falcatum	6
	Lycium barbarum	0
	Panax ginseng	125
	Perilla frutescens	44
	Saussurea lappa	41
	Schizonepeta tenuifolia	3
	Ziziphus jujuba	12
4	*Achyranthes bidentata*	3
	Angelica dahurica	15
	Atractylodes lancea	9
	Forsythia suspensa	27
	Gastrodia elata	18
	Ligusticum sinense	0
	Paeonia lactiflora	14
	Plantago asiatica	13
	Poria cocos	50
	Pueraria lobata	15
	Rehmannia glutinosa	26
	Rheum tanguticum	0
	Trichosanthes kirilowii	11
3	*Chaenomeles sinensis*	15
	Chrysanthemum indicum	24
	Chrysanthemum morifolium	1
	Eucommia ulmoides	70
	Gentiana macrophylla	3
	Juncus effusus	15
	Lithospermum erythrorhizon	20
	Lycium chinense	24
	Magnolia officinalis	28
	Paeonia suffruticosa	36
	Peucedanum praeruptorum	24
	Phellodendron chinense	8
	Polygonum multiflorum	16
	Sophora flavescens	168
	Uncaria gambir	22

(continued)

Table 18.3 (continued)

The number of GZs	Plants	The number of metabolites reported
2	*Akebia trifoliate*	2
	Anemarrhena asphodeloides	6
	Bambusa tuldoides	0
	Clematis chinensis	1
	Crataegus pinnatifida	10
	Lindera strychnifolia	40
	Notopterygium incisum	2
	Pinellia ternate	0
	Polyporus umbellatus	9
	Saposhnikovia divaricata	14
	Scrophularia ningpoensis	23

are reported in KNApSAcK World and KNApSAcK Core DB, respectively. A total of 87 plants, which correspond to 89% of plants utilized in both countries, are also utilized in other countries (the number of the geographic zones (GZs) > 2). Based on the number of GZs (>30), *Zingiber officinale, Foeniculum vulgare, Sesamum indicum, Cyperus rotundus, Curcuma longa, Citrus aurantium, Oryza sativa*, and *Hordeum vulgare* are the most popular. On the other hand, 11 plants, *Akebia trifoliate, Anemarrhena asphodeloides, Bambusa tuldoides, Clematis chinensis, Crataegus pinnatifida, Lindera strychnifolia, Notopterygium incisum, Pinellia ternate, Polyporus umbellatus, Saposhnikovia divaricata*, and *Scrophularia ningpoensis* are limited to Kampo and Jamu medicines. Metabolites reported in individual species, which are candidates for human efficacy, can be retrieved by the species–metabolite relation database, KNApSAcK Core, which contains 101,500 species–metabolite relationships involving 50,048 metabolites and 20,741 species (last update: March, 31, 2011). The number of metabolites reported in individual plants is also listed in the column denoted as '# of metabolites reported'. In conclusion, KNApSAcK Family enables global usage to be examined from various perspectives such as geographic zone and ingredients of medicinal plants. Because the data structure developed in the present study is based only on binary relations, extension of KNApSAcK Family DB by adding new attributes can be easily implemented.

18.5 Conclusion

To reduce and arrange the massive amount of the analytical chemistry data obtained from comprehensive metabolite analysis, multivariate analysis is essential for metabolomic studies. Additionally, metabolite databases are useful to process the metabolite fingerprints outputted by metabolome analysis, especially the KNApSAcK Family DB introduced in this chapter. The combination of these approaches and chemical analysis of metabolome demonstrates that metabolite profiling, comparison of metabolite fingerprints, and selection of markers can be simultaneously evaluated in a systematic and statistical fashion. Moreover, these approaches can be applied to multi-omics studies with other post-genomic research, such as transcriptomics and proteomics, to elucidate the regulation of gene-to-metabolite flow in metabolic pathways and the metabolic network in a cell and organism. Herein the utility, effectiveness, and potential of computational analysis in metabolomic studies are described.

Finally, the basal scheme described in this chapter, analytical chemistry to computational analyses, is a method aimed at computational systems biology. This approach should be applicable to metabolome analysis of other organisms and products, and their effects, leading to a comprehensive and global understanding of metabolism.

Acknowledgments This work was partially supported by the National Bioscience Database Center in Japan.

References

Adnyana IK, Soemardji AA (2008) Evaluation of pharmacological efficacy of Jamu medicine. University of Toyama, Institute of Natural Medicine, Toyama

Allwood JW, Clarke A, Goodacre R, Mur LA (2010) Dual metabolomics: a novel approach to understanding plant-pathogen interactions. Phytochemistry 71:590–597

Avula B, Wang YH, Moraes RM, Khan IA (2011) Rapid analysis of lignans from leaves of *Podophyllum peltatum* L. samples using UPLC-UV-MS. Biomed Chromatogr. doi:10.1002/bmc.1595

Bajpai V, Sharma D, Kumar B, Madhusudanan KP (2010) Profiling of *Piper betle* Linn. cultivars by direct analysis in real time mass spectrometric technique. Biomed Chromatogr 24:1283–1286

Cantarelli MA, Pellerano RG, Del Vitto LA, Marchevsky EJ, Camiña JM (2010) Characterisation of two South American food and medicinal plants by chemometric methods based on their multielemental composition. Phytochem Anal 21:550–555

Cardoso-Taketa AT, Pereda-Miranda R, Choi YH, Verpoorte R, Villarreal ML (2008) Metabolic profiling of the Mexican anxiolytic and sedative plant *Galphimia glauca* using nuclear magnetic resonance spectroscopy and multivariate data analysis. Planta Med 74:1295–1301

Chan CO, Chu CC, Mok DK, Chau FT (2007a) Analysis of berberine and total alkaloid content in *Cortex Phellodendri* by near infrared spectroscopy (NIRS) compared with high-performance liquid chromatography coupled with ultra-visible spectrometric detection. Anal Chim Acta 592:121–131

Chan EC, Yap SL, Lau AJ, Leow PC, Toh DF, Koh HL (2007b) Ultra-performance liquid chromatography/time-of-flight mass spectrometry based metabolomics of raw and steamed *Panax notoginseng*. Rapid Commun Mass Spectrom 21:519–528

Chapman JR (1996) Mass spectrometry: ionization methods and instrumentation. In: Chapman JR (ed) Protein and peptide analysis by mass spectrometry. Humana Press, Totowa

Chau FT, Chan HY, Cheung CY, Xu CJ, Liang Y, Kvalheim OM (2009) Recipe for uncovering the bioactive components in herbal medicine. Anal Chem 81:7217–7225

Chen Y, Wang J, Jia X, Tan X, Hu M (2011) Role of intestinal hydrolase in the absorption of prenylated flavonoids present in Yinyanghuo. Molecules 16:1336–1348

Chen M, Zhao L, Jia W (2005) Metabonomic study on the biochemical profiles of a hydrocortisone-induced animal model. J Proteome Res 4:2391–2396

Choe S, Kim S, Lee C, Yang W, Park Y, Choi H, Chung H, Lee D, Hwang BY (2011) Species identification of *Papaver* by metabolite profiling. Forensic Sci Int. doi:10.1016/j.forsciint.2011.04.015

Choi YH, Kim HK, Hazekamp A, Erkelens C, Lefeber AW, Verpoorte R (2004) Metabolomic differentiation of *Cannabis sativa* cultivars using ^{1}H NMR spectroscopy and principal component analysis. J Nat Prod 67:953–957

Cloarec O, Dumas ME, Craig A, Barton RH, Trygg J, Hudson J, Blancher C, Gauguier D, Lindon JC, Holmes E, Nicholson J (2005) Statistical total correlation spectroscopy: an exploratory approach for latent biomarker identification from metabolic 1H NMR data sets. Anal Chem 77:1282–1289

Dan M, Su M, Gao X, Zhao T, Zhao A, Xie G, Qiu Y, Zhou M, Liu Z, Jia W (2008) Metabolite profiling of *Panax notoginseng* using UPLC-ESI-MS. Phytochemistry 69:2237–2244

Everitt BS, Landau S, Leese M, Stahl D (2011) Cluster analysis, 5th edn. Wiley, New Jersey

Fan Q, Wang Y, Sun P, Liu S, Li Y (2010) Discrimination of *Ephedra* plants with diffuse reflectance FT-NIRS and multivariate analysis. Talanta 80:1245–1250

Fernandes MB, Scotti MT, Ferreira MJ, Emerenciano VP (2008) Use of self-organizing maps and molecular descriptors to predict the cytotoxic activity of sesquiterpene lactones. Eur J Med Chem 43:2197–2205

Fischedick JT, Hazekamp A, Erkelens T, Choi YH, Verpoorte R (2010) Metabolic fingerprinting of *Cannabis sativa* L., cannabinoids and terpenoids for chemotaxonomic and drug standardization purposes. Phytochemistry 71:2058–2073

Frédérich M, Jansen C, de Tullio P, Tits M, Demoulin V, Angenot L (2010) Metabolomic analysis of *Echinacea* spp. by ^{1}H nuclear magnetic resonance spectrometry and multivariate data analysis technique. Phytochem Anal 21:61–65

Ge GB, Zhang YY, Hao DC, Hu Y, Luan HW, Liu XB, He YQ, Wang ZT, Yang L (2008) Chemotaxonomic study of medicinal *Taxus* species with fingerprint and multivariate analysis. Planta Med 74:773–779

Gorzsás A, Stenlund H, Persson P, Trygg J, Sundberg B (2011) Cell-specific chemotyping and multivariate imaging by combined FT-IR microspectroscopy and orthogonal projections to latent structures (OPLS) analysis reveals the chemical landscape of secondary xylem. Plant J 66:903–914

Ioset KN, Nyberg NT, Van Diermen D, Malnoe P, Hostettmann K, Shikov AN, Jaroszewski JW (2011) Metabolic profiling of *Rhodiola rosea* rhizomes by ^{1}H NMR spectroscopy. Phytochem Anal 22:158–165

Jackson JE (1991) A user's guide to principal components. Wiley, New York

Jaiswal R, Deshpande S, Kuhnert N (2011) Profiling the chlorogenic acids of *Rudbeckia hirta, Helianthus tuberosus, Carlina acaulis, Symphyotrichum novae-angliae* leaves by LC-MSn. Phytochem Anal. doi:10.1002/pca.1299

Johnson HE, Broadhurst D, Kell DB, Theodorou MK, Merry RJ, Griffith GW (2004) High-throughput metabolic fingerprinting of legume silage fermentations via Fourier transform infrared spectroscopy and chemometrics. Appl Environ Microbiol 70:1583–1592

Kanaya S, Kinouchi M, Abe T, Kudo Y, Yamada Y, Nishi T, Mori H, Ikemura T (2001) Analysis of codon usage diversity of bacterial gene with a self-organizing map (SOM). Gene 276:89–99

Kanehisa M, Goto S (2000) KEGG: Kyoto encyclopedia of genes and genomes. Nucl Acids Res 28:27–30

Kanehisa M, Goto S, Furumichi M, Tanabe M, Hirakawa M (2010) KEGG for representation and analysis of molecular networks involving diseases and drugs. Nucl Acids Res 38:D355–D360

Kanehisa M, Goto S, Hattori M, Aoki-Kinoshita KF, Itoh M, Kawashima S, Katayama T, Araki M, Hirakawa M (2006) From genomics to chemical genomics: new developments in KEGG. Nucl Acids Res 34:D354–D357

Kang J, Choi MY, Kang S, Kwon HN, Wen H, Lee CH, Park M, Wiklund S, Kim HJ, Kwon SW, Park S (2008) Application of a ^{1}H nuclear magnetic resonance (NMR) metabolomics approach combined with orthogonal projections to latent structure-discriminant analysis as an efficient tool for discriminating between Korean and Chinese herbal medicines. J Agric Food Chem 56:11589–11595

Kim DH, Jarvis RM, Xu Y, Oliver AW, Allwood JW, Hampson L, Hampson IN, Goodacre R (2010) Combining metabolic fingerprinting and footprinting to understand the phenotypic response of HPV16 E6 expressing cervical carcinoma cells exposed to the HIV anti-viral drug lopinavir. Analyst 135:1235–1244

Kim HK, Choi YH, Erkelens C, Lefeber AW, Verpoorte R (2005) Metabolic fingerprinting of *Ephedra* species using ^{1}H-NMR spectroscopy and principal component analysis. Chem Pharm Bull 53:105–109

Kim HK, Choi YH, Verpoorte R (2011) NMR-based plant metabolomics: where do we stand, where do we go? Trends Biotechnol 29:267–275

Kohonen T (1990) Self-organizing map. Proc IEEE 78:1464–1480

Kokalj M, Kolar J, Trafela T, Kreft S (2011) Differences among *Epilobium* and *Hypericum* species revealed by four IR spectroscopy modes: transmission, KBr tablet, diffuse reflectance and ATR. Phytochem Anal. doi:10.1002/pca.1315

Lee EJ, Shaykhutdinov R, Weljie AM, Vogel HJ, Facchini PJ, Park SU, Kim YK, Yang TJ (2009) Quality assessment of ginseng by ^{1}H NMR metabolite fingerprinting and profiling analysis. J Agric Food Chem 57:7513–7522

Li P, Yang L, Gong Y (2009) Application of systems biology technology in research of traditional Chinese medicine. J Trad Chin Med 29:153–157

Liang Y, Xie P, Chan K (2004) Quality control of herbal medicines. J Chromatogr B 812:53–70

Liang X, Zhang X, Dai W, Lv Y, Yan S, Zhang W (2009) A combined HPLC-PDA and HPLC-MS method for quantitative and qualitative analysis of 10 major constituents in the traditional Chinese medicine Zuo Gui Wan. J Pharm Biomed Anal 49:931–936

Liu Y, Xu F, Zeng X, Yang L, Deng Y, Wu Z, Feng Y, Li X (2010) Application of a liquid chromatography/tandem mass spectrometry method to pharmacokinetic study of mangiferin in rats. J Chromatogr B Anal Technol Biomed Life Sci 878:3345–3350

Lu HM, Liang YZ, Wu XJ, Qiu P (2006) Tentative fingerprint-efficacy study of *Houttuynia cordata* injection in quality control of traditional Chinese medicine. Chem Pharm Bull 54:725–730

Lubbe A, Pomahacova P, Choi Y, Verpoorte R (2009) Analysis of metabolic variation and galanthamine content in *Narcissus* bulbs by ^{1}H NMR. Phytochem Anal 21:66–72

Matsumoto C, Kojima T, Ogawa K, Kamegai S, Oyama T, Shibagaki Y, Kawasaki T, Fujinaga H, Takahashi K, Hikiami H, Goto H, Kiga C, Koizumi K, Sakurai H, Muramoto H, Shimada Y, Yamamoto M, Terasawa K, Takeda S, Saiki I (2008) A proteomic approach for the diagnosis of '*Oketsu*' (blood stasis), a pathophysiologic concept of Japanese traditional (*Kampo*) medicine. Evid Based Complem Altern Med 5:463–474

Ni Y, Peng Y, Kokot S (2008) Fingerprinting of complex mixtures with the use of high performance liquid chromatography, inductively coupled plasma atomic emission spectroscopy and chemometrics. Anal Chim Acta 616:19–27

Okada T, Afendi FM, Altaf-Ul-Amin MD, Takahashi H, Nakamura K, Kanaya S (2010) Metabolomics of medicinal plants: the importance of multivariate analysis of analytical chemistry data. Curr Comput Aided Drug Des 6:179–196

Okada T, Nakamura Y, Kanaya S, Takano A, Malla KJ, Nakane T, Kitayama M, Sekita S (2009) Metabolome analysis of *Ephedra* plants with different contents of ephedrine alkaloids by using UPLC-Q-TOF-MS. Planta Med 75:1356–1362

Pichersky E, Gang DR (2000) Genetics and biochemistry of secondary metabolites in plants: an evolutionary perspective. Trends Plant Sci 5:439–445

Pongsuwan W, Bamba T, Harada K, Yonetani T, Kobayashi A, Fukusaki E (2008) High-throughput technique for comprehensive analysis of Japanese green tea quality assessment using ultra-performance liquid chromatography with time-of-flight mass spectrometry (UPLC/TOF MS). J Agric Food Chem 56:10705–10708

Qiao X, He WN, Xiang C, Han J, Wu LJ, Guo DA, Ye M (2011) Qualitative and quantitative analyses of flavonoids in *Spirodela polyrrhiza* by high-performance liquid chromatography coupled with mass spectrometry. Phytochem Anal 22:475–483

Qiao X, Zhang X, Ye M, Su YF, Dong J, Han J, Yin J, Guo DA (2010) Rapid characterization of triterpene saponins from *Conyza blinii* by liquid chromatography coupled with mass spectrometry. Rapid Commun Mass Spectrom 24:3340–3350

Qin XM, Dai YT, Zhang LZ, Guo XQ, Shao HX (2009) Discrimination of three medicinal materials from the *Citrus* genus by HPLC fingerprint coupled with two complementary software. Phytochem Anal 20:307–313

Qiu J (2007) Traditional medicine: a culture in the balance. Nature 448:126–128

Qiu Y, Lu X, Pang T, Zhu S, Kong H, Xu G (2007) Study of traditional Chinese medicine volatile oils from different geographical origins by comprehensive two-dimensional gas chromatography-time-of-flight mass spectrometry (GC×GC-TOFMS) in combination with multivariate analysis. J Pharm Biomed Anal 43:1721–1727

Saito K, Matsuda F (2010) Metabolomics for functional genomics, systems biology, and biotechnology. Annu Rev Plant Biol 61:463–489

Sangat HM, Zuhud EAM, Damayanti EK (2000) Kamus penyakit dan tumbuhan obat Indonesia: etnofitomedika. Yayasan Obor Indonesia, Jakarta

Sidhu OP, Annarao S, Chatterjee S, Tuli R, Roy R, Khetrapal CL (2011) Metabolic alterations of *Withania somnifera* (L.) Dunal fruits at different developmental stages by NMR spectroscopy. Phytochem Anal 22:495–502

Shinbo Y, Nakamura Y, Altaf-Ul-Amin MD, Asahi H, Kurokawa K, Arita M, Saito K, Ohta D, Shibata D, Kanaya S (2006) KNApSAcK: a comprehensive species-metabolite relationship database. In: Saito K, Dixon RA, Willmitzer L (eds) Plant metabolomics (Biotechnology in agriculture and forestry), vol 57. Springer-Verlag, Heidelberg

Smith RD, Light-Wahl KJ, Winger BE, Goodleftt DR (1994) Electronspray ionization. In: Matsuo T, Caprioli RM, Gross ML, Seyama Y (eds) Biological mass spectrometry: present and future. Wiley, New York, pp 41–74

Soares PK, Scarminio IS (2008) Multivariate chromatographic fingerprint preparation and authentication of plant material from the genus *Bauhinia*. Phytochem Anal 19:78–85

Takahashi H, Hirai A, Shojo M, Matsuda K, Parvin AK, Asahi H, Nakamura K, Altaf-Ul-Amin MD, Kanaya S (2011) Species-metabolite relation database KNApSAcK and its multifaceted retrieval system, KNApSAcK family. In: Casciano DA, Sahu SC (eds) Handbook of systems toxicology. Wiley, New York

Tan SN, Yong JW, Teo CC, Ge L, Chan YW, Hew CS (2011) Determination of metabolites in *Uncaria sinensis* by HPLC and GC-MS after green solvent microwave-assisted extraction. Talanta 83:891–898

Tanaka K, Kuba Y, Ina A, Watanabe H, Komatsu K (2008a) Prediction of cyclooxygenase inhibitory activity of curcuma rhizome from chromatograms by multivariate analysis. Chem Pharm Bull 56:936–940

Tanaka K, Kuba Y, Sasaki T, Hiwatashi F, Komatsu K (2008b) Quantitation of curcuminoids in curcuma rhizome by near-infrared spectroscopic analysis. J Agric Food Chem 56:8787–8792

Tanaka K, Tamura T, Fukuda S, Batkhuu J, Sanchir C, Komatsu K (2008c) Quality evaluation of Astragali Radix using a multivariate statistical approach. Phytochemistry 69:2081–2087

Tarachiwin L, Katoh A, Ute K, Fukusaki E (2008) Quality evaluation of *Angelica acutiloba* Kitagawa roots by ^{1}H NMR-based metabolic fingerprinting. J Pharm Biomed Anal 48:42–48

Tianniam S, Bamba T, Fukusaki E (2009) Non-targeted metabolite fingerprinting of oriental folk medicine *Angelica acutiloba* roots by ultra performance liquid chromatography time-of-flight mass spectrometry. J Sep Sci 32:2233–2244

Tianniam S, Tarachiwin L, Bamba T, Kobayashi A, Fukusaki E (2008) Metabolic profiling of *Angelica acutiloba* roots utilizing gas chromatography-time-of-flight-mass spectrometry for quality assessment based on cultivation area and cultivar via multivariate pattern recognition. J Biosci Bioeng 105:655–659

Tilton R, Paiva AA, Guan JQ, Marathe R, Jiang Z, van Eyndhoven W, Bjoraker J, Prusoff Z, Wang H, Liu SH, Cheng YC (2010) A comprehensive platform for quality control of botanical drugs (PhytomicsQC): a case study of *Huangqin Tang* (HQT) and PHY906. Chin Med 5:30

Tomer KB (2001) Separations combined with mass spectrometry. Chem Rev 101:297–328

Tong L, Wan M, Zhou D, Gao J, Zhu Y, Bi K (2010) LC-MS/MS determination and pharmacokinetic study of albiflorin and paeoniflorin in rat plasma after oral administration of *Radix Paeoniae Alba* extract and Tang–Min–Ling–Wan. Biomed Chromatogr 24: 1324–1331

Trygg J (2004) Prediction and spectral profile estimation in multivariate calibration. J Chemom 18:166–172

Trygg J, Wold S (2002) Orthogonal projections to latent structures (O-PLS). J Chemom 16:119–128

van der Kooy F, Maltese F, Choi YH, Kim HK, Verpoorte R (2009) Quality control of herbal material and phytopharmaceuticals with MS and NMR based metabolic fingerprinting. Planta Med 75:763–775

Verma RS, Padalia RC, Chauhan A, Verma RK, Yadav AK, Singh HP (2010) Chemical diversity in Indian oregano (*Origanum vulgare* L.). Chem Biodivers 7:2054–2064

Vestal ML (2001) Methods of ion generation. Chem Rev 101:361–375

Wang H, Hollywood K, Jarvis RM, Lloyd JR, Goodacre R (2010) Phenotypic characterization of *Shewanella oneidensis* MR-1 under aerobic and anaerobic growth conditions by using fourier transform infrared spectroscopy and high-performance liquid chromatography analyses. Appl Environ Microbiol 76:6266–6276

Wang M, Lamers RJ, Korthout HA, van Nesselrooij JH, Witkamp RF, van der Heijden R, Voshol PJ, Havekes LM, Verpoorte R, van der Greef J (2005) Metabolomics in the context of systems biology: bridging traditional Chinese medicine and molecular pharmacology. Phytother Res 19:173–182

Wang T, An Y, Zhao C, Han L, Boakye-Yiadom M, Wang W, Zhang Y (2011) Regulation effects of *Crataegus pinnatifida* leaf on glucose and lipids metabolism. J Agric Food Chem 59:4987–4994

Wang X, Lv H, Sun H, Liu L, Yang B, Sun W, Wang P, Zhou D, Zhao L, Dou S, Zhang G, Cao H (2008) Metabolic urinary profiling of alcohol hepatotoxicity and intervention effects of Yin Chen Hao Tang in rats using ultra-performance liquid chromatography/electrospray ionization quadruple time-of-flight mass spectrometry. J Pharm Biomed Anal 48:1161–1168

Wang Y, Tang H, Nicholson JK, Hylands PJ, Sampson J, Whitcombe I, Stewart CG, Caiger S, Oru I, Holmes E (2004) Metabolomic strategy for the classification and quality control of phytomedicine: a case study of chamomile flower (*Matricaria recutita* L.). Planta Med 70:250–255

Ward JH (1963) Hierarchical grouping to optimize an objective function. J Am Stat Assoc 58:236–244

Weljie AM, Newton J, Mercier P, Carlson E, Slupsky CM (2006) Targeted profiling: quantitative analysis of ^{1}H NMR metabolomics data. Anal Chem 78:4430–4442

Wharfe ES, Winder CL, Jarvis RM, Goodacre R (2010) Monitoring the effects of chiral pharmaceuticals on aquatic microorganisms by metabolic fingerprinting. Appl Environ Microbiol 76:2075–2085

Winder CL, Cornmell R, Schuler S, Jarvis RM, Stephens GM, Goodacre R (2011) Metabolic fingerprinting as a tool to monitor whole-cell biotransformations. Anal Bioanal Chem 399:387–401

Wishart DS, Knox C, Guo AC et al (2009) HMDB: a knowledgebase for the human metabolome. Nucl Acids Res 37(database issue):D603–D610

Wold S, Sjöström M, Eriksson L (2001) PLS-regression: a basic tool of chemometrics. Chemom Intell Lab Sys 58:109–130

Wu YW, Sun SQ, Zhou Q, Leung HW (2008) Fourier transform mid-infrared (MIR) and near-infrared (NIR) spectroscopy for rapid quality assessment of Chinese medicine preparation Honghua Oil. J Pharm Biomed Anal 46:498–504

Xiang Z, Wang XQ, Cai XJ, Zeng S (2011) Metabolomics study on quality control and discrimination of three *Curcuma* species based on gas chromatograph-mass spectrometry. Phytochem Anal 22:411–418

Xie G, Plumb R, Su M, Xu Z, Zhao A, Qiu M, Long X, Liu Z, Jia W (2008) Ultra-performance LC/TOF MS analysis of medicinal *Panax* herbs for metabolomic research. J Sep Sci 31: 1015–1026

Yan G, Sun H, Sun W, Zhao L, Meng X, Wang X (2010) Rapid and global detection and characterization of *aconitum* alkaloids in Yin Chen Si Ni Tang, a traditional Chinese medical formula, by ultra performance liquid chromatography-high resolution mass spectrometry and automated data analysis. J Pharm Biomed Anal 53:421–431

Yang S, Chen C, Zhao Y, Xi W, Zhou X, Chen B, Fu C (2011) Association between chemical and genetic variation of wild and cultivated populations of *Scrophularia ningpoensis* hemsl. Planta Med 77:865–871

Yi LZ, Yuan DL, Liang YZ, Xie PS, Zhao Y (2007) Quality control and discrimination of *Pericarpium Citri Reticulatae* and *Pericarpium Citri Reticulatae Viride* based on high-performance liquid chromatographic fingerprints and multivariate statistical analysis. Anal Chim Acta 588:207–215

Zhang S, Nagana Gowda GA, Ye T, Raftery D (2010) Advances in NMR-based biofluid analysis and metabolite profiling. Analyst 135:1490–1498

Zhu Z, Zhao L, Liu X, Chen J, Zhang H, Zhang G, Chai Y (2010) Comparative pharmacokinetics of baicalin and wogonoside by liquid chromatography-mass spectrometry after oral administration of *Xiaochaihu Tang* and *Radix scutellariae* extract to rats. J Chromatogr B Anal Technol Biomed Life Sci 878:2184–2190

Chapter 19
Genomic and Transcriptomic Profiling: Tools for the Quality Production of Plant-Based Medicines

Nikolaus J. Sucher, James R. Hennell and Maria C. Carles

19.1 Introduction

Humans and non-human primates have used medicinal plants for medication since pre-historic times (Halberstein 2005). To date, herbal medicines form the backbone of primary health care for 70–95 % of the population in the developing world. Over 70 % of the population in developed nations use some form of complementary/alternative medicines many of which are herbal medicines. The annual estimate of the global market for herbal medicines was US $83 billion and is expected to exponentially increase in the coming years (Robinson and Zhang 2011). The estimated world market for medicinal plants and their derivatives (raw material) was estimated at US $65 billion in 2003 with an expected annual growth rate of 7 % (Anonymous 2003). Medicinal plants have been and continue to be an important source for drug discovery (Balunas and Kinghorn 2005). At the same time, however, there is increasing evidence that the non-sustainable harvesting of medicinal plants from the wild, is threatening these irreplaceable resources (Cordell 2009; Canter et al. 2005; Chapman and Chomchalow 2005). It has been argued accordingly that cultivation and technology be used to turn medicinal

N. J. Sucher (✉)
Science, Technology, Engineering & Math, Roxbury Community College,
1234 Columbus Ave Roxbury Crossing, MA 02120, USA
e-mail: nsucher@rcc.mass.edu; nikolaussucher@me.com

J. R. Hennell
Centre for Complementary Medicine Research, University of Western Sydney,
Penrith, NSW 2751, Australia
e-mail: J.Hennell@uws.edu.au

M. C. Carles
Natural Sciences, Northern Essex Community College, 110 Elliot Street,
Building E, Room 367 Haverhill, MA 01830, USA
e-mail: mcarles@necc.mass.edu

S. Chandra et al. (eds.), *Biotechnology for Medicinal Plants*,
DOI: 10.1007/978-3-642-29974-2_19, © Springer-Verlag Berlin Heidelberg 2013

plants into sustainable commodities in order to meet future public health care needs (Cordell 2009; Canter et al. 2005; Chapman and Chomchalow 2005). A stable and controlled supply of herbal medicines depends on the availability of a high quality source of raw materials. Cultivation of medicinal plants used for the manufacture of traditional herbal medicines using good agricultural practices is considered to be a key element in the drive to modernize and internationalize traditional Chinese medicine (TCM) (Zhang et al. 2010) and has motivated the cultivation and breeding of selected Chinese medicinal plants in Germany, a major market for TCM (Heuberger et al. 2010). In another approach, several research groups have developed plant tissue culture and in vitro micropropagation techniques for the cultivation, rapid propagation, and metabolic engineering of medicinal plants (Zhou and Wu 2006; Saxena and Bopana 2009; Yang and Stockigt 2010; Chaudhuri et al. 2007; Lattoo et al. 2006; Ramesh et al. 2011).

Quality and efficacy of the herbal medicines are directly linked to the quality of the medicinal herb raw material. Medicinal herb species authentication traditionally consists in the matching of macroscopic and microscopic-type specimen descriptions with the starting herb material by an experienced taxonomist. Over the last two decades, genomic profiling or DNA fingerprinting has been developed for the identification of medicinal plants (Sucher and Carles 2008). A large number of molecular techniques have been used to authenticate plants based on species-specific variations in the sequences of various chloroplast and nuclear DNA regions (Sucher and Carles 2008). Polymerase chain reaction (PCR)-based methods only require minute amounts of DNA and can be used both from fresh and dried plant parts. Application of DNA fingerprinting to the authentication of medicinal plants has been the subject of two recent reviews (Heubl 2010; Sucher and Carles 2008). The use of DNA fingerprinting goes, however, far beyond its application in species identification. The plethora of applications of DNA fingerprinting techniques in plants and its historic development have been described in great detail in two books (Weising 1995, 2005).

DNA sequencing is increasingly used either in combination with or as a replacement for traditional DNA fingerprinting techniques. The availability of the sequences of whole genomes has led to the development of functional genomics, which aim to understand how an organism's genotype determines its phenotype (Holtorf et al. 2002; Iyer-Pascuzzi et al. 2009). Functional genomics-based technologies are now commonly referred to as 'omics' approaches and include transcriptomics, proteomics, and metabolomics (Fukushima et al. 2009).

Following in the footsteps of automated DNA sequencing (Smith et al. 1986), "next generation sequencing" (NGS) technology (Shendure and Ji 2008; Oetting 2010; Brautigam and Gowik 2010) is revolutionizing genomics and transcriptomics (Imelfort and Edwards 2009) as well as genome-wide genetic marker discovery and genotyping (Davey et al. 2011). Rapid advances in NGS technology are driving down the cost of sequencing and bringing large-scale sequencing projects into the reach of individual investigators (Zhang et al. 2011; Shendure and Ji 2008; Varshney et al. 2009). NGS technology is beginning to be used to target even the largest nuclear genomes of plants (You et al. 2011; Magbanua et al. 2011;

Steuernagel et al. 2009; Straub et al. 2011). NGS technology has the potential to complement or even replace traditional DNA fingerprinting techniques (Sucher et al. 2012). NGS can also facilitate and complement barcoding of plants. The ability to rapidly and cost-effectively sequence entire chloroplast genomes should be useful in resolving even the toughest barcoding challenges.

An important advantage of sequencing-based approaches compared to classical DNA fingerprinting, which is based on electrophoretic band patterns, is that the DNA sequence is independent of the particular method that was used to generate it. Furthermore, DNA sequences can be deposited into public repositories for genetic information such as GenBank (Benson et al. 2009) that can be searched easily by other scientists using specialized bioinformatics software such as BLAST (Mount 2007; Altschul et al. 1990; Hennell et al. 2012).

Here, we present an overview of recent publications that exemplify the use of genomic and transcriptomic profiling in medicinal plants. For easy reference and convenience of the reader, we have included here a brief description of the chemistry of the two most popular NGS platforms, which has been taken in slightly modified form from our recent review on the application of NGS in plants (Sucher et al. 2012).

19.2 NGS Technology

NGS is rapidly developing (Rothberg et al. 2011) and the manufacturers of NGS machines are competing fiercely in the market place (Metzker 2010). Major differentiating attributes between the various NGS approaches that are of most significance to the end-user are the average read length, run time and total number of bases sequenced per run, and the cost per million bases sequenced (Metzker 2010). Other differentiating factors such as the underlying NGS chemistry, template library preparation (emulsion or solid phase PCR; single molecule), and detection method (optical vs. non-optical) may need to be considered, however, depending on the particular application. To date, the two most commonly used NGS techniques are Pyrosequencing™ used by Roche Diagnostics (formally 454 Sequencing) and Illumina® sequencing (formally Solexa). Here, we provide a brief description of the underlying chemistry of these two NGS platforms. Videos and animations illustrating further these and other NGS platform methodology are available on YouTube (http://www.youtube.com).

19.3 Pyrosequencing™

Pyrosequencing™ is used to determine the sequence of a DNA strand based upon the release of pyrophosphate (PP_i) when a nucleotide is incorporated during DNA synthesis (Ronaghi et al. 1998). Pyrosequencing™ starts by breaking down the DNA template into many shorter sequences of about 300 base pairs

(Parchman et al. 2010). Next, the fragments are ligated to adapters that allow the small DNA fragments to bind to the complementary adapter strands bound to beads (one DNA fragment per bead). The bead-bound fragments are then subjected to emulsion PCR for clonal amplification. Beads containing the amplified clonal DNA are then randomly deposited into a microfabricated array of wells such that each well contains only one bead. Pyrosequencing™ is then performed on each bead simultaneously. The sequencing reaction takes place by adding sequentially the dNTPs (A, T, C and G) to the polymerase reaction. When a dNTP is successfully incorporated into a nascent DNA strand in a micro well, PP_i is released and converted to ATP using ATP sulfurylase and adenosine 5′ phosphosulfate. The ATP is subsequently used to drive the conversion of luciferin to oxyluciferin using luciferase, which releases an amount of light proportional to the original amount of PP_i produced (Ronaghi et al. 1996). An imaging device is used to record the chemiluminescent light signals emanating from each well. The remaining unincorporated dNTPs are degraded by apyrase to ensure that a reaction only occurs upon the addition of new dNTPs. The process is continually repeated until the template DNA strand has been copied and the intensity profiles of each bead are generated. The average read length is in the range of 300–400 bp (Brautigam and Gowik 2010; Metzker 2010).

19.4 Illumina® Sequencing

Illumina® sequencing is essentially a highly parallelized adaption of traditional Sanger sequencing (Illumina 2010). For Illumina® sequencing the DNA templates are broken down into short sequences (35–150 bp) to which oligonucleotides containing primer-binding sites are ligated at both ends. Two different adaptors for binding to a flowcell are subsequently attached to the DNA strand using PCR. The DNA is then denatured and run through a flowcell that contains the adapter complements bound to its surface. The single stranded DNA fragments anneal randomly throughout the flowcell to the complementary sites. The surface-bound adapter is then elongated by a polymerase reaction to form a surface-bound clone of the template DNA strand. The DNA is again denatured and washed so only surface-bound DNA strands remain. Bridge amplification is then performed, where the adaptor on the free end of the surface-bound DNA bends down to anneal to an opposite adapter (forming a bridge). The bridged DNA strand is then cloned, thus forming two strands of surface-bound DNA. This process is repeated until up to 1,000 copies are produced. Bridge amplification is used because the DNA strands are clonally amplified within a very localized area, forming 'hot spots' where each unique DNA fragment has been amplified. The sequencing reaction is performed in cycles by adding the dNTPs (all four at the same time) to the flowcell. Each dNTP has a unique, reversibly bound fluorescent label that also serves as a polymerase reaction terminator. Only a single dNTP is therefore incorporated during each polymerase reaction cycle. An imaging device is used to record the colored fluorescent signal generated in the flowcell upon excitation of each DNA 'hotspot' by a laser. The reversible fluorescent terminator on the incorporated nucleotide is then cleaved and the process is repeated until the template strands are copied. The color of each 'hotspot' during each nucleotide incorporation cycle is recorded to determine the sequence. The average read length is around 75 bp (Brautigam and Gowik 2010; Metzker 2010).

19.5 Data Processing

All NGS approaches require powerful computers to reconstruct the genome sequence from the small individual DNA fragments. Each array contains thousands of DNA templates, and consequently NGS typically yields hundreds of millions or even billions of reads during each run.

The relatively short sequence reads obtained by NGS are normally not sufficient to guide the unequivocal assembly of an entire genome de novo. Therefore, whole genome assembly from NGS data usually requires the availability of an existing reference sequence against which the short NGS reads are aligned. It has recently been reported, however, that a combination of traditional Sanger sequencing, which results in longer sequence reads, and NGS was successfully used to assemble the complete genomes of *Cucumis sativus* (Huang et al. 2009b) and *Vitis vinifera* (Velasco et al. 2007) in the absence of existing reference genomes.

19.6 Genome-Based Authentication of Medicinal Plants

Authentication of medicinal plants is traditionally the purview of expert taxonomists, who examine their macroscopic and microscopic morphology. Morphological identification is often not possible or very difficult following processing of the plants to generate herbal medicines. Therefore, additional authentication methods have been sought. A variety of genome-based methods have been developed for the identification and authentication of medicinal plants (Sucher and Carles 2008). Nuclear and chloroplast DNA have been used for DNA fingerprinting and DNA sequencing-based approaches. In principle, both approaches can be used to differentiate between individuals, species, and populations and to detect the presence of adulterants. We have previously argued that the generation of DNA sequence-based "barcodes" of medicinal plants will be worth the concerted effort of the medicinal plant research community within the wider effort of defining barcodes for every species on earth (Sucher and Carles 2008). DNA barcoding, the use of short, standardized regions of the genome as taxon barcodes for biological identification of species first proposed by Hebert et al. in (2003) is being recognized as a powerful addition to the tools of the taxonomist. While DNA barcoding has proven to be considerably more demanding in plants than in animals (where a single locus has proved sufficient) (Chase et al. 2007), a concerted effort of scientists worldwide has led to the identification of four regions that can be successfully used for DNA barcoding of land plants (Hollingsworth et al. 2009). Recent publications appear to indicate that barcoding is increasingly playing an important role in the characterization and authentication of medicinal plants (Zuo et al. 2011; Xue and Li 2011; Xiao et al. 2011; Sun et al. 2011; Quan and Zhou 2011; Li et al. 2010b, 2011; Hollingsworth et al. 2011; Gu et al. 2011; Fu et al. 2011; Tezcan et al. 2010; Ran et al. 2010; Pang et al. 2010; Ma et al. 2010; Lou et al. 2010; He et al. 2010; Guo et al. 2010; Gao et al. 2010; Chen et al. 2010; Schmiderer et al. 2010).

Three plastid markers, *rbcL*, *matK*, and the intergenic spacer *trnH–psbA*, have proven to be effective in a wide range of land plant taxa and have recently been used with good success in medicinal plants (He et al. 2010; Zuo et al. 2011; Gu et al. 2011; Fu et al. 2011; Yan et al. 2008; Yao et al. 2009). In addition, the nuclear ribosomal marker *ITS* (or a subset of *ITS* termed *ITS2*), which has been used for the genomic authentication of many medicinal plants, has also proved suitable for species discrimination in a large number of medicinal plant species (Sucher and Carles 2008; Pang et al. 2010; Gu et al. 2011; Fu et al. 2011; Xie et al. 2009; Howard et al. 2009; Ruzicka et al. 2009).

Twenty-five years after publication of the first complete sequence of a chloroplast genome by Shinozaki et al. (1986), several groups have used NGS to sequence the complete chloroplast genomes from multiple plants in parallel (Parks et al. 2009) (Doorduin et al. 2011). For example, Nock et al. (2011) used massively parallel sequencing of total DNA and reference-guided assembly using the existing rice genome sequence as a scaffold to assemble the complete chloroplast sequence of wild relatives of increasing evolutionary distance from rice. These authors suggested that the entire chloroplast genome rather than single genes might in fact serve as the "… the elusive universal single-locus plant barcode for plant species identification" (Nock et al. 2011).

19.7 Genomic Profiling of Medicinal Plants

Genetic markers in plants include microsatellites (SSR, simple sequence repeats; short tandem repeats) (Gardner et al. 2011; Michalczyk et al. 2011; Buehler et al. 2011; Delmas et al. 2011; Csencsics et al. 2010) and single nucleotide polymorphisms (Marroni et al. 2011; Myles et al. 2010; Barbazuk and Schnable 2011; Arai-Kichise et al. 2011; Hyten et al. 2010). These markers are used for genome mapping and the discovery of disease-associated alleles, population mapping, forensics, the analysis of germplasm, trait mapping, and marker assisted selection in plant breeding and others (Appleby et al. 2009). A number of different assays have been developed for high throughput genotyping in plants using these markers (Appleby et al. 2009; Jones et al. 2009). Genomic polymorphisms can also be harnessed for species authentication in the absence of prior sequence information using sequence-independent array technology (Niu et al. 2011; Jayasinghe et al. 2009). However, NGS is rapidly becoming the method of choice for the characterization of genetic markers (Myles et al. 2010; Marroni et al. 2011; Elshire et al. 2011; Arai-Kichise et al. 2011; Deschamps and Campbell 2010) and the unequivocal identification of multiple alleles of homologous loci in polyploid plants (Griffin et al. 2011). Notwithstanding the increasing use of DNA sequence-based approaches, fingerprinting techniques continue to be used for genomic profiling of medicinal plants (Zhou et al. 2008; Kumar et al. 2007; Techen et al. 2010; Xue and Xue 2008; Qiu et al. 2009; Diao et al. 2009; Ruzicka et al. 2009; Tamhankar et al. 2009; Gupta et al. 2007; Huang et al. 2009a; Li et al. 2009). An interesting novel approach is the use of real-time PCR for the quantitation of amplified markers (Xue and Xue 2008), which should allow to determine the extent of adulterant contamination in raw material samples.

A number of groups used DNA fingerprinting techniques to assess the genetic variation of wild and cultivated TCM plants. Qiu et al. (2009) used inter-simple sequence repeat (ISSR) markers to characterize the genetic variation in wild and cultivated *Rhizoma Corydalis*, a Chinese herbal medicine. Wild populations *Corydalis yanhusuo* W.T. Wang ex Z.Y. Su et C.Y. Wu (*Fumariaceae*) in southern China have been considerably depleted and have virtually disappeared from some locations due to human interference and environmental deterioration. These authors found that the average within-population diversity of ISSR markers was higher in wild than in cultivated populations and recommended that the extant wild populations should be given a high priority for in situ conservation as they may function as reservoirs of genetic variation in the species in the face of extinction (Qiu et al. 2009). Lata et al. (2010) reported that micro propagated plants of *Cannabis sativa* over 30 passages in culture and hardening in soil for 8 months were genetically stable when assessed using ISSR markers and showed a similar cannabinoid profile.

The genetic variation among wild and cultivated populations of the Chinese medicinal plant *Coptis chinensis* (*Ranunculaceae*) was investigated using ISSR by Shi et al. (2008). They found that the genetic diversity in wild and cultivated populations was overall similar but there was significant genetic variation between the wild populations.

Using amplified fragment length polymorphisms (AFLP), He et al. (2009) investigated the population structure and genetic diversity of wild and cultivated populations of the TCM plant *Magnolia officinalis* subsp. *biloba* (*Magnoliaceae*). Principal coordinates analysis of AFLP data did not differentiate between wild and cultivated populations, which led the authors to conclude that alleles from the wild population were maintained in the cultivated gene pool.

Ramesh et al. (2011) used a randomly amplified polymorphic DNA (RAPD) fingerprinting approach to compare the genetic stability of 19 different microporpagated *Bacopa monnieri* with the mother plant (wild type). Although this study illustrates the continued use of RAPD, comparative assessment of various DNA fingerprinting techniques indicates that this technique is less reliable and more difficult to implement consistently than AFLP, SSR, and ISSR (McGregor et al. 2000; Pejic et al. 1998).

19.8 Transcriptomic Profiling of Medicinal Plants

Transcriptomics aims at the genome-wide characterization of mRNA expression in cells or organisms as a function of developmental state, environment, or disease. Transcriptomic approaches have been used to discover the synthetic pathways of secondary metabolite production (Goossens et al. 2003). Transcriptomic data has also been used to identify the genetic basis of the variation in mRNA abundance that is due to genotypic variation in regulatory genes rather than exclusively the environment (Druka et al. 2010). The most widely used transcriptomics technique is based on the use of DNA microarrays (Gregory and Belostotsky 2009) and quantitative real-time PCR (Li et al. 2010a); however, a number of DNA fingerprinting techniques have also been adapted. For example, modifications of the AFLP protocol have been developed for transcriptomic (cDNA-AFLP) and epigenetic studies (methylation sensitive amplified polymorphism; MSAP) (Jones et al. 2009).

NGS is proving to be an efficient and cost-effective method for transcriptomic studies in plants including the discovery of SSR markers in expressed sequences, so-called expressed sequence tag (EST)-SSRs (Kaur et al. 2011; Zeng et al. 2010; Parchman et al. 2010; Molina et al. 2011; Hiremath et al. 2011; Franssen et al. 2011; Severin et al. 2010; Lu et al. 2010; Angeloni et al. 2011; Hao et al. 2011; Wu et al. 2011). Furthermore, NGS holds promise to be useful in the characterization of DNA methylation, histone modification, alternative splicing identification, small RNA profiling, DNA-protein, and potentially protein–protein interactions (Lister et al. 2009; Wei et al. 2011; Zhao et al. 2010; Mohorianu et al. 2011).

Two research groups used 454 pyrosequencing and Illumina paired end sequencing to investigate the transcriptome of *Taxus mairei* and *Taxus cuspidatus*, respectively (Wu et al. 2011; Hao et al. 2011). Wu et al. (2011) obtained 81,148 high quality reads from the needles of *Taxus cuspidata* with 20,557 unique sequences. Performing a similarity search in publicly accessible databases, they were able to annotate some 14,095 unique sequences corresponding to 11,220 unique genes, of which 2,403 transcripts could be mapped to 3,821 biochemical pathways using the two Kyoto Encyclopedia of Genes and Genomes (Kanehisa and Goto 2000). They also identified 753 SSR motifs, which they validated by Sanger sequencing.

In a parallel study, Hao et al. (2011) assembled 36,493 unique sequences of the *Taxus mairei* transcriptome using 13,737,528 sequencing reads (2.03 Gb). Using BLAST for a similarity search with known proteins they identified 23,515 UniGenes (cut-off E-value above 10^{-5}). They then performed a comparative study of the transcriptomes of three different tissues of *Taxus mairei* and identified a large number of genes associated with tissue-specific functions and the taxane biosynthetic pathway. The expression genes involved in the biosynthesis of taxane was found to be significantly higher in the root than in the leaf and the stem.

Li et al. (2010a) used 454 pyrosequencing to generate ESTs from the vegetative organs of *Glycyrrhiza uralensis*. These authors generated 59,219 ESTs with an average read length of 409 bp, which they combined with 50,666 *G. uralensis* ESTs that were identified in GenBank. They then assembled the combined EST data set into 27,229 unique sequences (11,694 contigs and 15,535 singletons). BLAST searches of multiple public databases were performed to annotate the ESTs. A total of 20,437 unique sequences representing approximately 10,000 independent transcripts could be annotated (e-value $\leq 10^{-5}$) including 16 genes encoding for enzymes that are part of the 18-step glycyrrhizin skeleton synthesis pathway.

The transcriptomes of the roots of *Panax ginseng* and *Panax quinquefolius* (American ginseng) were characterized using 454 pyrosequencing by Sun et al. (2010) and Chen et al. (2011). Sun and colleagues obtained more than 200,000 high quality reads from a cDNA library as template. The reads could be assembled into 31,088 unique transcripts (16,592 contigs and 14,496 singletons). Interestingly, the authors note that the average lengths of the reads and the contigs were comparable to those generated from an American ginseng root cDNA library in a previous study using the Sanger method. Bioinformatic analysis of the data indicated that all genes encoding enzymes involved in the biosynthesis of the ginsenoside backbone were present in their data set. Chen and colleagues produced a total of 217,529 high quality reads (ESTs), with an average length of 409 bases, generated from a one-quarter run to yield 31,741 unique sequences. Over 69 % (20,198) of the unique sequences could be annotated using BLAST similarity searches. A total of 16,810 and 16,577 unique sequences were assigned to functional classifications and biochemical pathways based on Gene Ontology analysis and the Kyoto Encyclopedia of Genes and Genomes assignment, respectively. Nine genes involved in the biosynthesis of ginsenoside skeletons and many candidate genes putatively responsible for modification of the skeletons were identified.

Zeng et al. (2010) used 454 pyrosequencing technology to generate 76,459 ESTs and 2,810 EST-SSRs from the TCM plant *Epimedium sagittatum* (Sieb. Et Zucc.) Maxim. They reported that some 1580 EST-SSR markers (86 %) were transferable across the *Epimedium* genus. These EST-SSR markers will therefore be useful in authentication and the marker-assisted selection breeding of *Epimedium* species, some of which are endangered.

19.9 Conclusions

NGS is increasingly being used for the genomic and transcriptomic profiling of medicinal plants and may in the future largely replace traditional DNA fingerprinting techniques. NGS can facilitate and complement barcoding of plants. The ability to rapidly and cost-effectively sequence entire chloroplast genomes should be useful in resolving even the toughest barcoding challenges. An important advantage of sequencing-based approaches compared to classical DNA fingerprinting, which is based on electrophoretic band patterns, is that the DNA sequence is independent of the particular method that was used to generate it. Furthermore, DNA sequences can be deposited into public repositories for genetic information such as GenBank that can be searched easily by other scientists using bioinformatics software that is easily accessible on the World Wide Web.

References

Altschul SF, Gish W, Miller W, Myers EW, Lipman DJ (1990) Basic local alignment search tool. J Mol Biol 215(3):403–410

Angeloni F, Wagemaker CAM, Jetten MSM, Op den Camp HJM, Janssen-Megens EM, Francoijs KJ, Stunnenberg HG, Ouborg NJ (2011) De novo transcriptome characterization and development of genomic tools for *Scabiosa columbaria* L. using next-generation sequencing techniques. Mol Ecol Resour 11(4):662–674

Anonymous (2003) Growth in global trade in medicinal plants (R&D News). Manuf Chemist (April 1)

Appleby N, Edwards D, Batley J (2009) New technologies for ultra-high throughput genotyping in plants. Plant Genomics 513:19–39

Arai-Kichise Y, Shiwa Y, Nagasaki H, Ebana K, Yoshikawa H, Yano M, Wakasa K (2011) Discovery of genome-wide DNA polymorphisms in a Landrace cultivar of Japonica rice by whole-genome sequencing. Plant Cell Physiol 52(2):274–282

Balunas MJ, Kinghorn AD (2005) Drug discovery from medicinal plants. Life Sci 78(5):431–441

Barbazuk WB, Schnable PS (2011) SNP discovery by transcriptome pyrosequencing. cDNA Libr 729:225–246

Benson DA, Karsch-Mizrachi I, Lipman DJ, Ostell J, Sayers EW (2009) GenBank. Nucleic Acids Res 37(Database issue):D26–D31

Brautigam A, Gowik U (2010) What can next generation sequencing do for you? Next generation sequencing as a valuable tool in plant research. Plant Biol (Stuttg) 12(6):831–841

Buehler D, Graf R, Holderegger R, Gugerli F (2011) Using the 454 pyrosequencing-based technique in the development of nuclear microsatellite loci in the alpine plant *Arabis alpina* (Brassicaceae). Am J Bot 98(5):e103–e105

Canter PH, Thomas H, Ernst E (2005) Bringing medicinal plants into cultivation: opportunities and challenges for biotechnology. Trends Biotechnol 23(4):180–185

Chapman K, Chomchalow N (2005) Production of medicinal plants in Asia. Acta Hort (ISHS) 679:45–59

Chase MW, Cowan RS, Hollingsworth PM, van den Berg C, Madrinan S, Petersen G, Seberg O, Jorgsensen T, Cameron KM, Carine M, Pedersen N, Hedderson TAJ, Conrad F, Salazar GA, Richardson JE, Hollingsworth ML, Barraclough TG, Kelly L, Wilkinson M (2007) A proposal for a standardised protocol to barcode all land plants. Taxon 56(2):295–299

Chaudhuri RK, Pal A, Jha TB (2007) Production of Genetically uniform plants from nodal explants of Swertia Chirata Buch.-Ham ex. wall.: an endangered medicinal herb. In Vitro Cellular Develop Biol Plant 43(5):467–472

Chen SL, Yao H, Han JP, Liu C, Song JY, Shi LC, Zhu YJ, Ma XY, Gao T, Pang XH, Luo K, Li Y, Li XW, Jia XC, Lin YL, Leon C (2010) Validation of the ITS2 region as a novel DNA barcode for identifying medicinal plant species. PLoS ONE 5(1):e8613

Chen S, Luo H, Li Y, Sun Y, Wu Q, Niu Y, Song J, Lv A, Zhu Y, Sun C, Steinmetz A, Qian Z (2011) 454 EST analysis detects genes putatively involved in ginsenoside biosynthesis in Panax ginseng. Plant Cell Rep 30:1593–1601

Cordell GA (2009) Sustainable drugs and global health care. Quim Nova 32:1356–1364

Csencsics D, Brodbeck S, Holderegger R (2010) Cost-effective, species-specific microsatellite development for the Endangered Dwarf Bulrush (*Typha minima*) using next-generation sequencing technology. J Hered 101(6):789–793

Davey JW, Hohenlohe PA, Etter PD, Boone JQ, Catchen JM, Blaxter ML (2011) Genome-wide genetic marker discovery and genotyping using next-generation sequencing. Nat Rev Genet 12(7):499–510

Delmas CEL, Lhuillier E, Pornon A, Escaravage N (2011) Isolation and characterization of microsatellite loci in *Rhododendron ferrugineum* (Ericaceae) using pyrosequencing technology. Am J Bot 98(5):e120–e122

Deschamps S, Campbell M (2010) Utilization of next-generation sequencing platforms in plant genomics and genetic variant discovery. Mol Breeding 25(4):553–570

Diao Y, Lin XM, Liao CL, Tang CZ, Chen ZJ, Hu ZL (2009) Authentication of Panax ginseng from its adulterants by PCR-RFLP and ARMS. Planta Med 75(5):557–560

Doorduin L, Gravendeel B, Lammers Y, Ariyurek Y, Chin AWT, Vrieling K (2011) The complete chloroplast genome of 17 individuals of pest species *Jacobaea vulgaris*: SNPs, microsatellites and barcoding markers for population and phylogenetic studies. DNA Res 18(2):93–105

Druka A, Potokina E, Luo Z, Jiang N, Chen X, Kearsey M, Waugh R (2010) Expression quantitative trait loci analysis in plants. Plant Biotechnol J 8(1):10–27

Elshire RJ, Glaubitz JC, Sun Q, Poland JA, Kawamoto K, Buckler ES, Mitchell SE (2011) A robust, simple genotyping-by-sequencing (GBS) approach for high diversity species. PLoS ONE 6(5):e19379

Franssen S, Shrestha R, Brautigam A, Bornberg-Bauer E, Weber A (2011) Comprehensive transcriptome analysis of the highly complex *Pisum sativum* genome using next generation sequencing. BMC Genomics 12(1):227

Fu YM, Jiang WM, Fu CX (2011) Identification of species within Tetrastigma (Miq.) Planch. (Vitaceae) based on DNA barcoding techniques. J Syst Evol 49(3):237–245

Fukushima A, Kusano M, Redestig H, Arita M, Saito K (2009) Integrated omics approaches in plant systems biology. Curr Opin Chem Biol 13(5–6):532–538

Gao T, Yao H, Song JY, Liu C, Zhu YJ, Ma XY, Pang XH, Xu HX, Chen SL (2010) Identification of medicinal plants in the family Fabaceae using a potential DNA barcode ITS2. J Ethnopharmacol 130(1):116–121

Gardner MG, Fitch AJ, Bertozzi T, Lowe AJ (2011) Rise of the machines—recommendations for ecologists when using next generation sequencing for microsatellite development. Mol Ecol Resour 11:1093–1101

Goossens A, Häkkinen ST, Laakso I, Seppänen-Laakso T, Biondi S, Sutter VD, Lammertyn F, Nuutila AM, Söderlund H, Zabeau M, Inzé D, Oksman-Caldentey K-M (2003) A functional genomics approach toward the understanding of secondary metabolism in plant cells. Proc Natl Acad Sci U S A 100(14):8595–8600

Gregory BD, Belostotsky DA (2009) Whole-genome microarrays: applications and technical issues. Plant Syst Biol 553:39–56

Griffin PC, Robin C, Hoffmann AA (2011) A next-generation sequencing method for overcoming the multiple gene copy problem in polyploid phylogenetics, applied to Poa grasses. BMC Biol 9:19

Gu J, Su JX, Lin RZ, Li RQ, Xiao PG (2011) Testing four proposed barcoding markers for the identification of species within *Ligustrum* L. (Oleaceae). J Syst Evol 49(3):213–224

Guo HY, Wang WW, Yang N, Guo BL, Zhang S, Yang RJ, Yuan Y, Yu JL, Hu SN, Sun QS, Yu J (2010) DNA barcoding provides distinction between *Radix Astragali* and its adulterants. Sci China-Life Sci 53(8):992–999

Gupta S, Pandey-Rai S, Srivastava S, Naithani SC, Prasad M, Kumar S (2007) Construction of genetic linkage map of the medicinal and ornamental plant *Catharanthus roseus*. J Genet 86(3):259–268

Halberstein RA (2005) Medicinal plants: historical and cross-cultural usage patterns. Ann Epidemiol 15(9):686–699

Hao DC, Ge G, Xiao P, Zhang Y, Yang L (2011) The first insight into the tissue specific *Taxus* Transcriptome via Illumina second generation sequencing. PLoS ONE 6(6):e21220

He J, Chen L, Si Y, Huang B, Ban X, Wang Y (2009) Population structure and genetic diversity distribution in wild and cultivated populations of the traditional Chinese medicinal plant *Magnolia officinalis* subsp. biloba (Magnoliaceae). Genetica 135(2):233–243

He J, Wong KL, Shaw PC, Wang H, Li DZ (2010) Identification of the medicinal plants in *Aconitum* L. by DNA barcoding technique. Planta Med 76(14):1622–1628

Hebert PD, Cywinska A, Ball SL, deWaard JR (2003) Biological identifications through DNA barcodes. Proc Biol Sci 270(1512):313–321

Hennell JR, D'Agostino PM, Lee S, Khoo CS, Sucher NJ (2012) Using GenBank® for genomic authentication: a tutorial. Methods Mol Biol 862:181–200

Heuberger H, Bauer R, Friedl F, Heubl G, Hummelsberger J, Nogel R, Seidenberger R, Torres-Londono P (2010) Cultivation and breeding of Chinese medicinal plants in Germany. Planta Med 76(17):1956–1962

Heubl G (2010) New aspects of DNA-based authentication of Chinese medicinal plants by molecular biological techniques. Planta Med 76(17):1963–1974

Hiremath PJ, Farmer A, Cannon SB, Woodward J, Kudapa H, Tuteja R, Kumar A, BhanuPrakash A, Mulaosmanovic B, Gujaria N, Krishnamurthy L, Gaur PM, KaviKishor PB, Shah T, Srinivasan R, Lohse M, Xiao Y, Town CD, Cook DR, May GD, Varshney RK (2011) Large-scale transcriptome analysis in chickpea (*Cicer arietinum* L.), an orphan legume crop of the semi-arid tropics of Asia and Africa. Plant Biotechnol J 9:922–931

Hollingsworth P, Forrest L, Spouge J, Hajibabaei M, Ratnasingham S, Bank M, Chase M, Cowan R, Erickson D, Fazekas A (2009) A DNA barcode for land plants. Proc Natl Acad Sci U S A 106:12794–12797

Hollingsworth PM, Graham SW, Little DP (2011) Choosing and using a plant DNA barcode. PLoS ONE 6(5):e19254

Holtorf H, Guitton M-C, Reski R (2002) Plant functional genomics. Naturwissenschaften 89(6):235–249

Howard C, Bremner PD, Fowler MR, Isodo B, Scott NW, Slater A (2009) Molecular identification of *Hypericum perforatum* by PCR amplification of the ITS and 5.8S rDNA region. Planta Med 75(8):864–869

Huang H-R, Zhou G, Ge X-J, Wei X, Jiang Y-S, Tang H (2009a) Eight polymorphic microsatellite loci for the Chinese medicinal plant *Artemisia annua* L. (Asteraceae). Conserv Genet 10(3):593–595

Huang S, Li R, Zhang Z, Li L, Gu X, Fan W, Lucas WJ, Wang X, Xie B, Ni P, Ren Y, Zhu H, Li J, Lin K, Jin W, Fei Z, Li G, Staub J, Kilian A, van der Vossen EA, Wu Y, Guo J, He J, Jia Z, Tian G, Lu Y, Ruan J, Qian W, Wang M, Huang Q, Li B, Xuan Z, Cao J, Asan WuZ, Zhang J, Cai Q, Bai Y, Zhao B, Han Y, Li Y, Li X, Wang S, Shi Q, Liu S, Cho WK, Kim JY, Xu Y, Heller-Uszynska K, Miao H, Cheng Z, Zhang S, Wu J, Yang Y, Kang H, Li M, Liang H, Ren X, Shi Z, Wen M, Jian M, Yang H, Zhang G, Yang Z, Chen R, Ma L, Liu H, Zhou Y, Zhao J, Fang X, Fang L, Liu D, Zheng H, Zhang Y, Qin N, Li Z, Yang G, Yang S, Bolund L, Kristiansen K, Li S, Zhang X, Wang J, Sun R, Zhang B, Jiang S, Du Y (2009b) The genome of the cucumber *Cucumis sativus* L. Nat Genet 41(12):1275–1281

Hyten D, Song Q, Fickus E, Quigley C, Lim J-S, Choi I-Y, Hwang E-Y, Pastor-Corrales M, Cregan P (2010) High-throughput SNP discovery and assay development in common bean. BMC Genomics 11(1):475

Illumina (2010) Technology spotlight: Illumina® sequencing. http://www.illumina.com/documents/products/techspotlights/techspotlight_sequencing.pdf

Imelfort M, Edwards D (2009) De novo sequencing of plant genomes using second-generation technologies. Briefings Bioinformatics 10(6):609–618

Iyer-Pascuzzi A, Simpson J, Herrera-Estrella L, Benfey PN (2009) Functional genomics of root growth and development in *Arabidopsis*. Curr Opin Plant Biol 12(2):165–171

Jayasinghe R, Hai Niu L, Coram TE, Kong S, Kaganovitch J, Xue CC, Li CG, Pang EC (2009) Effectiveness of an innovative prototype subtracted diversity array (SDA) for fingerprinting plant species of medicinal importance. Planta Med 75(10):1180–1185

Jones N, Ougham H, Thomas H, Pašakinskienė I (2009) Markers and mapping revisited: finding your gene. New Phytol 183(4):935–966

Kanehisa M, Goto S (2000) KEGG: kyoto encyclopedia of genes and genomes. Nucleic Acids Res 28(1):27–30

Kaur S, Cogan NO, Pembleton LW, Shinozuka M, Savin KW, Materne M, Forster JW (2011) Transcriptome sequencing of lentil based on second-generation technology permits large-scale unigene assembly and SSR marker discovery. BMC Genomics 12:265

Kumar J, Verma V, Shahi AK, Qazi GN, Balyan HS (2007) Development of simple sequence repeat markers in cymbopogon species. Planta Med 73(3):262–266

Lata H, Chandra S, Techen N, Khan IA, ElSohly MA (2010) Assessment of the genetic stability of micropropagated plants of *Cannabis sativa* by ISSR markers. Planta Med 76(1):97–100

Lattoo S, Bamotra S, Sapru Dhar R, Khan S, Dhar A (2006) Rapid plant regeneration and analysis of genetic fidelity of in vitro derived plants of *Chlorophytum arundinaceum* Baker—an endangered medicinal herb. Plant Cell Rep 25(6):499–506

Li L, Yao X, Chen X, Huang H (2009) Development and characterization of microsatellite loci in Chinese medicinal plant *Akebia trifoliate* ssp. australis and cross-species amplification in closely related taxa. Conserv Genet 10(4):959–962

Li Y, Luo HM, Sun C, Song JY, Sun YZ, Wu QO, Wang N, Yao H, Steinmetz A, Chen SL (2010a) EST analysis reveals putative genes involved in glycyrrhizin biosynthesis. BMC Genomics 11:268

Li YH, Ruan JL, Chen SL, Song JY, Luo K, Lu D, Yao H (2010b) Authentication of *Taxillus chinensis* using DNA barcoding technique. J Medicinal Plants Res 4(24):2706–2709

Li M, Cao H, But PPH, Shaw PC (2011) Identification of herbal medicinal materials using DNA barcodes. J Syst Evol 49(3):271–283

Lister R, Gregory BD, Ecker JR (2009) Next is now: new technologies for sequencing of genomes, transcriptomes, and beyond. Curr Opin Plant Biol 12(2):107–118

Lou SK, Wong KL, Li M, But PPH, Tsui SKW, Shaw PC (2010) An integrated web medicinal materials DNA database: MMDBD (Medicinal Materials DNA Barcode Database). BMC Genomics 11:402

Lu T, Lu G, Fan D, Zhu C, Li W, Zhao Q, Feng Q, Zhao Y, Guo Y, Li W, Huang X, Han B (2010) Function annotation of the rice transcriptome at single-nucleotide resolution by RNA-seq. Genome Res 20(9):1238–1249

Ma XY, Xie CX, Liu C, Song JY, Yao H, Luo K, Zhu YJ, Gao T, Pang XH, Qian J, Chen SL (2010) Species identification of medicinal pteridophytes by a DNA barcode marker, the chloroplast psbA-trnH intergenic region. Biol Pharm Bulletin 33(11):1919–1924

Magbanua ZV, Ozkan S, Bartlett BD, Chouvarine P, Saski CA, Liston A, Cronn RC, Nelson CD, Peterson DG (2011) Adventures in the enormous: a 1.8 million clone BAC library for the 21.7 Gb genome of loblolly pine. PLoS ONE 6(1):e16214

Marroni F, Pinosio S, Di Centa E, Jurman I, Boerjan W, Felice N, Cattonaro F, Morgante M (2011) Large scale detection of rare variants via pooled multiplexed next generation sequencing: towards next generation Ecotilling. Plant J 67:736–745

McGregor CE, Lambert CA, Greyling MM, Louw JH, Warnich L (2000) A comparative assessment of DNA fingerprinting techniques (RAPD, ISSR, AFLP and SSR) in tetraploid potato (*Solanum tuberosum* L.) germplasm. Euphytica 113(2):135–144

Metzker ML (2010) Sequencing technologies—the next generation. Nat Rev Genet 11(1):31–46

Michalczyk IM, Schumacher C, Mengel C, Leyer I, Liepelt S (2011) Identification and characterization of 12 microsatellite loci in *Cnidium dubium* (Apiaceae) using next-generation sequencing. Am J Bot 98(5):e127–e129

Mohorianu I, Schwach F, Jing R, Lopez-Gomollon S, Moxon S, Szittya G, Sorefan K, Moulton V, Dalmay T (2011) Profiling of short RNAs during fleshy fruit development reveals stage-specific sRNAome expression patterns. Plant J 67:232–246

Molina C, Zaman-Allah M, Khan F, Fatnassi N, Horres R, Rotter B, Steinhauer D, Amenc L, Drevon J–J, Winter P, Kahl G (2011) The salt-responsive transcriptome of chickpea roots and nodules via deepSuperSAGE. BMC Plant Biol 11(1):31

Mount DW (2007) Using the Basic Local Alignment Search Tool (BLAST). CSH Protoc 2007:pdb top17

Myles S, Chia J-M, Hurwitz B, Simon C, Zhong GY, Buckler E, Ware D (2010) Rapid genomic characterization of the genus *vitis*. PLoS ONE 5(1):e8219

Niu L, Mantri N, Li CG, Xue C, Pang E (2011) Array-based techniques for fingerprinting medicinal herbs. Chin Med 6(1):18

Nock CJ, Waters DLE, Edwards MA, Bowen SG, Rice N, Cordeiro GM, Henry RJ (2011) Chloroplast genome sequences from total DNA for plant identification. Plant Biotechnol J 9(3):328–333

Oetting WS (2010) Impact of next generation sequencing: the 2009 Human Genome Variation Society Scientific Meeting. Hum Mutat 31(4):500–503

Pang X, Song J, Zhu Y, Xie C, Chen S (2010) Using DNA barcoding to identify species within Euphorbiaceae. Planta Med 76(15):1784–1786

Parchman TL, Geist KS, Grahnen JA, Benkman CW, Buerkle CA (2010) Transcriptome sequencing in an ecologically important tree species: assembly, annotation, and marker discovery. BMC Genomics 11:180

Parks M, Cronn R, Liston A (2009) Increasing phylogenetic resolution at low taxonomic levels using massively parallel sequencing of chloroplast genomes. BMC Biol 7(1):84

Pejic I, Ajmone-Marsan P, Morgante M, Kozumplick V, Castiglioni P, Taramino G, Motto M (1998) Comparative analysis of genetic similarity among maize inbred lines detected by RFLPs, RAPDs, SSRs, and AFLPs. Theor Appl Genet 97(8):1248–1255

Qiu YX, Zong M, Yao Y, Chen BL, Zhou XL, Chen ZL, Fu CX (2009) Genetic variation in wild and cultivated rhizoma *corydalis* revealed by ISSRs markers. Planta Med 75(1):94–98

Quan XQX, Zhou SL (2011) Molecular identification of species in *Prunus Persia* (Rosaceae), with emphasis on evaluation of candidate barcodes for plants. J Syst Evol 49(2):138–145

Ramesh M, Vijayakumar K, Karthikeyan A, Pandian S (2011) RAPD based genetic stability analysis among micropropagated, synthetic seed derived and hardened plants of *Bacopa monnieri* (L.): a threatened Indian medicinal herb. Acta Physiologiae Plantarum 33(1): 163–171

Ran JH, Wang PP, Zhao HJ, Wang XQ (2010) A test of seven candidate barcode regions from the Plastome in *Picea* (Pinaceae). J Integr Plant Biol 52(12):1109–1126

Robinson MM, Zhang X (2011) The world medicines situation 2011. Traditional medicines: global situation, issues and challenges. WHO. Accessed WHO/EMP/MIE/2011.2.3. Downloaded on 11 July 2011

Ronaghi M, Karamohamed S, Pettersson B, Uhlen M, Nyren P (1996) Real-time DNA sequencing using detection of pyrophosphate release. Anal Biochem 242(1):84–89

Ronaghi M, Uhlen M, Nyren P (1998) A sequencing method based on real-time pyrophosphate. Science 281(5375):363–365

Rothberg JM, Hinz W, Rearick TM, Schultz J, Mileski W, Davey M, Leamon JH, Johnson K, Milgrew MJ, Edwards M, Hoon J, Simons JF, Marran D, Myers JW, Davidson JF, Branting A, Nobile JR, Puc BP, Light D, Clark TA, Huber M, Branciforte JT, Stoner IB, Cawley SE, Lyons M, Fu Y, Homer N, Sedova M, Miao X, Reed B, Sabina J, Feierstein E, Schorn M,

Alanjary M, Dimalanta E, Dressman D, Kasinskas R, Sokolsky T, Fidanza JA, Namsaraev E, McKernan KJ, Williams A, Roth GT, Bustillo J (2011) An integrated semiconductor device enabling non-optical genome sequencing. Nature 475(7356):348–352

Ruzicka J, Lukas B, Merza L, Gohler I, Abel G, Popp M, Novak J (2009) Identification of *Verbena officinalis* based on ITS sequence analysis and RAPD-derived molecular markers. Planta Med 75(11):1271–1276

Saxena S, Bopana N (2009) In vitro clonal propagation of *Asparagus racemosus*, a high value medicinal plant. Methods Mol Biol 547:179–189

Schmiderer C, Mader E, Novak J (2010) DNA-based identification of *Helleborus niger* by high-resolution melting analysis. Planta Med 76(16):1934–1937

Severin A, Woody J, Bolon Y-T, Joseph B, Diers B, Farmer A, Muehlbauer G, Nelson R, Grant D, Specht J, Graham M, Cannon S, May G, Vance C, Shoemaker R (2010) RNA-Seq Atlas of Glycine max: a guide to the soybean transcriptome. BMC Plant Biol 10(1):160

Shendure J, Ji H (2008) Next-generation DNA sequencing. Nat Biotechnol 26(10):1135–1145

Shi W, Yang CF, Chen JM, Guo YH (2008) Genetic variation among wild and cultivated populations of the Chinese medicinal plant *Coptis chinensis* (Ranunculaceae). Plant Biol (Stuttg) 10(4):485–491

Shinozaki K, Ohme M, Tanaka M, Wakasugi T, Hayashida N, Matsubayashi T, Zaita N, Chunwongse J, Obokata J, Yamaguchi-Shinozaki K, Ohto C, Torazawa K, Meng BY, Sugita M, Deno H, Kamogashira T, Yamada K, Kusuda J, Takaiwa F, Kato A, Tohdoh N, Shimada H, Sugiura M (1986) The complete nucleotide sequence of the tobacco chloroplast genome: its gene organization and expression. EMBO J 5(9):2043–2049

Smith LM, Sanders JZ, Kaiser RJ, Hughes P, Dodd C, Connell CR, Heiner C, Kent SB, Hood LE (1986) Fluorescence detection in automated DNA sequence analysis. Nature 321(6071):674–679

Steuernagel B, Taudien S, Gundlach H, Seidel M, Ariyadasa R, Schulte D, Petzold A, Felder M, Graner A, Scholz U, Mayer KF, Platzer M, Stein N (2009) De novo 454 sequencing of barcoded BAC pools for comprehensive gene survey and genome analysis in the complex genome of barley. BMC Genomics 10:547

Straub S, Fishbein M, Livshultz T, Foster Z, Parks M, Weitemier K, Cronn R, Liston A (2011) Building a model: developing genomic resources for common milkweed (*Asclepias syriaca*) with low coverage genome sequencing. BMC Genomics 12(1):211

Sucher NJ, Carles MC (2008) Genome-based approaches to the authentication of medicinal plants. Planta Med 74(6):603–623

Sucher NJ, Hennell JR, Carles MC (2012) DNA fingerprinting, DNA barcoding, and next generation sequencing technology in plants. Methods Mol Biol 862:13–22

Sun C, Li Y, Wu Q, Luo H, Sun Y, Song J, Lui E, Chen S (2010) De novo sequencing and analysis of the American ginseng root transcriptome using a GS FLX Titanium platform to discover putative genes involved in ginsenoside biosynthesis. BMC Genomics 11(1):262

Sun ZY, Gao T, Yao H, Shi LC, Zhu YJ, Chen SL (2011) Identification of *Lonicera japonica* and its related species using the DNA barcoding method. Planta Med 77(3):301–306

Tamhankar S, Ghate V, Raut A, Rajput B (2009) Molecular profiling of "Chirayat" complex using Inter Simple Sequence Repeat (ISSR) markers. Planta Med 75(11):1266–1270

Techen N, Chandra S, Lata H, Elsohly MA, Khan IA (2010) Genetic identification of female *Cannabis sativa* plants at early developmental stage. Planta Med 76(16):1938–1939

Tezcan M, Vlachonasios K, Aki C (2010) DNA barcoding study on *Sideritis trojana* bornm: an endemic medicinal plant of IDA mountain, Turkey. Fresenius Environ Bull 19(7):1352–1355

Varshney RK, Nayak SN, May GD, Jackson SA (2009) Next-generation sequencing technologies and their implications for crop genetics and breeding. Trends Biotechnol 27(9):522–530

Velasco R, Zharkikh A, Troggio M, Cartwright DA, Cestaro A, Pruss D, Pindo M, FitzGerald LM, Vezzulli S, Reid J, Malacarne G, Iliev D, Coppola G, Wardell B, Micheletti D, Macalma T, Facci M, Mitchell JT, Perazzolli M, Eldredge G, Gatto P, Oyzerski R, Moretto M, Gutin N,

Stefanini M, Chen Y, Segala C, Davenport C, Demattè L, Mraz A, Battilana J, Stormo K, Costa F, Tao Q, Si-Ammour A, Harkins T, Lackey A, Perbost C, Taillon B, Stella A, Solovyev V, Fawcett JA, Sterck L, Vandepoele K, Grando SM, Toppo S, Moser C, Lanchbury J, Bogden R, Skolnick M, Sgaramella V, Bhatnagar SK, Fontana P, Gutin A, Van de Peer Y, Salamini F, Viola R (2007) A high quality draft consensus sequence of the genome of a Heterozygous Grapevine variety. PLoS ONE 2(12):e1326

Wei LQ, Yan LF, Wang T (2011) Deep sequencing on genome-wide scale reveals the unique composition and expression patterns of microRNAs in developing pollen of *Oryza sativa*. Genome Biol 12(6):R53

Weising K (1995) DNA fingerprinting in plants and fungi. CRC Press, Boca Raton

Weising K (2005) DNA fingerprinting in plants: principles, methods, and applications, 2nd edn. Taylor & Francis Group, Boca Raton

Wu QO, Sun C, Luo HM, Li Y, Niu YY, Sun YZ, Lu AP, Chen SL (2011) Transcriptome analysis of *Taxus cuspidata* needles based on 454 pyrosequencing. Planta Med 77(4):394–400

Xiao WL, Motley TJ, Unachukwu UJ, Lau CBS, Jiang B, Hong F, Leung PC, Wang QF, Liyingston PO, Cassileth BR, Kennelly EJ (2011) Chemical and genetic assessment of variability in commercial *Radix Astragali* (Astragalus spp.) by Ion Trap LC-MS and nuclear ribosomal DNA barcoding sequence analyses. J Agric Food Chem 59(5):1548–1556

Xie H, Huo KK, Chao Z, Pan SL (2009) Identification of crude drugs from Chinese medicinal plants of the genus *Bupleurum* using ribosomal DNA ITS sequences. Planta Med 75(1):89–93

Xue CY, Li DZ (2011) Use of DNA barcode sensu lato to identify traditional Tibetan medicinal plant *Gentianopsis paludosa* (Gentianaceae). J Syst Evol 49(3):267–270

Xue CY, Xue HG (2008) Application of real-time scorpion PCR for authentication and quantification of the traditional Chinese medicinal plant *Drynaria fortunei*. Planta Med 74(11):1416–1420

Yan P, Pang QH, Jiao XW, Zhao X, Shen YJ, Zhao SJ (2008) Genetic variation and identification of cultivated *Fallopia multiflora* and its wild relatives by using chloroplast matK and 18S rRNA gene sequences. Planta Med 74(12):1504–1509

Yang L, Stockigt J (2010) Trends for diverse production strategies of plant medicinal alkaloids. Nat Prod Rep 27(10):1469–1479

Yao H, Song JY, Ma XY, Liu C, Li Y, Xu HX, Han JP, Duan LS, Chen SL (2009) Identification of Dendrobium species by a candidate DNA barcode sequence: the chloroplast psbA-trnH intergenic region. Planta Med 75(6):667–669

You FM, Huo N, Deal KR, Gu YQ, Luo MC, McGuire PE, Dvorak J, Anderson OD (2011) Annotation-based genome-wide SNP discovery in the large and complex *Aegilops tauschii* genome using next-generation sequencing without a reference genome sequence. BMC Genomics 12:59

Zeng S, Xiao G, Guo J, Fei Z, Xu Y, Roe BA, Wang Y (2010) Development of a EST dataset and characterization of EST-SSRs in a traditional Chinese medicinal plant, *Epimedium sagittatum* (Sieb. Et Zucc.) Maxim. BMC Genomics 11:94

Zhang B, Peng Y, Zhang Z, Liu H, Qi Y, Liu S, Xiao P (2010) GAP production of TCM herbs in China. Planta Med 76(17):1948–1955

Zhang J, Chiodini R, Badr A, Zhang G (2011) The impact of next-generation sequencing on genomics. J Genet Genomics 38(3):95–109

Zhao C-Z, Xia H, Frazier T, Yao Y–Y, Bi Y-P, Li A-Q, Li M-J, Li C-S, Zhang B-H, Wang X-J (2010) Deep sequencing identifies novel and conserved microRNAs in peanuts (*Arachis hypogaea* L.). BMC Plant Biol 10(1):3

Zhou LG, Wu JY (2006) Development and application of medicinal plant tissue cultures for production of drugs and herbal medicinals in China. Nat Prod Rep 23(5):789–810
Zhou X, Li Q, Yin Y, Chen Y, Lin J (2008) Identification of medicinal Ganoderma species based on PCR with specific primers and PCR-RFLP. Planta Med 74(2):197–200
Zuo Y, Chen Z, Kondo K, Funamoto T, Wen J, Zhou S (2011) DNA barcoding of Panax species. Planta Med 77(2):182–187

Index

S. Chandra et al. (eds.), *Biotechnology for Medicinal Plants*,
DOI: 10.1007/978-3-642-29974-2,

MIX
Papier aus verantwortungsvollen Quellen
Paper from responsible sources
FSC® C105338
FSC
www.fsc.org

If you have any concerns about our products,
you can contact us on
ProductSafety@springernature.com

In case Publisher is established outside the EU,
the EU authorized representative is:
Springer Nature Customer Service Center GmbH
Europaplatz 3, 69115 Heidelberg, Germany

Printed by Libri Plureos GmbH
in Hamburg, Germany